Lecture Notes in Mechanical Engineering

Lecture Notes in Mechanical Engineering (LNME) publishes the latest developments in Mechanical Engineering—quickly, informally and with high quality. Original research reported in proceedings and post-proceedings represents the core of LNME. Volumes published in LNME embrace all aspects, subfields and new challenges of mechanical engineering. Topics in the series include:

- Engineering Design
- Machinery and Machine Elements
- Mechanical Structures and Stress Analysis
- Automotive Engineering
- Engine Technology
- Aerospace Technology and Astronautics
- Nanotechnology and Microengineering
- Control, Robotics, Mechatronics
- MEMS
- Theoretical and Applied Mechanics
- Dynamical Systems, Control
- Fluid Mechanics
- Engineering Thermodynamics, Heat and Mass Transfer
- Manufacturing
- Precision Engineering, Instrumentation, Measurement
- Materials Engineering
- Tribology and Surface Technology

To submit a proposal or request further information, please contact the Springer Editor of your location:

China: Ms. Ella Zhang at ella.zhang@springer.com
India: Priya Vyas at priya.vyas@springer.com
Rest of Asia, Australia, New Zealand: Swati Meherishi at swati.meherishi@springer.com
All other countries: Dr. Leontina Di Cecco at Leontina.dicecco@springer.com

To submit a proposal for a monograph, please check our Springer Tracts in Mechanical Engineering at http://www.springer.com/series/11693 or contact Leontina.dicecco@springer.com

Indexed by SCOPUS. All books published in the series are submitted for consideration in Web of Science.

More information about this series at http://www.springer.com/series/11236

Andrii Bieliatynskyi · Vera Breskich
Editors

Safety in Aviation and Space Technologies

Select Proceedings of the 9th World Congress "Aviation in the XXI Century"

Springer

Editors
Andrii Bieliatynskyi
North Minzu University
Yinchuan, China

Vera Breskich
Peter the Great St. Petersburg Polytechnic
University
St. Petersburg, Russia

ISSN 2195-4356 ISSN 2195-4364 (electronic)
Lecture Notes in Mechanical Engineering
ISBN 978-3-030-85056-2 ISBN 978-3-030-85057-9 (eBook)
https://doi.org/10.1007/978-3-030-85057-9

This Springer imprint is published by the registered company Springer Nature Switzerland AG
The registered company address is: Gewerbestrasse 11, 6330 Cham, Switzerland

Contents

Observations of LED Side Lights at the Stage of Visual Piloting

Yuliia Kvach

Abstract The side lights of the light signal system indicate the side boundaries of the runway and its color marking by distance. To encode sections of the runway, depending on where they are installed, the lights emit white or yellow light. The fire must be in the pilot's field of vision from the moment of recognition until the moment of flight near it, only in this case the aerodrome lights give a clear idea of the location of the aircraft. The development of LED technologies has allowed the Ukrainian Association of Lighting Enterprises of Vatra Corporation to develop a design of LED equipment for aerodrome light signal systems, namely a universal fire that performs the functions of bidirectional fire and roundabout fire. Features of fire design, angles of convergence of side lights, difficulties in observing LED airfield lights affect on the visual contact. If the light signal system fails, the crew will not be able to establish its location or visual contact may be erroneous. To solve the complex problem of visibility of light signals from LED light signal aerodrome side lights, a tool using the MatLab interface is proposed. The tool allows to model a fragment of a light signal picture in the conditions of deficit of time at a stage of visual piloting. The tool simulates the illumination created on the retina of the pilot's eye from aerodrome side lights in the MatLab interface. Lighting simulation will manage risk in order to prevent aviation events.

Keywords LED fire · Observation of LED signals · Modeling of a fragment of a light signal picture · Illumination on a retina of a pilot's eye

1 Introduction

The side aerodrome lights of the light signal system limit the runway on two sides with the appropriate lighting characteristics and color of radiation.

Y. Kvach (✉)
National Aviation University, Liubomyra Huzara Ave., Kyiv 03058, Ukraine
e-mail: juli_k74@ukr.net

A. Bieliatynskyi and V. Breskich (eds.), *Safety in Aviation and Space Technologies*, Lecture Notes in Mechanical Engineering,
https://doi.org/10.1007/978-3-030-85057-9_1

The lighting characteristics and location of the side lights of aerodrome lights must be such that in difficult meteorological conditions the pilot can clearly see the required number of lights of the light signal system. The aerodrome lights of the light signaling system must be visible within a horizontal angle so that the pilot can observe them while piloting the aircraft both during the day and at night.

1.1 Side Lights of the Runway

Each side light emits light of different intensity, depending on which part of the light beam is their line of sight. Taking into account the characteristics of the side fire, setting the angle of elevation and the angle of convergence, you can determine the intensity of the light beam directed at the pilot. The angles of convergence of the side lights are 4.50 with a runway width of 60 m and 3.50 with a runway width of 45 m. The angle of increase of the side lights is 3.50.

A tool using the MatLab interface is offered to solve the problem of monitoring light signals from LED light signal aerodrome side lights. The tool allows to model a fragment of a light signal picture in the conditions of deficit of time at a stage of visual piloting. The tool simulates the illumination created on the pilot's retina by side LED aerodrome lights in the MatLab interface. Such lighting modeling will allow to manage risk in order to prevent aviation events.

1.2 Review of Research and Publications

The current regulations [1–6] contain requirements for the design and lighting characteristics of the side aerodrome runway fire.

In [7], the author presents the patterns of visual observation of objects depending on the conditions of observation and the characteristics of the object.

The study of the influence of blue light on physiological and subjective human reactions was considered by the authors in [7].

The definition of visual search of objects as airfield lights, constantly emitting light on the runway depending on the transparency of the atmosphere is considered in the simulation in the created tools of the MatLab environment in [8–10].

2 Problem Statement

The side lights emit a constant white light of constant light. In the presence of a shifted runway threshold, the side lights emit red light in the direction of approach to the landing and emit yellow light on the section that is one third of the length of the runway.

Side lights should be visible from all directions to help orient the pilot who is approaching or taking off in any direction. In the case where the side lights are designed to control the flight in a circle, it is necessary to observe them from all directions.

According to the requirements, the side lights emit light in all directions at angles of elevation up to 15є above the horizon with a light intensity of at least 50 cd. Except in the absence of external lighting, the light intensity of the side lights at the aerodrome shall be not less than 25 cd to prevent the pilot from being blinded.

The side lights of the runway, equipped for accurate landing, meet the technical requirements specified in Figs. 1 and 2.

Given the fact that the observation of the side light must be provided from a certain area of space, it is clear that the photometric body should be normalized in accordance with the recommendations of ICAO [1]. Spatial parameters of radiation (Figs. 1 and 2) provide the necessary guidance during the approach.

According to the perception of light color by the pilot, it is necessary that the technical conditions in determining the chromaticity of the colors of airfield lights meet the current conditions of the International Commission on Lighting (ICL) (Fig. 3) [3, 4].

But it is impossible to set the technical conditions for colors so as to completely eliminate the possibility of their misperception. For sufficiently reliable recognition, it is important that the illumination of the eye significantly exceeds the threshold of perception, that the color does not undergo significant changes due to selective (selective) atmospheric attenuation (absorption) and the color vision of the observer

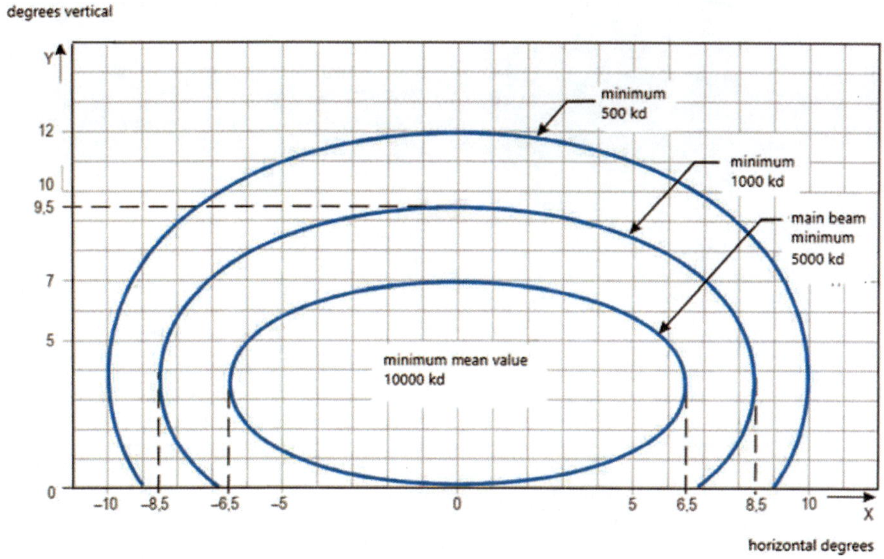

Fig. 1 Diagram of isocandles of side lights at a runway width of 45 m (White lights)

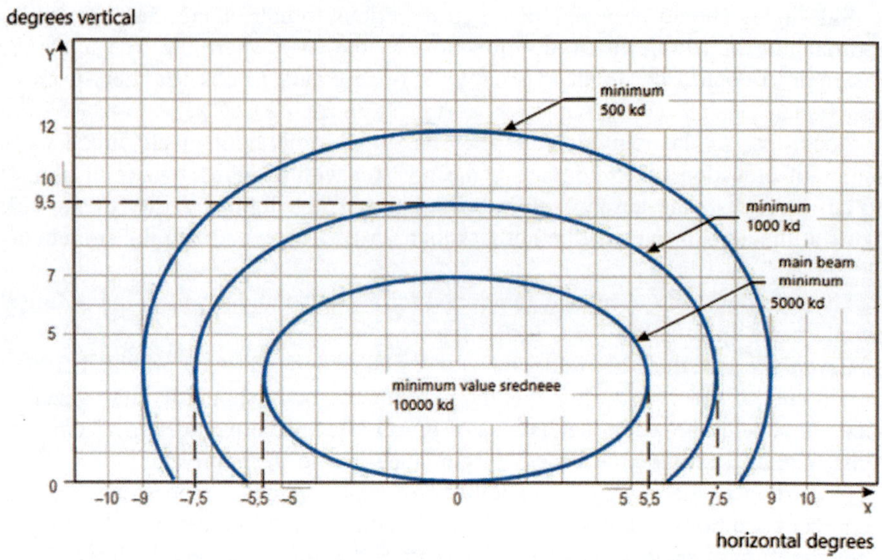

Fig. 2 Diagram of isocandles of side lights at a runway width of 60 m (White lights)

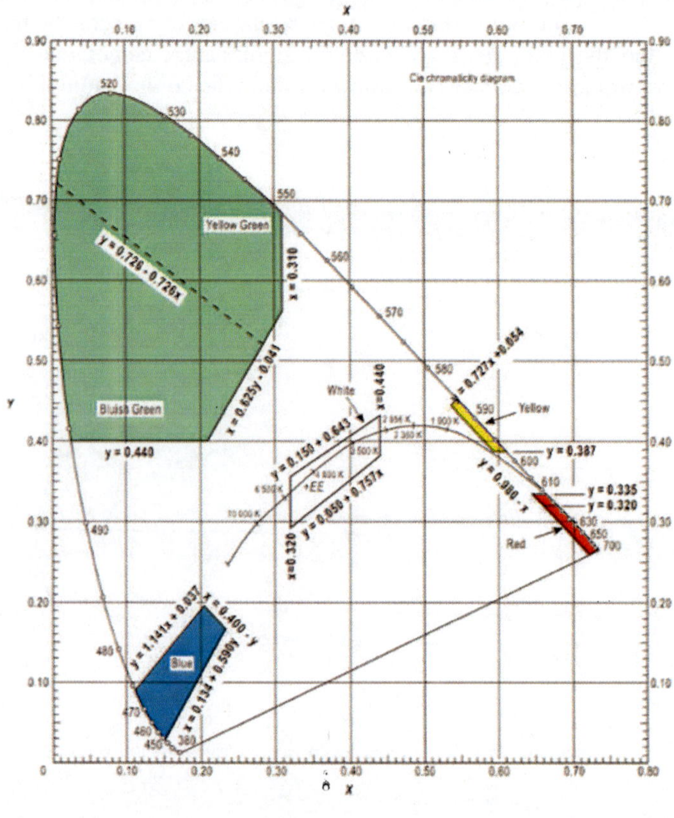

Fig. 3 Colors of LED aerodrome light signals

was normal. There is also a risk of distorted color perception in very bright light, which can be caused by a high-intensity light source at close range. Experience has shown that satisfactory recognition is possible when these parameters are given due attention.

Characteristics of chromaticity are made on the basis of research of visual perception and correspond to the coordinate system accepted by ICL. The parameters of the fire color are compared with the parameters determined on the outer curve of the isocandel, in order to ensure the absence of such a color change that could lead to incorrect perception of the light signal by the pilot.

3 Computational Methodology

The development of LED technologies has allowed the Ukrainian corporation "Fire" to develop a design of LED aerodrome universal side light NL 02V-RWY 45-R (L)-W/Y, NL 02V-RWY 60-R (L)-W/Y (Fig. 4). The use of LED modules in the construction of airfield lights has a number of advantages. The most significant advantages of LED airfield light include: low power consumption, instant on and off, vibration resistance, long service life, high light quality with a color rendering index in the range of 80–95 and an optimal radiation pattern with the ability to create directional light.

The design of the ground side runway light consists of two oppositely directed LED assemblies of eight LEDs mounted on a board complete with special optics to assemble and maximize light flux. Each module is represented by eight Cree LEDs (Fig. 5). The design of the bidirectional LED side light of the runway is additionally

Fig. 4 Ground side fire of the runway NL 02V-RWY 45-R (L)-W/Y, NL 02V-RWY 60-R (L)-W/Y

Fig. 5 Ground bidirectional side fire of the runway

supplemented by a reflector that forms the output light in the upper hemisphere. Thus, the improved design of the bidirectional side light of the runway can be considered a universal fire, which is a replacement for the installation of two ground lights and a roundabout type "bush". This design of universal fire must emit light intensity in the upper hemisphere of not less than 50 cd, performing the function of circular fire (Fig. 6).

In the study of bidirectional side LED runway light, the light radiation has parameters that meet (Fig. 7):

- chromaticity coordinates: $x = 0.5455$ $y = 0.4531/u' = 0.2970$ $v' = 0.5551$
- dominant wavelength 585.2 nm with 99.8% color purity
- luminous flux 138.2 lm
- RGB color ratio: R = 5.4% G = 94.6% B = 0.0%

To solve the problem of observing light signals from LED side lights of the runway, a toolkit using the MATLAB interface is proposed [8, 9]. The toolkit

Fig. 6 Improved design of bidirectional runway side lights

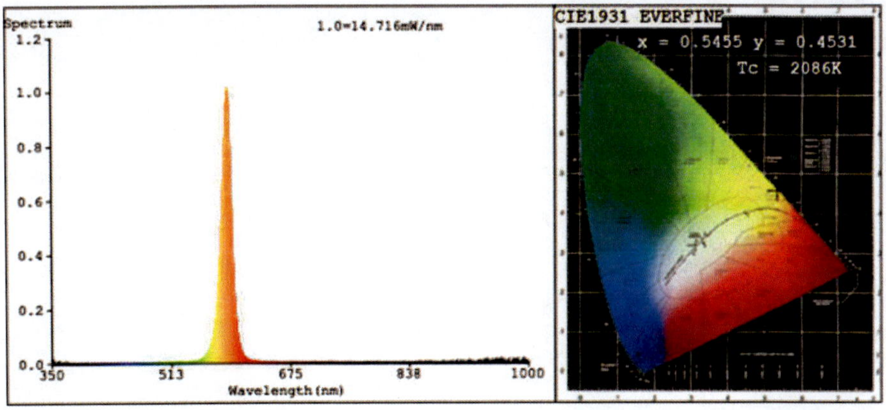

Color Parameters:

Chromaticity Coordinate:x=0.5455 y=0.4531/u'=0.2970 v'=0.5551

Tc=2086K(Duv=0.0118) Dominant WL:Ld =585.2nm Purity=99.8%

Ratio:R=5.4% G=94.6% B=0.0%;;Peak WL:Lp=587.5nm HWL:Lhd=14.1nm

Render Index:Ra=9.4

R1 =0	R2 =37	R3 =25	R4 =0	R5 =0	R6 =12	R7 =2	
R8 =0	R9 =0	R10=0	R11=0	R12=0	R13=0	R14=49	R15=0

Photo Parameters:

Flux = 138.2 lm Fe = 256.0 mW

Fig. 7 The result of the study of the chlorine characteristics of the side fire of the runway

allows to model fragments of a light signal picture of a runway in the conditions of deficit of time at a stage of visual piloting.

Using the tools in the environment MatLab [11–15] simulates the illuminance created on the retina of the pilot from the system of side LED lights of the runway, depending on the input to the input field (Fig. 8):

- coordinates (X, Y, Z) and angles (V, G) of glide firets on the runway;
- photometric data of glide fires by means of the respective ies-files of corresponding fires [11];
- the coordinates of the pilot's finding in relation to the glide fire system to determine the length of the trajectory between the light emitter and the receiver;
- parameters characterizing the complexity of meteorological conditions with indicating the background brightness (Lback), atmospheric transparency and meteorological of visibility range (MVR).

Toolkit allows to get:

- illumination taking into account the cosine dependence created by glide fires on the retina of the pilot's eye (E0);
- illumination by Allard, generated by glide fires on the retina of the pilot's eye;

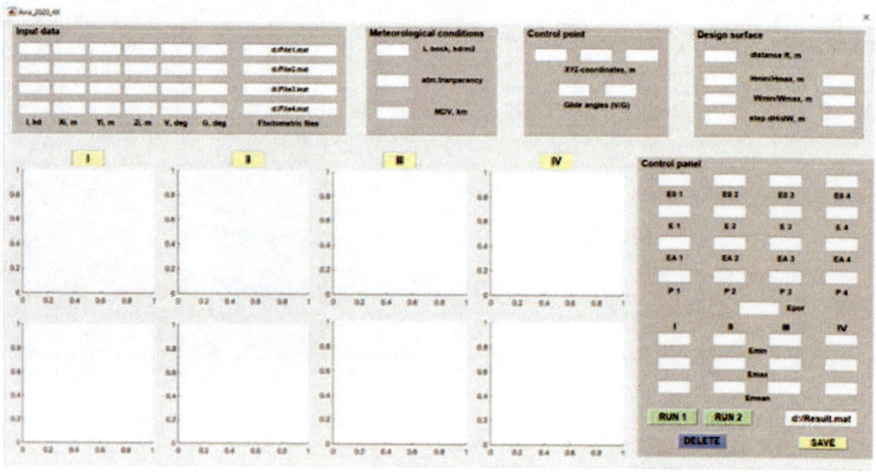

Fig. 8 Tools for determining the visual recognition of LED side lights of the runway de-pending on the angle of flight, individual characteristics of the observer, the transparency of the atmosphere

- total illumination generated by glide fires on the retina of the pilot's eye (E) with the determination of minimum, average and maximum values;
- the threshold of illumination, which allows to estimate the visibility of glide fires depending on difficult meteorological conditions;
- a graphic representation of illumination from each individual glide fire (Fig. 9);
- graphical representation of illumination from the system of glide fires;
- graphical representation of the radiation direction of glide fires (Fig. 10).

Graphical representations of the direction of a single runway side light or a group of runway side lights with appropriate light intensity on the pilot's retina allow conclusions to be drawn as to their contribution to the overall visual observation picture perceived by the pilot.

4 Results and Discussion

Universal runway side lights help the pilot follow the correct descent trajectory during approach and accurately control the distance from the runway even in difficult weather conditions. Therefore, it is important that the aerodrome side lights of the runway emit light both in the direction of approach to the landing and in the upper hemisphere. Thus, there is a designation of the geometric parameters of the artificial surface of the runway, i.e. a picture of observation is formed, which allows the pilot to navigate in three-dimensional space.

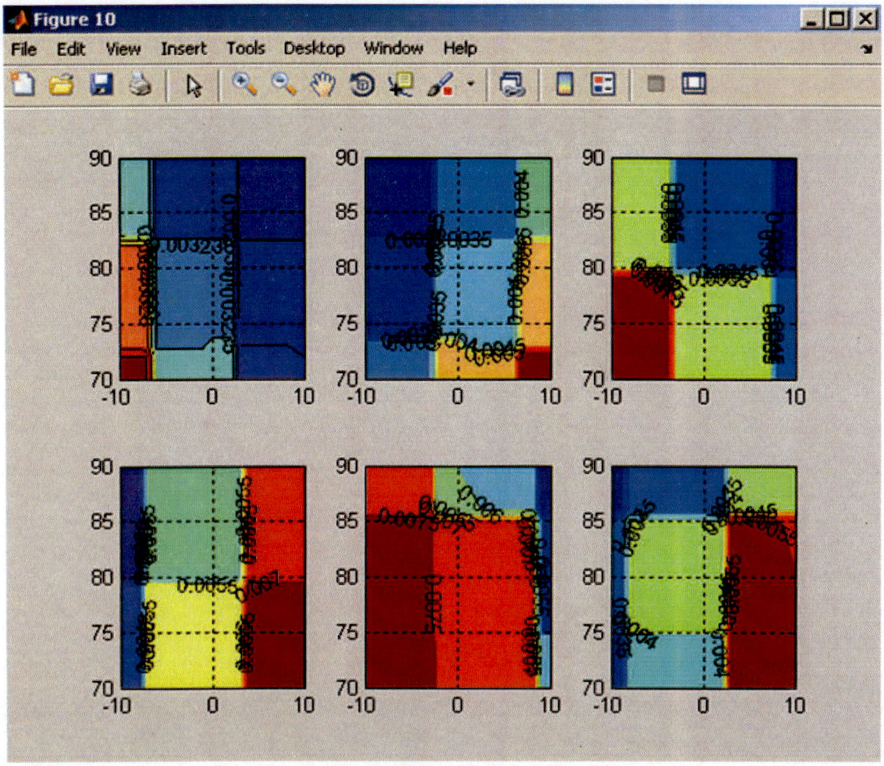

Fig. 9 Graphic representation of illumination of LED glide lights

In the calculations of the visibility of light signals it is necessary to take into account the light threshold, i.e. the lowest value of illumination on the pupil of the observer, when the light signal is confidently perceived.

Taking into account the design of the ground side fire, the modeling of the illumination created on the retina of the pilot's eye in different meteorological conditions was carried out.

When simulating the pilot's perception of the light signal from the group of side lights of the runway in fog (Fig. 11), the lights are not recognized, which is confirmed by the calculation of the probability of observation.

When modeling the pilot's perception of the light signal from the subsystem of the side lights of the runway in a haze (Fig. 12), but at dusk and against the background of snow—universal lights are not recognized.

When modeling the pilot's perception of the light signal from the subsystem of the side lights of the runway in conditions of good visibility (Fig. 13) during the day against the background of grass—the whole group of side lights of the runway is confidently different.

Thus, the simulation of the illumination created on the retina of the pilot takes into account the determination of the brightness of the background, which

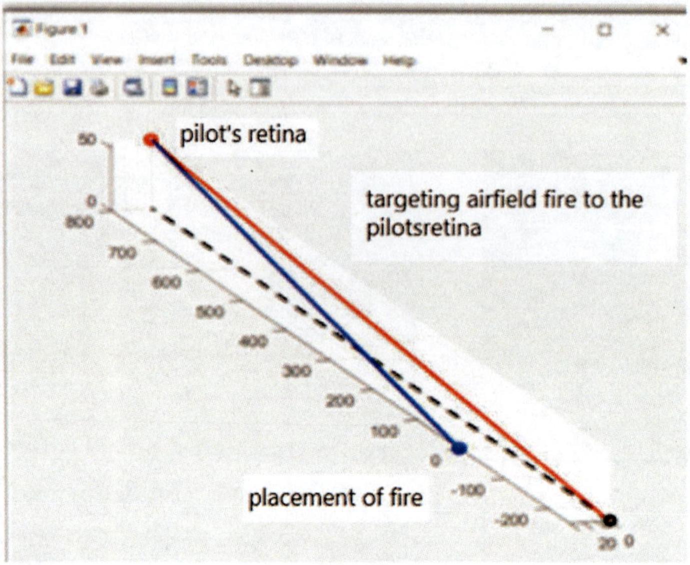

Fig. 10 Graphic representation of the radiation direction of the LED side lights of the runway with the appropriate light intensity on the retina of the pilot

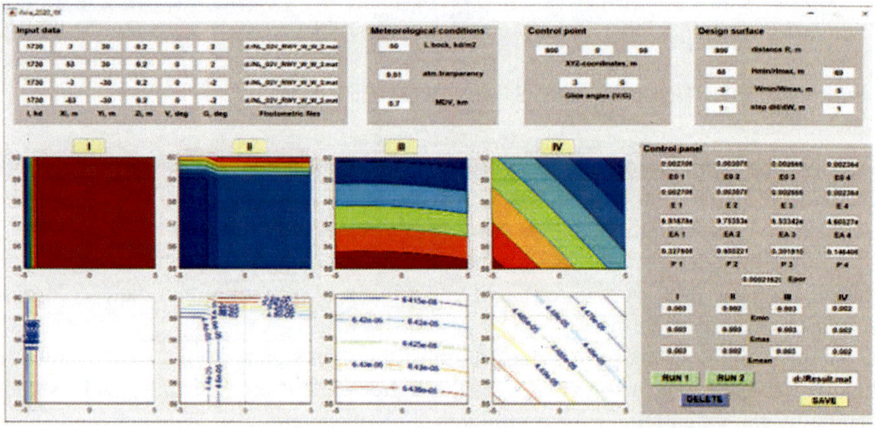

Fig. 11 Modeling of determination of visual recognition of LED side lights in conditions of visible fog

recognizes the individual ground sideways of the runway or subsystem of side lights of the runway depending on location coordinates and their photometric characteristics taking into account complex meteorological conditions.

Simulation of the light signal picture and observation of the LED side lights of the runway will allow to manage the risk in order to prevent aviation events.

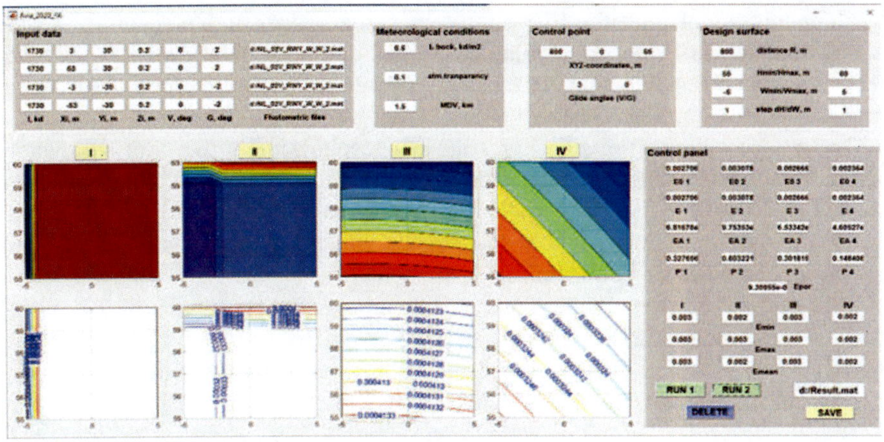

Fig. 12 Simulation of the definition of visual recognition of LED side lights in a haze on a background of snow

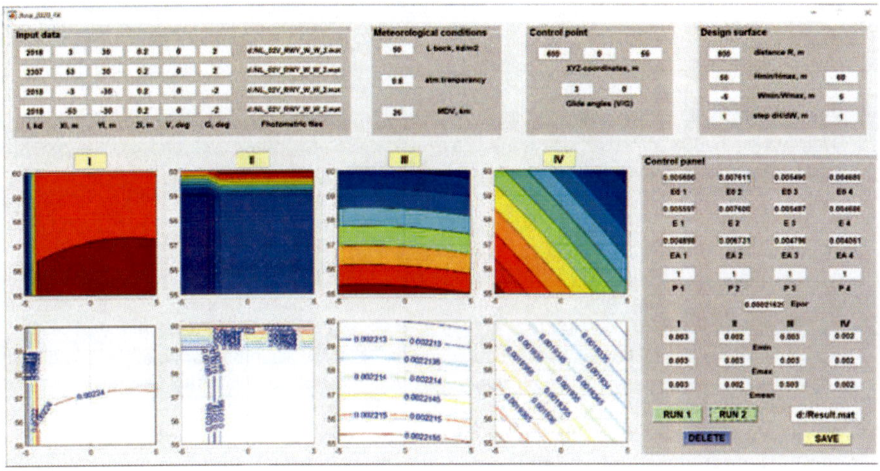

Fig. 13 Simulation of visual recognition of LED side lights in conditions of good visibility against grass background

5 Conclusions

According to the simulation of the visibility of the LED side lights of the runway by the pilot when visually guiding the observation at a distance of 800 m, which corresponds to the complexity of the category I aerodrome in conditions of visible fog, the decision error is quite high.

In conditions of light fog, the pilot does not recognize lights according to the simulation, which is not favorable. Therefore, the universal side LED light NL 02V-RWY 45-R(L)-W/Y and NL 02V-RWY 60-R(L)-W/Y in fog conditions is not recommended for use.

In haze conditions, the side lights are not recognized by the pilot simulation, which is not acceptable. Therefore, the universal side LED lights NL 02V-RWY 45-R(L)-W/Y and NL 02V-RWY 60-R(L)-W/Y in haze conditions both on the background of grass in the light of stars and on the background of snow when stars appear lights are not recommended for use.

The study of universal side lights in difficult weather conditions allows to say that the universal side LED lights NL 02V-RWY 45-R(L)-W/Y and NL 02V-RWY 60-R(L)-W/Y can be reliably recognized if illuminance on the retina of the pilot exceeds the threshold illuminance and the probability of observation is not less than one.

References

1. Doc 9157-AN/901 (2004) ICAO aerodrome design guide. 4. Visual aids, 4th ed
2. Doc 9157 (2017) Part 5—Electrical systems, Aerodrome design manual, pp 226
3. CIE—Publication (1971) Colorimetry, 15
4. ISO/CIE 11664-4: (2019) (E) Colorimetry. Part 4. CIE color space 1976 L * a * b *
5. Doc 9328-AN/908 ICAO (2005) A guide to the practice of tracking and reporting on visibility observation to runways, 3rd ed., pp 124
6. Price JC, Forrest JS (2016) Chapter 8—Airport safety programs. In: Practical airport operations, safety, and emergency management, pp 273–338. https://doi.org/10.1016/B978-0-12-800515-6.00008-1. https://www.sciencedirect.com/science/article/pii/B9780128005156000081
7. Scheuermaier K, Münch M, Ronda JM, Duffy JF (2018) Improved cognitive morning performance in healthy older adults following blue-enriched light exposure on the previous evening. Behav Brain Res 348:267–275. https://doi.org/10.1016/j.bbr.2018.04.021
8. Kvach Y, Zelenkov A (2018) Mathematical modeling of the visibility range of the aerodromie fires of the light-signal systems from the meteorological visibility range in Matlab environment. Aviation in the XXI
9. Kvach Y (2019) Modeling illumination on the retina of the pilot's eye during in complicated meteorological conditions in MATLAB environment. Svitlotekhnika Elektroenergetika 55. https://doi.org/10.33042/2079-424X-2019-2-55-59-62
10. Kvach Y, Shkvarnytska T (2020) Monitoring of performance of aerodrome lights in operating conditions. EasyChair Preprint, pp 3762. https://doi.org/10.13140/RG.2.2.24916.42882
11. Buffolo M, Meneghini M, De Santi C, Felber H, Renso N, Meneghesso G, Zanoni E (2016) Experimental observation of TDDB-like behavior in reverse-biased green InGaN LEDs. Microelectron Reliab 64:610–613. https://doi.org/10.1016/j.microrel.2016.07.103
12. Lindfield G, Penny J (2019) Chapter 1—An introduction to Matlab. In: Numerical methods, 4th ed., pp 1–72. https://doi.org/10.1016/B978-0-12-812256-3.00010-5
13. Radionova LV, Chernyshev AD (2015) Mathematical model of the vehicle in MATLAB simulink. Procedia Eng 129:825–831. https://doi.org/10.1016/j.proeng.2015.12.114

14. Šalík P, Róka R, Gorazd T (2018) Simulation platform of optical transmission system in Matlab simulink. Procedia Comput Sci 134:196–203. https://doi.org/10.1016/j.procs.2018.07.162
15. Suthanthira Vinita N, Manivannan L, Meenakshi T, Radhika K (2020) Stability analysis of Quadrotor using state space mathematical modeling. Mater Today: Proc 33(7):4040–4043. https://doi.org/10.1016/j.matpr.2020.06.428

Middleware Service for the Integration of Control Systems of the Aviation Engine and Aircraft

Serhii Tovkach

Abstract The integration of an aircraft engine and an aircraft is such a combination of their structural forms, aerodynamic and propulsive-economic characteristics and control systems (including thrust vector), which provides an improvement in certain flight performance characteristics of the aircraft. A program structural service of multi-mode aircraft aviation engine integration has been considered. It can be effectively applied for ensuring adaptation to changing flight modes by determining the shape of an aircraft and its aviation engine, from one side, for determining the purpose of aircraft, its parameters and aviation engine parameters, as well as flight modes by correct choice of control laws, from another side. The considered technique and service, based on mathematical modeling of the AIRCRAFT-AVIATION ENGINE-FUEL system, in combination with the use of multiparameter optimization methods, is an effective tool for choosing design parameters and forming the technical appearance of an aviation engine.

Keywords Aviation engine · Fuel system · Optimization · Integration · Computer technologies · Flight programs · Thrust

1 Introduction

During an aircraft designing, the most important tasks can be defined as the choice of the type of engine, the parameters of its working process, dimensions, which are determined, as a rule, by the required thrust-to-weight ratio, as well as the layout of the control system and its location on the aircraft [1–8], which are estimated primarily by the advantages obtained from their mutual integration.

S. Tovkach (✉)
National Aviation University, Liubomyra Huzara Ave. 1, Kyiv 03058, Ukraine

A. Bieliatynskyi and V. Breskich (eds.), *Safety in Aviation and Space Technologies*, Lecture Notes in Mechanical Engineering,
https://doi.org/10.1007/978-3-030-85057-9_2

For better adaptation the aviation engine (AE) characteristics to the tasks solved by the aircraft in flight, it is necessary to integrate control systems. Integrated control systems are especially effective for the control systems of modern multi-mode aircraft. On the basis of such systems, optimal control programs for the control system are formed using the criteria for estimation the effectiveness of the aircraft.

As a result, the main task of the automatic control system (ACS) is formed, which boils down to the maximum adaptation of the thrust and economic characteristics and the engine stability reserves to the needs of the aircraft. The optimal control program for the AE is a program for changing the controlled parameters of the engine, air intake and jet nozzle, in which the selected criterion of optimality reaches an extreme value.

In this case, matching the characteristics of the engine and the aircraft in the take-off areas and at transonic and supersonic flight speeds is achieved by increasing the steepness of the increase in thrust with an increase in flight speed in the modes of full engine uprating.

For this purpose, combined control programs are used to increase the gas temperature T_g^* in front of the turbine and in the afterburner T_{af}^* and to "spin up" the engine rotors according to the flight M number. The "temperature uprating" can reach 150–200 K. A typical example is the RD-33 engine control program [1–3]. Due to the presence of large excess thrust in aircraft with a high thrust-to-weight ratio, their integration with the aviation engine can be expanded by using the engine as an energy source for direct improvement of the aircraft's flight characteristics, for example, by taking compressed air from the engine and blowing it out onto the bearing surfaces of the airframe.

Thus, blowing out thin flat jets of air through slots in the area of the trailing edges of the cantilever part of the wing (jet flaps) or on the upper surface of the wing in the area of mechanical flaps allows, due to the effect of supercirculation and deflection downward from the axial direction of the air jets, to increase the lift of the wing and, thus, improve the takeoff and landing and maneuverability of the aircraft. Investigations [1–4] show that with numbers $M < 0.5$–0.7 the increase in lifting force can reach 20–30%.

The implementation of such systems should provide for the optimization of the amount of bleed air, sampling points and parameters of compressed air transportation systems to the places of its blowing on the aircraft, since air bleeding from the engine, contributing to the improvement of the aerodynamic characteristics of the aircraft, leads to a decrease in the engine thrust. To reduce thrust losses during air intake, it may be necessary to adjust the elements of the engine flow path.

An investigation [1–14] of the engine by Boeing, Pratt & Whitney and Rolls Royce showed that the creation of an aircraft-aviation engine-fuel system will significantly improve fuel efficiency, reliability and flight safety.

2 Aviation Engine Efficiency of High-Speed Aircraft

For choosing the type and parameters of engines and control systems of high-speed aircraft, the type of engine and its position scheme on the aircraft, the used type of fuel, the influence of the control system and fuel on the scheme, the volumetric layout and aerodynamic characteristics of the aircraft, etc. should be taken into account [8]. A schematic diagram of such investigations is shown in Fig. 1.

The choice of parameters and the estimation of the efficiency of the AE are made on the basis of a system approach, when the engine and the power plant are considered as subsystems of a higher level complex. To solve the problems of multi-parameter optimization, complex mathematical models of the entire system, consisting of aircraft and control systems, are developed, taking into account the properties of the fuels used. The indicators of the functional efficiency of the aircraft are used as criteria for the selection of parameters and the formation of the technical appearance of the AE. In the considered investigation, the aviation engine is represented by its effective characteristics, obtained from the conditions for considering the joint operation of the elements, and the specific mass characteristics, and the aircraft is represented by the volumetric layout, aerodynamic and specific mass characteristics of the glider. The flight is performed according to a given or optimized program.

The integration of the aviation engine and the aircraft is carried out on the basis of the conditions for ensuring the mass balance, volumetric layout and starting thrust-to-weight ratio. The practical flight range L is considered as an efficiency

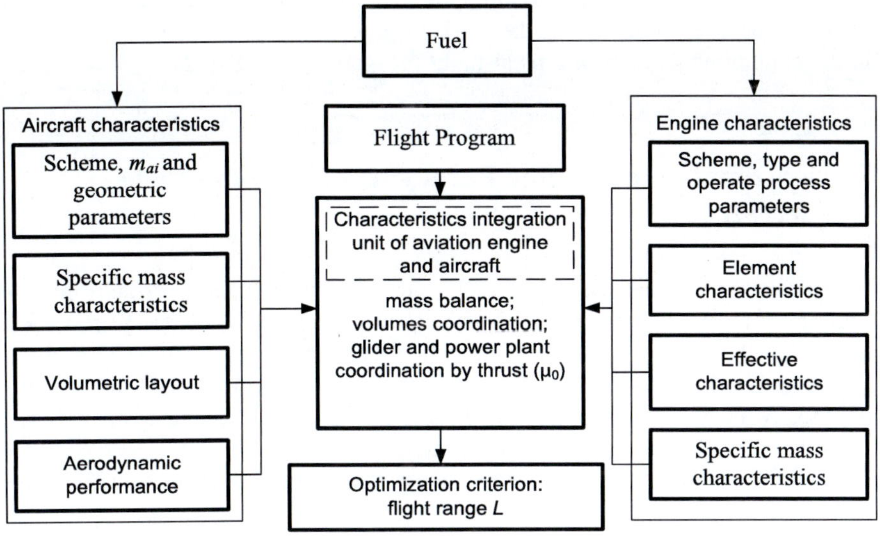

Fig. 1 Investigation of the efficiency of aviation engine of a high-speed aircraft [1]

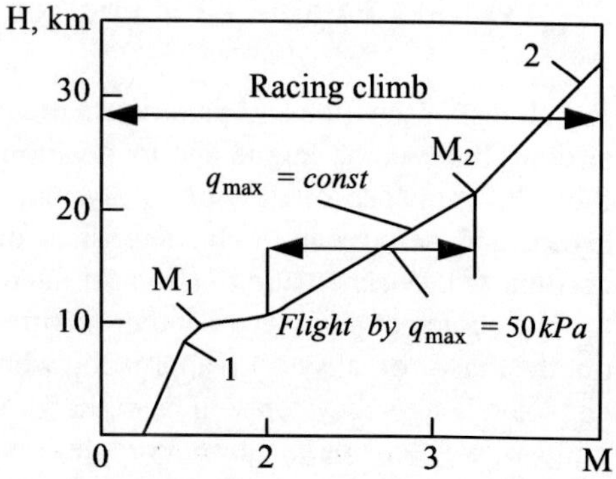

Fig. 2 Typical hypersonic flight program: 1 is subsonic cruise; 2 is hypersonic cruise flight

criterion. These conditions allow for the interconnection of quantitative indicators of the entire system.

The mathematical models of engines [9] (from the inlet to the compressor or to the combustion chamber and to the throat section of the jet nozzle) are based on methods traditional for the theory of aviation engines, but taking into account the real properties of the gas, dissociation and uneven flow at the inlet.

In parallel with the mathematical model of the engine to determine the effective traction and economic characteristics of an aviation engine, a technique for calculating the mass and size characteristics of engines and all other elements should be developed [4, 5]. Such techniques can be created on the basis of draft design, taking into account statistical data on completed and designed structures.

In combined power plants (PP), engines of various types can be considered, for example, TLP with a parallel or sequential (tandem) arrangement of circuits, turbofan and ramjet engines and steam-hydrogen rocket-turbine engines (RTEsh) of several types, the parameters of the working process of which are optimized based on the conditions L_{max} for a typical flight program [5, 6] (Fig. 2).

3 Coordination of Engine Parameters for Multi-mode Aircraft

In a certain respect, any aircraft is multi-mode, but the property of multi-mode is most characteristic of fighter aircraft, which must perform a wide range of various tasks (Fig. 3): intercept enemy aircraft at the maximum possible (supersonic) speed; conduct a maneuverable air battle in a wide range of altitudes and flight speeds; accompany attack aircraft in flight to distant targets; carry out long-term air patrols in a given area, attack various ground targets; make throws at the surface of the earth with flying around obstacles or following the terrain, based on poorly prepared airfields [11].

It is obvious that the requirements for the engines imposed by different flight modes of a multi-mode aircraft are extremely contradictory. For ensuring high

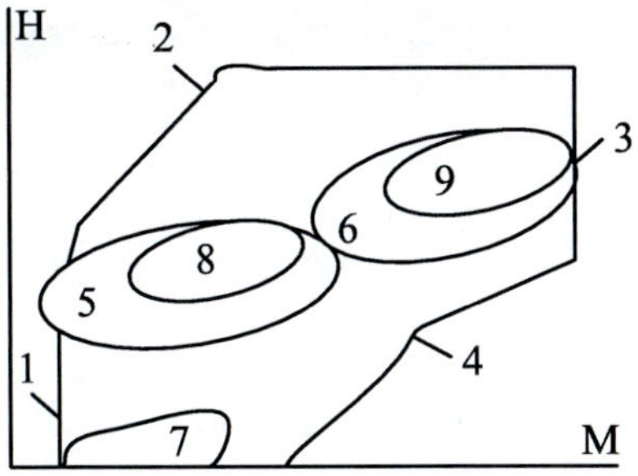

Fig. 3 Areas of the main flight modes of a multi-mode aircraft with restrictions: 1 is by wing bearing properties; 2 is by engine traction capabilities; 3 is by kinematic heating; 4 is by aircraft strength; 5 is subsonic maneuverable battle; 6 is interception; 7 is ground target attacks; 8 is subsonic cruising; 9 is supersonic cruising

limiting values of altitude and speed, maneuverability, short take-off distance, first of all, high engine thrust is required, even at the expense of efficiency, which is provided with great difficulty with a high thrust-to-weight ratio of the aircraft. When performing a number of tasks, the dynamic properties of the engine [11–13] are more important.

The main problems solved at the stage of conceptual design of an aviation engine of a multi-mode aircraft include the following:

- choice of engine type is with or without afterburner; traditional schemes or with a variable workflow;
- selection of the size of the engine (thrust-to-weight ratio of the aircraft) and another interconnected parameter of the aircraft tie is the specific wing load;
- selection of the degree of bypass of the engine on the assumption that other parameters of the working process ($T^*_{g\,\max}, \pi^*_{c\Sigma}$) are determined by the level of capabilities achieved in this generation;
- selection of the estimated number M of the air intake;
- selection of the laws of engine regulation, which determine the nature of the flow of its altitude-speed and throttle characteristics, for example, the law of change T^*_g in flight speed (braking temperature).

4 Middleware Service

For creation the information systems (IS) of aviation engine (AE) and aircraft, it is necessary to use ready-made software platforms for group work that already have powerful tools for multi-platform software development in their arsenal, ready-made technologies for deploying and managing the security subsystem.

Control programs (flight programs) must perform a number of sequential steps before each call to the integration central unit:

- Determine the number of physical databases and their names installed on the integration central unit;
- Determine the ability to access each database separately;
- Execute the corresponding query for each database;
- Process and fill in the received response; (repeat steps 2–4 for each database; add up the accumulated responses and process them as a single package of information).

For effectively solve such a problem, it is necessary to use special middleware services, that is a general-purpose service that sits between platforms and applications of distributed system of aviation engine and aircraft (Fig. 4). By platform, can be meant a set of low-level services and processing elements defined by a pair of processor architecture and operating system APIs such as Intel × 86 and Win32, Sun SPARCstation and SunOS, IBM and AIX, Alpha AXP.

Middleware integrated environment is a programming environment that is designed to simplify the development and management of applications for the specialized application domain (Fig. 4). An Integrated Environment (IE) is defined by an application programming interface, user interface, and toolkit. In addition to middleware services that an IE imports from other products, it can also include its own services. These are the so-called transaction monitors (IBM CICS, BEA Systems Tuxedo, Transarc Encina), fourth generation languages (Uniface, Cognos and Focus), CAD systems (Mentor Graphics Falcon, DEC Powerframe), CASE systems (HP SoftBench, Texas Instrument Composer), and distributed resource management systems of aviation engine and aircraft.

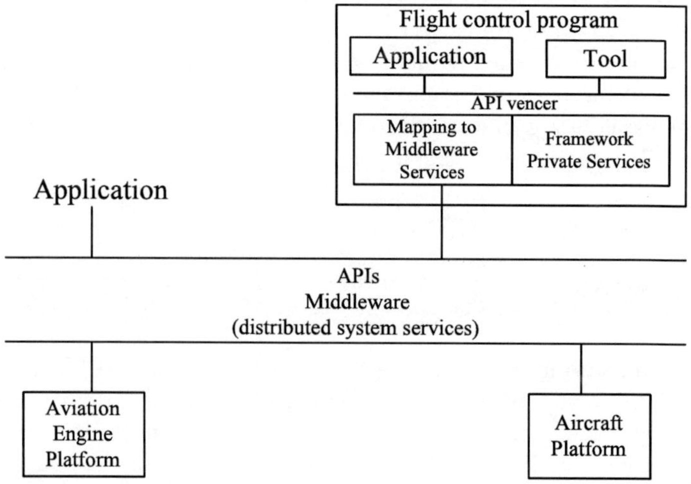

Fig. 4 Middleware integrated environment

During the organization of the interaction of information systems (α_1 and α_2), it is required to implement the rules for mutual data conversion [13, 14]:

$$\Omega = \begin{cases} \Psi\ \alpha_1 \Rightarrow \alpha_2 \\ \Psi\ \alpha_2 \Rightarrow \alpha_1 \end{cases} \quad (1)$$

where Ω is the space of conversion rules; $\Psi\ \alpha_i \Rightarrow \alpha_j$ is the rule of conversion from one information system to another, $i, j = 1, 2, i \neq j$.

Each rule for converting an information system contains many corrections for converting individual objects:

$$\Psi\ \alpha_i \Rightarrow \alpha_j = \begin{cases} Y_1(\omega_1\ \alpha_1 \mapsto \omega_1\alpha_2); \\ Y_2(\omega_2\ \alpha_1 \mapsto \omega_2\alpha_2); \\ \vdots \\ Y_n(\omega_n\ \alpha_1 \mapsto \omega_n\alpha_2), \end{cases} \quad (2)$$

where $\omega_1\alpha_i \ldots \omega_n\alpha_i$ are the objects of the information system for which the conversion rules are determined; $Y_1 \ldots Y_n$ is conversion rule for a separate object, i, $j = 1, 2, i \neq j$.

During the organization of the information interaction between n information systems, it is required to determine a total $n^2 - n$ of conversion rules Ω while the total number of object conversion rules Ψ cannot be calculated.

5 Conclusions

The analysis of the scheme, parameters and appearance of the engine are determined according to the relevant criteria for the efficiency of the aircraft based on the use of multi-parameter optimization methods. The development of mathematical models for optimization studies is based on the use of characteristics of aviation engine elements obtained experimentally, as well as on the basis of simplified (one-dimensional) methods for their calculation. Computer technologies are becoming more widespread for 2D and 3D modeling and calculation of the characteristics of aviation engine elements for an aircraft of a given intended purpose.

Taking into account the peculiarities of information resources and services of automated control systems for aviation engines and aircraft, consisting in ultra-high volumes of information arrays, their distributed nature, heterogeneity and different conditions of maintenance, control and access policy, one of the key tasks for achieving the functional properties of qualitatively improved versions of information systems the next generation should include:

- the need to use ready-made software platforms for group work with powerful tools already in their arsenal for multi-platform software development, ready-made technologies for deploying and managing security subsystems;
- the use of common Middleware services, which allows the Spiral Model evolutionary method to be applied to introduce new versions of the information system by replacing basic services with new versions with a decrease in their unit cost;
- implementation of the possibilities of information exchange between functional subsystems by integrating heterogeneous information flows on the basis of bringing them to a single universal presentation format.

Large aviation organizations already rely on middleware, the trend towards simplifying middleware and expanding its functionality in aircraft and aircraft engine applications will make this dependence even stronger.

References

1. Solonin VI, Palkin VA (2014) Development of the united states of technology for creating engines of the sixth generation for military aviation. Engine 4(94):2–7
2. Seabridge A, Moir I (2020) Design and development of aircraft systems. Wiley. 3, 400
3. Etzkorn LH (2017) Introduction to middleware. Web services. In: Object components and cloud computing, pp 688
4. Tovkach SS (2020) The Liebman process for distribution of the information flows of the engine automatic control systems. J Nano- Electron Phys 12(1):01003. https://doi.org/10.21272/jnep.12(1).01003
5. Ayoade G, Chamry A (2018) Secure data processing for IoT middleware systems. J Supercomput 75(7):1. https://doi.org/10.1007/s11227-018-26868
6. Bagai IE, Roedig U, Hollick M (2015) Using channel state information for tamper detection in the internet of things. In: Proceedings of the 31st Annual Computer Security Applications Conference, pp 131. https://doi.org/10.1145/2818000.2818028
7. Mechitov K, Gul A (2012) Building portable middleware services for heterogeneous cyber-physical systems. 31. https://doi.org/10.1109/SESENA.2012.6225732
8. Marcin Z (2017) Reconfiguration of unmanned aircraft control systems. Trans Aerosp Res 2:80. https://doi.org/10.2478/tar-2017-0017
9. Wirma Y, Maharani A, Hidayah Z (2020) Indonesia's civil servants' performance at aviation engineering: exploration study. Integr J Bus Econ 4(2):160. https://doi.org/10.33019/ijbe.v4i2.279
10. Ab N, Bhardwaj AK (2014) Important pilot static system in aircraft control system. Am J Eng Res 1 03(10):138
11. Mao L, Xianlian X (2019) Design on four-axis aircraft control system based on Somatosensory interaction. IOP Conf Ser Mater Sci Eng 042052. https://doi.org/10.1088/1757-899x/569/4/042052
12. Vinnichenko AV, Nazarevich SA, Karpova IR (2020) Automated reverse engineering control system model. Issues Radio Electron. https://doi.org/10.21778/2218-5453-2020-4-39-49
13. Gryadunov KI, Kozlov AN, Nemchikov ML, Melnikova IS (2019) Aviation engines diagnostics by estimating the metal contamination in oils. Civil Aviation High Technol 22 (3):35. https://doi.org/10.26467/2079-0619-2019-22-3-35-44
14. Terentev M, Karpenko S, Zylin E (2018) Nonparametric method for failures diagnosis in the actuating subsytem of aircraft control system. IOP Conf Ser Mater Sci Eng 312(1):012025. https://doi.org/10.1088/1757-899X/312/1/012025

Application of Fiber-Optic Sensors for the Aircraft Structure Monitoring

Ruslan Zakirov and Feruza Giyasova

Abstract There is a transition from periodic forms of aircraft maintenance to the "on condition" forms performing currently can be observed. Aircraft Health Monitoring system is a very important tool in the formation of "on condition" based maintenance schedule for modern commercial aircraft, which allows collecting and analyzing information about aircraft systems condition. One of the most important components of Aircraft Health Monitoring is the Aircraft Structure Condition Monitoring system, which allows detecting and conditions monitoring of various types of internal damages and mechanical stresses arising in the aircraft structure during its operation. The main advantage of the Aircraft Structure Condition Monitoring system is the absence of its influence on the mechanical and strength characteristics of the aircraft structural elements. Implementation of the system is very important due to constantly expanding possibilities of using composite materials in aircraft structure. This article proposes a method for implementing Aircraft Structure Condition Monitoring system by using fiber–optic sensors. A feature of the proposed system is the integration of sensors into the structure of the aircraft. A review and analysis of the applicability of photodetectors for the proposed system is also given.

Keywords Aircraft · Monitoring · Fiber-optic · Structure · Sensor · Maintenance · Bragg grating

1 Introduction

A significant contribution to the issue of flight safety may be created by onboard systems of a new generation, which not only monitor the technical condition of the aircraft, but can also predict the moment when a problem unit will require

R. Zakirov (✉) · F. Giyasova
Semiconductor Physics and Microelectronics Research Institute, National University of Uzbekistan named after Mirzo Ulugbek, Universitet Str., Building 4, Tashkent 100174, Uzbekistan

A. Bieliatynskyi and V. Breskich (eds.), *Safety in Aviation and Space Technologies*, Lecture Notes in Mechanical Engineering,
https://doi.org/10.1007/978-3-030-85057-9_3

replacement. Such systems can diagnose episodic failures which can appear during flight. The transition from the routine maintenance to aircraft health monitoring (AHM) expected soon. These systems predict failures and degradation of unit performance and automatically create maintenance and repair schedule, which in the long term can lead to a complete rejection of periodic forms of maintenance.

Increasing of composite materials using in aircraft structure, including in aerospace engineering, as well as in other areas requiring increased reliability of structures [1], makes the task of structure health monitoring (SHM technology—one of the most significant part of AHM technology) very important. The creation and application of "smart" materials, including those with adaptive properties, also requires the use of a more advanced system of SHM. Such a system should promptly detect defects in the structure, adequately respond to their presence and issue appropriate influences or recommendations.

One of the most promising approaches is the using of fiber–optic sensors as part of the structure health monitoring system. Fiber–optic sensors have the following significant advantages over classical ones:

- low weight;
- high sensitivity;
- electromagnetic compatibility;
- possibility of networking, as well as multiplexing;
- compatibility with composite structures.

A great advantage of fiber optic sensors is also the ability to measure many parameters necessary for aircraft structure diagnostics:

- deformation, pressure and force;
- electric and magnetic fields;
- sound and vibration;
- pH and viscosity; etc.

2 Fiber Optic Sensor Classification

Fiber optic sensors use Interferometry principles of operation and divided as follow:

- Fabry–Perot interferometer;
- Michelson interferometer;
- Sagnac interferometer;
- Mach-Zender interferometer.

Sensors can also be built on a grating basis. These sensors are divided as follow:

- long period grating (LPG);
- fiber Bragg grating (FBG);
- tilted Fiber Bragg grating;
- chirped Fiber Bragg grating.

Distributed optical fiber sensor (DOFS) are also used in SHM. The following principles are used in these sensors:

- Raman scattering;
- Rayleigh scattering;
- Brillouin scattering.

Let us consider the described types of fiber optic sensors in detail.

2.1 *Interferometric Sensors*

The most widespread items are the fiber-optic interferometers of Mach–Zehnder, Meikelson, Fabry–Perot, Sagnac. Using of a resonator in the path of transmitted radiation is a distinctive feature of interference sensors. Combination or separation of radiation is performed by using fiber-optic splitters and mirrors, which are dielectric coatings, including multilayer ones, applied to the end of the fiber.

Due to the emerging phenomenon of light interference, it is possible to determine the corresponding parameters of the material or processes occurring in it [2]. This method makes it possible to measure ultra-small deformations with a high frequency, which makes it possible to register acoustic waves in the material.

The resolution of Fabry–Perot sensors is up to 0.000015%, the range of working deformations is $\pm 0.1\%$ and temperatures are from –40 to +250 °C [3]. Fabry–Perot sensors have small linear dimensions as a rule from 1 to 20 mm long. With these sensors it is possible to register both ultrasonic vibrations and acoustic emission. To detect structural damage in composite aircraft structures, it is possible to use such sensors for recording acoustic waves.

2.2 *Fiber Bragg Sensing Technology*

The sensing element of a point fiber–optic sensor is a fiber–Bragg grating (FBG). A fiber Bragg grating reflects radiation at a certain wavelength and is transparent for other wavelengths. This selective reflection is achieved by creating a periodic structure in the fiber core. The reflected signal is recorded by the receiving equipment.

An external influence on the FBG leads to a change in the parameters of the Bragg grating, which in turn leads to a change in the wavelength reflected from it. This change can be used to make decision about the required characteristics—deformation, temperature, pressure.

FBG–based sensors do not have any electronic components, i.e., they are passive. This opens wide possibilities for the using of such sensors in hazardous areas and in areas with strong electromagnetic interference.

Multiple Bragg gratings can be created on a single fiber. Each of them will reflect radiation at its own wavelength. This allows us to create a distributed monitoring system with wavelength multiplexing.

Advantages of FBG-based sensors:

- wide range of measurements;
- ability to integrate the sensor into the control object;
- explosion and fire safety;
- easy way of multiplexing;
- insensitivity to electromagnetic interference;
- resistance to the environmental impact.

2.3 Distributed Sensing Technology

In the distributed measurement method, the fiber optic cable itself is a distributed sensing element, that is, a sensor (Fig. 1). The method is based on registration of stimulated Raman scattering. The scattered signal is recorded by the receiving equipment [4].

There are several ways to process signals when using a distributed sensor: Optical time–domain reflectometry (OTDR), Raman optical time-domain reflectometry (ROTDR), and reflectometry based on Brillouin scattering (Brillouin optical time–domain reflectometry—BOTDR).

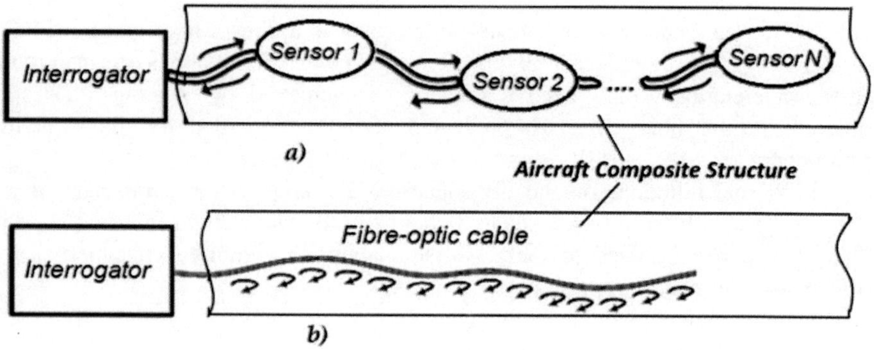

Fig. 1 Fiber-optic sensor system: **a**—fiber-optic sensors based on a Bragg grating; **b**—distributed fiber-optic sensors

The OTDR method is based on Rayleigh scattering caused by the transmission of light through an optical fiber. The scattered signal is registered by a photodetector.

The ROTDR method is based on Raman scattering. In this case, spectrum components called Stokes and anti–Stokes components that were missed in the emitted signal appear in the scattered signal. By the ratio of the intensities of these components, it is possible to determine the temperature of the fiber at any point. The BOTDR method is based on Brillouin scattering. The deformation and temperature of an optical fiber can be determined by the frequency change of the scattered light described in details in [5].

The properties of the fiber-optic cable can be varied over a wide range, resulting in fiber with properties that meet the requirements of the application.

3 Application of Fiber Optic Methods for Aircraft Structures Integrity of Monitoring

3.1 Acoustic Emission

Currently, there is a continuous trend towards an increasing of using of composite materials in aircraft construction. However, it is much more difficult to detect damage in composite structures than in metal ones. Cracks and delamination in the material can occur without visible damage to the surface layers. By monitoring acoustic emission, it is possible to detect the appearance of such internal damage by analyzing the sound waves emanating from them, and also to determine their location.

Conventional piezo-acoustic sensors are not very useful for the aircraft structure control due to their sensitivity to electromagnetic interference, massive connecting cables, complexity with multiplexing and impossibility of using in hazardous areas.

Using of fiber–optic distributed systems for monitoring acoustic emission is free from the above disadvantages. Localization of the source of acoustic emission is also possible. For these purposes, in [6], one of the methods of artificial intelligence was applied, namely, the Support Vector Regression (SVR) method.

3.2 Localization of Defects

Structural health monitoring is critical for loop safety and predicting the life-of-structure—especially in aviation and aircraft production industry. Among the various methods, Lamb waves have great opportunities for detecting damage and defects due to their sensitivity to small damage and the possibility of propagation over relatively large distances.

Using of fiber–optic sensors based on Bragg gratings for detecting Lamb waves, including a phased array, is shown in [7]. Such sensors have small size and weight, and can operate in hard operating conditions—at high humidity, under water, at high temperatures, and can also be built into the structure of a composite material. It is also possible to generate sound using optical fiber and a laser source.

3.3 *Temperature Measurement*

Along with the monitoring of aircraft structure mechanical stresses, temperature measurement is one of the main tasks solved using fiber-optic sensors in AHM. Temperature measurement is performed using both fiber-optic sensors based on Bragg gratings and distributed fiber–optic sensors [8].

Typical applications for fiber optic temperature sensors are fire alarm systems, thermal monitoring of power cables and power lines, temperature monitoring of chemical processes, and pipeline leak detection. It is also possible to embed such a sensor into composite structure and measure its temperature field.

3.4 *Stresses and Deformations Monitoring*

Monitoring of mechanical stresses inside structures is one of the most frequently solved problems using fiber–optic sensors [9]. Currently, the most accessible are sensors based on a fiber Bragg grating, the principle of which is based on changing its period under the influence of external factors. This change is caused both by a change in the deformation of the lattice and by a change in its temperature. Without information about the temperature, it is impossible to give an unambiguous assessment of the deformation type and size.

Using of an FBG embedded in a composite material as a sensor is associated with certain difficulties associated primarily with the fact that both axial and radial components contribute to the resulting deformation of the lattice.

According to the values of stresses inside the structure, it is possible to assess its shape, which is especially important for use in "smart" structures [10]. In the aircraft production industry using of "smart" designs will make it possible to create unique devices for the active control of aerodynamic flows, which will significantly increase the efficiency and safety of aircraft composite structure diagnostics technology.

A femtosecond laser can also be used to create various structures inside a fiber—waveguides, Bragg gratings, interferometers, spectrometers, polarizing elements, mirrors, etc. (Fig. 2). These structures may have different orientations and can help to solve the problem of three–dimensional voltage measurement and determination of the shape of the structure using a fiber Bragg grating.

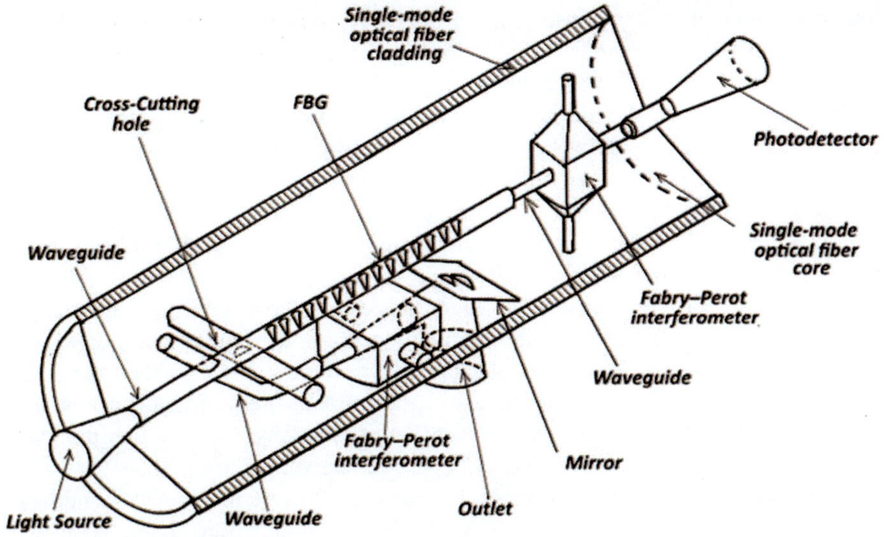

Fig. 2 Variants of structures formed in fiber using a femtosecond laser. *Source* https://dspace.susu.ru/xmlui/bitstream/handle/0001.74/29171/2019_464_vakhitovadv.pdf?sequence=1

Due to the waveguides created inside the fiber, it becomes possible to output radiation not only through the ends of the fiber. This opens up wide possibilities for creating distributed networks of such sensors with different characteristics.

The complex development of such systems can lead to the creation of a fiber–optic sensor that does not require an interrogator or contains it inside the fiber. Such a sensor can be built into the material and not have any physical connections with the outside components, i.e., there will be no need to bring the sensor outside and process the results of its operation on bulky equipment, which will improve the technology and reduce the cost of using such sensors. The measurement results will be transmitted over a radio channel, which will also be built into the fiber.

4 Analysis of Photodetectors for Using in SHM System

Engineers designing optoelectronic devices for optical signal detection and transmission recently started to pay particular attention to the development of photodiodes operable in the spectral range of 0.85 and 1.31 μm. Photodiodes for the above spectral ranges are mostly designed and manufactured on the basis of gallium arsenide, indium phosphide, cadmium sulfide and telluride, as well as on solid solutions based on them. For example, photodiodes with a p–n junction based on gallium arsenide make it possible to obtain photosensitivity in the spectral range of 250–900 nm. Review of the existing experimental results in [11] indicates that the maximum photosensitivity in the gallium arsenide photodiode occurs near 800 nm.

The maximum photo–response reaches 0.5 A/W, and at 300 nm appeared to be 0.04 A/W. The material most suitable for the spectral range of 0.9–1.7 μm is zinc doped indium phosphide with a photo–separating p–n junction.

In heteroepitaxial N^{+}–InP/n_0–$In_{0.53}Ga_{0.47}As$/n^{+} InP photodiode structures low capacitance values (1.6 pF at 5 V) and high photosensitivity at a maximum of 0.9 A/W were reported [12]. The current in the measuring circuit did not exceed 100 μA. $In_xGa_{1-x}As$ is believed to be the most suitable material for high–speed photodiodes designed for fiber–optic communication systems operable in the spectral range of 1.3 and 1.55 μm.

It is also important to ensure the highest possible sensitivity, minimum capacitance, minimum dark current, minimum equivalent noise power while choosing a photodiode for the required spectral range. To solve this issue, researchers apply methods of manufacturing photodiodes in the form of mesa structures as well as building corresponding guard ring to rectifying transitions, even though the above techniques are not widely believed to be very effective.

The most suitable photodetectors for fiber optic systems are semiconductor p–i–n and avalanche photodiodes (APD). In p–i–n–photodiodes, a layer with intrinsic conductivity (i–region) is usually introduced in between layers with polar conductivity, and when the reverse bias voltage is applied to this region, the region will subsequently be depleted (less free carriers), whereas the strong electric field in it will accelerate carriers that will be generated as a result of absorption of light. They are more sensitive due to reduced losses from recombination. The barrier capacitance is negligent, due to which a satisfactory frequency characteristics are provided. These diodes require a small reverse bias voltage (5 V or even less), which promotes their predominant application in LANs and other terminal devices.

Avalanche photodiodes have internal amplification capacity and differ from p–i–n photodiodes by the presence of another additional layer. At high reverse bias voltages (near 100 V), a strong accelerating field is formed inside them, and avalanched multiplication of charge carriers occurs, i.e., an increase in the photocurrent. These devices are characterized by high sensitivity, high gain and high speed, however, their widespread application is hindered by their complexity, high cost, high operating voltages, the need to stabilize voltages and temperatures and they can operate only in the mode of weak signal amplification.

It's important to mention that at significantly larger areas of 4–40 mm^2, when the following conditions are met:

- material of the base region was properly selected;
- physical properties of the formation of the metal—semiconductor transition in gallium arsenide were applied;
- integration of the metal—semiconductor transition with a rectifying p–n junction was ensured during the technological cycle;
- the closure of adjacent potential barriers was achieved one could obtain photodiode structures with low dark currents (10–(6–7) A/cm^2) and high photosensitivity (2–20 A/W) [9], which are designed to detect optical signals [13].

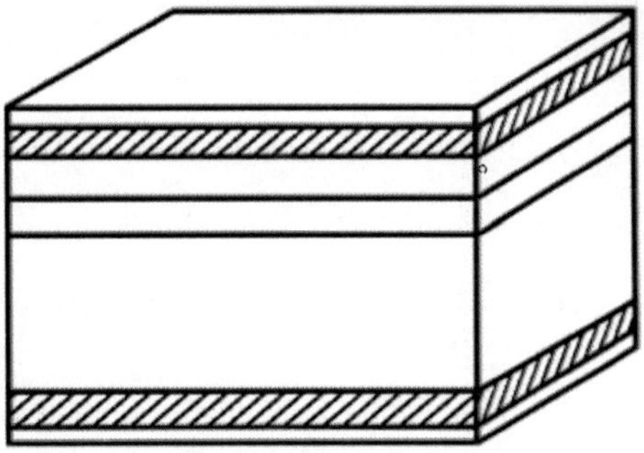

Fig. 3 Cross-section of photosensitive multilayered semiconductor Au–νGaAs:O–nCdS–nInP–Au of structure

Meanwhile, experts in photosensitive devices witnessed increased application of film structures, since this technology allows relative ease in building the films with required dimensions and the technique itself is easy to deploy in comparison to the technique of bulk structures.

In the aircraft SHM systems using of a multilayer photosensitive Au–νGaAs:O–nCdS–nInP–Au–structure based on gallium arsenide with embedded potential barriers photodetector (Fig. 3) is proposed.

The Au–translucent rectifier layer; RSC–region of space charge; nInP–sprayed layers of indium phosphide and nCdS–cadmium sulfide with a thickness 0.4 μm; νGaAs:O is gallium arsenide substrate semi–insulating with a thickness of 350 μm.

5 Conclusion

Using of fiber-optic sensors for aircraft structure health monitoring, including structures made of composite materials, is very promising. The increasing number of aircraft elements from composite materials, as well as the creation and use in the future of "smart" materials and structures, including those with adaptive properties, will require continuous improvement of the system for monitoring the conditions of the structure, as well as assessing its geometric shape, residual life and other parameters.

Promising complex systems created inside the fiber, combining elements of electronics, optoelectronics, micromechanics and photonics, will radically change the ways of using fiber–optic sensors in aviation materials and structures.

References

1. Zhu P, Xie X, Sun X, Sotoac MA (2019) Distributed modular temperature–strain sensor based on optical fiber embedded in laminated composites. Compos B Eng 168:267–273
2. Yu H, Wang Y, Ma J, Zheng Z, Luo Z, Zheng Y (2018) Fabry-Perot interferometric high-temperature sensing up to 1200 °C based on a silica glass photonic crystal fiber. Sensors 18:273
3. Yoshino T, Kurosawa K, Itoh K, Ose T (1982) Fiber–optic Fabry-Perot interferometer and its sensor applications. IEEE J Quantum Electron 4:626–665
4. Hartog AH (2017) An introduction to distributed optical fiber. Sensors 442
5. Wei H, Zhao X, Kong X et al (2014) The performance analysis of distributed Brillouin corrosion sensors for steel reinforced concrete structures. Sensors 14:431–442
6. Sai Y, Zhao X, Hou D, Jiang M (2017) Acoustic emission localization based on FBG sensing network and SVR algorithm. Photonic Sens 7(1):48–54
7. Tian Z, Yu L, Sun X, Lin B (2019) Damage localization with fiber Bragg grating Lamb wave sensing through adaptive phased array imaging. SAGE Publ Struct Health Monit 17(1):334–344
8. Yu F, Okabe Y (2017) Fiber-optic sensor–based remote acoustic emission measurement in a 1000 °C environment. Sensors 17:1–14
9. Ramakrishnan M, Rajan G, Semenova Y, Farrell G (2016) Overview of fiber optic sensor technologies for strain/temperature sensing applications in composite materials. Sensors 16:1–27
10. Sun J, Guan Q, Liu Y, Leng J (2016) Morphing aircraft based on smart materials and structures: a state–of–the–art review. J Intell Mater Syst Struct 27(17):2289–2312
11. MacDougal M, Hood A et al (2011) Part of the SPIE infrared technology and applications XXXVII. V 8012:21
12. Dhar NK, Dat R, Sood AK (2013) In: Pyshkin SL, Ballato JM (eds) Optoelectronics—advanced materials and devices, pp 149–186
13. Karimov AV, Yodgorova DM, Abdulkaev OA (2011) Physical principles of photo–current generation in multi–barrier punch–through–structures. Second chapter of book "Photodiodes", pp 23–36

Iterative-Analytical Model of Nonlinear Deformation and Fracture of Multilayer Composite Panels

Aleksandr Gondlyakh, Andrey Chemeris, Aleksandr Kolosov, Aleksandr Sokolskiy, Valeriy Scherbina, and Sergiy Antonyuk

Abstract This work is devoted to the issues of numerical modeling of the processes of occurrence and propagation of cracks and delamination zones in spatial multilayer systems. To simulate the processes of delamination and destruction of composite structures, an iterative-analytical theory is used. A special eight-node isoparametric multilayer finite element has been developed and numerically implemented. Nonlinear problems are solved on the basis of the developed universal algorithm, which includes an iterative-analytical method of variable approximations. Comparison of the results of numerical solutions of test problems with analytical solutions of the data indicates their good agreement. Numerical modeling of the process of propagation of main cracks and delamination zones of a nine-layer metal-composite plate with a central hole has been performed. The developed methods make it possible to reliably estimate the change in the stress–strain state of multilayer structures depending on the accumulation of damage during their operation and can be used in the development of information support systems for the life cycle of multilayer aircraft panels.

Keywords Refined iterative-analytical theory · Fracture · Delamination zone · Numerical simulation · Finite element method · Multilayer structures

1 Introduction

For structures made of composite materials, the presence of more or less significant zones of destruction, if not a normal phenomenon, at least does not always lead to fatal consequences. However, when designing critical facilities, the presence of

A. Gondlyakh (✉) · A. Chemeris · A. Kolosov · A. Sokolskiy · V. Scherbina
National Technical University of Ukraine "Igor Sikorsky Kyiv Polytechnic Institute", Prospect Peremohy 37, Kyiv 03056, Ukraine

S. Antonyuk
Institute of Particle Process Engineering, Technical University of Kaiserslautern, 67653 Kaiserslautern, Germany

A. Bieliatynskyi and V. Breskich (eds.), *Safety in Aviation and Space Technologies*, Lecture Notes in Mechanical Engineering,
https://doi.org/10.1007/978-3-030-85057-9_4

such zones should be taken into account, and the possibility of their further development and impact on the bearing capacity of the structure as a whole should be analyzed. Due to the complexity of physical processes occurring at the micro-level during the destruction of composite materials (destruction of the matrix, fibers, boundary layer, the influence of structural defects, dislocation phenomena, etc.), the analysis of the reliability of structures is carried out on their basis, with the necessary for engineering practice the degree of credibility is possible with the help of phenomenological strength criteria.

The exhaustion of the strength of individual layers of the composite does not always affect the current stress state of the entire structure, and sometimes it is even difficult to detect experimentally, but it can significantly affect the further behavior of the object under study, provided that the fracture zone progresses. This determines the relevance of the development of an apparatus for numerical modeling of the evolution of layer-by-layer destruction, both of individual elements of a multilayer package of layers, and of the entire structure as a whole.

It is necessary to note the important fact that the history of cracking in the layers, as a rule, significantly affects the final picture of the structure failure. Therefore, the analysis of the strength of structures made of composites must be carried out on the basis of refined theories of multilayer systems, as well as using special multilayer finite elements (FEs), which make it possible to take into account, in the process of deformation, the change in boundary conditions between layers due to the appearance of delamination zones in them.

2 Previous Studies

One of the most widespread types of destruction of aircraft construction objects is the destruction of elements due to the formation and development of cracks and zones of delamination in them. So, for example, statistics show that more than 82% of failures occur precisely due to the development of cracks in aircraft structural elements, and less than 18% of failures arise as a result of manufacturing defects or disruption of operation processes. In the classification of causes that led to the development of cracks, more than 50% of failures are accounted for by technological and fatigue cracks.

Theories for calculating multilayer composite plates and shells currently offer a wide range of hypotheses for reducing the three-dimensional problem of mechanics of a deformable solid to two-dimensional problems of the theory of shell systems. The mathematical models for calculating shells developed within this area, in contrast to the classical theory of Kirchhoff–Love shells and the theory of S. P. Timoshenko, are designed to investigate the specifics of the three-dimensional stress state of structures made of composite materials, due to:

- multilayer;
- inhomogeneity of the distribution of physical and mechanical characteristics over the thickness of the structure, arising in the technological process of forming a product when laying layers or winding fibers;
- the design features of the layers and their structure (honeycombs, partitions, tubular structures, etc.);
- the need for a detailed study of the stress state in the contact zone of the layers, in the zones of stress concentrators, such as inclusions, reinforcing elements (frame, stiffener), as well as in the zones of end edge effects;
- the possibility of changing the mechanical characteristics of the layers during deformation of the product during operation due to the formation of cracks, delamination zones, fiber break, fracture and plastic deformation of the matrix, etc.

This incomplete list of factors necessary for the correct analysis of the stress state of composite shell systems indicates the complex problems that arise in the development of appropriate mathematical models for a reliable assessment of their strength and reliability. As is known, the great difficulties arising in the direct integration of the differential equations of three-dimensional mechanics of a deformable solid as applied to the study of the stress–strain state of shell systems, due to the presence of a small parameter, lead to the need to develop special mathematical models of calculation. These mathematical models make it possible, albeit with a certain degree of approximation, but very effectively to assess the strength properties of homogeneous in thickness and multilayer spatial structures.

In this regard, a number of authors have recently proposed a large number of different refined theories of shells. An analysis of survey papers [1–4] shows that all these approaches can be grouped as follows:

- hypothesis method;
- the thickness series expansion method;
- the method of asymptotic integration of three-dimensional equations.

Each of these approaches has advantages and disadvantages. Due to the fact that the most important stage in the construction of any theory of shells is the problem of choosing the basis for the expansion of the vector of displacements (stresses) over the thickness of the layer packet, a successful choice of functions for the transition from three-dimensional equations of elasticity theory to two-dimensional relations of the theory of shells largely determines the field of application of this or a different theory. The use of the thickness series method, despite the possibility of obtaining in some cases an almost exact three-dimensional solution, is hampered by the need to retain a sufficiently large number of expansion terms for a detailed description of the internal stress–strain state of multilayer spatial structures.

Similar difficulties arise when using the asymptotic method. The method of hypotheses, despite the clearness of its physical interpretation, the relative simplicity of deriving resolving equations, effective numerical implementation, as well as the possibility of obtaining closed analytical solutions for certain classes of problems, does not allow detailing the concentration of the stress state in regions of singularities when solving a number of applied problems due to the unavoidable the error hypothetically included in the calculation. So, for example, in the study of deformation processes of both uniform thickness and multilayer systems of thin shells, the physical and mechanical characteristics of the layers of which differ by less than an order of magnitude, the application of the classical theory of laminated plates (CLPT) Love Kirchhoff is the most widespread.

Remaining one of the most studied methods for solving the problems of statics and dynamics of thin shells, which makes it possible to analyze the stress state of a wide class of real structures with a sufficient degree of accuracy, the classical theory can lead not only to quantitative, but also to qualitatively incorrect results when modeling the deformation processes of multilayer shell systems, made of composites. For these systems, it is necessary to take into account the specifics of the operation of these materials, which consists in low shear stiffness and strong anisotropy of the physicomechanical and strength properties of individual layers of the package. Therefore, the features of deformation of structures made of composite materials cannot be taken into account on the basis of the classical theory of shells.

Taking the shear into account as a generalized unknown significantly expands the class of the problems under consideration, allows one to study the stress–strain state of the average thickness of isotropic and also anisotropic multilayer shell systems, the physical and mechanical characteristics of the layers of which differ by less than an order of magnitude. In this regard, Timoshenko's hypothesis was widely used in engineering practice to assess the strength and analyze the stress–strain state of shells of medium thickness (straight line hypothesis—The First Order Shear Deformation Theory (FSDT)).

Replenishment of the expansion basis for one more function makes it possible to construct refined models of the theory of shells, taking into account the shift and curvature of the element initially normal to the middle surface of the shell. Thus, the class of problems under consideration extends to the region of essentially inhomogeneous multilayer shell systems, the physical and mechanical characteristics of which differ by several orders of magnitude. The works [5–15] are devoted to the solution of this problem. Similarly, the refinement of the sought solution based on the expansion of the basis of the expansion of the components of the displacement vector by introducing an additional function characterizing the inhomogeneous distribution of compression deformations in different layers makes it possible to analyze the stress–strain state of thick plates and shells.

Recently, a number of works have appeared that use iterative processes to refine the components of the stress–strain state of multilayer shells. For example, in [16–21], the theory of shells is proposed, based on the refinement of the sought solution by integrating the equilibrium equations with respect to transverse tangential stresses for plates and shallow shells, based on the hypothesis method. When applying iterative schemes for refining solutions of differential equations of equilibrium (motion), it should be emphasized that the analysis of the convergence of the integration process indicates a fairly rapid achievement of acceptable accuracy of the solution both for a uniform thickness and for multilayer plates with symmetric or skew-symmetric stacking of layers.

In the case of a complex scheme of layer reinforcement or strong anisotropy, the convergence of the iterative procedures significantly deteriorates and a significant number of iterations is required to achieve the required calculation accuracy. This is adequate to the method of series expansion with a large number of series members. This fact is due to the fact that, as a rule, it is not possible to identify in advance the dominant functions in the expansion of the displacement vector over the thickness of the structure, which determine a qualitatively correct description of the distribution of certain components of the stress–strain state of a complex multilayer structure.

Despite the fact that many authors have proposed various refined models of the deformation of shell systems, at present there is a limited number of works devoted to the construction of mathematical models of deformation of multilayer shells for the analysis of various evolutionary processes occurring in the material shell, depending on the level of its components of the stress–strain state. The construction of this kind of theories of shells is necessary in the case of studying the nonlinear deformation of structures taking into account the change in the physicomechanical parameters of the layers due to the appearance and development of plastic zones and, especially, fracture zones.

The process of destruction of shells made of composite materials, especially reinforced ones, is characterized by the gradual nucleation of microdefects, an increase in their concentration in the bulk of the material, and the subsequent formation of a fracture zone. In addition to this process, which can be described in sufficient detail on the basis of standard phenomenological methods of mechanics, composite structures are inherent in the possibility of the appearance of delamination zones during their deformation.

It should be noted that modeling this effect is a stumbling block for many refined shell theories based on the hypothesis method, since it requires taking into account the change in boundary conditions at the contact boundary of delaminated layers, the initial location of which is unknown. So, for example, as shown in [17–20], the use of the hypotheses of a straight line, a straight normal and a static hypothesis of continuous change in the diagram of transverse shear stresses does not allow obtaining results that are even qualitatively close to the reference solution. This situation is aggravated by the emergence of new additional delamination zones. The authors of [22] argue that a unified universal refined model of analysis of the stress–strain state of multilayer composite plates and shells does not currently exist. Therefore, depending on the specific problem, it is necessary to use different formulations of hypotheses for reducing the three-dimensional equations of mechanics to the equations of the theory of shells.

This situation becomes much more complicated in the case of solving the problem of nonlinear deformation and fracture of multilayer structures. Since even a structure that is uniform in thickness can turn into a multilayer system when zones of delamination appear during its operation, the solution to the problem of nonlinear deformation and fracture in this case becomes much more complicated. This fact implies an automatic redefinition of the basis for the expansion of the components of the displacement vector over the thickness of the structure, since the initially continuous reduction functions in this case inevitably undergo a discontinuity, which is caused by a change in the boundary conditions of contact at the boundary of the delaminated sections.

This task is of an evolutionary nature, therefore, in this case, it is necessary to take into account the entire history of changes in the stress state of a multilayer object before and after the appearance of delamination zones. In this regard, in [17–19], an iterative-analytical theory of multilayer structures is proposed. A distinctive feature of this theory from the known models is that, on its basis, at each step of integration over time/load, not only the determination of the components of the stress–strain state of a multilayer system, but also a physically justified determination of the reduction functions depending on the physical processes occurring in the material of the layers in any moment in time. The use of this methodology for practical purposes makes it possible to effectively build systems of information support for the life cycle of structural elements in aircraft construction, mechanical engineering, building engineering etc. The basic procedure for constructing a resolving system of equations of motion for multilayer spatial systems is briefly discussed below.

3 Materials and Methods

3.1 Theoretical Provisions

The theoretical provisions of this work are based on the iterative-analytical theory of spatial systems [15, 17–19], the essence of which is as follows. The components of the displacement vector of a multilayer structure are represented as:

$$u_i = \sum_{S=1}^{S} F_S^i(x^3, t) v_S^i(x^1, x^2, t), \tag{1}$$

where v_i^S are the components of the generalized displacement vector of the middle surface of the structure x^1, x^2;
F_S^i are functions to be determined ($F_1^i = 1$; $F_2^i = x^3$);
t is time.

In order for the vector $\vec{u}$ to describe the actual deformed state of the shell, it is necessary to fulfill the Hamilton-Ostrogradsky variational principle, namely:

$$\delta\int_{t_0}^{t_1}(W-K-A)dt=\delta_{v_i^s}\int_{t_0}^{t_1}(W-K-A)dt+\delta_{F_i^s}\int_{t_0}^{t_1}(W-K-A)dt=0 \quad (2)$$

- here the notation of type $\delta_{v_i^s}$ and $\delta_{F_i^s}$ means that the variation of the functional is made by v_i^S or F_S^i, respectively;
- W is the internal energy of the system;
- K is kinetic energy;
- A is the work of external forces.

Taking into account (1) and the independence of the variations δv_j^r and δF_r^j, we obtain the resolving system of equations:

$$\begin{cases}\int_{t_0}^{t_1}\left\{\int_V A_j^r\left(v_i^s;\frac{\partial v_i^s}{\partial x^\alpha};\frac{\partial^2 v_i^s}{\partial x^\alpha \partial x^\beta};\ddot{v}_i^s;F_s^i;\frac{dF_s^i}{dx^3};\frac{d^2F_s^i}{(dx^3)^2};\ddot{F}_i^s;p_j^r\right)\delta v_j^r dV\right\}dt=0,\\ \int_{t_0}^{t_1}\left\{\int_V B_j^r\left(v_i^s;\frac{\partial v_i^s}{\partial x^\alpha};\frac{\partial^2 v_i^s}{\partial x^\alpha \partial x^\beta};\ddot{v}_i^s;F_s^i;\frac{dF_s^i}{dx^3};\frac{d^2F_s^i}{(dx^3)^2};\ddot{F}_i^s;p_j^r\right)\delta F_j^r dV\right\}dt=0\end{cases} \quad (3)$$

and the corresponding boundary conditions:

$$\begin{cases}\int_{t_0}^{t_1}\left\{\int_S a_j^r\left(v_i^s;\frac{\partial v_i^s}{\partial x^\alpha};F_s^i;\frac{dF_s^i}{dx^3};q_j^r\right)\delta v_j^r dS\right\}dt=0,\\ \int_{t_0}^{t_1}\left\{\int_S b_j^r\left(v_i^s;\frac{\partial v_i^s}{\partial x^\alpha};F_s^i;\frac{dF_s^i}{dx^3};q_j^r\right)\delta F_r^j dS\right\}dt=0.\end{cases} \quad (4)$$

A distinctive feature of expression (3) is that the solution of the resolving system of equations ensures the determination of not only the vector of generalized unknowns v_i^s, but also the determination of physically substantiated functions F_s^i, depending on the physical processes occurring in the material of the layers at any time, for example, plastic deformation, delamination or accumulation defects and destruction of layers.

Solving the system of Eq. (3) by direct methods can be difficult. At the same time, the use of iterative methods, consisting in the sequential refinement of components v_i^s or functions F_s^i, proceeding from the condition of minimizing the residual vector of the three-dimensional operator of elasticity theory, makes it possible to significantly simplify the obtaining of the desired solution. In this case, the second subsystem of Eq. (3), as can be seen, is represented as a system of ordinary differential equations for functions F_s^i in the sense of the Krylov-Kantorovich method, i.e.:

$$\mathrm{D}_j^{r(n-1)}\left(F_s^{i(n)};\frac{dF_s^{i(n)}}{dx^3};\frac{d^2F_s^{i(n)}}{(dx^3)^2};p_j^r\right)=0 \quad (5)$$

with appropriate boundary conditions:

$$\mathrm{d}_j^{r(n-1)}\left(F_s^{i(n)};\frac{dF_s^{i(n)}}{dx^3};p_j^r\right)=0 \quad (6)$$

The coefficients in Eqs. (5) and (6) are integral characteristics that depend on the components of the stress–strain state obtained at iteration n − 1, namely:

$$\mathrm{D}_j^{r(n-1)}=\mathrm{D}_j^{r(n-1)}\left\{\int\limits_S (V_j^{r(n-1)};p_j^{r(n-1)})\right\}dS. \quad (7)$$

This technique for solving the main system of Eq. (3) greatly simplifies the solution of the problem, since relations (5) are a system of ordinary differential equations. This system can be easily solvable by known methods of mathematical physics, and with an appropriate grouping of unknowns, it also allows obtaining an analytical solution of an exponential form, namely:

$$\begin{aligned} u_\alpha^m &= v_\alpha^m+\left\{\sum_{i=1}^{4}C_i^{\alpha m}e^{r_\alpha^{im}x^3}+G_\alpha^m/\psi_\alpha^m\right\}\xi_\alpha^m \\ u_3^m &= \left\{\sum_{\alpha=1}^{2}\left[\sum_{i=1}^{4}\varphi_{\alpha i}^{3m}C_i^{\alpha m}e^{r_\alpha^{im}x^3}+C_i^{3m}e^{r_\alpha^{im}x^3}\right]+\Upsilon_3^m\right\}\omega^1 \end{aligned} \quad (8)$$

where v_α^m, ξ_α^m, ω^1 are components of the vector of tangential displacements, generalized shear and deflection of the middle layer, respectively;
$C_i^{\alpha m}$ is the constants of integration, determined from the conditions of contact between the layers and on the front surfaces of the layer package;
r_α^{im} is integral characteristics of the package deformation, which depend not only on the physical and mechanical characteristics of its layers, but also on the components of the deformed state of the layer package, determined at iteration n − 1;
G_α^m/ψ_α^m, Υ_3^m is particular solutions of characteristic equations.

A feature of the obtained solutions is that on their basis it is possible to simulate not only the effect of slipping of layers, but also the effect of compression (separation) of layers from each other.

3.2 *Procedure for Accounting for Cracks in a Multilayer Refined FE*

In this work, the process of crack formation is modeled by analogy with [15, 17–19] and is interpreted as the acquisition of specific anisotropic properties by the material in a certain local area due to the influence of cracks on the stress–strain state of the structure (Fig. 1). A characteristic feature of this technique for modeling cracks is not a simple zeroing of certain components of the stress tensor, but the correction of the tensor of the physicomechanical constants of the fractured region on the basis of static hypotheses adopted depending on the type of fracture. This approach is more justified in comparison with the known approaches to zeroing one or another component of the tensor of the physicomechanical characteristics of the layer material, since in this case the physical meaning of the process of formation of destruction zones is completely preserved.

3.3 *Correction of the Reduction Functions F_s^i and the FE Stiffness Matrix Taking into Account the Formed Cracks*

Due to the absence of a universal strength criterion and a significant difference in the strength properties of composite materials, an orientation towards any one

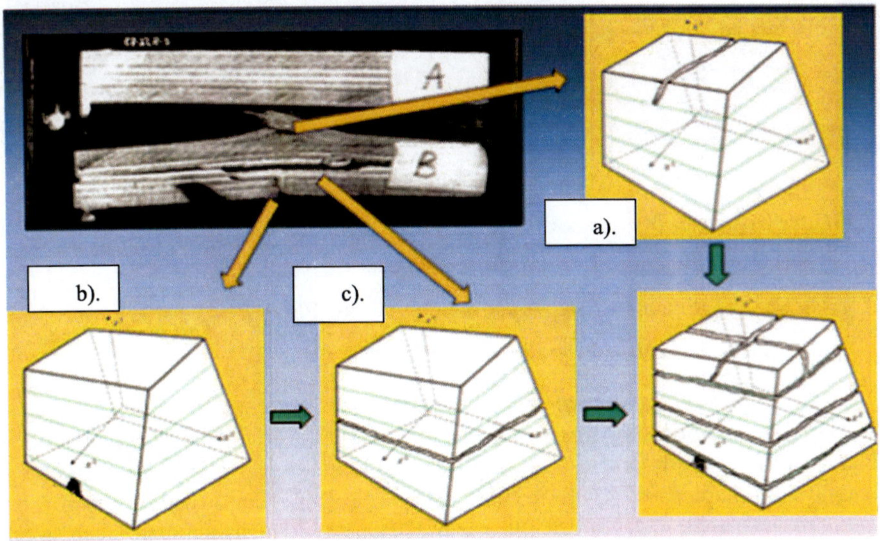

Fig. 1 Schemes of fracture of multilayer FE: **a**—layer rupture (breaking); **b**—rushing of the layer; **c**—layer delamination

theory of strength can unreasonably narrow the class of problems under consideration. Therefore, it is assumed that any fracture surface can be used, the equation of which is uniquely formulated in the space of stresses or deformations. This approach allows you to quickly change the strength conditions depending on the type of material used, especially since the package may contain layers whose strength properties are described by different strength criteria. When constructing a model of physically nonlinear deformation of structural layers, we will proceed from the main provisions of the theory of plastic flow with translational isotropic hardening. An important independent stage in the numerical implementation of the study of the evolution of physically nonlinear processes and fracture occurring in composite multilayer systems, based on the FEM positions, is the linearization of the stiffness matrices (SM) at each loading step, taking into account the real diagram of the material operation of the layers and the formed cracks. In this work, the formation of an SM element, the layers of which undergo plastic deformation, is carried out using the instant tensor of deformation constants C_{cm}^{ijkl}.

Moreover, the latter is calculated taking into account the secant values of the physicomechanical characteristics of the layers, taken from real diagrams obtained through a series of experiments. When forming a SM element with cracks, we will proceed from the following assumptions. Suppose that when individual layers of FE reach the limiting stress values (after fulfilling the condition of layer destruction), they cease to resist stretching and bending. The complete derivation of the expression for calculating the coefficients of the stiffness matrix of a multilayer eight-node isoparametric FE with cracks in separate layers is given in [15, 17, 19].

Let us write the final expression for the variation of the deformation energy functional of a multilayer FE:

$$\delta \overline{w}_p^t = \sum_{m=1}^{M} \sum_{t=1}^{8} \sum_{r=1}^{8} \overset{\mathrm{T}_{k'q'm}}{G_{rtsf}} u_{k'}^{s(t)} \delta u_{q'}^{f(r)} = \sum_{t=1}^{8} \sum_{r=1}^{8} \left[\sum_{m=1}^{M-M_t} G_{rtsf}^{k'q'm} + \sum_{m=M_t}^{M} \mathrm{T}_{rtsf}^{k'q'm} \right] u_{k'}^{s(t)} \delta u_{q'}^{f(r)}, \tag{9}$$

where Mt is the number of destroyed layers;

$G_{rtsf}^{k'q'm}$ is SM coefficients of non-destroyed layers of p-th FE;

$\mathrm{T}_{rtsf}^{k'q'm}$ is SM coefficients of destroyed layers of p-th FE;

$u_{k'}^{s(t)}$ is components of displacements in FE nodes.

As you know, the most effective in solving nonlinear problems is the use of step-iterative methods. In this paper, to solve physically and geometrically nonlinear problems with crack formation, a step-iterative algorithm for sequential loading is implemented, based on the modified Newton–Kantorovich method. The modeling of the plastic work of the material is carried out in the process of iterations by the method of additional loads. Reforming the SM (variable stiffness method) is performed at the beginning of the next loading step (taking into account the design scheme of the deformed structure) and after each stage of “applying” cracks (taking into account the anisotropy caused by cracking). In this case, the adaptive change in the approximating FE functions is taken into account on the basis of the iterative-analytical method of variable approximations (to redefine the functions of the expansion of the displacement vector over the thickness of the layer package, taking into account the formed cracks) (Fig. 2).

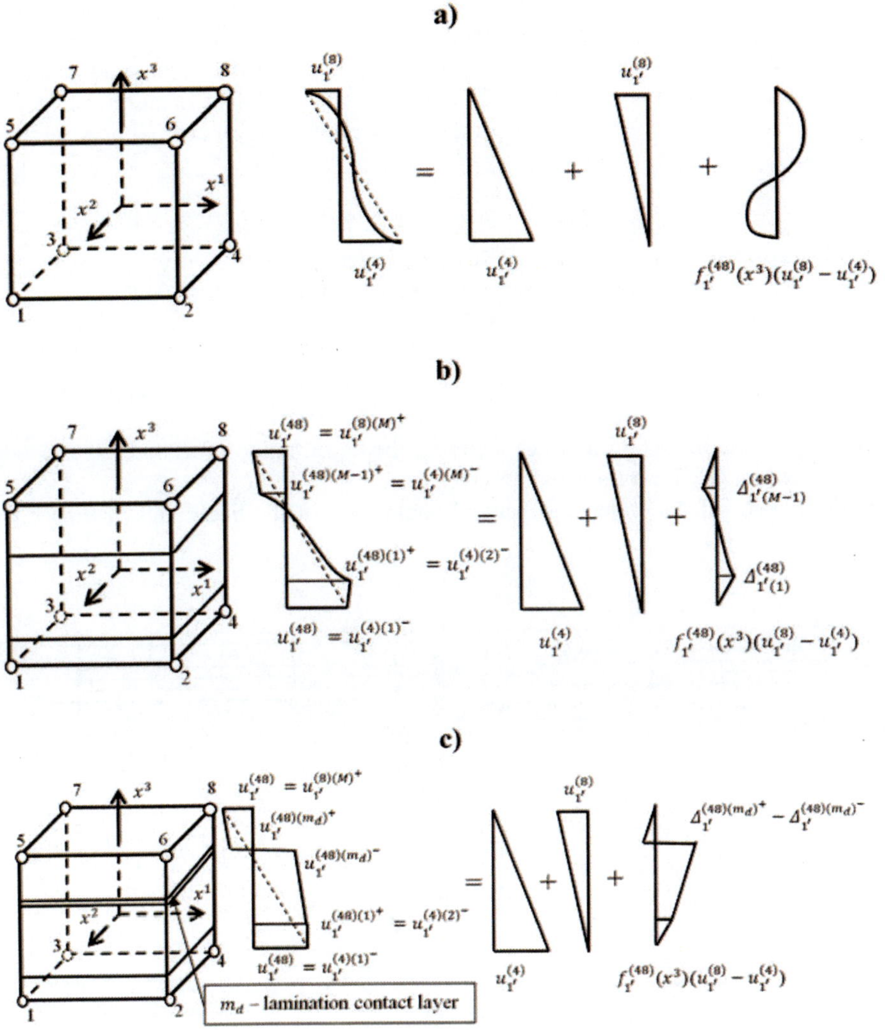

Fig. 2 Iterative-analytical correction of the approximation functions of the components of the FE displacement vector depending on the physically nonlinear processes occurring in the material of the layers during deformation of the structure: **a**—approximating FE functions in the initial state of the layer material; **b**—approximating FE functions when plastic zones appear in layers; **c**—approximating FE functions when zones of delamination appear in layers

3.3.1 Determination of the Area of Application of the Developed FE in the Study of the Processes of Destruction and Plastic Deformation of Multilayer Structures

This FE was developed in the Automation System for Calculating the Strength of Shell and Combined Systems (APROKS). The practice of using this FE when solving a large number of test and applied problems has shown its effectiveness in studying the processes of nonlinear deformation and fracture of homogeneous thickness and multilayer spatial structures [19, 23–25]. Currently, on the basis of the above relations in the ABAQUS environment, the user end element USER_IAT_3D and an algorithm for modeling the processes of nonlinear deformation and progressive destruction of the material of the layers of spatial structures are developed based on this FE.

3.3.2 Approbation of the Possibility of Modeling Delamination Processes During Bending and Separation of a Layer Package Based on USER_IAT_3D FE

The possibility of modeling the process of delamination of multilayer structures based on the developed refined USER_IAT_3D element is considered on the example of the model problem of studying the deformation of a three-layer freely supported beam with a thin middle layer.

Geometric, physical and mechanical parameters of the beam and loading conditions are shown in Fig. 3.

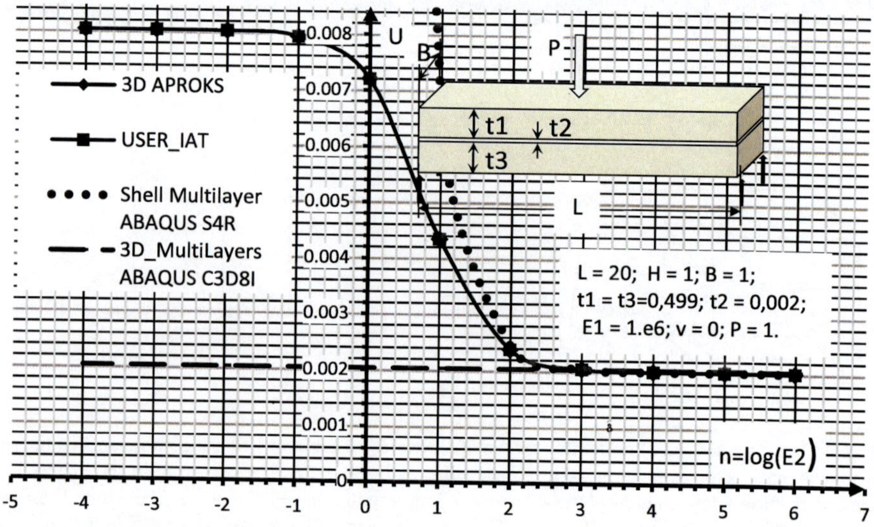

Fig. 3 Change in the deflection at the center of a freely supported beam depending on the level of the elastic modulus of the middle thin layer

The separation process is considered on the example of the deformation of a five-layer rigidly restrained cantilever beam with a thin middle layer, which is under the action of a load uniformly distributed on the front surfaces (Fig. 4). To assess the reliability of the results obtained on the basis of the developed refined USER_IAT_3D element, a series of numerical experiments was carried out in parallel to calculate this structure based on three-dimensional FE within the APROKS system. The beam calculations were performed independently on the basis of the ABAQUS Multilayer S4R multilayer shell FE and the ABAQUS Multilayer C3D8I three-dimensional multilayer element.

Figure 4 shows the results of studies reflecting the dependence of the change in the deflection of the center point of the beam depending on the level of the elastic modulus of the middle layer. As a result of calculations, it was found that when the stiffness of the middle layer changes by two orders of magnitude with respect to the stiffness of the outer layers, the determination of the stress–strain state of such structures must be made taking into account the factor of inhomogeneous distribution of transverse shear deformations over the thickness of the layer package.

When the stiffness of the middle layer changes by six orders of magnitude or more, the outer layers begin to work independently of each other. The calculation data for the stress–strain state of the beam, obtained from the proposed USER_IAT_3D element, is in full agreement with the results of three-dimensional and theoretical calculations. At that time, the use of multilayer shell and three-dimensional elements from the ABAQUS library reflected the correct pattern of beam deformation only in the range of small values of the transverse shear deformation inhomogeneity parameter.

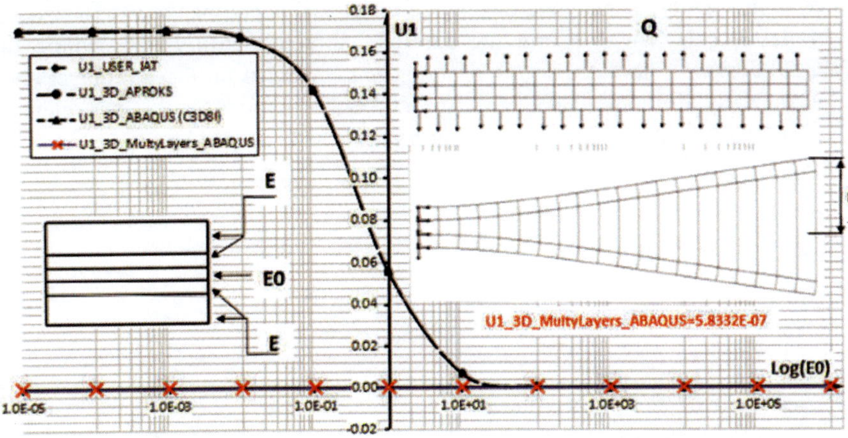

Fig. 4 Change in the opening of the free end of a rigidly restrained five-layer beam depending on the level of the elastic modulus E0 of the middle thin layer (L = 10; H = 1; B = 1; E1 = 1 106; v = 0; E0 = 1.0• 〖10〗 ^6 ÷ 1.0• 〖10〗 ^(−5); v = 0.; h1 = h5 = 1/3; h2 = h3 = h4 = 1/6; Q = 1)

4 Results and Discussion

4.1 Approbation of the Algorithm for Modeling the Destruction of a Nine-Layer Plate

The approbation of modeling the fracture processes of multilayer structures on the basis of the developed refined USER_IAT_3D element is considered on the example of a model problem of nonlinear deformation at rupture of a nine-layer metal-composite compact specimen with a central hole. The specimen consists of three aluminum layers, two composite layers and three adhesive layers (Fig. 5).

The specimen is loaded with a uniformly distributed tensile load at the ends in the direction of the specimen length. Geometrical parameters of the specimen: length L = 101.6 mm; width W = 50.8 mm; thickness of the aluminum layer is 0.34 mm; composite layer thickness is 0.458 mm; thickness of the adhesive layer is 0.034 mm; the diameter of the central hole is 12.7 mm.

Physical and mechanical characteristics of aluminum layers: E = 70,000 MPa; Poisson's ratio v = 0.33. Physical and mechanical characteristics of composite layers: E_1 = 46,400 MPa, E_{22} = 46,400 MPa, E_{33} = 63,430 MPa, $\nu_{12} = 0.01$, $\nu_{13} = 0.02$, $\nu_{23} = 0.318$, $G_{12} = 12,000\,\text{MPa}$, $G_{13} = 12,000\,\text{MPa}$, $G_{23} = 16,410\,\text{MPa}$. Strength limits of composite layers: $\sigma_{11} = 402\,\text{MPa}$, $\sigma_{22} = 402\,\text{MPa}$, $\sigma_{33} = 550\,\text{MPa}$, $\sigma_{12} = 400\,\text{MPa}$, $\sigma_{13} = 400\,\text{MPa}$, $\sigma_{23} = 400\,\text{MPa}$. Physical and mechanical characteristics of adhesive layers: $\text{E} = 46,400\,\text{MPa}$, $\nu = 0.39$. Strength limits of adhesive layers: $\sigma_{11} = 402 MPa$, $\sigma_{22} = 402\,\text{MPa}$, $\sigma_{33} = 402\,\text{MPa}$, $\sigma_{12} = 150\,\text{MPa}$, $\sigma_{23} = 150\,\text{MPa}$.

All planes of symmetry are taken into account in the design scheme (Fig. 6).

To assess the reliability of the results obtained on the basis of the developed refined USER_IAT_3D element, a series of numerical experiments was carried out in parallel to calculate this structure based on three-dimensional FE within the APROKS system. The calculations of the specimen based on the multilayer shell FE ABAQUS Multilayer_S4R and the three-dimensional multilayer element ABAQUS Multilayer_C3D8I (Fig. 7) were performed independently.

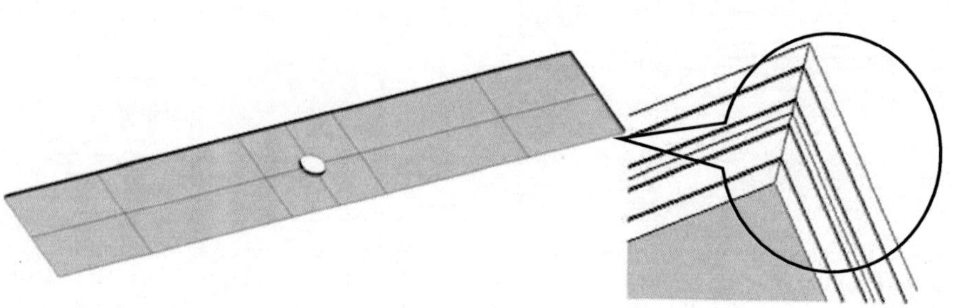

Fig. 5 Nine-layer composite specimen

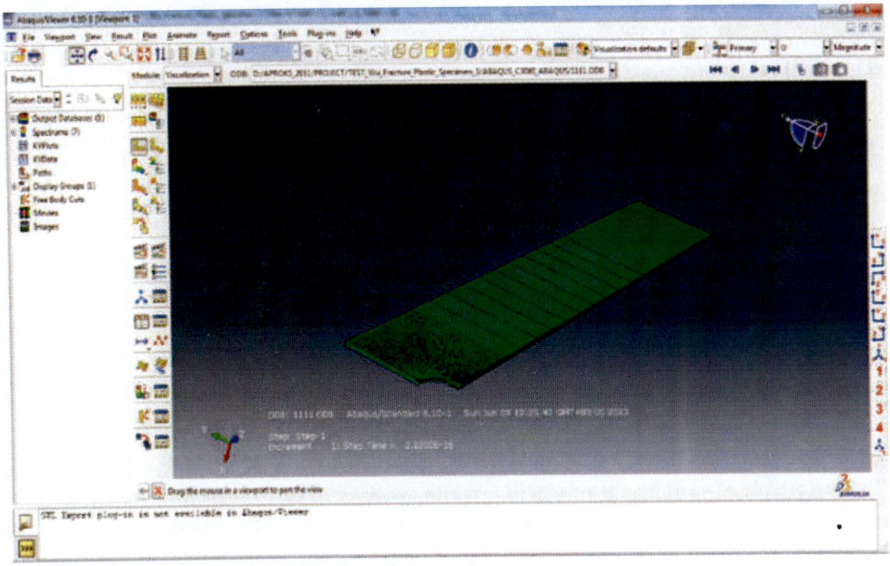

Fig. 6 Calculated FE scheme of a nine-layer composite specimen

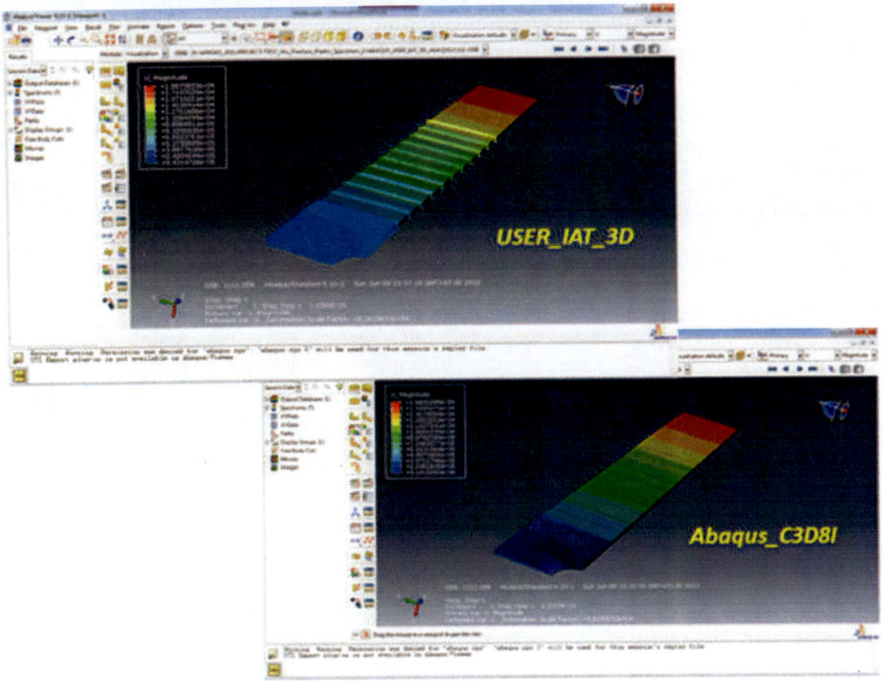

Fig. 7 Results of specimen calculation using USER_IAT_3D and C3D8I FE

Fig. 8 Nucleation of delamination zone in the adhesive layer

As a result of numerical modeling of the process of destruction of the specimen, it was found that the initiation of damage begins in the middle adhesive layers at the boundary of the hole in the zone of maximum shear stresses (Fig. 8). The development of these zones leads to the appearance of a delamination zone in the area of the hole and the appearance of plastic deformation zones in the aluminum layers (Fig. 9).

The evolution of the further development of delamination zones in the adhesive layers and plastic deformation zones in aluminum layers leads to the appearance of a main crack in the composite layer (Fig. 10). Further development of main cracks in the composite, delamination in the adhesive layers and the achievement of critical deformations in the zones of physically nonlinear deformation of aluminum layers leads to the appearance of through cracks in the aluminum layers (Fig. 11). Ultimately, these processes lead to the complete exhaustion of the bearing capacity of the compact multilayer specimen.

Fig. 9 Formation of plastic zones in aluminum layers

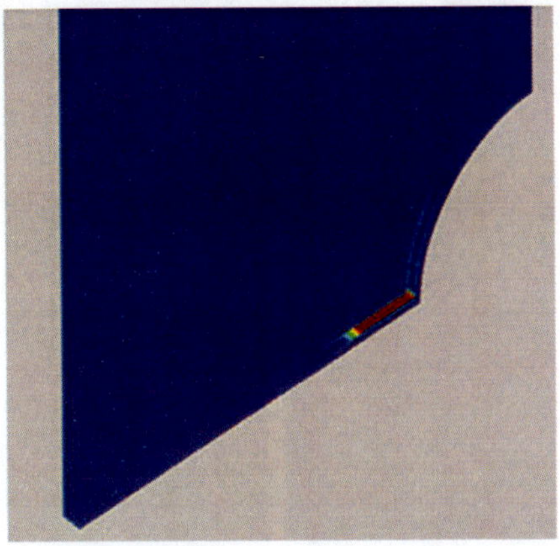

Fig. 10 Through-hole pull-off crack in a composite layer

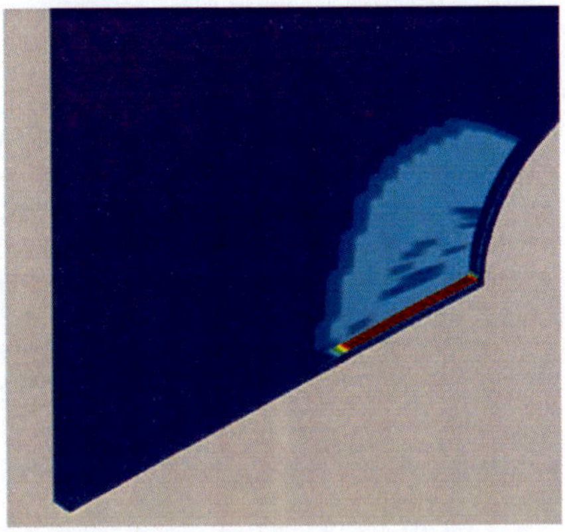

Fig. 11 A typical picture of the destruction of the specimen during the propagation of a through crack in the composite

5 Conclusions

In this work, the main relations of the iterative-analytical theory of spatial multilayer structures are presented. A feature of this theory is that, on its basis, it is possible to efficiently simulate the evolutionary processes of initiation and propagation of cracks and delamination zones in multilayer composite systems, ranging from a structure that is uniform in thickness up to independently deformable layers with slip. An eight-node isoparametric multilayer user finite element (USER_IAT_3D) has been developed for modeling the processes of delamination and destruction of composite structures.

The refined multilayer finite element makes it possible to simulate the inhomogeneity of the distribution of transverse shear and tensile deformations over the thickness of the structure, as well as to avoid the occurrence of numerical instability of solutions during the formation of thin layers in the process of cracking. The FE was adapted to the ABAQUS environment. A universal algorithm has been developed, consisting of four main blocks: (1) block that implements integration over the loading parameter; (2) a block providing the "additional loads" method; (3) a block that implements the method of variable stiffness; (4) a block that implements the iterative-analytical method of variable approximations.

Control and test problems are solved and their good agreement with analytical solutions is shown.

The evolutionary process of destruction of a nine-layer composite plate with a central hole is investigated. The developed methods make it possible to reliably assess the change in the stress–strain state of multilayer structures, depending on the accumulation of damage during their operation.

The use of the proposed methods is promising in the development of information support systems for the life cycle of multilayer aircraft panels.

References

1. Abrate S, Sciuva MD (2017) Equivalent single layer theories for composite and sandwich structures: a review. Compos Struct 179(1):482–494
2. Teotia M, Soni RK (2018) Applications of finite element modelling in failure analysis of laminated glass composites: a review. Eng Failure Anal 94:412–437
3. Sayyad SA, Ghugal YM (2017) Bending, buckling and free vibration of laminated composite and sandwich beams: a critical review of literature. Compos Struct 171:486–504
4. Kreja I (2011) A literature review on computational models for laminated composite and sandwich panels. Cent Eur J Eng 1(1):59–80. https://doi.org/10.2478/s13531-011-0005-x
5. Wu Z, Chen W (2006) An efficient higher-order theory and finite element for laminated plates subjected to thermal loading. Compos Struct 73:99–109
6. Sarfaraz KR, Dehghan MB, Abedian A (2009) Nonlinear analysis of FGM plates under pressure loads using the higher-order shear deformation theories. Compos Struct 89:333–344
7. Moleiro F, Franco Correia VM, Ferreira AJM, Reddy JN (2019) Fully coupled thermo-mechanical analysis of multilayered plates with embedded FGM skins or core layers using a layerwise mixed model. Compos Struct 210:971–996
8. Amabili M, Reddy JN (2020) The nonlinear, third-order thickness and shear deformation theory for statics and dynamics of laminated composite shells. Compos Struct 244:112265
9. Trabelsi S, Frikha A, Zghal S, Dammak F (2019) A modified FSDT-based four nodes finite shell element for thermal buckling analysis of functionally graded plates and cylindrical shells. Eng Struct 178:444–459
10. Cho M, Oh J (2007) Higher order zig-zag theory for fully coupled thermo-electric-mechanical smart composite plates. Int J Solids Struct 44:100–127
11. Mantari JL, Oktem AS, Guedes SC (2012) A new higher order shear deformation theory for sandwich and composite laminated plates. Compos B Eng 43(3):1489–1499. https://doi.org/10.1016/j.compositesb.2011.07.017
12. Wu Z, Chen R, Chen W (2005) Refined laminated composite plate element based on global–local higher order shear deformation theory. Compos Struct 70:135–152
13. Jin Q, Yao W (2020) An accurate zigzag theory for bending and buckling analysis of thick laminated sandwich plates with soft core. J Compos 54(18):2473–2488. https://doi.org/10.1177/0021998319899138
14. Akavci SS (2010) Two new hyperbolic shear displacement models for orthotropic laminated composite plates. Mech Compos Mater 46(2):215–226

15. Sakharov AS, Gondlyakh AV, Mel'nikov SL, Snitko AN (1989) Numerical modeling of processes of failure of multilayered composite shells. Mech Compos Mater 25(3):337–343
16. Massabo R (2020) Mechanics based modeling of composite and sandwich structures in the naval environment: elastic behavior, fracture and damage evolution. In: Advances in thick section composite and sandwich structures, pp 347–386
17. Sakharov AS, Kozak AL, Gondlyakh AV, Melnikov SL (1984) Mathematical model of the deformation of multilayer composite shell systems. Strength Mater Theory Struct 44:13–16
18. Gondlyakh AV (2012) Refined model of multilayer structures deformation for progressive destruction processes study. Eastern-European J Enterpr Technol 2, 7, 56, 52–57
19. Sakharov AS, Gondlyakh AV, Strizhalo AV (1997) On features of numerical integration for the equations of motion of laminated shell systems in the iterative analytic theory. Int Appl Mech 33(9):713–718
20. Piskunov VG (2003) An iterative analytical theory in the mechanics of layered composite systems. Mech Compos Mater 39:1–16
21. Piskunov VG, Rasskazov AO (2002) Evolution of the Theory of Laminated Plates and Shells. Int Appl Mech 38:135–166
22. Kulkarni P, Dhoble A, Padole P (2018) A review of research and recent trends in analysis of composite plates. Indian Academy of Sciences Sadhana 43–96
23. Gondlyakh AV (2012) Adaptation in ABAQUS of the iterated-analytical multilayer user finite element. Eastern-European J Enterpr Technol 37(57):62–68
24. Sakharov AS, Kolosov AE, Sokolskii AL, Sivetskii VI (2013) Modeling of polymer melting processes in screw extruder channels. Chem Petrol Eng 49(5, 6):357–363
25. Gondlyakh O, Krytskyi V, Onopriienko V, Chemerys A, Krytska N (2017) Computer analysis of thermomechanical state of sealing steel lining for containment of NPPs with VVER-1000/V-320 in emergencies. Nucl Rad Saf 4(76):28–39

Vibration Measurement Technologies and Systems

Volodymyr Kvasnikov and Anzhelika Stakhova

Abstract One of the current problems of modern technology is the measurement of vibration parameters. Increased vibration in machines, for example due to bearing wear, can cause a serious damage. Most often and rather quickly the destruction of the object occurs under vibration with conditions of resonance. Stresses caused by vibration contribute to the accumulation of damage in materials, the appearance of cracks and destruction. Therefore, lowering the level is an urgent task and is a prerequisite for the design and manufacture of process equipment. Such problems can be detected early by using vibration measurement techniques. Thus, there is a great need for measuring, evaluating and monitoring vibration in industrial systems—not only for process machines, but increasingly also for industrial machines. A study of oscillatory processes is great interest to all sectors of the economy—metallurgy, power engineering, rocketry, etc. So, the paper reviews the existing methods and means of measuring vibration parameters and identifies ways to further develop this issue.

Keywords Vibration · Vibration measurement · Measuring sensor · Vibration parameters · Vibration protection

1 Assignment Formulation

The decision making process on the state of the machine is usually assisted by technical diagnostic methods and determined by different types of attributes. One defect can be observed by more than one symptom (e.g. in electrical machines by changes of characteristic of vibration, temperature and electromagnetic field). In the case of choosing the vibration as a physical phenomenon that reflects changes in the machine can be determined by observing many of the symptoms in both the time and the frequency domains [1, 2].

V. Kvasnikov (✉) · A. Stakhova (✉)
National Aviation University, Liubomyra Huzara Ave., 1, Kyiv 03058, Ukraine
e-mail: sap@nau.edu.ua

A. Bieliatynskyi and V. Breskich (eds.), *Safety in Aviation and Space Technologies*, Lecture Notes in Mechanical Engineering,
https://doi.org/10.1007/978-3-030-85057-9_5

Dangerous vibrations in machines and moving mechanisms can be caused by a variety of factors: imbalance, alignment errors, shaft curvature, foundation problems, cavitation, gearing errors in gears, adhesion to fan blades, vortex vibrations in gases and liquids, electrical or magnetic disturbances and many other reasons. Virtually every application requires a competent analysis of vibration problems [3].

Vibrations in large rotating machines such as turbines and generators, compressors or large compressors are especially important. If the vibration amplitude becomes too large, there is a risk of extremely serious and very costly damage. In order to ensure particular safety in these cases, the corresponding requirements for protective systems have been defined, for example in ISO 10816 (new 20816) [4, 5] and API 670 [6]. Typical monitoring parameters in these machines are the measurement of the absolute vibration of the bearing (using acceleration or vibration sensors), relative vibration of the shaft (using non-contact displacement sensors) and measuring the axial displacement of the shaft (also using non-contact displacement sensors).

The oscillations occur at different phase positions. Only phase position measurement allows analysis of the causes of vibration. For this reason, the position of the shaft is usually also recorded with reference sensors (and the speed is measured at the same time). Characteristic values are then monitored that estimate the amount of oscillation (such as oscillation width Smax, oscillation amplitude vRMS, RMS values for harmonic oscillation components).

To register vibration values, contactless displacement sensors, piezo acceleration sensors, linearized (electrodynamic) vibration speed sensors and air gap sensors are used.

2 Equipment Control According to ISO

Machine control according to ISO 10816/20816 [4, 5] is often used as a basis for fans, motors and pumps. A characteristic feature of this standard is the identification of vibration conditions using four assessment zones A, B, C and D:

- A = new machine
- B = suitable for continuous operation
- C = can only work for a limited time. Take action as soon as possible
- D = possible damage to the machine.

For this type of condition-oriented monitoring, it is important that the monitoring device actually conforms to the specifications of the standard with respect to signal processing. Fixed and modular systems are available on the market that can scale to a single channel. Stationary systems are suitable for rail mounting and have interfaces (fieldbus, 4–20 mA, digital outputs) to a PLC or control system.

The effective value of the vibration velocity is often measured in a fixed frequency range from 3 to 1000 Hz as an aggregate parameter. This parameter is especially important because the vibration speed level is also a measure of the vibration energy level generated by the machine. Vibration speed is usually more important for monitoring than acceleration measurement.

On the other hand, the standard itself is not enough, the current monitoring parameters contained in the standard requires to hold or construction of easily configurable system for measuring and analyzing.

2.1 *Vibration Measurement Methodology Based on ISO*

Factors affecting measurement accuracy in the time and frequency domain are as follows:

- the frequency band depends on the type and individual characteristics of the machine, but most often 10–1000 Hz is used;
- the speed is measured in three orthogonal directions X-Y-Z (An approximate distribution of the measuring points is shown in Fig. 1);
- measured speed when the machine is running;
- vibration measuring devices that meet the conditions of the effective speed of the frequency range;
- allowable error 10%;

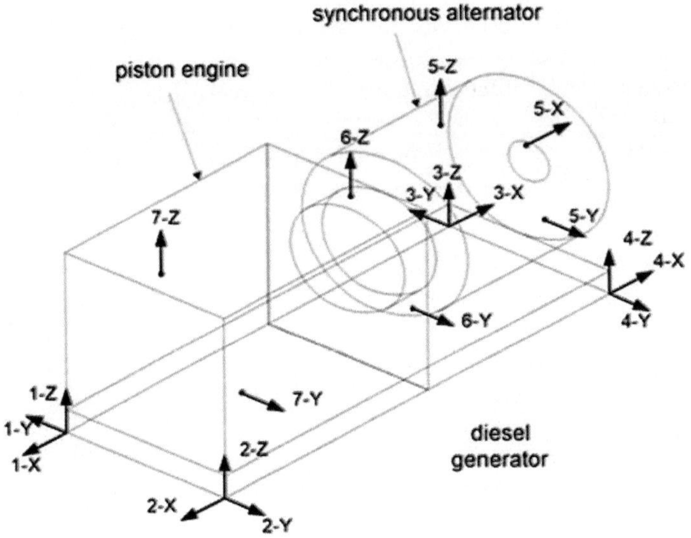

Fig. 1 Approximate distribution of measurement points during vibration diagnostics of a generator set

- averaging measurement time is not less than 1 s;
- measuring instruments worthy with the recommended instructions;
- measurement is performed along the axis of the machine shaft;
- vibration transducers are mounted in such a way as to ensure linear signal processing;
- the measured values are read at maximum speed (in real time).

The vibration sensors are installed correctly, which does not introduce additional errors.

2.2 *Recommendations for the Selection of Measuring Sensor Parameters*

The quality of the converters affects the quality of the data. Results vary with air temperature, humidity, magnetic fields, vibration, other equipment, and installation errors. To obtain correct results, correctly prepared parts of the measuring circuit are used to eliminate factors influencing the measurement. An important element is the correct installation and placement of the sensors on the machine body. The transducers are fixed and located in three directions in accordance with the X-Y-Z—coordinate system on the bed of the testing machine (Fig. 1). The transducers should provide a linear transformation of the signal generated by the machine in the range from 10 to 1000 Hz (or in another frequency range depending on the machine being evaluated and the experience of the researcher). The mass of the transducer should be no more than 1/10 of the mass of the vibration machine.

The most commonly used vibration sensors are accelerometers (Fig. 2) [7, 8]. Accelerometers are mainly used for two specific types of acceleration measurement—impact (shock) and vibration. Impact is effectively a large acceleration over a short period of time, while vibration is a small, repeatable acceleration. Accelerometers are also used to measure the position, velocity, or acceleration of bodies, such as aircraft and ships. Accelerometers are normally mechanically

Fig. 2 Typical accelerometers

bonded to the object whose acceleration is to be measured. The accelerometer detects acceleration along one axis and is insensitive in orthogonal directions.

The operating principle of an accelerometer is based on the inertial effects associated with a mass connected to a moving object through a spring and damper. When the moving object accelerates, there is a relative displacement between the object and the mass. The relative displacement is either directly measured through a position sensor, such as a linear potentiometer (seismic accelerometer), or indirectly sensed by the output voltage of a piezoelectric crystal (piezoelectric accelerometer). See Fig. 3 for a schematic illustration of the two types of accelerometers. Whereas seismic accelerometers are intended mainly for low-frequency accelerations, piezoelectric accelerometers do not give an output for constant or slow-varying accelerations due to the basic characteristics of piezoelectric motion transducers. Instead, they are excellent for dynamic measurements, such as in high-frequency vibrations and impacts.

Each accelerometer used in a measurement procedure must have a calibration table. The most important parameters contained in the calibration table are: name and type of accelerometer, serial number, reference sensitivity, frequency range, set resonant frequency, measurement range, sensitivity characteristic (amplitude and phase) depending on frequency and environmental conditions during use.

We have the following notation during the measurement and analyses:

acceleration of vibration (as function of time)—*a(t)*: Peak: A_p.
velocity (as function of time)—*v(t):* Peak: V_p;
RMS: V_{RMS}.
displacement (as function of time)—*d(t)*: Peak: D.

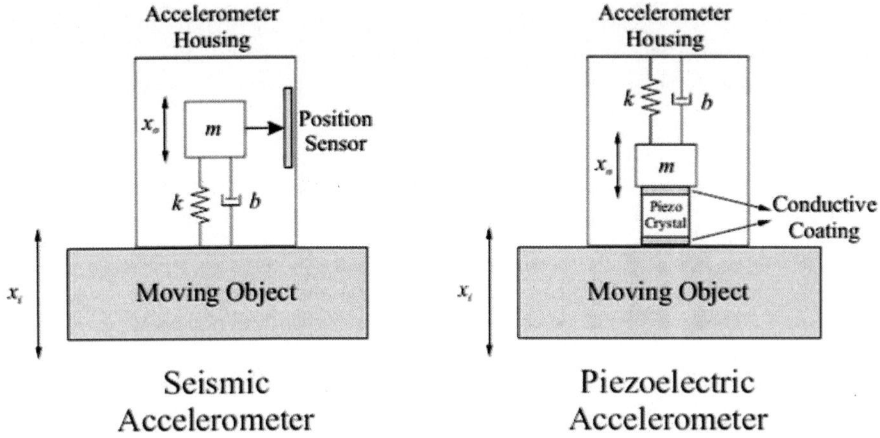

Fig. 3 Schematic representation of accelerometers

3 Vibration Protection

General methods of vibration control are based on the analysis of equations that describe the oscillations of machines in production conditions [9, 10] and are classified as follows:

- reduction of vibrations in the source by reducing or eliminating excitatory forces;
- debugging from resonant modes by rational choice of the reduced mass or rigidity of the oscillating system;
- vibration damping—reduction of vibrations due to the friction force of the damping device, i.e. the transfer of vibrational energy into heat;
- dynamic damping—introduction into the oscillating system of additional masses or increasing the rigidity of the system;
- vibration isolation—introduction into the oscillating system of additional elastic connection, in order to reduce the transmission of vibrations, adjacent structural element or workplace;
- use of personal protective equipment.

Reduction of vibration in the source of its occurrence is achieved by reducing the force that causes oscillations. Therefore, even at the design stage of machines and mechanical devices, it is necessary to choose kinematic schemes, and whose dynamic processes caused by shocks and accelerations would be eliminated or reduced. Reduction of vibration can be achieved by balancing the masses, changing the mass or stiffness, reducing the technological tolerances in the manufacture and assembly, the use of materials with high internal friction. It is important to increase the accuracy of processing and reduce the roughness of friction surfaces.

Debugging from resonance mode. To reduce vibrations, it is essential to prevent resonant modes of operation in order to eliminate resonance with the frequency of the forcing force. The natural frequencies of individual structural elements are determined by the calculation method according to known values of mass and stiffness or experimentally on stands.

Resonant modes during the operation of technological equipment are eliminated in two ways: by changing the characteristics of the system (mass or stiffness) or by setting another mode of operation (debugging the resonant value of the angular frequency of the forcing force).

4 Condition Monitoring of Rolling Bearings

The condition monitoring of rolling bearings is an established technology. Motors, pumps and fans with roller bearings are either checked with handheld measuring devices at regular intervals or monitored with permanently installed systems [11]. In this context, bearing damage parameters such as the BCU or k(t) value, so-called shock impulse parameters, are important. These characteristic values summarize the state of health of a rolling bearing into a total value. This gives maintenance staff a reliable basis for making decisions about a preventive replacement of the bearing. From the point of view of many maintenance managers, online monitoring of the BCU value is desirable, even for medium-sized and smaller units.

Due to their sensitivity and high bandwidth, acceleration sensors are usually used for rolling bearing diagnosis. In addition to the characteristic values shown above, a common evaluation is the calculation of an envelope frequency spectrum. This evaluation method enables the assessment of various, so-called damage symptom frequencies. Rolling bearings have design-related, fixed damage symptom frequencies for damage to the inner and outer ring, the rolling elements (balls or rollers) and the cage. Each damage symptom frequency can be assessed individually. This enables precise statements to be made about the condition of the bearing. The basis of this measurement technology is that bearing damage generates high-frequency periodic shock excitations. The envelope frequency spectrum enables precise evaluation through skillful signal processing (high pass + rectification + low pass).

5 Intelligent Signal Analysis and Industrial Analytics

In some cases, the methods described so far are not sufficient to find and monitor the causes of vibration problems. In such cases, further measured values and system parameters are recorded and analyzed together with the vibrations. Only the correlation of several influencing factors delivers a meaningful value that can be monitored. This requires methods of intelligent signal analysis, also known as 'Industrial Analytics' or 'Machine Learning'. Whether correlation procedures or more detailed statistical procedures are necessary must be decided depending on the application.

With the help of industrial analytics methods, it is also possible to uncover optimization potential in machines, analyze quality problems or even increase the efficiency of the machine. The correlation process is based on complex and computationally intensive algorithms. Especially when the number of input variables is high, one PC is no longer sufficient to process the data and the calculation has to be moved to the cloud. Well-known cloud platforms have long offered options for statistical analysis of the measured values—including measured vibration values. The necessary evaluation methods, rules and algorithms are created by data analysts and integrated into the cloud systems. If the use of cloud-based systems is planned, it must be ensured that the vibration measurement technology has suitable IIoT interfaces. MQTT is one of the standard protocols that can be used for communication between field devices and the cloud.

6 Systems for Web-Based Vibration Monitoring

Web technology has long since found its way into vibration measurement. Acquisition and monitoring devices are available that can be integrated directly into the cloud as IIoT devices with dynamic input channels and integrated, web-based visualization and analysis software. With these devices, the configuration and online display of vibrations and characteristic values is carried out directly via sophisticated analysis diagrams and practical diagnostic functions with the web browser [12]. Such devices often have an internal data memory that can buffer data —independently from the cloud.

7 Vibration Monitoring in the Ex Area

Wireless monitoring systems are often used in Ex areas or with distributed machines. Such systems can measure machine vibrations and temperature synchronously and monitor the machine condition. A wireless system consists of three individual components: wireless sensors, wireless data acquisition station and analysis software. The software is used to monitor the machine status in real time and enables users to recognize machine errors in advance.

In widely distributed industrial areas, Long Range WAN, or LoRaWAN for short, is increasingly being used as a transmission medium for the measurement signals. LoRaWAN offers the possibility to cover large industrial areas with self-sufficient, battery-powered sensors. LoRaWAN is particularly suitable for vibration measurement in the sense of condition monitoring and it can be assumed that intelligent sensors will be available shortly.

8 Motor Current Analysis as a Monitoring Tool

The motor current analysis is based on the fact that mechanical defects such as bearing damage, imbalance or cavitation are mapped directly in the motor current and can be detected by precise current and voltage measurement—virtually from a distance [13, 14]. Such CMS systems [15] are suitable for most electric motor driven devices and generators. A very wide range of electrical and mechanical faults is diagnosed, including not only the motor or generator but also the connected machines such as pumps, fans, compressors and turbines. Such systems provide information on the type and severity of the development of errors. For an investigation, a motor is connected to clamps and voltage converters. An evaluation device records the measured values over a period of around 30 min, creates a machine analysis and uses it to generate a clear report in which the damage symptoms and the health status of the machine are assessed. Corresponding solutions are available for both stationary and mobile use.

9 Summary

Vibration is the most dangerous of all types of mechanical influences for technical objects. The alternating stresses caused by vibration contribute to the accumulation of damage in materials, the appearance of cracks and destruction. Most often and rather quickly the destruction of the object occurs under vibration under conditions of resonance. Vibrations also cause failures of machines and devices.

The paper reviewed and analyzed the current state of vibration measurement, methods and means of measuring vibroacoustic parameters, as well as methods of vibration control. New detection, storage and analytics systems need to be developed to monitor vibrations. For complete information, in order to link the measurement data of the sensors with real conditions and consequences.

References

1. Listewnik K, Polanowski S, Dereszewski M, Żukowski J (2011) Preliminary evaluation of effective vibration analysis for the fault diagnosis of natural gas engine-driven compressor. J KONES 18:243–250
2. Żółtowski B, Niziński S (2002) Modeling process of machine operation. Manual, Bydgoszcz-Sulejówek
3. ISO 4S66 (2010) Mechanical vibration and shock. Vibration of fixed structures: guidelines for the measurement of vibrations and evaluation of then-effects on structures, pp 48
4. ISO 10816-4 (2009) Mechanical vibration. In: Evaluation of machine vibration by measurements on non-rotating parts
5. ISO 20816-4 (2018) Mechanical vibration. In: Measurement and evaluation of machine vibration
6. API STD 670, 5th Edition (2014) Machinery protection systems
7. Elgar P (1998) Sensors for measurement and control. Addison-Wesley Longman, Essex, England
8. Histand MB, Alciatore DG (1999) Introduction to mechatronics and measurement systems. WCB/McGraw-Hill, Boston, MA
9. Veprik AM, Babitsky VI (2000) Vibration protection of sensitive electronic equipment from harsh harmonic vibration. J Sound Vib 238(1):19–30. https://doi.org/10.1006/jsvi.2000.3098
10. Karnovsky IA, Lebed E (2016) Theory of vibration protection. Springer International Publishing
11. Mathew J, Alfredson RJ (1984) The condition monitoring of rolling element bearings using vibration analysis. https://doi.org/10.1115/1.3269216
12. Barszcz T (2009) Architecture and performance of RPC-based distributed vibration monitoring and diagnostic system. Measurement 42(4):532–541

13. Firmansah A, Mufti N, Affandi AN, Zaeni I AE (2019) Self-powered IoT based vibration monitoring of induction motor for diagnostic and prediction failure. IOP Conf Ser: Mater Sci Eng 588(1):012016. https://doi.org/10.1088/1757-899X/588/1/012016
14. Kryter RC, Haynes HD (1989) Condition monitoring of machinery using motor current signature analysis. CONF-890555-3. Oak Ridge National Lab
15. Hmida MA, Braham A (2018) An on-line condition monitoring system for incipient fault detection in double-cage induction motor. IEEE Trans Instrum Meas 67(8):1850–1858. https://doi.org/10.1109/TIM.2018.2806009

Methods of Statistical Modeling of Atmospheric Turbulence

Anatolii Pashko and Tetiana Oleshko

Abstract The paper studies the methods of statistical modeling of Gaussian random processes used in the modeling of atmospheric turbulence. The need to model atmospheric turbulence arises in many applied problems. Atmospheric turbulence is the main factor that determines the aircraft design resource, affects the assessment of aviation resource. Atmospheric turbulence affects the flight trajectory of unmanned aerial vehicles, the assessment of the variance of the deflection of ballistic projectiles during firing. For practical calculations, we use the implementations of the random field obtained as a result of statistical modeling. The study is based on the methods of statistical modeling of stationary Gaussian random processes and fields. When modeling stationary Gaussian processes, the shaping filter method can be used. The paper proposes a modeling method that uses spectral representations of a stationary Gaussian process and field. Using the spectral representation allows you to build models with a given accuracy and reliability. The paper presents new estimates of accuracy and reliability in the space of integrated functions. The obtained estimates depend on the parameters of the spectral decompositions. This allows us to take into account the properties of atmospheric turbulence in the construction of statistical implementations. The use of spectral images of Gaussian random processes and fields allows to obtain simpler statistical modeling algorithms. This is very important when organizing and conducting computational experiments to solve applied problems.

Keywords Statistical simulation · Gaussian random processes · Atmospheric turbulence

A. Pashko
Taras Shevchenko National University of Kyiv, Volodymirskaya St., 64, Kyiv 01033, Ukraine

T. Oleshko (✉)
National Aviation University, Liubomyra Huzara Ave., 1, Kyiv 03058, Ukraine
e-mail: 11lota@ukr.net

A. Bieliatynskyi and V. Breskich (eds.), *Safety in Aviation and Space Technologies*, Lecture Notes in Mechanical Engineering,
https://doi.org/10.1007/978-3-030-85057-9_6

1 Introduction

Monte Carlo methods, or statistical modeling methods, are often used in the design and construction of intelligent systems to solve complex problems. Their feature is that instead of using combinatorial methods, you can perform a computational experiment a large number of times. The method is well suited for problems that cannot be solved by classical mathematical methods. Methods of machine learning in the fields of meteorology, economics, finance and others face the problem of having large data for their study. With the help of statistical modeling is the accumulation of characteristics and parameters that reflect the behavior of complex systems, taking into account the influence of possible external factors. When using statistical modeling methods, the system is endowed with real properties and modeling is provided under different initial conditions. The accumulated data is, in fact, a source of data for learning. This is especially important in reinforcement learning tasks. Learning based on a model experiment gives good results [1, 2]. The Monte Carlo method solves the problem of reinforcement learning by averaging the results of the experiment. Among the methods of statistical modeling a special place is occupied by methods of modeling of Gaussian stationary processes [3, 4]. In [5–7], methods for modeling Gaussian processes based on spectral images of these processes were investigated. The use of the developed algorithms allows to build implementations of Gaussian random processes with given accuracy and reliability.

An important area of application of statistical modeling methods is atmospheric turbulence modeling [8]. Estimate of influence of atmospheric turbulence is an important issue in many applications. So, when flying in a turbulent atmosphere, the aircraft experiences overloads from wind gusts directed along all axes—longitudinal, normal and transverse [9, 10]. The most significant overloads are directed along the normal axis. Atmospheric turbulence is the main factor determining the aircraft design resource. Various approaches to the description of turbulence in different applied problems are presented in papers [11–13].

When assessing the impact of wind effects on the movement of the aircraft, the wind speed is considered as the sum of two components.

$$W(x) = W_0 + W_T(x),$$

where $W_0 \in R^3$ is constant component, $W_T(x)$ is atmospheric turbulence.

Atmospheric turbulence is assumed to be a three-dimensional Gaussian random field $W_T(x) = (X1(x), X2(x), X3(x))^T$ of argument $x \in R^3$. The matrix correlation function of the field is described by the longitudinal correlation function.

$$R_t(x) = EW_t(x+y)W_t(y),$$

$$R_t(x) = \sigma^2 \exp\left(-\frac{|x|}{L}\right),$$

where L is turbulence scale, $W_t(y)$ is projection of vector $W(y)$ onto the direction x, along which the correlation moment is calculated.

Matrix spectral density corresponding to the correlation function $R_t(x)$, has the form

$$F_{k,l}(\lambda) = \frac{2\sigma^2|\lambda|^2\left(\delta_{k,l} - \frac{\lambda_k\lambda_l}{|\lambda|^2}\right)}{L\pi^2\left(|\lambda|^2 + L^{-2}\right)^3},$$

where $\delta_{k,l}$ is Kronecker symbol.

Wind loading and turbulence in the atmosphere have a significant impact on the dynamics of movement of unmanned aerial vehicles (UAV), as well as unguided and corrected aerial vehicles (UCAV). Regular movements of air masses and random gusts of wind, their unsteadiness caused by atmospheric turbulence, are sources of additional forces and moments acting on UAV and UCAV in flight. They lead to disturbed motion, complicate the trajectory correction process, create loads on the aircraft elements, which contribute to the damage to the structure of the deployable stabilizers, etc. [14]. In connection with the above, studies aimed at studying the effect of atmospheric turbulence on the dynamics of motion of aircraft for various purposes are very relevant and have great practical importance.

One of the ways to improve the accuracy of shooting in ballistic problems is to correct the trajectories of the projectile in order to parry the disturbing factors acting on it and, consequently, to compensate for the resulting misses.

When solving ballistic problems in the process of creating corrected munitions, it is necessary to take into account the influence of meteorological factors on the dynamics of their flight. Of all the meteorological factors, wind has the greatest influence on the movement of corrected munitions. When analyzing the effect of wind, a systematic component is usually identified that characterizes the large-scale displacement (kilometers and tens of kilometers) of air masses, at which the average value of the wind speed remains constant or almost constant during the flight of the projectile. This component is measured by sounding the atmosphere and is taken into account when calculating firing settings. At the same time, gusts of wind act on the projectile, caused by the movement of much smaller air masses, covering areas of tens and hundreds of meters. These gusts determine the turbulent random component of the wind speed, which is not predicted when the incoming trajectory is formed [15, 16].

One of the most attractive aspects of the theory of stability of turbulent flows is its intermediate position between the deterministic description using macroscopic equations (such as the Navier-Stokes equation) and the theory of random processes.

The aim of this work is to develop methods of statistical modeling of atmospheric turbulence based on spectral representations of Gaussian stationary processes and comparison of the obtained results with the classical shaping filter algorithm.

2 Shaping Filter Method

The models for the formation of a random atmosphere are usually based on realizable algorithms of canonical expansions, forming filters, or special empirical orthogonal functions. When solving the problems of the dynamics of the flight of corrected ammunition in the atmosphere, in the mathematical description of the process, it is necessary to sufficiently fully take into account the model of the meteorological field and its spatio-temporal variability. There are various ways of accounting for atmospheric turbulence in a general mathematical model describing the movement of corrected munitions. The model of the wind field has become widespread under the assumption of the homogeneity and isotropy of the turbulent air environment. In this case, the spatial distribution of the wind speed field is characterized by two correlation functions: longitudinal R_t and lateral R_n. They describe the statistical characteristics of two components of the random wind speed: one is directed tangentially to the trajectory, the other is normal.

For correlation functions we use expressions:

$$R_t(r) = \sigma^2 \exp\left(-\frac{|r|}{L}\right),$$

$$R_n(r) = \sigma^2 \left(1 - \frac{|r|}{2L}\right) \exp\left(-\frac{|r|}{L}\right),$$

where σ^2 is dispersion of wind speed components, L is turbulence scale, $|r|$ is modulus of the radius vector determining the position of the points of the wind field relative to the considered point, $r = Vt$.

A random process of wind disturbances, the correlation functions of which have the form (1), can be represented by means of white noise that has passed a linear shaping filter, the structure and parameters of which are determined by the following equations [10]:

$$\frac{dW(t)}{dt} = -W(t)\frac{V}{L} + \sigma\left(\sqrt{\frac{2V}{L}}\right)\xi(t),$$

$$\frac{dW_n'(t)}{dt} = W_n(t) + \sigma\left(\sqrt{\frac{3V}{L}}\right)\xi(t),$$

$$\frac{dW_n(t)}{dt} = -\left(\frac{V}{L}\right)^2 W_n'(t) - \frac{2V}{L}W_n(t) + \sigma\left(\sqrt{\left(\frac{3V}{L}\right)^3}\right)\left(1 - 2\sqrt{3}\right)\xi(t),$$

The system of difference equations for modeling has the form

$$W_{k+1} = W_k + h\left(-W_k\frac{V}{L} + \sigma\left(\sqrt{\frac{2V}{L}}\right)\xi_k\right),$$

$$W'_{k+1} = W'_k + h\left(W^n_k + \sigma\left(\sqrt{\frac{3V}{L}}\right)\xi_k\right),$$

$$W^n_{k+1} = W^n_k + h\left(-\left(\frac{V}{L}\right)^2 W'_k - \frac{2V}{L}W_k + \sigma\left(\sqrt{\left(\frac{3V}{L}\right)^3}\right)\left(1-2\sqrt{3}\right)\xi_k\right).$$

The simulation results are shown in Figs. 1 and 2.

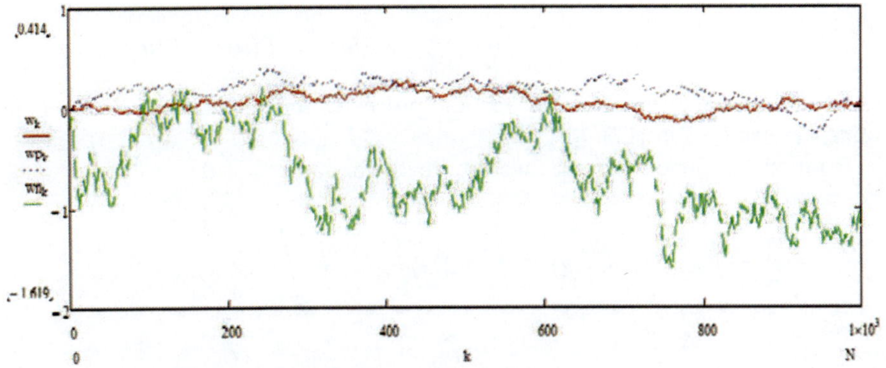

Fig. 1 Wind component implementations for L = 150, $\sigma = 5$

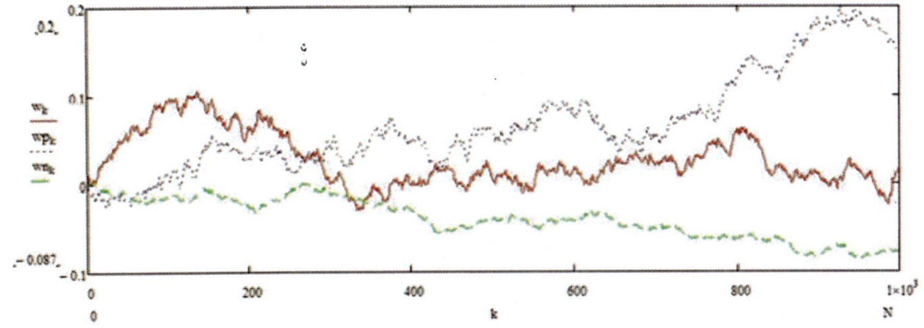

Fig. 2 Wind component implementations for L = 2000, $\sigma = 5$

3 Spectral Model of Turbulence

Let (T, B, μ) be some measurable space with $\mu(T) = 1$.

Let $X(t), t \in R^1$ be Gaussian stationary random process, $EX(t) = 0$. Suppose $B(\tau) = EX(t+\tau)X(t) = \int_0^\infty \cos(u\tau)dF(u)$ is correlation function of the process, $F(u)$ is spectral function.

The random process has an image.

$$X(t) = \int_0^\infty \cos(tu)d\zeta_1(u) + \int_0^\infty \sin(tu)d\zeta_2(u),$$

where $\zeta_1(u)$ and $\zeta_2(u)$ are centered uncorrelated random processes with uncorrelated increments such that for arbitrary $0 \leq u_1 < u_2$ the following equality takes place

$$E(\zeta_1(u_2) - \zeta_1(u_1))^2 = E(\zeta_2(u_2) - \zeta_2(u_1))^2 = F(u_2) - F(u_1).$$

Images of stationary Gaussian random processes for different correlation functions were studied in [17].

To model a random process, we use image for some $\Lambda > 0$.

$$X(t) = \int_0^\infty \cos(tu)d\zeta_1(u) + \int_0^\infty \sin(tu)d\zeta_2(u) = \int_0^\Lambda \cos(tu)d\zeta_1(u)$$
$$+ \int_0^\Lambda \sin(tu)d\zeta_2(u) + \int_\Lambda^\infty \cos(tu)d\zeta_1(u) + \int_\Lambda^\infty \sin(tu)d\zeta_2(u).$$

Definition 1 Let $\Lambda > 0$, $D(\Lambda, N) : 0 = u_0 < u_1 < \ldots < u_N = \Lambda$ be splitting of the interval $[0, \Lambda]$. Random process $X_N(\Lambda, t) = \sum_{i=0}^{N-1} (\cos(u_i t)(\zeta_1(u_{i+1}) - \zeta_1(u_i)) + \sin(u_i t)(\zeta_2(u_{i+1}) - \zeta_2(u_i)))$ is called an approximation model of the process $X(t)$ (A is model).

Therefore, model A of process $X(t)$ can be obtained by modeling the sum

$$X_N(\Lambda, t) = \sum_{i=0}^{N-1} \left(\cos(u_i t)\eta_i^1 + \sin(u_i t)\eta_i^2\right)$$

where $\{\eta_i^1, \eta_i^2\}$ are independent strictly subgaussian random variables with $E\eta_i^1 = E\eta_i^2 = 0$ and $E(\eta_i^1)^2 = E(\eta_i^2)^2 = F(u_{i+1}) - F(u_i)$.

4 Estimates of Accuracy and Reliability of Models

Let some random process $X(t)$ and all $X_n(t,\Lambda)$ belong to some functional Banach space $A(T)$ with norm $\|\cdot\|$. Let two numbers $\delta > 0$ and $0<\alpha<1$ be given. The model $X_n(t,\Lambda)$ approximates the process $X(t)$ with reliability $1-\varepsilon$ and accuracy δ in the norm of space $A(T)$, if there is an inequality $P\{\|X(t) - X_n(t,\Lambda)\| > \delta\} \le \varepsilon$.

Modeling of atmospheric turbulence is reduced to building a model of a stationary Gaussian process. The following theorems hold.

Theorem 1 *The model $X_n(t,\Lambda)$ approximates the process $X(t)$ with reliability $1-\varepsilon$ and accuracy δ in the norm of space $L_2(T)$, if for numbers Λ and n the following inequalities hold*

$$B2_{n,\Lambda} < \delta^2 \text{ and } \exp\left\{\frac{1}{2}\right\}\frac{\delta}{\sqrt{B2_{n,\Lambda}}}\exp\left\{-\frac{\delta^2}{2B2_{n,\Lambda}}\right\} \le \varepsilon,$$

where $B2_{n,\Lambda} = \int_T E(X(t) - X_n(t,\Lambda))^2 d\mu(t)$.

Theorem 2 *The model $X_n(t,\Lambda)$ approximates the process $X(t)$ with reliability $1-\varepsilon$ and accuracy δ in the norm of space $L_2(T)$, if for numbers Λ and n the following inequalities hold*

$$B3_{n,\Lambda} < \delta^2 \text{ and } \exp\left\{\frac{1}{2}\right\}\frac{\delta}{\sqrt{B3_{n,\Lambda}}}\exp\left\{-\frac{\delta^2}{2B3_{n,\Lambda}}\right\} \le \varepsilon,$$

where $B3_{n,\Lambda} = 2T\sum_{i=0}^{n-1}\int_{\lambda_i}^{\lambda_{i+1}}\left(1 - \frac{\sin(T(\lambda-\lambda_i))}{T(\lambda-\lambda_i)}\right)f(\lambda)d\lambda + T\left(\int_{\Lambda}^{\infty} f(\lambda)d\lambda\right)$.

Let the condition $T(\lambda_{i+1} - \lambda_i) \le 1$ be satisfied for splitting D_Λ. Suppose the splitting is chosen uniform, that is $\lambda_{i+1} - \lambda_i = \frac{\Lambda}{n}$ and $\frac{T\Lambda}{n} \le 1$.

For the correlation function $R_t(r) = \sigma^2 \exp\left(-\frac{|r|}{L}\right)$ spectral density is $f(x) = \frac{\sigma^2 L}{\pi(1+L^2x^2)}$.

Corollary 1 *The model $X_n(t,\Lambda)$ approximates the process $X(t)$ with reliability $1-\varepsilon$ and accuracy δ in the norm of space $L_2(T)$, if for numbers Λ and n the following inequalities hold*

$$G1_{n,\Lambda} < \delta^2 \text{ and } \exp\left\{\frac{1}{2}\right\} \frac{\delta}{\sqrt{G1_{n,\Lambda}}} \exp\left\{-\frac{\delta^2}{2G1_{n,\Lambda}}\right\} \le \varepsilon,$$

where $G1_{n,\Lambda} = \frac{T\Lambda^2}{6n^2} + \frac{T}{L}\left(\frac{\pi}{2} - arctg(L\Lambda)\right)$.

For the correlation function $R_n(r) = \sigma^2\left(1 - \frac{|r|}{2L}\right)\exp\left(-\frac{|r|}{L}\right)$ *spectral density is* $f(x) = \frac{2\sigma^2 L^3 x^2}{\pi(1+L^2x^2)^2}$.

Corollary 2 *The model* $X_n(t, \Lambda)$ *approximates the process* $X(t)$ *with reliability* $1 - \varepsilon$ *and accuracy* δ *in the norm of space* $L_2(T)$, *if for numbers* Λ *and* n *the following inequalities hold*

$$G2_{n,\Lambda} < \delta^2 \text{ and } \exp\left\{\frac{1}{2}\right\} \frac{\delta}{\sqrt{G2_{n,\Lambda}}} \exp\left\{-\frac{\delta^2}{2G2_{n,\Lambda}}\right\} \le \varepsilon,$$

where $G2_{n,\Lambda} = \frac{T^3\pi\sigma^2\Lambda^2 L}{2n^2} + \frac{TL^2}{2}\left(\frac{\pi}{2} - arctg\left(L^2\Lambda^2\right)\right)$.

The simulation results are shown in Figs. 3 and 4.

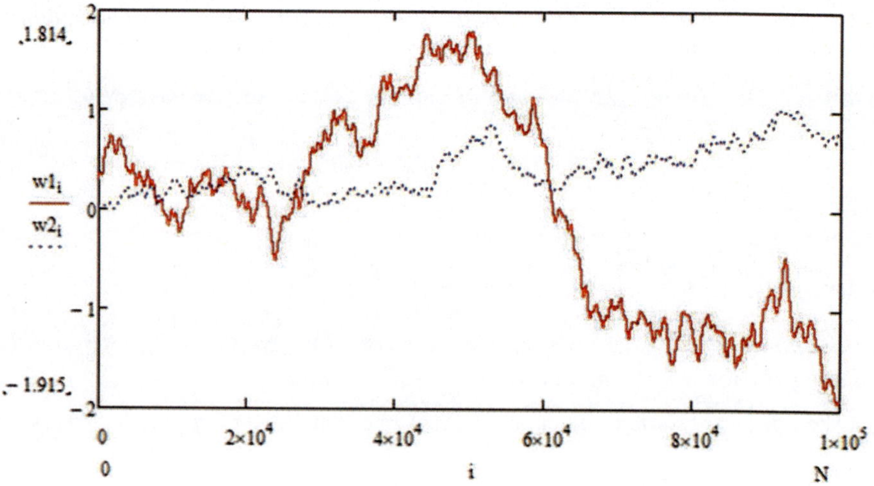

Fig. 3 Wind component implementations for L = 150

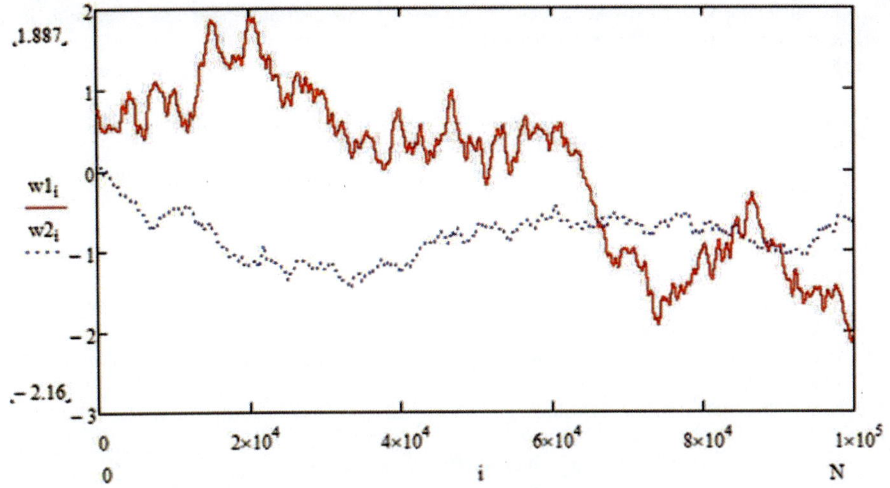

Fig. 4 Wind component implementations for L = 1500

5 Conclusions

The paper studies methods of statistical modeling of atmospheric turbulence based on the spectral representation of Gaussian stationary processes. For comparison the modeling filter method is considered. The use of spectral representation makes it possible to build models with a given accuracy and reliability. The paper gives estimates of accuracy and reliability in the space of integrable functions. Shaping filter modeling requires solving a system of stochastic differential equations.

The obtained models can be used to study the effect of atmospheric turbulence on the reliability of aircraft, to plot the trajectories of unmanned aerial vehicles, and to assess the dispersion of ballistic projectiles.

References

1. Lyuger D (2003) Artificial intelligence: strategies and techniques for complex problem solving. 864
2. Pashko AO, Lukovych OV, Rozora IV, Oleshko TA, Vasylyk OI (2019) Analysis of simulation methods for fractional Brownian motion in the problems of intelligent systems design. IEEE Int Conf Adv Trends Inf Theory (ATIT), pp 373–378. https://doi.org/10.1109/ATIT49449.2019.9030478
3. Kozachenko Y, Pogorilyak O, Rozora I, Tegza A (2016) Simulation of stochastic processes with given accuracy and reliability. ISTE Press, pp 181–250. https://doi.org/10.1016/C2016-0-02585-8
4. Kozachenko Y, Rozora I (2003) Simulation of Gaussian stochastic processes. Random Oper Stochas Eq 3:275–296

5. Kozachenko Y, Pashko AO (1989) Estimates on the distribution of the supremum of a stationary Gaussian. Ukrain Math J 41:279–286
6. Kozachenko YV, Pashko AO, Rozora IV (2007) Modeling of random processes and fields. 230
7. Pashko A, Vasylyk O (2018) Simulation of fractional brownian motion basing on its spectral representation .Theory Stochastic Process 23(39):1, 73–81
8. Optis M, Perr-Sauer J (2019) The importance of atmospheric turbulence and stability in machine-learning models of wind farm power production. Renew Sustain Energy Rev 112:27–41
9. Dobrolyenskiy YP (1969) Flight dynamics in a turbulent atmosphere. Mechanical Engineering, Moscow, pp 11–46
10. Shalygin AS and Palagin YI (1986) Applied statistical modeling methods. Mechanical Engineering, Leningrad. 320
11. Chen CJ, Jaw SY (1998) Fundamentals of turbulence modelling. Taylor & Francis, DC, Washington
12. Durbin PA, Reif BAP (2001) Statistical theory and modelling for turbulent flows. John Wiley & Sons Ltd, West Sussex, England
13. Monin AS, Yaglom AM (1965) Statistical fluid mechanics 1–2
14. Dmitriyevskiy AA, Lysyenko LN (1978) Applied problems of optimal control theory of motion of unmanned aerial vehicles. Mechanical engineering, Moscow 328
15. Weinacht P, Oberlin G (2018) Effect of crosswind on direct-fire projectile dispersion and hit probability and considerations for crosswind estimation. 30
16. MacEachern C, Yıldız I (2018) Wind energy comprehensive energy systems. 1: 665–701. https://doi.org/10.1016/B978-0-12-809597-3.00118-8
17. Yaglom AM (1981) Correlation theory of stationary random functions. 280

Designing Integrated Cargo Delivery Systems on Logistics Principles

Viktor Voitsehovskiy, Mariya Grygorak, Sergiy Lytvynenko, Tetiana Gabrielova, and Larysa Lytvynenko

Abstract In the paper an integrated cargo delivery systems on logistics principles was proposed. In order to optimize the cargo business of the air carrier as an additional source of income and to enable the cargo transportation development on the network of airlines effective cargo traffic management system should be implemented. The principal scheme of operating activities for a network air carrier in relation to the cargo component was designed. Implementation of cargo traffic management system of the network air carrier was considered on logistics principles. To synchronize and optimize cargo traffic flows during the planned period on the airline network to obtain profit from cargo transportation mathematical modelling of cargo traffic of a network air carrier, including a model of operational management of cargo flows in the real time mode and a nonlinear dynamic traffic flow planning model can be applied.

Keywords Cargo · Delivery systems · Logistics principles · Network air carrier · Cargo traffic

1 Introduction

Issues on the designing integrated cargo delivery systems were considered by a large number of researchers. In particular, Khalipova et al. [1] proposed a model for the integrated management of the international delivery chains formation providing integrated multicriteria analysis of potential cargo transportation routes and cost of delivery using the systemic approach. It is aimed at the rational allocation and optimal servicing of traffic flows. In turn, Delgado et al. [2] focused on ensuring optimal allocation of cargo to the passenger network for profit maximization through application of a multistage stochastic programming model.

The paper by Pimentel and Alvelos [3] emphasized the expediency of integration of the existing passenger transport network with the urban freight process. For this

V. Voitsehovskiy · M. Grygorak · S. Lytvynenko (✉) · T. Gabrielova · L. Lytvynenko
National Aviation University, Liubomyra Huzara Ave. 1, Kyiv 03058, Ukraine

A. Bieliatynskyi and V. Breskich (eds.), *Safety in Aviation and Space Technologies*, Lecture Notes in Mechanical Engineering,
https://doi.org/10.1007/978-3-030-85057-9_7

purpose the mathematical model was developed allowing minimization of the total service time, network services' synchronization and providing a balance between loads and system capacities.

Van et al. [4] described main directions of implementing cargo hitching project providing social, environmental impacts along with sustainable business model for the system.

The air cargo schedule recovery problem for air carriers to adapt to sudden demand changes giving recovery savings was introduced by Delgado et al. [5]. Zhang et al. [6] constructed two mixed integer programming models and a two-stage hybrid algorithm for the capacitated air-cargo network planning and scheduling problems.

Another aspect of additional economic effect obtaining by air carriers was considered by Yu et al. [7]. In order to gain more revenues differential pricing strategies of air freight transport carriers in the spot market should by applied.

The paper by Wang [8] analyses factors of an airline's optimal network selection showing comparison of 1-hub, 2-hub and fully connected networks.

Fageda and Flores-Fillol [9] investigated the interrelations between airline network structure and airport congestion, ways of adjusting frequencies to delays depending on the network type operating by airlines.

Gao et al. [10] designed a fast Dijkstra-based all-pair shortest path algorithm with time and budget requirements and apply Fenwick Tree to improve the time efficiency, also studied the reachability for airline networks through the implementation of Analytic Hierarchy Process to measure their service qualities.

Air carriers select hub-strategies when costs are sufficiently low, and asymmetric equilibrium where one air carrier chooses a fully-connected strategy and the other chooses a hub-and-spoke strategy may occur, as discussed by Flores-Fillol [11].

Temporal and spatial analysis was made by Jiang et al. [12] for studying features of the network structure on the example of Chinese airline. Chinese market airline network where nodes and edges denote cities and direct flights was also assessed by Liu et al. [13]. Another practical study based on activities of Iranian airlines was carried out by Omrani and Soltanzadeh [14] with use of dynamic DEA models to evaluate network structure.

Kölker and Lütjens [15] offered an approach to integrate network planning and scheduling, solve minor effects of aircraft rotation and passenger demand by applying genetic algorithms. Logothetis and Miyoshi [16] proposed a new model for evaluating connectivity at hub airports, using the Hub Connectivity Performance Analyser to derive conclusions about the hub performance and efficiency of airlines being influential connectors.

Particular attention within the traffic flows management should be paid modelling aimed at the minimization of airport slots for airlines [17]. Lurkin et al. [18] used discrete choice models under competition among airline itineraries, adding non-stop versus connecting level of service, carrier, and time of day using nested logit and ordered generalized extreme value models.

Optimizing of the network can be provided by airline partnerships in alliances and implementation of hybrid cooperation agreements, as stated by Casanueva et al. [19] and Fageda et al. [20] respectively.

In general, despite the various-aspect studies of the traffic flows management the issue of application a nonlinear multiproduct model of cargo traffic management of the network air carrier is not sufficiently solved.

In general, despite the various-aspect studies of the traffic flows management the issue of application a nonlinear multiproduct model of cargo traffic management of the network air carrier is not sufficiently solved.

2 Methodology

2.1 Operating Activities for a Network Air Carrier in Relation to the Cargo Component

Currently, the limit between profitability and loss to air carriers is very small, therefore, no matter how hard it sounds, but only those carriers have the right to survive who will effectively form the rotation of air planes, have a wide network of directions and minimum possible service delays at airports. Domestic carriers usually have problems with all these aspects, which greatly affects their competitiveness in the world transportation markets. Especially it becomes noticeable when competing with recognized world aviation brands. The principal scheme of operating activities for a network air carrier in relation to the cargo component is represented in Fig. 1.

Dynamic pricing is based on a number of factors that need to be analyzed in real time mode, at the same time, primarily, a flexible tariff line for all possible sales routes with the possibility of their combination with the products of other airlines should be developed or the construction of intermodal cargo transportation with the participation of different transport modes, which is now becoming more popular, should be done. This module corresponds to the tariff management systems. Distribution and tariff policy are closely interconnected, which served as the basis for a close cooperation of suppliers of these solutions in the cargo transportation market.

The modular architecture of the system provides an ability to manage effectively all processes in the field of cargo transportation. One of the most difficult iterations in the airline is the priority of cargo selection. Availability of a hub airport means the domination of many segmental transportations, that is, those involving several carriers' flights in different directions.

Costs for reloading cargo on a passenger flight are very difficult to accurately measure, since it is practically impossible to separate the costs of cargo loading transportation from the passenger one. In general, when transporting additional cargo loading, the costs will change slightly, fuel consumption will increase somewhat due to the increase in commercial loading. Revenues from cargo loading

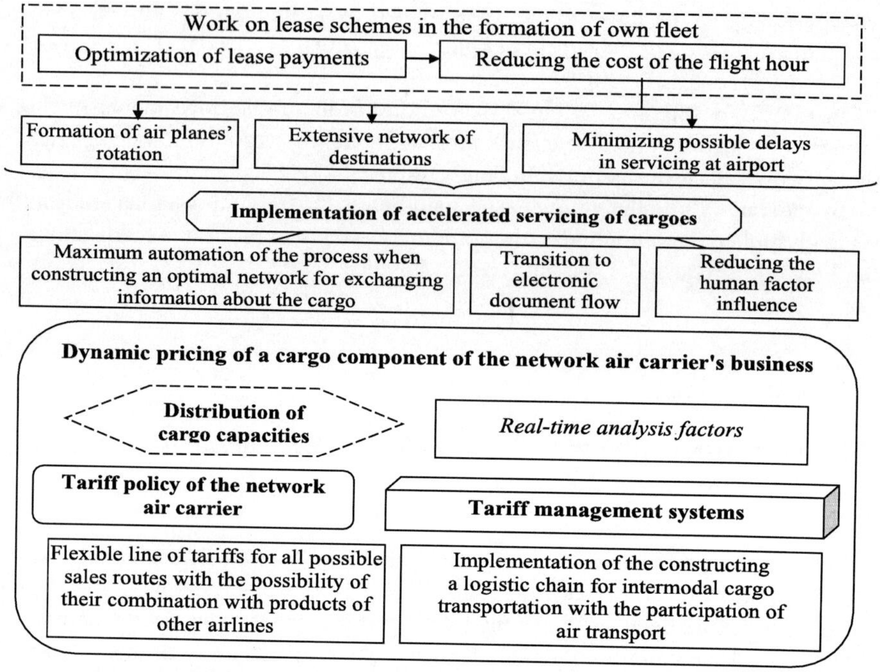

Fig. 1 The principal scheme of operating activities for a network air carrier in relation to the cargo component

are measured more clearly knowing cargo tariffs and loading volume. An increase in cargo loading of a passenger flight will allow an air carrier to obtain a significant net profit, since if costs increase by a small amount, and revenues from cargo loading transportation increase on the contrary, the general economic effect is clearly visible.

Cargo traffic management will enable a network carrier to structure and direct cargo flows in different seasons of the year, attracting or restricting commercial loading of different values in "low" and "high" seasons. This will optimize the cargo business of the air carrier as an additional source of income and will enable the cargo transportation development on the network of airlines. When considering cargo transportations as an additional business, they are developing on the existing network of airlines.

2.2 *Cargo Traffic Distribution on the Part of Airline Network*

The carrier's airline network is formed by passenger transportation, cargo traffic does not affect the development of the general network. Therefore, it is important to synchronize and optimize cargo traffic flows during the planned period on the airline's network in order to obtain profit from cargo transportation. For this purpose, the use of mathematical modelling of cargo traffic of a network air carrier was offered. This research is devoted to the implementation of a nonlinear multiproduct model of cargo traffic management of a network air carrier.

The scheme of cargo traffic distribution on the part of airline's network is given in Fig. 2. The cargo traffic can be divided: according to the time frame—the one that arrives in a year, in half a month, in a month; by structure—permanent, direct, transfer; by category of cargo—special and general.

In addition, there is an opportunity to provide an intermodal transportation in the airport of the flight start, for which special ground deliveries are organized (in the case under consideration, from Karaganda). Also the use of the network of other air carriers is possible (in our example on the New York-Seattle route and New York-Toronto route). Ground carriers interacting with air carriers organize servicing of supply chains of goods and generate cargo traffic of multimodal and intermodal carriage of cargoes. Air transportation interacting with road transportation on the start–end segments of the path is becoming particularly in demand.

2.3 *Main Strategic and Tactical Tasks of Air Carriers*

In accordance with the analysis carried out, the authors identified main strategic and tactical tasks of air carriers, represented in Fig. 3. So, reduction of costs (prime cost) by optimizing and technological improvement of the process as a whole, as well as the selection of optimal cargo traffic for the network will be key strategic tasks. Moreover, additional services will be important to ensure the reliability and convenience of transportation.

In turn, tactical tasks will include the construction of the logic of cargo priority and the system of through accounting of capacities, technologicality of the process

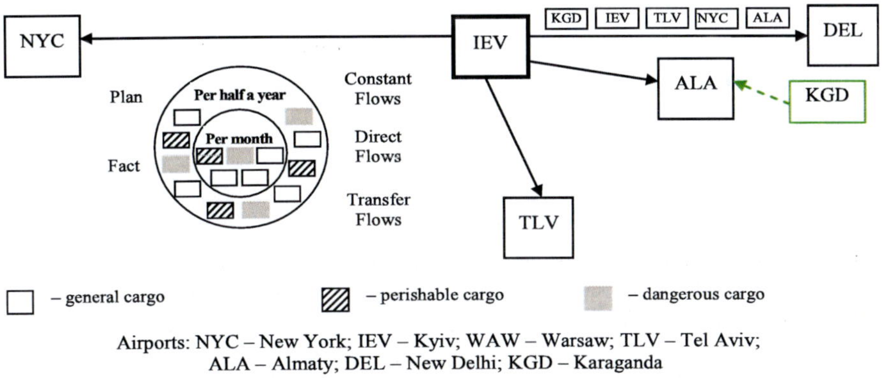

Fig. 2 The scheme of cargo traffic distribution on the part of airline network

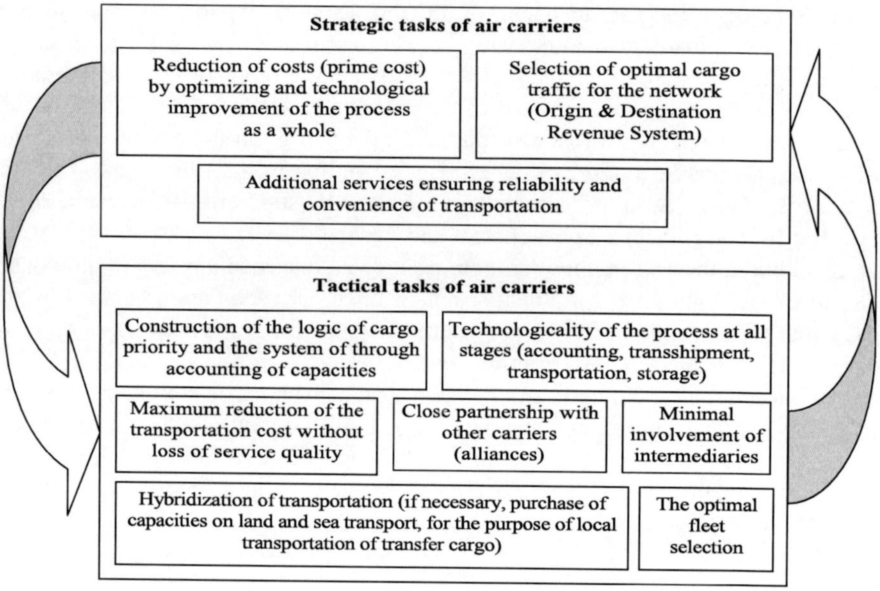

Fig. 3 Main strategic and tactical tasks of air carriers

at all stages (accounting, transshipment, transportation, storage), maximum reduction of the transportation cost without loss of service quality, close partnership with other carriers and joining a variety of alliances, minimal involvement of intermediaries, hybridization of transportation (if necessary, purchase of capacities on land and sea transport, for the purpose of local transportation of transfer cargo), as well as the optimal fleet selection.

3 Results and Discussion

3.1 Forming Integrated Cargo Delivery Systems with the Participation of Air Carriers on Logistics Principles

The airline logistics involves considering the cargo delivery from airport to airport, in a larger chain of goods delivery. Air transportation is one of the main elements of logistics chain: manufacturer—seller—transportation to the airport—air transportation—transportation from the airport—buyer, in case complete door-to-door delivery process is considered.

All aspects of air transportation are handled by the airline providing the service. The air carrier must minimize the cost of delivering the cargo for maximum profit without losing the quality and speed of cargo delivery. Different types of cargo delivery can be compared by such key characteristics as transportation costs, risks of cargo loss and delivery time. Shipping by air is more expensive but has advantages in terms of speed and integrity of cargo. In a large number of cases, air transportation is the only possible way of transportation in the logistics supply chain. One of the issues in air transportation is to organize cargo delivery by air as a logistics system.

Air carrier is a part of the cargo delivery chain and performs logistics functions on its own airline network. But the successful implementation of the entire delivery chain depends largely on it. Implementation of cargo traffic management system of the network air carrier based on logistics is presented in Fig. 4.

For the optimal operating of the air carrier, air transportation services must be provided in accordance with logistical requirements and the creation of logistical flows of goods and services should be provided. Air carriers should monitor the situation in markets for the consumption of cargo transportation services, as the fluctuations in cargo flows depend on it. Optimizing the cargo delivery on the network of airlines creates conditions for obtaining income of all participants of the cargo delivery chain. Minimum transit time providing minimum delivery time of products can be considered as air transportation advantage. In addition, the cargo integrity is another advantage when transported by air. Air transport is one of the most reliable because airlines pay a lot of attention to safety issues. This is particularly relevant in transporting high value products.

Increase in profits from the cargo loading transportation, increase of the load factor of flights, increase of network connectivity, the carrier's participation in global cargo delivery networks, improvement of the interaction between participants of the cargo delivery chain are benefits of optimizing the cargo delivery on the network of airlines for the carrier. The customer benefits from increased cargo transportation capacities and transportation cost optimization. Freight forwarders, agents and logistics operators benefit from attracting and maintaining additional cargo flows on the network of airlines. Thus, the cargo traffic management on the network of airlines is closely linked to the production activities of all participants in the cargo delivery chain.

A two-stage model of cargo traffic management of the network air carrier includes model of operational management of cargo flows in the real time mode and nonlinear dynamic traffic flow planning model. It should be noted that conditions for the use of penalties in a nonlinear multiproduct model are correct only if full payment is made at the time of calculation. In this case, the "minus" means that the air carrier returns the prepayment, and the "plus" means that agents pay fines to the carrier for not using the hard block. If the carrier simply returns the prepayment, then there is no need to take into account anything at all, since the model does not consider the dynamics of money. According to the rules of the hard block all capacities bought by the customer must be used. They may be returned only with a fine, then, in this case, the capacities are returned with the return of part of the funds paid for them. There is a possibility that these capacities will fly empty, but in this case the customer will receive nothing for them.

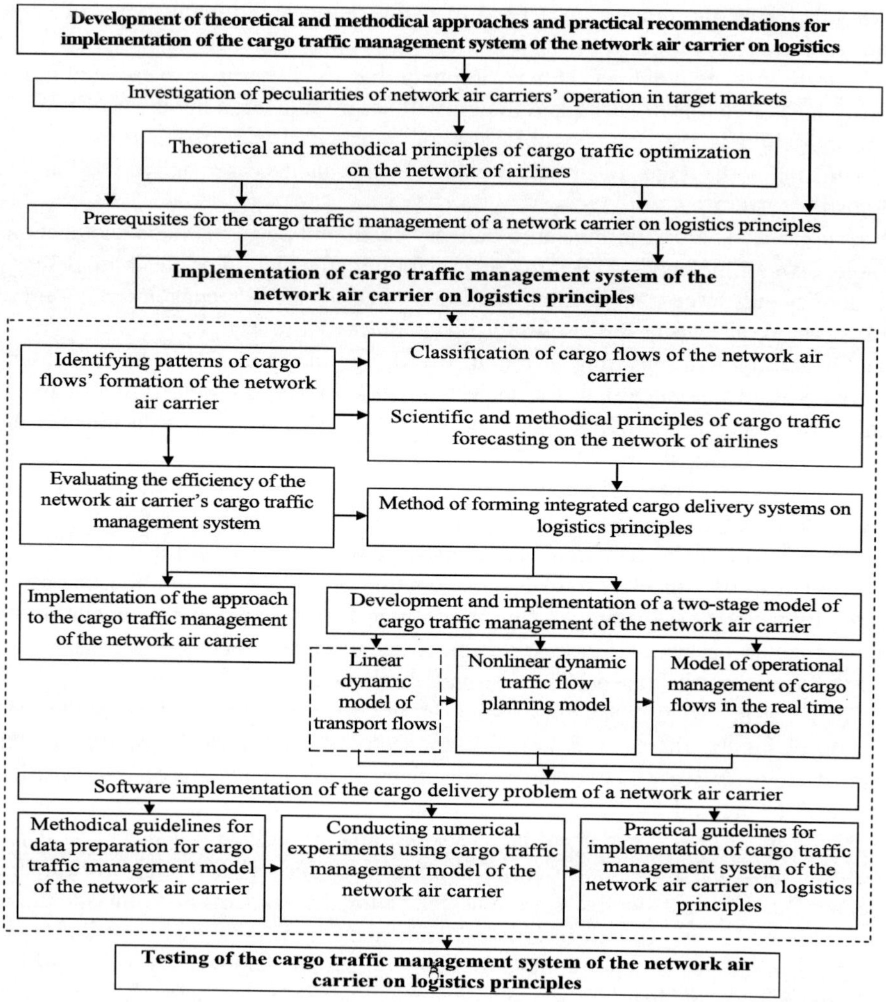

Fig. 4 Implementation of cargo traffic management system of the network air carrier on logistics principles

For any planning, extraordinary demand is possible. These problems arise during the implementation of the plan. The air carrier starts planning in a year and in 2 weeks something extraordinary happens, for example, a board manufacturing plant in China, concentrated production for 30% of the countries, is flooded, and this immediately prompted reorientation of cargo traffic. At the same time, the transportation had to be urgent, since contracts were broken, these fees rose sharply several times, which means that if the added value of these components used to be within US $1 per kg and these cargo flows were served by other modes of transport, then under these conditions the added value increased to US $4 per kg and it became profitable to deliver this cargo by air. This means that the carrier must be prepared for the necessity to reorient its own cargo traffic and urgently look for free capacities for this cargo.

In the nonlinear multiproduct model such cases are taken into account as follows. The air carrier has a definite plan, however then the carrier has an unexpected shortage of shipping capacities, which means he has problems on performing carriage, therefore due to the fact that somewhere the air carrier refuses someone from customers and pays a fine, he increases own carrying capacities, freeing them up for another cargo, more expensive and one that spontaneously arose. In this model, special categories of cargoes are considered in detail.

4 Conclusions

The authors determined that today the boundary between profitability and losses for air carriers is very small, and therefore only those that effectively form rotation of airplanes, have an extensive network of directions and the minimum possible delays in servicing at airports are actively working in the markets. The proposed principal scheme of operational activities for a network air carrier in relation to the cargo component should allow the maximum balance of areas of passenger and cargo air transportation in the overall activity of the air carrier. Cargo traffic management has made it possible for a network air carrier to implement structuring and directing cargo flows in different seasons of the year, to attract or restrict commercial loading of different values in "low" and "high" seasons, depending on the situation. This research is devoted to the implementation of a nonlinear multiproduct model of cargo traffic management of a network air carrier, which is characterized by continuous variables. In this nonlinear model, the options that the airline will choose regarding certain storage and delivery airports options are specifically identified. Terms of use of penalties in a nonlinear multiproduct model are correct only if full payment at the time of calculation is made. The model determines the availability of ascertainment a number of cargoes' categories, which is an extremely important criterion.

The proposed recommendations can make it possible to take into account the capabilities of the considered model as much as possible when carrying out their practical implementation by the network air carriers within the scope of their economic activity.

References

1. Khalipova N, Bosov A, Prohoniuk I (2018) Development of a model for the integrated management of the international delivery chains formation. Eastern-European J Enterp Technol 3(93):59–72
2. Delgado F, Trincado R, Pagnoncelli KB (2019) A multistage stochastic programming model for the network air cargo allocation under capacity uncertainty. Transp Res PE: Logis Transp Rev 131:292–307
3. Pimentel C, Alvelos F (2018) Integrated urban freight logistics combining passenger and freight flows—mathematical model proposal. Transp Res Proc 30:80–89
4. Van DR, Wiegmans B, Tavasszy L, Hendriks B, He Y (2019) Evaluating new participative city logistics concepts: the case of cargo hitching. Transp Res Proc 39:565–575
5. Delgado F, Sirhan C, Katscher M, Larrain H (2020) Recovering from demand disruptions on an air cargo network. J Air Transp Manage 85:101799
6. Zhang C, Xie F, Huang K, Wu T, Liang Z (2017) MIP models and a hybrid method for the capacitated air-cargo network planning and scheduling problems. Transp Res PE: Logis Transp Rev 103:158–173
7. Yu S, Yang Z, Zhang W (2019) Differential pricing strategies of air freight transport carriers in the spot market. J Air Transp Manage 75:9–15
8. Wang X (2016) 1-Hub, 2-hub or fully connected network? A theoretical analysis of the optimality of airline network structure. Econ Transp 5:12–23
9. Fageda X, Flores-Fillol R (2016) How do airlines react to airport congestion? The role of networks. Reg Sci Urban Econ 56:73–81
10. Gao X, Xianzang Y, You X, Dang Y, Chen G, Wang X (2018) Reachability for airline networks: fast algorithm for shortest path problem with time windows. Theoret Comput Sci 749:66–79
11. Flores-Fillol R (2009) Airline competition and network structure. Transp Res PB: Methodol 43(10):966–983
12. Jiang Y, Yao B, Wang L, Feng T, Kong L (2017) Evolution trends of the network structure of Spring Airlines in China: a temporal and spatial analysis. J Air Transp Manage 60:18–30
13. Liu HK, Zhang XL, Zhou T (2010) Structure and external factors of Chinese city airline network. Phys Proc 3(5):1781–1789
14. Omrani H, Soltanzadeh E (2016) Dynamic DEA models with network structure: an application for Iranian airlines. J Air Transp Manage 57:52–61
15. Kölker K, Lütjens K (2015) Using genetic algorithms to solve large-scale airline network planning problems. Transp Res Proc 10:900–909
16. Logothetis M, Miyoshi C (2018) Network performance and competitive impact of the single hub—a case study on Turkish Airlines and Emirates. J Air Transp Manage 69:215–223
17. Babić D Kalić M (2012) Airline route network expansion: modelling the benefits of slot purchases. J Air Transp Manage 23:25–30
18. Lurkin V, Garrow LA, Higgins MJ, Newman JP, Schyns M (2018) Modeling competition among airline itineraries. Transp Res PA: Policy Prac 113:157–172
19. Casanueva C, Gallego Á, Castro I, Sancho M (2014) Airline alliances: mobilizing network resources. Tour Manage 44:88–98
20. Fageda X, Flores-Fillol R, Theilen B (2017) Hybrid cooperation agreements in networks: the case of the airline industry. Int J Ind Org 62:194–227

Optimization of Parameters for the Printing Process of Adhesive Application in Honeycomb Core Manufacturing

Andrii Kondratiev, Sergiy Melnikov, Tetyana Nabokina, and Anton Tsaritsynskyi

Abstract Thickness of the adhesive strip, mass application of the adhesive, graphic accuracy of the adhesive strip location, and strength of the adhesive bond at delamination are among the main parameters of the adhesive joint affecting the honeycomb core characteristics. The most advanced technology which provides obtaining high-quality honeycomb core is considered to be the technology where the adhesive is applied by the intaglio method. One of the ways to improve the adhesive application procedure by this method is the validation and selection of the conditions of printing process. The regression mathematical model is developed in order to establish the optimal relationship between the surface application of the adhesive and parameters of the process of adhesive strip application. The proposed model of the process under study, as a result of evaluation according to a number of criteria, allowed obtaining the minimum applied amount of adhesive with guaranteed uneven tearing strength. Analysis of the resulting model shows that the depth of printing elements has the most significant effect on the mass application of the adhesive. For example, with the decrease in the depth of printing elements, mass application of the adhesive is reduced as well. Speed of printing affects the parameter of adhesive application only in the interaction with the adhesive viscosity. Results of the work allow significantly reducing the procedures of preparation of the considered stages of technological process for the manufacturing of honeycomb cores by means of reduction of the experimental research.

Keywords Adhesive joint · Tearing strength · Process deviations

A. Kondratiev (✉)
O M Beketov National University of Urban Economy, Marshal Bazhanov St., 17, Kharkiv 61002, Ukraine
e-mail: andrii.kondratiev@kname.edu.ua

S. Melnikov · T. Nabokina · A. Tsaritsynskyi
National Aerospace University "Kharkiv Aviation Institute", Chkalova St., 17, Kharkiv 61070, Ukraine

A. Bieliatynskyi and V. Breskich (eds.), *Safety in Aviation and Space Technologies*, Lecture Notes in Mechanical Engineering,
https://doi.org/10.1007/978-3-030-85057-9_8

1 Introduction

The modern structural materials and design and technological solutions where these materials are used reveal the unique reserves for reducing the weight of load-bearing structures in almost all areas of technology and construction [1, 2].

Owing to more stringent requirements to the modern products and constructions, the structures with high specific strength consisting of fewer parts, joined together by bonding, become common use [3]. Solution to this issue is promoted by the use of sandwich structures with a core [4, 5]. These structures represents the sheets (skins) with the core placed between them; this core ensures the joint work of the skins and the required stiffness of the structure. Composite panels comprising thin outer layers of high strength and honeycomb core between them are ones of the most promising structural elements in the modern products and constructions [6]. This is due to the fact that the use of such panels allow solving two main problems, namely, the optimal involvement of the material in the work on accommodation of external loads and provision of a number of other functional characteristics, such as noise absorption, thermal insulation, etc. [7]. The advanced feature in the creation of honeycomb structures is the use of polymeric composite materials in them [8].

Among many types of honeycomb cores, hexagonal honeycombs are the most widespread ones. The main features of such cores are high specific strength and stiffness, high thermal conductivity and fatigue resistance, good anti-vibration characteristics, and the ability to handle complex surfaces [1, 8, 9]. Polymeric paper "Nomex", glass cloth, polymeric films, aluminum foil, carbon filler and other materials can serve as a base of the honeycomb core [10].

More than 90% of honeycomb cores are manufactured in the production environment by stretching or corrugation [1, 11]. Besides, thickness of the adhesive strip, mass application of the adhesive (ratio of the mass of adhesive to the area of adhesive strips), graphic accuracy of the adhesive strip location, and strength of the adhesive bond at delamination are among the main parameters of the adhesive joint affecting the honeycomb core characteristics.

Currently, the following methods of adhesive strip application are used for the honeycomb manufacturing: flow method and intaglio printing method [1, 8, 11]. Comparative characteristics of these methods for applying adhesive strips are shown in Table 1. These data are taken from the paper [12] for the honeycomb core of aluminium foil of 23 μm thick, bonded with VK-25 adhesive.

The most advanced technology which provides obtaining high-quality honeycomb core is considered to be the technology where the adhesive is applied by the intaglio method derived from the printing industry [1, 8, 11].

The advantages of the intaglio printing method include [12]:

- possibility of obtaining the endless impression print with continuous application of adhesive strips along and across the roll, and at an angle;
- accelerated modes of drying of adhesive strips (up to 50 m/min) due to the use of low viscosity adhesive (i.e. less solvent is removed).

Table 1 Comparative characteristics of various methods of adhesive strip application in the process of bonding of honeycomb cores

Adhesive application method	Adhesive thickness (μm)	Mass application of adhesive· (10^{-5} g/cm^2)	Range of deviations of the strip width from the nominal value (μm)	Speed of application of adhesive strips (m/min)
Intaglio printing method	3…5	15…17	±0.1	5…50
Relief printing method	8…18	50…85	±0.6	0.5…1.0
Flow method	10…50	60…230	±0.4	1.5…3.4
Screen printing method	10…17	60…80	±0.4	6.0…30

Application of adhesive strips by the intaglio method, compared to the flow method used in industrial production, makes it possible to reduce mass application of adhesive 3.5–15 times, and also to increase the productivity several times [11, 12].

One of the ways to improve the adhesive application procedure by this method is the validation and selection of the conditions of printing process, which provide the minimum mass application of the adhesive with the necessary tolerances for the width of adhesive strip and delamination strength of the adhesive joint of honeycomb faces.

The objective of this work consists in establishing the optimal relationship between the surface application of the adhesive and parameters of the process of adhesive strip application on the foil by intaglio printing method.

To achieve this objective, the following problem was solved: determination of the optimal conditions of the printing process that ensure the minimum mass application of adhesive at a given tolerance for the width of the adhesive strip and guaranteed delamination strength of the adhesive joint.

2 Literature Review

The papers [12, 13] include the comparative characteristics of various methods for applying adhesive strips on the foil. These papers show a decrease in the adhesive thickness to 3–5 μm achieved by intaglio printing at the reduction of tolerance for the width of the adhesive strip to 0.1 μm and high speed of application.

The studies [14, 15] show that the amount of VK-25 adhesive applied by the intaglio printing method on the aluminum foil is directly proportional to the adhesive viscosity, the foil feed rate and the depth of printing elements. These papers, on the basis of their findings, give a list of defects of the honeycomb cores and their effect on physical and mechanical characteristics of the cores. Analysis of the influence of the parameters of technological operations on the occurrence of defects allowed identifying a number of process parameters and formulating the requirements for the tolerances on their values.

Peculiarities of application of adhesive strips onto polymeric paper and assessment of the impact of the direction of their location on the mechanical properties of polymeric honeycomb core are studied in [16]. Conclusions made in [14, 15] with regard to the choice of the method of application of VK-25 adhesive strips onto aluminium foil refer to the polymeric paper as well. For this material, the intaglio printing method of application of VK-25 adhesive strips is more rational than others (flow method, relief and screen printing methods). The experiments on application of strips of VK-25 adhesive of 2.1 mm wide and pitch of 10 mm onto polymeric paper of 55 μm thick were carried out by varying the following parameters of the printing process: speed of printing, adhesive viscosity, screen line number and depth of printing elements—in total, 87 variants of parameter combinations.

The quality of bonding of honeycomb cores is ensured by the geometrical parameters of the adhesive joint and its continuity. The works [17, 18] deal with the problem of permissible dimensions of disbonds in the honeycomb structures. However, the issues related to specification of manufacturing tolerances were not reflected in these studies.

The paper [19] gives the results of studies aimed at taking into consideration the technological imperfections of the honeycomb core manufacturing in designing of sandwich panels with the honeycomb core. This paper offers a technique for optimization of tolerances for the manufacture of honeycomb core of a panel with the provision of specified bearing capacity of the structure. However, the proposed approach to standardization of tolerances on honeycomb cells' bonding parameters is focused only on the ensuring of the regulated deviation of the bearing capacity of a particular panel within the object under consideration.

Technological methods for correcting the physical and mechanical characteristics of honeycomb cores are proposed in [20]. This paper deals with the ranges of possible changes in the pitch of adhesive strips depending on the method of adhesive application and geometrical parameters of the honeycomb core. It is stated that the intaglio printing method is the only method acceptable in terms of ensuring maximum physical and mechanical characteristics. However, the method developed in [20] does not allow establishing the optimal parameters for the intaglio printing method.

In the number of works [21, 22] a conclusion is made that an effective alternative to field tests and analytical models is the use of information technologies of the finite element analysis. At the same time, the dependence of the results of calculation under the finite element method on the choice of size of finite elements, and degree of irregularity of the zone for determination of stresses in a particular problem, is well known.

Analysis of the above publications indicates that the problem of improving the quality of bonding of honeycomb cores is already solved to a greater or lesser degree. However, for one of the most advanced technologies of adhesive application by intaglio printing, a number of issues were not taken into account. Solving of these issues can have significant effect on the end results and requires a separate study.

3 Materials and Methods

Experimental studies of the process of application of adhesive strips by intaglio printing were carried out on the aluminium foil, widely used in the manufacturing of honeycomb cores. Printing was carried out using VK-25 adhesive on the printing equipment of the test section of the Ukrainian Institute of Mechanical Engineering. Experimental data were processed following the technique of the experimental planning. To build a model of the process of application of adhesive strips, we used the scheme of complete factorial experiment 2^3. The Cochran's, Student's, and Fisher's criteria were used to check the consistency of variance estimates.

4 Results

Intaglio printing differs from the other printing methods in that the printing elements on the form cylinder are located below spacing elements (Fig. 1) [14, 15]. The system of raster base lines and deep printing elements 7 is located on the surface of form cylinder 1. The adhesive in form cylinder 1 is removed by blade 6, remaining only in the deep printing elements. At the time of printing, printing material 4, passing between form cylinder 1 and printing cylinder 2, comes into contact with plate 1, from which adhesive 5 is transferred onto printing material.

Parameters of the printing process of adhesive strip application are as follows: shape of printing elements; depth of printing elements h_p; screen line number (the number of printing elements per 1 cm segment); adhesive viscosity μ_η; printing

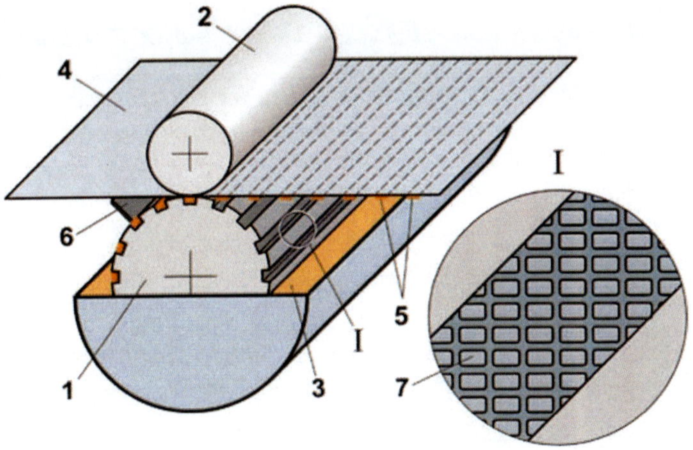

Fig. 1 Scheme of application of adhesive strips by intaglio printing: 1—form cylinder; 2—printing cylinder; 3—container with adhesive; 4—printing material; 5—adhesive strips; 6—blade; 7—deep printing elements

speed V_p; pressure in the printing area; angle of sharpening and setting of the blade; temperature of drying cylinder etc.

On the basis of the studies [14, 15], a rhombic shape of the printing elements was chosen with the ratio of spaces to printing elements 1:3; screen line number 34 lines/cm; pressure in the printing area 200…250 N/cm. The angle of sharpening and setting of the blade was chosen so as to ensure the effective removal of adhesive from the spaces of the printing cylinder, and the temperature of the drying cylinder should provide the drying of adhesive strips of 2.1 mm wide with the pitch of 10 mm.

In the course of the factorial experiment, three factors were varied:

- depth of printing elements h_p, μm—X_1;
- adhesive viscosity μ_η, s—X_2;
- printing speed V_p, mm/min—X_3.

The levels, intervals and ranges of determination of the limits of variation for the variable parameters $h_p(X_1)$, $\mu_\eta(X_2)$ and $V_p(X_3)$ are presented in Table 2.

Mass application of adhesive y_η was chosen as a main parameter of optimization. Amount of adhesive on the unit of foil surface was understood as mass application of adhesive y_η.

The limiting parameters are the tolerance for the width of the adhesive strip not exceeding ±0.1 mm, and delamination strength of the adhesive joint S ≥ 6 N/cm.

For the identified technological factors of the printing process of application of adhesive strips, the problem of establishing the optimal relationship between them and the surface application of adhesive can be solved by the multiple regression method [23, 24]. For constructing a model of the process of application of adhesive strips, we used the scheme of complete factorial experiment 2^3, which allows, based on eight experiments, obtaining the model of the form [24, 25]

$$Y = b_0 + b_1X_1 + b_2X_2 + b_3X_3 + b_{12}X_1X_2 + b_{13}X_1X_3 + b_{23}X_2X_3 + b_{123}X_1X_2X_3 \quad (1)$$

Matrix of the experiment, its results and data for their processing are given in Table 3. As can be seen from Table 3, the limiting parameters (graphic accuracy and delamination strength) in all eight experiments did not go beyond the permissible values.

Table 2 Levels, intervals and ranges of determination of the limits of variation for the variable parameters

Variable parameters	Levels			Variation intervals	Range of determination of the limits of variation
	Basic	Upper	Lower		
h_p (μm)	60	70^+	50^-	10	10…150
μ_η (s)	65	50^+	80^-	15	25…130
V_p (mm/min)	17.5	30^+	5^-	12.5	5…100

Table 3 Matrix and results of the experiment on application of VK-25 adhesive strips by intaglio printing method onto foil AMg-2N of 30 μm thick

b_i	b_0	b_1	b_2	b_3	b_4	b_5	b_6	b_7	Deviations of adhesive strip width from the nominal value (mm)	Delamination strength S (N/cm)	Mass application of adhesive (10^5 g/cm^2)		
											$y_{u1\ \min}$	$y_{u1\ \max}$	y_{mid}
1	+	–	–	–	+	+	+	–	±0.1	9	10.3	12.4	11.35
2	+	+	–	–	–	–	+	+	±0.07	10.2	25.3	22.9	24.1
3	+	–	+	–	–	+	–	+	±0.08	9.2	16.6	14.1	15.35
4	+	+	+	–	+	–	–	–	±0.03	9.9	24.3	19.7	22.0
5	+	–	–	+	+	–	–	+	±0.05	7	14.1	13.8	13.95
6	+	+	–	+	–	+	–	–	±0.05	10.7	24.1	32.5	28.3
7	+	–	+	+	–	–	+	–	±0.05	9.5	6.1	12.0	9.05
8	+	+	+	+	+	+	+	+	±0.08	10.6	17.2	18.5	17.85

To construct the model of the process, coefficients of the regression equation were determined by the formula [25]

$$b_i = \sum_{u=1}^{N} \left(X_{iu}\bar{y}_{\eta u}\right) \cdot N^{-1}, \quad i = 0.1, \ldots, 7; u = 1.2, \ldots, 8; N = 8 \tag{2}$$

The regression equation with computational coefficients is written as

$$\begin{aligned} y = {} & 17.74 + 5.32X_1 - 1.68X_2 - 0.456X_3 - 1.46X_1X_2 + 0.47X_1X_3 \\ & - 2.16X_2X_3 + 0.07X_1X_2X_3. \end{aligned} \tag{3}$$

Processing of experimental data from Table 3 is presented in Table 4.

The estimates of line-by-line variances of the optimization parameter obtained in the experiment (see Table 4) were determined according to the formula

$$S_u^2\{y_\eta\} = \sum_{q=1}^{2} \left(y_{\eta uq} - \bar{y}_{\eta u}\right)^2 = \left(y_{\eta u1} - \bar{y}_{\eta u}\right)^2 + \left(y_{\eta u2} - \bar{y}_{\eta u}\right)^2 = 2\left(y_{\eta u1} - \bar{y}_{\eta u}\right)^2 \tag{4}$$

For checking the consistency of variance estimates, the Cochran's criterion was used

$$G^{ex} = \frac{S_{u\ \max}^2\{y_\eta\}}{\sum_{u=1}^{q} S_u^2\{y\}} = 0.486. \tag{5}$$

Table 4 Processing of results of experiment on application of VK-25 adhesive strips by intaglio printing onto foil AMg-2N of 30 μm thick

b_i	b_0	b_1	b_2	b_3	b_4	b_5	b_6	b_7	Evaluation of line-by-line variances of mass application of adhesive 10^5 g/cm^2, S_u^2	Value of mass application of adhesive calculated using the model $y_{\eta u}$	Squared difference $(\bar{y}_{\eta u} - y_{\eta u})^2$
1	+	−	−	−	+	+	+	−	2.205	10.26	1.881
2	+	+	−	−	−	−	+	+	2.88	20.9	10.24
3	+	−	+	−	−	+	−	+	3.125	12.42	8.5849
4	+	+	+	−	+	−	−	−	10.58	15.22	10.3684
5	+	−	−	+	+	−	−	+	0.045	12.42	2.3409
6	+	+	−	+	−	+	−	−	35.28	25.22	9.4864
7	+	−	+	+	−	−	+	−	17.405	10.26	1.4641
8	+	+	+	+	+	+	+	+	0.845	20.9	9.3023

The table value of *G*-criterion at $\alpha = 0.05$; $f_1 = 1$; $f_2 = 2$ is equal to $G^{th} = 0.6798$. $G^{ex} < G^{th}$ [24], therefore, with the confidence level of 0.95 the estimates of variances $S_u^2\{y_\eta\}$ are consistent.

Variance of the experiment was calculated by the formula [25]

$$S^2\{y_\eta\} = \frac{\sum_{u=1}^{N} S_u^2\{y_\eta\}}{N} = 9.046. \quad (6)$$

Taking into account two repeated experiments, the variance of the mean value is [25]

$$S^2\{\bar{y}_\eta\} = \frac{S^2\{y_\eta\}}{n} = 4.52. \quad (7)$$

The variance of coefficients of the regression equation is [25]

$$S_u^2\{b_i\} = \frac{S_u^2\{\bar{y}_\eta\}}{N} = 0.57, \quad (8)$$

and the mean square error

$$S\{b_i\} = \sqrt{S^2\{b_i\}}\frac{1}{2} = 0.75. \quad (9)$$

The value of Student's *t*-criterion at $\alpha = 0.05$ and $f_y = 8$, $t_{\alpha,f_y} = 2.306$ [25]. Then the confidence interval for the coefficients of the regression equation is

$$\Delta b_i = \pm t_{\alpha,f_y} S\{b_i\} = \pm 2.306 \cdot 0.752 = \pm 1.73 \quad (10)$$

Coefficients b_i are significant if $|b_i| > \Delta b_i$. Equation (3) is written as

$$y_\eta = 17.74 + 5.319X_1 - 2.156X_2X_3 \quad (11)$$

Values y_η for each of the eight experiments determined by Eq. (11) are presented above in Tables 3 and 4. The variance of the adequacy of obtained model according to the results of the experiment was calculated by the formula

$$S_{ad}^2 = \frac{n}{f_{ad}} \sum_{u=1} \left(\bar{y}_{\eta u} - y_{\eta u}\right)^2 \quad (12)$$

of degrees of freedom f_{ad} = 5; S_{ad}^2 = 21.19.

Value of the Fisher's criterion is equal to [25]

$$F^{ex} = \frac{S_{ad}^2}{S^2\left(y_\eta\right)} = 2.34. \quad (13)$$

Table value of F-criterion at α = 0.05 and $f_1 = f_{ad}$ = 5; $f_2 = f_y$ = 8 is equal to F^{th} = 3.69. Since $F^{ex} < F^{th}$, then with a confidence level of 0.95 Eq. (11) adequately describes the process.

As a result of each of eight experiments in addition to the characteristics given in Table 3, the thickness of the adhesive strip was determined, which varied from 2.6 to 4.5 μm. Due to the fact that aluminum foil has the thickness tolerance limit of ±3 μm [14, 26], further decrease in the thickness of the adhesive strips and, therefore, mass application of adhesive is not advisable. At the same time, there is growing possibility of disbond in the honeycomb pack when it is pressed in places where the variations in thickness of the foil exceed the thickness of adhesive.

The regression dependence (11) characterizes the degree of impact of the relative parameters h_p, μ_η and V_p, referred to their upper (X_i = 1) and lower (X_i = − 1) values on the amount of adhesive applied. For practical use (11) and for establishing the reasonable tolerance limits for these parameters, it is necessary to transform it into a dependence connecting the adhesive application with the absolute values of h_p, μ_η and V_p.

For this purpose, we use the system of linear equations with the process parameters included in the experiment plan [27]:

$$
\begin{aligned}
y_{\eta 1} &= a_0 + 50a_1 + 1400a_6 = 11.35 \times 10^{-5}; \\
y_{\eta 2} &= a_0 + 70a_1 + 1400a_6 = 24.1 \times 10^{-5}; \\
y_{\eta 3} &= a_0 + 50a_1 + 875a_6 = 15.35 \times 10^{-5}; \\
y_{\eta 4} &= a_0 + 70a_1 + 875a_6 = 22 \times 10^{-5}; \\
y_{\eta 5} &= a_0 + 50a_1 + 2400a_6 = 13.95 \times 10^{-5}; \\
y_{\eta 6} &= a_0 + 70a_1 + 2400a_6 = 28.3 \times 10^{-5}; \\
y_{\eta 7} &= a_0 + 50a_1 + 1500a_6 = 9.05 \times 10^{-5}; \\
y_{\eta 8} &= a_0 + 70a_1 + 1500a_6 = 17.85 \times 10^{-5}.
\end{aligned} \tag{14}
$$

The linear function of its parameters was considered as a mathematical model [24, 28]

$$y_\eta^m = \sum_{i=1}^{n} a_i f_i(X), \tag{15}$$

where X is vector of independent variables; $f_i(X)$ is functions; α_i is parameters of the model (the required coefficients of system (14)).

The condition for the minimum of the root-mean-square deviation of model values y_η^{ex} from the vector of experimental data obtained at fixed values of the vector of independent variables X is written as

$$[F]^{th} \cdot [F] \cdot \{a\} = [F]^{th} \cdot \left\{ y_\eta^{ex} \right\}, \tag{16}$$

from which the vector of model parameters $\{a\}$ is defined as

$$\{a\} = \left([F]^{th} \cdot [F] \right)^{-1} \cdot [F]^{th} \cdot \left\{ y_\eta^{ex} \right\}, \tag{17}$$

where $[F]$ is model matrix:

$$[F] = \begin{vmatrix} f_1(x_1) & f_2(x_1) & \cdots & f_n(x_1) \\ \cdots & \cdots & \cdots & \cdots \\ f_1(x_m) & f_2(x_m) & \cdots & f_n(x_m) \end{vmatrix} \tag{18}$$

As a result, we obtained the desired dependence of the adhesive application on independent variables h_p, μ_η and V_p.

$$y_\eta = 1.732 \times 10^{-4} \left(3.07 \times 10^{-2} h_p + 1.18 \times 10^{-4} \mu_\eta V_p - 1 \right). \tag{19}$$

5 Discussion

Therefore, we established the optimal relationship between the surface application of adhesive y_η and parameters of the process of adhesive strip application—adhesive viscosity μ_η, speed of movement of the foil roll V_p, depth of printing elements of the application cylinder h_p. For this purpose we developed the mathematical model based on the scheme of planning of the complete factorial experiment 2^3. As a result of implementation of the model in eight experiments, conclusions of [14, 15] on the decisive role of these process parameters on the amount of adhesive applied were confirmed. Resulting formula (19) reflects the experimentally established consequence of the influence of h_p, μ_η and V_p on the adhesive application y_η. Analysis of the obtained model shows that the depth of printing elements h_p has the most significant effect on mass application of adhesive (with the decrease in the depth of printing elements, mass application of adhesive is reduced as well). Speed of printing V_p affects the parameter of adhesive application y_η only in interaction with the adhesive viscosity μ_η. Tables 3 and 4, and data of analysis of the model (19), show that minimum application of adhesive is achieved in the experiment No. 7. It is equivalent to the printing element depth of 50 μm, adhesive viscosity of 5 s and printing speed of 30 m/min. In this experiment, delamination strength (9.5 N/s) and tolerance for adhesive strip width (±0.05 mm) is more tight that the limiting parameters. These conditions can be recommended for application of the adhesive in the industrial production of honeycomb cores of aluminium foil.

However, it should be noted that formula (19) reflects only experimental consequence of the effect of h_p, μ_η and V_p on adhesive application y_η. The relationship between the thickness of adhesive strip and adhesive application, as well as the geometric parameters (the width of adhesive strip equal to the side of honeycomb core cell, pitch of adhesive strips and adhesive density) is not identified.

6 Conclusion

In order to establish the optimal relationship between the surface application of the adhesive and parameters of the process of adhesive strip application on the foil by intaglio printing, the mathematical model based on the scheme of planning of the complete factorial experiment 2^3 is developed. As a result of implementation of the model in eight experiments the prevailing role of the amount of adhesive applied and depth of printing elements is established, as well as interrelationship of the adhesive viscosity and speed of application of adhesive strips. Results of the work allow significantly reducing the procedures of preparation of the considered stages of technological process for the manufacturing of honeycomb cores by means of reduction of the experimental research.

References

1. Nunes JP, Silva JF (2016) Sandwiched composites in aerospace engineering. Adv Compos Mater Aerospace Eng 129–174
2. Fomin O, Lovska A, Kulbovskyi I, Holub H, Kozarchuk I, Kharuta V (2019) Determining the dynamic loading on a semi-wagon when fixing it with a viscous coupling to a ferry deck. East Eur J Enterp Technol 2, 7, 98:6–12. https://doi.org/10.15587/1729-4061.2019.160456
3. Bitzer T (1997) Honeycomb technology: materials, design, manufacturing, applications and testing, p 243
4. Slyvyns'kyy V, Gajdachuk V, Kirichenko V, Kondratiev A (2012) Basic parameters' optimization concept for composite nose fairings of launchers. In: 62nd international astronautical congress, IAC 2011, vol 9, pp 5701–5710
5. Rodichev YuM, Smetankina NV, Shupikov OM, Ugrimov SV (2018) Stress-strain assessment for laminated aircraft cockpit windows at static and dynamic loads. Strength Mater 50, 6:868–873. https://doi.org/10.1007/s11223-019-00033-4
6. Kondratiev A, Slivinsky M (2018) Method for determining the thickness of a binder layer at its non-uniform mass transfer inside the channel of a honeycomb filler made from polymeric paper. East Eur J Enterp Technol 6, 5, 96:42–48. https://doi.org/10.15587/1729-4061.2018.150387
7. Slyvynskyi VI, Sanin AF, Kharchenko ME, Kondratyev AV (2014) Thermally and dimensionally stable structures of carbon-carbon laminated composites for space applications. Proc Int Astronaut Congr IAC 65, 8:5739–5751
8. Bychkov AS, Kondratiev AV (2019) Criterion-based assessment of performance improvement for aircraft structural parts with thermal spray coatings. J Superhard Mater 41(1):53–59. https://doi.org/10.3103/S1063457619010088
9. Fomin O, Logvinenko O, Burlutsky O, Rybin A (2018) Scientific substantiation of thermal leveling for deformations in the car structure. Int J Eng Technol 7, 4, 3:125–129. https://doi.org/10.14419/ijet.v7i4.3.19721
10. Herrmann AS (1999) Design and manufacture of monolithic sandwich structures with cellular cares, p 274
11. Astrom BT (1999) Sandwich manufacturing: past, present and future, p 198
12. Slivinsky M, Slivinsky V, Gajdachuk V, Gajdachuk A, Kirichenko V (2004) New possibilities of creating efficient honeycomb structures for rockets and spacrafts. In: 55th international astronautical congress, pp 1–7. https://doi.org/10.2514/6.IAC-04-I.3.A.10
13. Slyvyns'kyy V, Gajdachuk V, Gajdachuk A, Slyvyns'ka N (2005) Weight optimization of honeycomb structures for space applications. In: 56th international astronautical congress, pp 1–10. https://doi.org/10.2514/6.IAC-05-C2.3.07
14. Slyvyns'kyy V, Slyvyns'kyy M, Gajdachuk A, Gajdachuk V, Melnikov S, Slyvynska N, Kirichenko V (2007) Technological possibilities for increasing quality of honeycomb cores used in aerospace engineering. In: 58th international astronautical congress Hyderabad C-07-C2.I.08
15. Melnikov S (2006) The relationship of the tolerance fields on the strength of the adhesive with uneven tearing and its deposition on the foil in the production of honeycomb core. Issues Des Manuf Aircraft Struct 44(1):114–119
16. Slyvyns'kyy V, Slyvyns'kyy M, Polyakov N, Gajdachuk A, Gajdachuk V, Kirichenko V (2008) Scientific fundamentals of efficient adhesive joint in honeycomb structures for aerospace applications. In: 59th international astronautical congress, C2, 1, 13, pp 1–8
17. Huang SJ, Chiu LW (2008) Modeling of structural sandwich plates with 'through-the-thickness' inserts: five-layer theory. Comput Model Eng Sci 34(1):1–32. https://doi.org/10.3970/cmes.2008.034.001
18. Okada R, Kortschot MT (2002) The role of the resin fillet in the delamination of honeycomb sandwich structures. Compos Sci Technol 62(14):1811–1819. https://doi.org/10.1016/S0266-3538(02)00099-4

19. Gaydachuk V, Koloskova G (2016) Mathematical modeling of strength of honeycomb panel for packing and packaging with regard to deviations in the filler parameters. East Eur J Enterp Technol 6, 1, 84:37–43. https://doi.org/10.15587/1729-4061.2016.85853.
20. Kondratiev A, Gaidachuk V, Nabokina T, Tsaritsynskyi A (2020) New possibilities of creating the efficient dimensionally stable composite honeycomb structures for space applications. Adv Intell Syst Comput 1113:45–59. https://doi.org/10.1007/978-3-030-37618-5_5
21. Mackerle J (2002) Finite element analyses of sandwich structures: a bibliography (1980–2001). Eng Comput 19(2):206–245. https://doi.org/10.2514/2.991
22. Allegri G, Lecci U, Marchetti M, Poscente F (2002) FEM simulation of the mechanical behavior of sandwich materials for aerospace structures. Exp Techn Des Composite Mater 5:209–220. https://doi.org/10.4028/www.scientific.net/KEM.221-222.209
23. Grove SM, Popham E, Miles ME (2006) An investigation of the skin/core bond in honeycomb sandwich structures using statistical experimentation techniques. Compos Part A Appl Sci Manuf 37, 5:804–812. https://doi.org/10.1016/j.compositesa.2005.07.005
24. Ayranci C, Carey JP (2011) Experimental validation of a regression–based predictive model for elastic constants of open mesh tubular diamond–braid composites. Polym Compos 32:243–251. https://doi.org/10.1002/pc.21042
25. Garson GD (2014) Multiple Regression, p 462
26. Sypeck DJ (2005) Wrought aluminum truss core sandwich structures. Metall Mater Trans B 36:125–131. https://doi.org/10.1007/s11663-005-0012-5
27. Seber George AF, Lee AJ (2003) Linear regression analysis, p 582
28. Kutner M, Nachtsheim C, Neter J, Li W (2004) Applied linear statistical models, p 1396

Rationalization of the Topology of Steel Combined Truss

Myron Hohol, Uliana Marushchak, Ivan Peleshko, and Dmytro Sydorak

Abstract The rationalization of topology of steel combined trusses is offered in the work. The search for a rational topology of steel combined trusses based on structural synthesis with the energy criterion of rationality is presented. The rational constructive form of a steel combined truss is offered. According to research, it is established that the most rational is the topology of trusses, in which the gratings have descending slopes and regular placement of racks. A rational angle of inclination of truss racks is obtained, which provides the minimum value of mass of combined trusses. The evaluation of rational parameters of combined steel trusses by the method of mathematical planning of the experiment is carried out. The diagram of maximum normal stresses in the stiffening beam of the combined truss is obtained. The potential of the calculated regulation of the stress–strain state in the stiffening beam, characteristic for such combined trusses, is revealed.

Keywords Truss · Rationality · Topology

1 Introduction

Currently, steel trusses are widespread in many branches of construction [1–3]. Modern process of designing rational building load-bearing structures of engineering structures—including trusses, most closely related to saving material and minimizing its volume or weight [4–6].

The new idea of the constructive decision of object can give bigger effect, than optimization of a design of the established geometry on its parameters [7]. This includes the synthesis of load-bearing structures at the topology level. However, the theory of trusses synthesis remains a complex problem. The variational principle of structural synthesis says that the potential energy of the system in the position of

M. Hohol (✉) · U. Marushchak · I. Peleshko · D. Sydorak
Lviv Polytechnic National University, 12 Bandera str., Lviv 79013, Ukraine
e-mail: gogolmyron@i.ua

A. Bieliatynskyi and V. Breskich (eds.), *Safety in Aviation and Space Technologies*, Lecture Notes in Mechanical Engineering,
https://doi.org/10.1007/978-3-030-85057-9_9

stable equilibrium reaches an absolute minimum by moving in the functional space, extended due to the fields of configuration functions.

The best trusses must most fully meet the predetermined conditions, and have a minimum cost of material and labor for its manufacture.

2 Presentation of the Main Material

An important point in the design of statically indeterminate trusses (combined steel trusses) is the establishment of a standard—uniformly stressed statically indeterminate trusses, which has the smallest volume under certain specific conditions [3]. Recently, as one of the approaches to the formation of rational structures, in addition to the classical criterion (minimum volume or weight), an approach based on energy principles (minimum potential deformation energy) is used [8].

To solve the problems of structural-parametric synthesis, it is proposed to use the material utilization factor (MUF), which is rationally used in rational design. The value of MUF KV is the ratio of the energy of the actual deformation of the structure to the maximum allowable potential deformation energy. It can serve as a criterion for the quality of design, including optimal, individual elements, components and structures in general.

Also, the problem of synthesis of topology (formation) of rod systems in the general case is proposed to be solved from the standpoint of generalized variational principles formulated for design tasks [9, 10].

Rational designs of trusses with the criterion of minimum costs, taking into account the complexity of manufacture, unification of structural elements and seriality, do not always coincide with the optimal designs of such trusses, developed on the criterion of minimum weight [7, 11].

Thus, we can identify the basic principles underlying the design of rational building rod structures: saving material, reducing installation time and reducing production complexity. These principles are contradictory and their successful solution is possible when creating structures with minimal costs, meeting the requirements of operation, reliability and durability [1, 12].

Therefore, the development of a methodology for selecting the topology of steel, including combined trusses, which would correspond to the basic principles of designing rational building rod structures, is an urgent problem today [8, 12–14].

The aim of the work is to determine the rational topology of steel combined trusses on the basis of structural synthesis taking into account the energy criterion of rationality.

2.1 *Evaluation of the Efficiency of the Topology of Steel Combined Trusses*

In order to choose a rational topology of steel combined trusses with a span of 30 m, performed their calculation on a PC with the program “LIRA-SAPR 2016 R5” for q = 12.75 kN/m: he most common trusses in the former Soviet Union, the Molodechno Union (Fig. 1a); combined trusses used in Russia (Fig. 1b); combined trusses used in Germany (Fig. 1c); optimized combined trusses in Israel (Fig. 1d).

Restrictions on the conditions of strength, rigidity and stability, formulated by norms and design practice, were formulated for trusses.

Synthesis of load-bearing structures—trusses should have a single methodological basis, as such we can take a generalized characteristic of the potential deformation energy (1), (2) and (3) [9].

The optimization criterion is key. This leads to difficulties in finding the global extremum of the function (functional) of the goal [15].

Potential deformation energy of combined trusses for stiffening beams [8]:

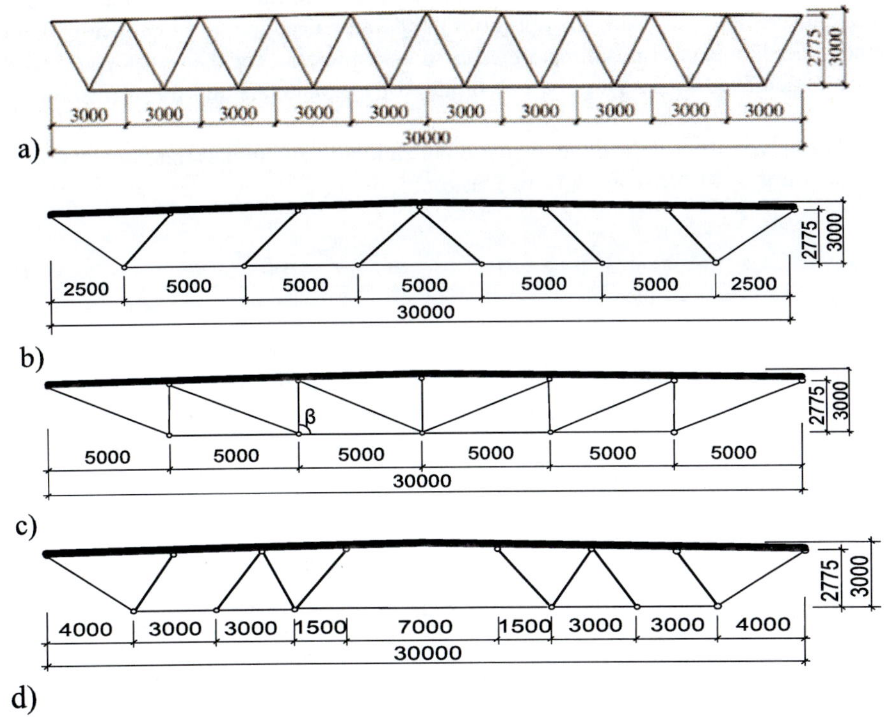

Fig. 1 Variants of truss topology

$$U_{bs} = \sum_{i=1}^{n} \int_{0}^{l_i} \frac{M_i^2}{2E_i I_i} dx + \sum_{i=1}^{n} \int_{0}^{l_i} \frac{N_i^2}{2E_i A_i} dx \tag{1}$$

Potential deformation energy of combined trusses for compressed lattice rods [8]:

$$U_c = \frac{R}{2E} \sum_{i=1}^{n} \frac{N_i l_i}{\varphi_i} \tag{2}$$

Potential deformation energy of combined trusses for stretched lattice rods [8]:

$$U_s = \frac{R}{2E} \sum_{i=1}^{n} N_i l_i \tag{3}$$

In Eqs. (1), (2) and (3): M is the bending moment; N—longitudinal force; E is the modulus of longitudinal elasticity of the material; A—cross-sectional area; I—moment of inertia of the section; li is the length of the i-th rod; n is the number of rods.

The energy criterion provides the minimum cost in the case of a homogeneous material. Here, first of all, the optimum lattice and a rational arrangement of struts and racks is meant. The ultimate goal is to save material for the developed design. This is achieved by varying the functional U by parameters and the corresponding formation of cross sections of the elements.

The second approach to the formation of rational structures is their evaluation by the amount of material used V, or mass m:

$$V = \sum_{i=1}^{n} A_i l_i \quad \text{or} \quad m = \sum_{i=1}^{n} \gamma A_i l_i \tag{4}$$

The criterion for the optimality of a truss made of a homogeneous linearly elastic material is the equivalence of a virtual system with internal forces N_i/φ_i, where φ_i is the coefficient of reduction of the calculated resistance *R* [16].

In the case of a homogeneous linear material having a modulus of longitudinal elasticity *E*, the mentioned criterion is expressed by the minimum potential deformation energy:

$$U = \sum_{i=1}^{n} \frac{N_i^2 l_i}{2E\varphi_i^2 A_i}, \tag{5}$$

where *n* is the number of rods of length l_i, N_i is the force in the rod.

Consider an example of evaluating the efficiency of the topology of steel combined trusses with a span of 30 m, which are shown in Fig. 1. The criterion of

Table 1 Potential deformation energy and mass of truss material

Variants of truss topology	Figure 1 (a)	Figure 1 (b)	Figure 1 (c)	Figure 1 (d)
U (J)	8205	10,839	8168	8285
Weight (kg)	2070.7	3206.0	1842.3	2489.9

Fig. 2 Combined trusses in aircraft: light, resilient, ultra-thin inclined wing supports—Boeing and NASA, USA

efficiency was the minimum potential deformation energy of the steel combined truss. Table 1 presents the value of the potential deformation energy of the truss *U* and the mass of the truss material—*m*, calculated by Formula (4 and 5) for four variants of steel combined trusses.

As can be seen, the smallest mass has a truss (Fig. 1c) with descending braces and vertical struts—$\beta = 90°$. The discrepancy between the extreme mass and potential energy is due to limitations in the choice of rolling profiles, which affects the accuracy of the problem.

The efficiency of all considered combined steel trusses also depends on the angle of inclination of the compressed rods β (Fig. 1c). Numerical and analytical calculations showed [15], hat the assumption of work in the elastic–plastic region (in supercritical behavior) gives an increase in the bearing capacity of combined trusses up to 20%, which depends on the magnitude of the angle of inclination of the compressed rods and the direction of their inclination and is confirmed by recent studies by Boeing and NASA, USA, Fig. 2 [17].

2.2 *Estimation of Rational Parameters of the Combined Steel Truss*

In order to assess, in terms of material consumption, rational parameters, such as the height of the combined steel truss, angles of inclination of the compressed rods of the truss lattice, a numerical experiment was conducted for a combined steel truss with a span of 30 m (Fig. 1c). Experimental studies were performed in accordance

Table 2 Characteristics of experiment planning

Characteristic	Planning options	
	Tilt angle β (deg) (X_1)	Truss height h (m) (X_2)
Main level "0"	45	2
	51.25	2.5
Lower level "−1"	67.5	3
	73.75	3.5
Top level "+1"	90	4
Interval of variation	6.25	0.5

with the plan of a two-factor three-level experiment (Table 2). For the changing factors, the height of the combined truss (h = 2.0 m, h = 2.5 m, h = 3.0 m, h = 3.5 m, h = 4.0 m), the angle of inclination of the compressed rods truss gratings (β = 45°, 56.25°, 67.5°, 78.75°, 90°). The evaluation criterion was the minimum weight of the truss m.

Analysis of the obtained mathematical dependences, as well as their graphical interpretation allow to determine the rational parameters of the combined steel combined truss with a run l = 30 m for the running load q = 12.75 kN/m.

As a result of processing the plans and the corresponding numerical data by the method of least squares, the regression equation (6) is obtained, which adequately describes the dependence of the values of the mass of the truss, as a criterion of a rational system on variables.

$$Y = b_0 + b_1x_1 + b_2x_2 + b_{11}x_1^2 + b_{22}x_2^2 + b_{12}x_1x_2 \tag{6}$$

where x_1, x_2—the values of the changing factors; Y is the response function of a two-factor experiment; b_0, b_1, b_2, b_{12}, b_{11}, b_{22}—regression coefficients.

Based on the obtained coefficients, the regression equation of the studied function is compiled. The planning matrix and the results of the complete two-factor experiment are shown in Table 3.

According to the obtained regression equations, isoparametric diagrams are constructed, which graphically describe the dependence of the criteria on the variable factors (Figs. 3 and 4).

In Figs. 3 and 4 shows that the weight of the truss m varies in the range from 1690 to 2107 kg when changing the height of the truss from 2.0 to 4.0 m, respectively. The angle of inclination β, at the same time, varies in the range from 45 to 90°.

The minimum weight of the truss m = 1640 kg was obtained at a height of the truss h of approximately 3 m, accordingly, at an angle of inclination of the racks β equal to approximately 79°–82°, and at an angle of 90°, the weight of the truss was 1842.3 kg, ie 12.3% more.

Table 3 Planning matrix and results of a complete two-factor experiment

№	Matrix planning in codes		Matrix planning in natural terms		*m* (kg)
	X_1	X_2	Angle of inclination β (deg)	Truss height h (m)	
1	−1	−1	67.5	3	1688.67
2	−1	0	67.5	2	2050.30
3	−1	+1	67.5	4	1726.63
4	0	−1	45	3	1758.36
5	0	0	45	2	2077.55
6	0	+1	45	4	2107.55
7	+1	−1	90	3	1692.75
8	+1	0	90	2	2074.28
9	+1	+1	90	4	1780.30

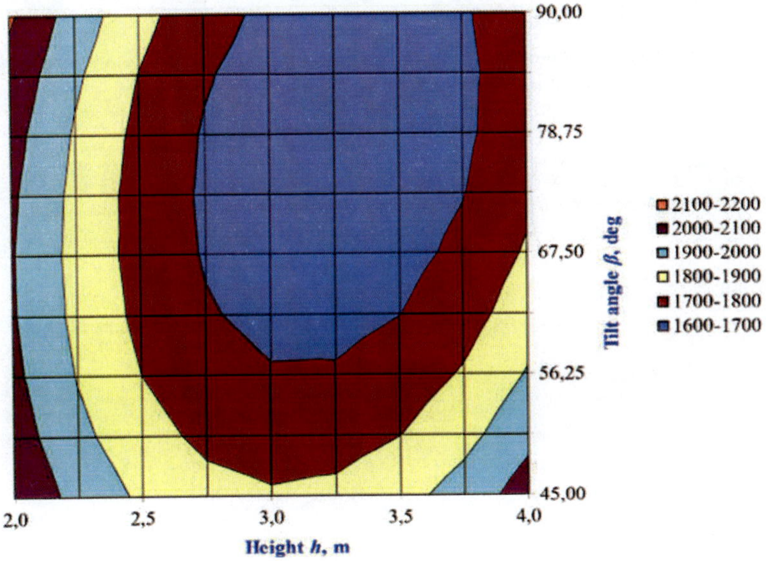

Fig. 3 Diagrams of the change in the mass of the truss *m* depending on the angle *β* of the compressed rods of the truss lattice and the height of the truss *h*

Such a rational combined steel truss with a run of 30 m (Fig. 5a) was calculated on a PC with the program "LIRA-SAPR 2016 R5" for q = 12.75 kN/m at an angle of inclination of compressed rods β = 80° and according to these results, a diagram of normal stresses *σ* in the cross sections of the stiffening beam is obtained (Fig. 5b).

The diagram is characterized by an uneven distribution of stresses along the length of the stiffening beam, they vary from 84 to 219 MPa. Of particular note is

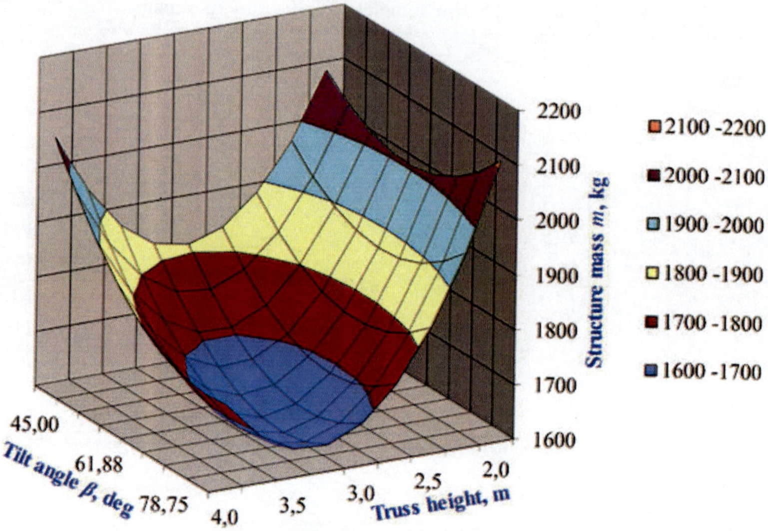

Fig. 4 Isoparametric diagram of the change in the mass of the truss m depending on the angle β of the compressed rods of the truss lattice and the height of the truss h

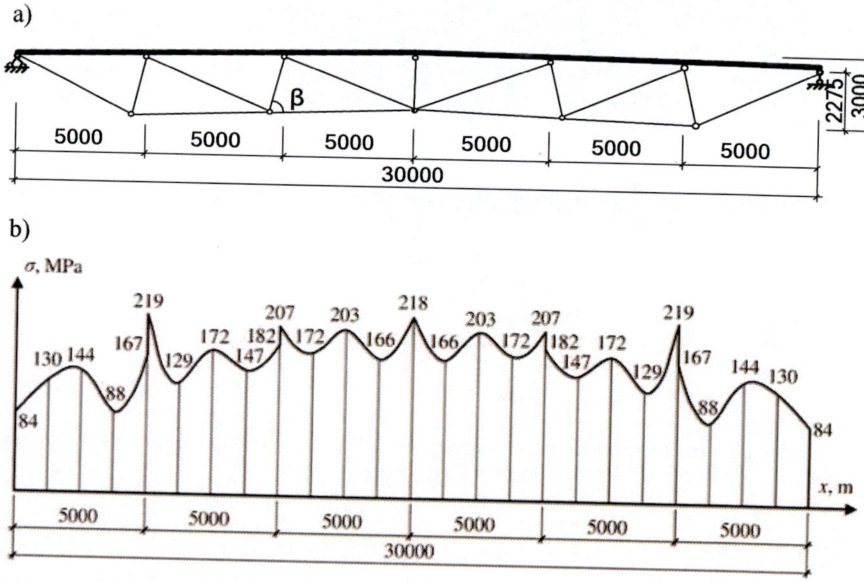

Fig. 5 **a** Scheme of a rational steel combined truss with a span of 30 m; **b** is a diagram of normal stresses σ in the cross sections of the stiffening beam of this truss

the underload of the extreme panels of the beam stiffness, normal stresses in which are only 84 MPa.

This indicates the presence of a large reserve in the extreme panels of the combined truss to ensure uniformity by the method of calculated control of the stress–strain state in the stiffening beam according to the technology described in [2].

Therefore, the disclosure of the settlement regulation potential inherent in such combined trusses is already at the design stage, by equalizing the stress values in the calculated cross sections of the stiffening beam, on the indicators of which the technical and economic indicators of the whole system largely depend, will increase their efficiency and increase the economy of steel.

3 Conclusions

1. A rational topology of combined steel trusses with a span of 30 m is chosen, which corresponds to the regular placement of vertical (or at an angle) compressed struts and descending stretched struts.
2. Rational topology of steel combined trusses and the angle of inclination of the truss racks is defined as 79°–82°, provides a reduction in the weight of the combined truss to 12.3%.
3. The uneven distribution of normal stresses in the stiffening beam of combined trusses is shown.
4. The potential of calculated stress regulation in the stiffening beam, characteristic for such combined trusses, is revealed.
5. Theoretical bases of increase of efficiency of the combined steel trusses are resulted.

References

1. Gasii G (2020) Testing of the combined structural elements of support of a mine opening. E3S Web Conf 168:00028. https://doi.org/10.1051/e3sconf/202016800028
2. Ruiz-Teran A, Aparicio A (2010) Developments in under-deck and combined cable-stayed bridges. In: Proceedings of ICE bridge engineering. Special issue about bridge engineering: cable-supported structures, vol 163, pp 67–78. https://doi.org/10.1680/bren.2010.163.2.67
3. Storozhenko L (2018) Investigation of the deformation state of a composite cable space frame structures with a photogrammetric method. Int J Eng Technol 7, 3, 2:442–446. https://doi.org/10.14419/ijet.v7i3.2.14568
4. Bendsoe M (2004) Topology optimization: theory, methods, and applications. Springer, Berlin, p 370
5. Bendsøe MP, Ben-Tal A, Zowe J (1994) Optimization methods for truss geometry and topology design. Struct Optimization 7:141–159. https://doi.org/10.1007/BF01742459
6. Kirsch U (1989) Optimal topologies of truss structures. Comput Methods Appl Mech Eng 72 (1):15–28. https://doi.org/10.1016/0045-7825(89)90119-9

7. Gogol M, Kropyvnytska T, Galinska T, Hajiyev M (2020) Ways to improve the combined steel structures of coatings. Lect Notes Civil Eng 73. https://doi.org/10.1007/978-3-030-42939-3_6
8. Toklu YC (2004) Nonlinear analysis of trusses through energy minimization. Comput Struct 1581–1589. https://doi.org/10.1016/j.compstruc.2004.05.008
9. Achtziger W (2007) On simultaneous optimization of truss geometry and topology. Struct Multidiscipl Optimization 33:285–304. https://doi.org/10.1007/s00158-006-0092-0
10. Kaveh A, Zolghadr A (2013) Topology optimization of trusses considering static and dynamic constraints using the CSS. Appl Soft Comput 13(5):2727–2734. https://doi.org/10.1016/j.asoc.2012.11.014
11. He L, Gilbert M (2015) Rationalization of trusses generated via layout optimization. Struct Multidiscip Optim 52(4):677–694. https://doi.org/10.1007/s00158-015-1260-x
12. Gogol M (2018) New effective combined steel structures. Int J Eng Technol 7, 3.2:343–348. https://doi.org/10.14419/ijet.v7i3.2.14432
13. Kalpakides VK, Patsouras CA, Hadjigeorgiou EP, Stavroulakis GE (2012) Optimal design of plane trusses using material forces. Compos B Eng 43(6):2669–2675. https://doi.org/10.1016/j.compositesb.2012.03.020
14. Michálek P, Bujňák J (2019) Experimental and theoretical study of composite trusses. Transport Res Procedia 40:815–822. https://doi.org/10.1016/j.trpro.2019.07.115
15. Aliawdin P, Urbańska K (2013) Limit analysis of geometrically hardening rod systems using bilevel programming. Procedia Eng 57:89–98. https://doi.org/10.1016/j.proeng.2013.04.014
16. Ruiz-Teran A, Aparicio A (2008) Structural behaviour and design criteria of underdeck cable-stayed bridges and combined cable-stayed bridges. Can J Civ Eng 35:938–950. https://doi.org/10.1139/L08-033
17. Cavallaro R, Demasi L (2016) Challenges, ideas, and innovations of joined-wing configurations: a concept from the past, an opportunity for the future. Prog Aerosp Sci 87:1–93. https://doi.org/10.1016/j.paerosci.2016.07.002

Construction of Analytical Coupling Model in Reinforced Concrete Structures in the Presence of Discrete Cracks

Ihor Yakovenko, Yevhen Dmytrenko, and Valentina Bakulina

Abstract It is presented the analytical method for determination of refined compliance (and secant rigidity C_{sm}) reinforcing communication, crossing crack λ_{sm}—significant parameters necessary for simulation of reinforced concrete constructions on the model of discrete cracks, including through the FEM. This method is based on a mathematical model, which is a nonlinear boundary value problem consisting of four equations, two of which are differential of the first order. Solution of the problem was carried out by numerical methods in mathematical calculation programs MathCAD and Wolphram Mathematica. Amongst the main assumptions of the model are bilinear and linear diagrams for description of behavior under the loading of concrete and reinforcement and the elastic–plastic law of adhesion between them applied. There were obtained the dependences of the secant stiffness Csm on the distribution of longitudinal forces $N_s(x)$ and displacements $U_s(x)$ in the reinforcement for different varied parameters, such as the areas, classes of concrete and diameters of reinforcement. The results clearly demonstrate the non-linear nature of the work of reinforcing communication that crosses a crack, which corresponds to the actual operation of the communication, but for greater reliability require experimental confirmation.

Keywords Rigidity · Coupling model · Reinforced concrete structures · Cracks

1 Introduction

When calculating statically definable and indeterminate reinforced concrete structures in terms of strength, rigidity and crack resistance, physical dependences (dependences of stresses on deformations, deformations on stresses) are of great importance. For reinforced concrete structures, they depend, in turn, on the level

I. Yakovenko (✉) · Y. Dmytrenko · V. Bakulina
National University of Life and Environmental Sciences of Ukraine,
Heroyiv Oborony St., 15, Kyiv 03041, Ukraine
e-mail: i2103@ukr.net

A. Bieliatynskyi and V. Breskich (eds.), *Safety in Aviation and Space Technologies*,
Lecture Notes in Mechanical Engineering,
https://doi.org/10.1007/978-3-030-85057-9_10

and mode of loading (hard or soft, active loading or unloading, single or multiple repetitive, etc.), as well as on the stage of operation of the structure—before or after the formation of cracks [1]:

- with the formation of a more or less clear system of parallel or orthogonal networks of cracks with an almost constant step;
- with the formation of one, two, three cracks or a system with non-parallel cracks.

2 Analysis of Recent Research

The work of structures in the first case was investigated in sufficient detail (first of all, in the works of prof. Karpenko [2], who developed a model of dispersed cracks), which cannot be said about the studies of the second case [3]. At the same time, this case is quite common (for example, near places of excitation of stress and strain fields—local loads, openings, holes, etc.). This must be taken into account when calculating reinforced concrete buildings and structures with load-bearing walls, beams and columns with holes, structures of any objects in the presence of defects and damage in them.

3 Statement of the Goal and Objectives of the Study

Models for Calculating Reinforced Concrete Structures After Cracking. When modeling cracked concrete, there are always two main models used.

1. Model of dispersed ("smeared") cracks. This model is based on the assumption that after the main tensile stress reaches the tensile strength of concrete, a series of parallel cracks (Fig. 1) will form in the concrete member in a direction perpendicular to the direction of the main tensile stress.

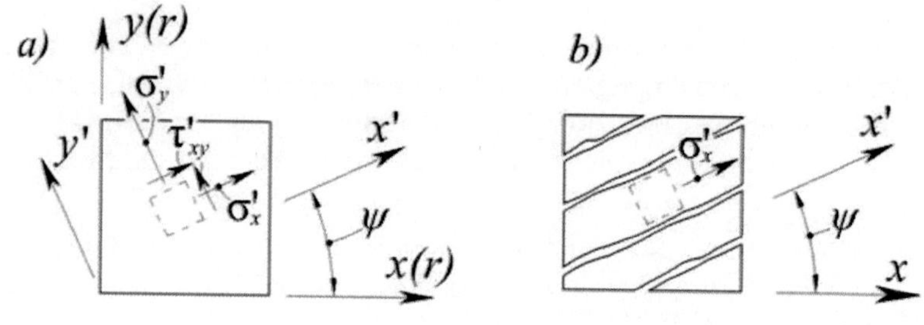

Fig. 1 Crack model to represent concrete work: immediately before the formation of cracks (**a**); immediately after cracking (**b**)

The presence of a crack [1] is modeled by a sharp decrease in the stiffness of the material in the system of two orthogonal axes of orthotropy—normal and tangent to the plane of the crack.

After the destruction of concrete from tension (crack formation) has occurred, at each subsequent step of the solution, it is checked whether the average normal deformation (across the cracks) remains tensile. If not, then it is considered that the "smeared cracks" are closed and can again transmit compressive forces [4].

2. Model of discrete fractures. An alternative to the "smeared" cracks model is the concept of discrete cracks, including in finite element mesh, which can be produced by sequential disconnection of common nodes of those elements in which stresses exceeded the tensile strength of concrete (Fig. 2).

The first published works on the application of this technique relate to the linear elastic analysis of beams with a previously known crack pattern [4]. Pairs of nodes with the same coordinates were introduced along these cracks (Fig. 2). For concrete and longitudinal reinforcement, quadrangular elements were used, including those consisting of two jointly working triangular elements with a linear deformation field, and for clamps—truss elements (Fig. 3). To simulate the phenomena of engagement of the crack faces, specially developed connecting elements were used (Fig. 4).

4 Statement of the Problem and Initial Preconditions for Mathematical Modeling of the Behavior of the Bond Crossing the Crack

For the design of structures using the discrete crack model, a necessary parameter is the secant stiffness of the reinforcement ties, the definition of which is considered in this work.

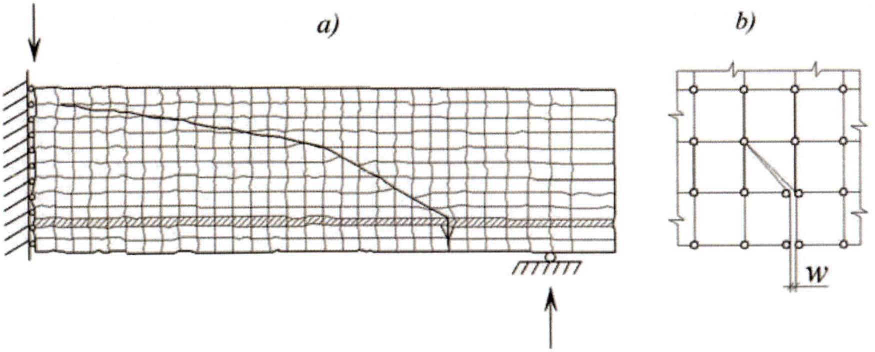

Fig. 2 Modeling the propagation of cracks using their predefined trajectories (**a**) and the introduction of pairs nodes with the same coordinates (**b**)

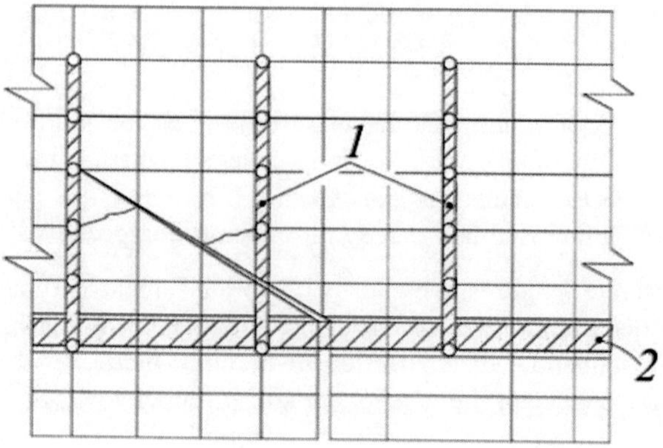

Fig. 3 General scheme of the finite element method for modeling reinforced concrete (for concrete and longitudinal reinforcement, quadrangular elements were used, including those consisting of two jointly working triangular elements with a linear deformation field, and for clamps—truss elements): *1*—cross clamps; *2*—longitudinal working reinforcement

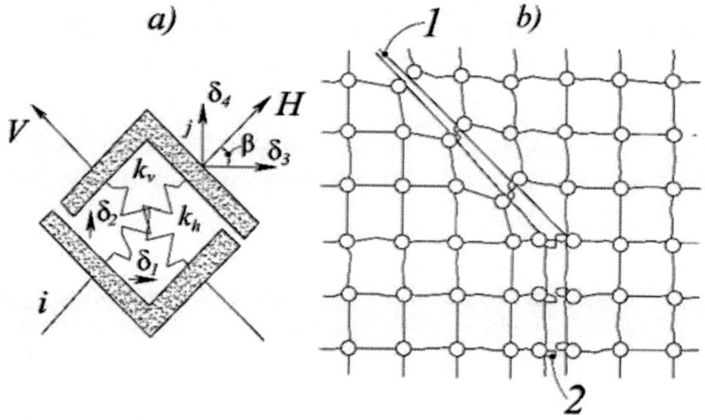

Fig. 4 Connecting element for modeling the meshing phenomenon (**a**) and modeling the meshing phenomenon using connecting elements (**b**): *1*—cracks; *2*—connections

The study of the dependence of the stiffness of the connection on the level of the stressed and deformed state was carried out by mathematical modeling of the bar [5]. Longitudinal compliance λsm (bond stiffness Csm) is determined by the ratio of the displacement Usm to the reaction of the bond Nsm in it:

$$\lambda_{sm} = \frac{1}{C_{sm}} = \frac{U_{sm}}{N_{sm}} \quad (1)$$

This reaction (it is also the force at the end of the rod) depends on the boundary conditions, which we vary during the study.

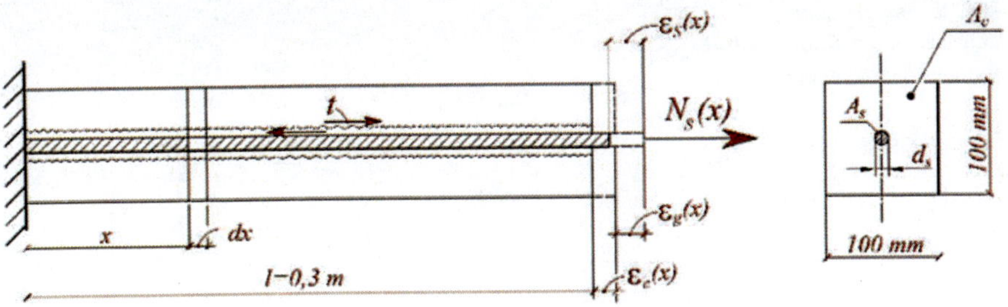

Fig. 5 Calculation scheme of a model element for modeling the behavior of links crossing the crack

In the model is considered a reinforced concrete element (representative volume) with a single central reinforcement. The left end of the element is rigidly fixed against any movements, the right end is free. A tensile force Ns is applied to the reinforcing bar, causing the bar and the element end to move Us and Ub, respectively, along the entire length of the reinforced concrete bar (Fig. 5).

The law of adhesion between concrete and reinforcement in the model under study is elastic–plastic, is described using a bilinear diagram τbond—εg(x) that takes into account the experimental data of Golyshev [6, 7], Yakovenko [8], Kolner [9], Kholmyanskiy [10], Babich [11], Kolchunov [12, 13] and others [14–16].

$$\tau_{bond} = k \cdot \varepsilon_g(x) = 0.4 \cdot E_{cm} \cdot [\varepsilon_s(x) - \varepsilon_c(x)] \tag{2}$$

when

$$\varepsilon_g(x) = [\varepsilon_s(x) - \varepsilon_c(x)] \leq \varepsilon_g^*(x) = 4.95 \cdot \frac{f_{ctm}}{E_{cm}} \tag{3}$$

$$\tau_{bond} = 0.0232 \cdot E_{cm} \cdot [\varepsilon_s(x) - \varepsilon_c(x)] + 1.866 \cdot f_{ctm} \tag{4}$$

when

$$\varepsilon_g(x) = [\varepsilon_s(x) - \varepsilon_c(x)] > \varepsilon_g^*(x) = 4.95 \cdot \frac{f_{ctm}}{E_{cm}} \tag{5}$$

where $\varepsilon_g(x)$ is the relative mutual displacement of concrete and reinforcement, in the section x (Fig. 6); the accumulation of relative mutual displacements of concrete and reinforcement in the area between the cracks and is the value of the crack opening width wk; $\varepsilon_g^*(x)$ is the boundary relative mutual displacement of concrete and reinforcement, corresponding to the end point of the first section of the bond diagram (Fig. 6).

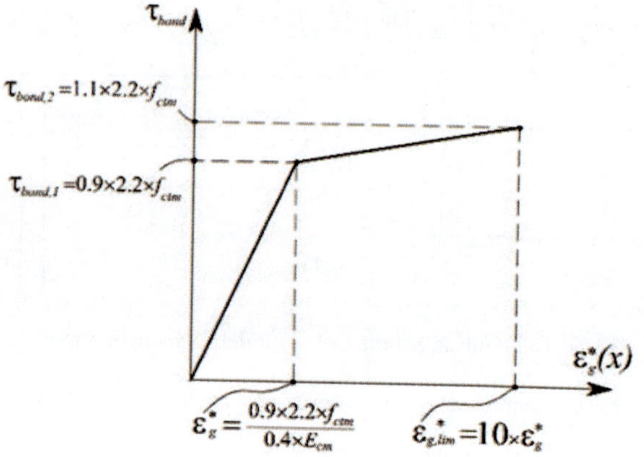

Fig. 6 Dependence of adhesion stresses on relative mutual displacements

The work of concrete in this model is also described using a bilinear diagram, shown in Fig. 7. For concrete, we use the following dependence that simulates the elastic–plastic work of the material (Fig. 7):

$$\varepsilon_c(x) = \begin{cases} \frac{N_c(x)}{E_{cm} \cdot A_c}, & if \quad \frac{N_c(x)}{A_c} \leq 0.9 \cdot f_{ctm}, \\ \frac{18 \cdot N_c(x)}{E_{cm} \cdot A_c} - 15.3 \cdot \frac{f_{ctm}}{E_{cm}}, & if \quad \frac{N_c(x)}{A_c} > f_{ctm}; \end{cases} \tag{6}$$

The reinforcement in this model is described using a line diagram.

Using the equilibrium conditions of the concrete and reinforcing bars, we obtain the following two differential equations connecting the forces in the bars and the shear stresses of adhesion (Fig. 8).

Using the equilibrium conditions of the concrete and reinforcing bars, we obtain the following two differential equations connecting the forces in the bars and the shear stresses of adhesion (Fig. 8).

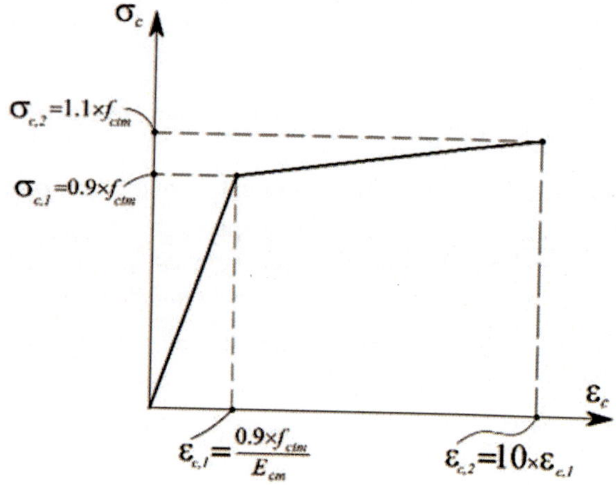

Fig. 7 Diagram of concrete deformation in the analytical model

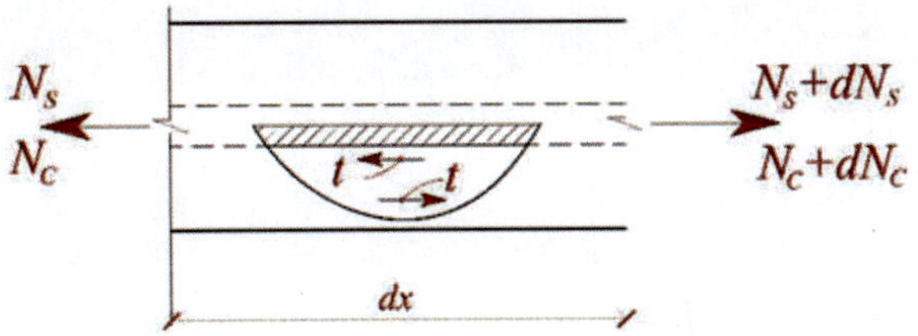

Fig. 8 Work of reinforcement in concrete

for reinforcement:

$$-N_s + N_s + dN_s - t \cdot dx = 0 \tag{7}$$

for concrete:

$$-N_c + N_c + dN_c + t \cdot dx = 0 \tag{8}$$

After the corresponding algebraic transformations, we will have:
for reinforcement:

$$\frac{dN_s}{dx} = \tau_{bond} \cdot \pi d_s \tag{9}$$

for concrete:

$$\frac{dN_c}{dx} = -\tau_{bond} \cdot \pi d_s \tag{10}$$

Hooke's law is valid for reinforcement:

$$\sigma_s = E_s \cdot \varepsilon_s; \Rightarrow \varepsilon_s = \frac{\sigma_s}{E_s} = \frac{N_s(x)}{E_s \cdot A_s} \tag{11}$$

So, we get:

$$\varepsilon_s(x) = \frac{1}{E_s A_S} \cdot N_s(x) \tag{12}$$

Thus, we have obtained a nonlinear boundary value problem consisting of four equations, two of which are differential of the first order, has the following form:

$$\begin{cases} \varepsilon_s(x) = \frac{1}{E_s \cdot A_S} \cdot N_s(x); \\ \varepsilon_c(x) = \begin{cases} \frac{N_c(x)}{E_{cm} \cdot A_c}, & if \quad \frac{N_c(x)}{A_c} \le 0.9 \cdot f_{ctm}, \\ \frac{18 \cdot N_c(x)}{E_{cm} \cdot A_c} - 15.3 \cdot \frac{f_{ctm}}{E_{cm}}, & if \quad \frac{N_c(x)}{A_c} > f_{ctm}; \end{cases} \\ \frac{dN_s(x)}{dx} = \begin{cases} \pi \cdot d_s \cdot 0.4 \cdot E_{cm} \cdot [\varepsilon_s(x) - \varepsilon_c(x)], & if \quad \varepsilon_g(x) \le \varepsilon_g^*(x) = 4.95 \cdot \frac{f_{ctm}}{E_{cm}}, \\ \pi \cdot d_s \cdot \{0.0232 \cdot E_{cm} \cdot [\varepsilon_s(x) - \varepsilon_c(x)] + 1.866 \cdot f_{ctm}\}, & if \quad \varepsilon_g(x) > \varepsilon_g^*(x) = 4.95 \cdot \frac{f_{ctm}}{E_{cm}}; \end{cases} \\ \frac{dN_c(x)}{dx} = \begin{cases} -\pi \cdot d_s \cdot 0.4 \cdot E_{cm} \cdot [\varepsilon_s(x) - \varepsilon_c(x)], & if \quad \varepsilon_g(x) \le \varepsilon_g^*(x) = 4.95 \cdot \frac{f_{ctm}}{E_{cm}}, \\ -\{0.0232 \cdot E_{cm} \cdot [\varepsilon_s(x) - \varepsilon_c(x)] + 1.866 \cdot f_{ctm}\}, & if \quad \varepsilon_g(x) > \varepsilon_g^*(x) = 4.95 \cdot \frac{f_{ctm}}{E_{cm}}. \end{cases} \end{cases} \tag{13}$$

The boundary conditions of the problem are written in the following form:

$$N_c(x = 1) = 0, \ldots N_S(x = 1) = C \tag{14}$$

5 Research Results

The solution of the problem was carried out by numerical methods in mathematical calculation programs MathCAD and Wolphram Mathematica [4].

The resulting graphs of changes in the required parameters U_s, U_c, N_s, N_c along the length of the bar for different diameters of the reinforcement are shown in Figs. 9 and 10.

By varying the values of the force at the end of the reinforcing bar, graphs of the distribution of longitudinal forces in reinforcement $N_s(x)$ and concrete $N_c(x)$ along the length of the bar were obtained under the action of the force at the end

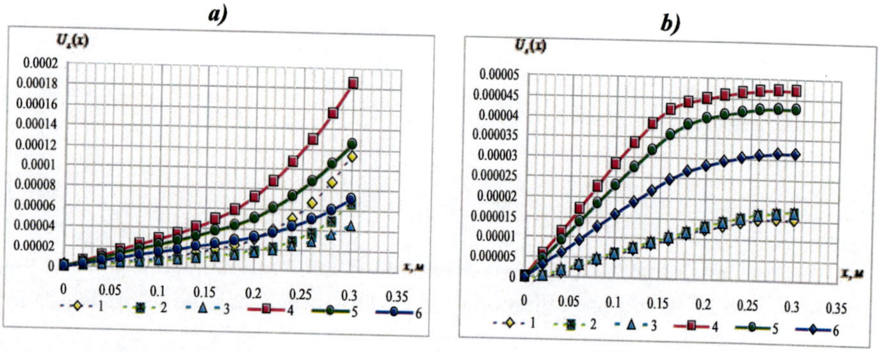

Fig. 9 Graphs of the distribution of relative longitudinal deformations in reinforcement U_s (**a**) and concrete U_c (**b**) along the length of the bar x under the action of a force at the end $N_s(x) = 0.025MN$: *1*—analytical model using the dependence of M. Kholmyanskiy, with a reinforcing bar Ø10A400C; *2*—the same as *1*, but with Ø12A400C; *3*—the same as *1*, but with Ø16A400C; *4*—analytical model with elastic–plastic law, Ø10A400C; *5*—the same as *3*, but with Ø12A400C; *6*—the same as *3*, but with Ø16A400C

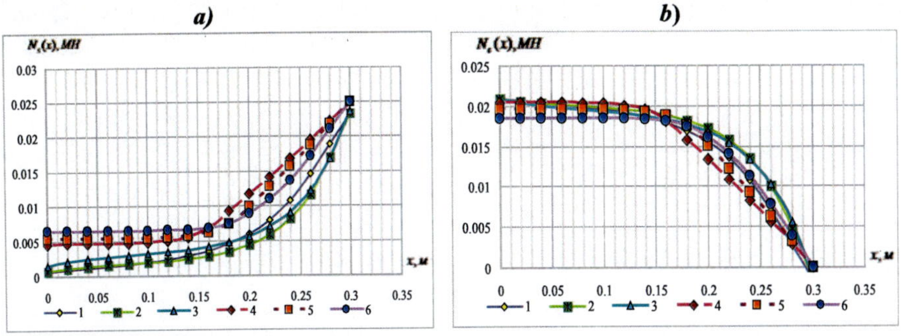

Fig. 10 Graphs of disribution of longitudinal forces in reinforcement $N_s(x)$ (**a**) and concrete $N_c(x)$ (**b**) along the length of the bar under the action of the force at the end $N_s(x) = 0.25MN$: *1*—analytical model using the dependence of M. Kholmyanskiy, with a reinforcing bar Ø10A400C; *2*—the same as *1*, but with Ø12A400C; *3*—the same as *1*, but with Ø16A400C; *4*—analytical model with elastic–plastic law, with reinforcing bar Ø10A400C; *5*—the same as 4, but with Ø12 A400C; *6*—the same as *4*, but with Ø16A400C

$N_s(x) = 0.25MN$ (Fig. 10) for various reinforcement diameters, concrete classes and cross-sectional areas.

The limiting factors were:

- tensile strength of reinforcemet f_t;
- the average tensile strength of concrete f_{ctm}, which was used to determine the stage of operation of the reinforced concrete part of the element section;
- the limiting relative displacements of the reinforcement relative to the concrete at the end at which the bond between the reinforcement and the concrete is broken, according to the accepted dependence of the work of the reinforcement-concrete contact (Fig. 6);
- ultimate relative deformations of concrete in tension according to the accepted dependence of concrete deformation (Fig. 7).

After the results obtained, the dependences of the secant stiffness on the distribution of longitudinal forces and displacements in the reinforcement were built (Fig. 11) for different areas of concrete, for different classes of concrete (Fig. 12), as well as for different diameters of reinforcement (Fig. 13).

6 Conclusions

1. With a change in the diameter of the reinforcement, the displacement of the end of the pulled out reinforcing bar increases to 61% with a decrease in the diameter of the reinforcement (by 60% over its cross-sectional area); at the same time, taking into account the elastic–plastic work of concrete and the bond of cohesion also noticeably increases these displacements up to 39% in comparison

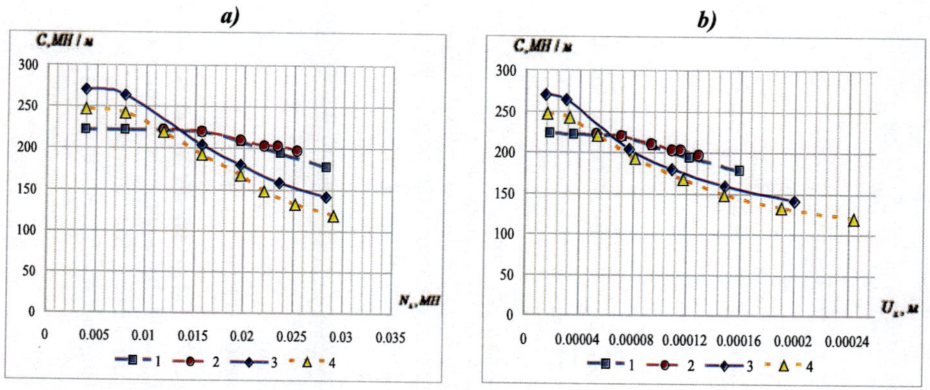

Fig. 11 Graphs of the dependence of the secant stiffness C_{sm} on the forces N_s (**a**) and displacements U_s (**b**) in the reinforcement for various areas of concrete: *1*—analytical model using M. Kholmyanskiy's dependence, at $A_c = 0.02\,m^2$; *2*—the same as *1*, but with $A_c = 0.01\,m^2$; *3*—an analytical model with elastic–plastic law $A_c = 0.02\,m^2$; *4*—the same as *3*, but with $A_c = 0.01\,m^2$

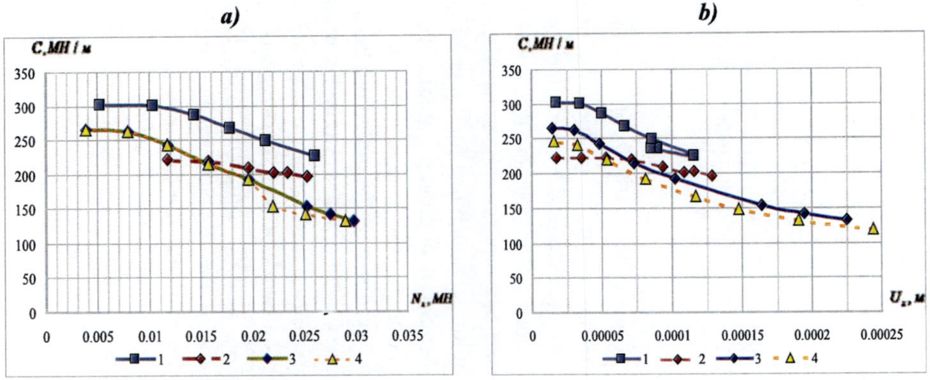

Fig. 12 Graphs of the dependence of the secant rigidity C_{sm} on the forces N_s (**a**) and displacements U_s (**b**) in the reinforcement for different classes of concrete: *1*—analytical model using the dependence of M. Kholmyanskiy, with the class of heavy concrete C20/25; *2*—the same as *1*, but with concrete class C16/20; *3*—analytical model with elastic–plastic law, with concrete class C20/25; *4*—the same as *3*, but with concrete class C16/20

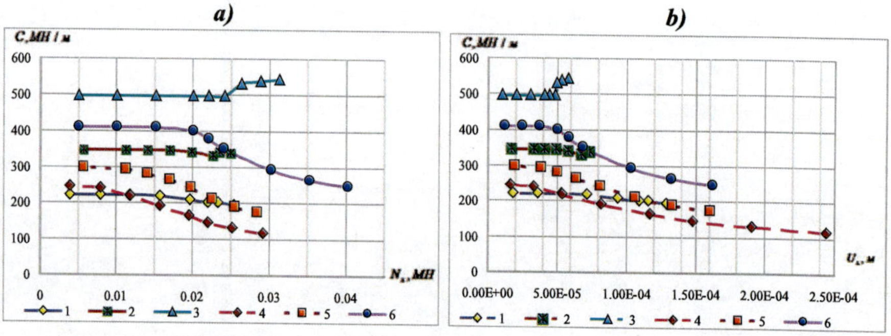

Fig. 13 Graphs of the dependence of the secant stiffness on the forces (**a**) and displacements (**b**) in the reinforcement for different diameters of the reinforcement: *1*—analytical model using the dependence of M. Kholmyanskiy, with a reinforcing bar Ø10A400C; *2*—the same as *1*, but with Ø12A400C; *3*—the same as *1*, but with Ø16A400C; *4*—analytical model with elastic–plastic law, Ø10A400C; *5*—the same as *3*, but with Ø12A400C; *6*—the same as *3*, but with Ø16A400C

with the model of M. Kholmyanskiy [10]. It should be noted that the displacements of the end of the concrete prism (at the point of intersection with the axis of the reinforcement) are characterized by the same tendency (the difference between the elastic–plastic work of concrete and the model of M. Kholmyanskiy is 68%) as for the reinforcing bar, however, the concavity, the characteristic for the graphs of reinforcement turns into a bulge for the graphs of concrete, i.e., the nature of the increase in displacement changes qualitatively.

2. With an increase in the distance from the restrained end of the reinforced concrete prism, the longitudinal force in the pulled out reinforcing bar increases to 80%, starting from half the length of the reinforced concrete prism, remaining constant in the first half adjacent to the restrained end; at the same time, taking into account the elastic–plastic work of concrete does not change the observed tendency (the deviation is up to 35% compared to the Kholmyanskiy model), only increasing the gradient of the increase in the longitudinal force in the second half of the prism; for concrete, the tendency to change the longitudinal force along the length of the reinforced concrete prism is the opposite in comparison with the revealed tendency for the longitudinal force of reinforcement (the maximum deviation is up to 29% compared to the Kholmyanskiy model).
3. With a decrease in the cross-sectional area of the prism by 50%, the following tendency is observed: with an increase in the longitudinal force in the reinforcing bar by 725%, the secant stiffness of the bond C_{sm} decreases to 96%; at the same time, taking into account the elastic–plastic work of concrete changes the observed tendency—instead of convex graphs, they are concave and the gradient of decreasing secant stiffness increases; It should be noted that for the relationship secant bond stiffness C_{sm}—displacement, the observed tendency is repeated (with an increase in displacements by a factor of 10, the secant bond stiffness C_{sm} decreases to 120%).

4. With a decrease in the class of concrete of the prism by 20% and with an increase in the longitudinal force in the reinforcing bar (up to 600%), the secant stiffness of the connection C_{sm} decreases (up to 34%); at the same time, taking into account the elastic–plastic work of concrete changes the observed trend, significantly reducing it to 40% (i.e., the effect of the concrete class is significantly reduced); for the dependence secant stiffness of the bond C_{sm}—displacement, the observed tendency is repeated (with an increase of 5.4 times of displacement U_s the secant stiffness of the bond C_{sm} decreases by 54%, taking into account the elastic–plastic work of concrete changes the observed tendency, decreasing it by up to 30%).
5. With a decrease in the diameter of the reinforcement (by 60% in its cross-sectional area) and with an increase in the longitudinal force in the reinforcing bar (up to 400%), the secant stiffness of the bond decreases (up to 75%); at the same time, not taking into account the elastic–plastic work of concrete according to the dependencies of M. Kholmyanskiy changes the observed tendency to the opposite (the deviation is up to 12% in the direction of increase). Note that an increase (by 60% in its cross-sectional area) in the reinforcement diameter changes the current stiffness by 150% in the direction of increase. At the same time, taking into account the elastic–plastic work of concrete does not change the observed tendency (the increase in stiffness is up to 220%). It should also be noted that for the dependence secant stiffness of the bond—displacement, the observed tendencies are qualitatively preserved, and quantitatively, the increase in stiffness is up to 250% with an increase of 60% in the cross-sectional area of the reinforcement.
6. With an increase in the longitudinal forces in the reinforcement, the displacement at the end of the reinforcing bar increases; at the same time, taking into account the elastic–plastic work of materials (nonlinear work of concrete, nonlinear adhesion resistance) affects the displacements (the maximum deviation is up to 140% relative to linear displacements) and the graph is convex; at the same time, for the curve constructed with linear work of concrete and nonlinear adhesion, the convex character remains, however, taking into account the elastic work of concrete with nonlinear work of adhesion insignificantly affects the presented dependence (the maximum deviation is up to 18% relative to linear displacements); a graph that takes into account the linear work of adhesion and nonlinear work of concrete takes on a constant value of the longitudinal forces in the reinforcement in the first quarter of the section, however, with a further increase in displacements, the curve increases and has a concave character (the maximum divergence from the curve constructed with nonlinear adhesion and nonlinear work of concrete is 45%); a graph that takes into account the elastic work of concrete and the elastic work of the adhesion elements is linear (the maximum divergence from the curve plotted with nonlinear adhesion resistance and nonlinear concrete work is up to 43% in longitudinal forces and up to 95% in displacements).

References

1. Iakovenko I, Kolchunov V, Lymar I (2017) MATEC web of conferences. In: 6th international scientific conference «reliability and durability of railway transport engineering structures and buildings», p 0216 12. https://doi.org/10.1051/matecconf/201711602016
2. Karpenko N (1996) General models of reinforced concrete mechanics, p 416
3. Levin V, Rogozhin N (2015) Mathematical modelling of joint deformation of reinforcing bar and concrete in the vicinity of an isolated crack (in relation to the model of discrete cracks). Bull Donbass National Acad Civil Eng Archit 3(113):5–7. http://nbuv.gov.ua/UJRN/vdnaba_2015_3_3
4. Emelyanov S, Nemchinov Y, Kolchunov V, Yakovenko I (2016) Details of large-panel buildings seismic analysis Enfoque. UTE 7(2):120–134. https://doi.org/10.29019/enfoqueute.v7n2.100
5. Kolchunov V, Yakovenko I, Dmitrenko E (2016) The analytical core model formation of the nonlinear problem bond armature with concrete. Acad J Ser Ind Mach Build Civil Eng 2,47:125–132. http://journals.nupp.edu.ua/znp/article/view/76/60
6. Golyshev A, Kolchunov Vl (2009) Resistance of reinforced concrete, p 432
7. Golyshev A, Kolchunov Vl, Yakovenko I (2015) Resistance of reinforced concrete structures, buildings and structures erected in complex engineering-geological conditions, p 371
8. Iakovenko I, Kolchunov Vl (2017) The development of fracture mechanics hypotheses applicable to the calculation of reinforced concrete structures for the second group of limit states. J Appl Eng Sci 15, 3, 455:366–375. https://doi.org/10.5937/jaes15-14662
9. Kolner V, Aliev Sh, Goldfine B (1965) Coupling with concrete and the strength of the sealing of bar reinforcement of periodic profile. Concrete Reinf Concr 11:15–19
10. Kholmyanskiy M (1981) Contact of reinforcement with concrete, p 184
11. Babych Y, Kochkarev D, Filipchuk S (2018) Mathematical model of armature adhesion with high-strength concrete. Build Constr Theory Pract 2, 21:154–161. http://nbuv.gov.ua/UJRN/budktp_2018_2_21
12. Kolchunov Vl, Yakovenko I (2016) About the violation solid effect of reinforced concrete in reconstruction design of textile industry enterprises. Proc Higher Educ Inst Textile Technol Ser 3, 363:258–263. https://www.researchgate.net/publication/316687268
13. Dem'yanov A, Kolchunov Vl, Iakovenko I, Kozarez A (2019) Load bearing capacity calculation of the system "reinforced concrete beam—deformable base" under torsion with bending. E3S Web Conf 97. https://doi.org/10.1051/e3sconf/20199704059
14. Jendele L, Cervenka J (2006) Finite element modelling of reinforcment with bond. Comput Struct 84(28):1780–1791
15. Pokorný P, Kouřil M, Stoutil J, Bouška P, Simon P, Juránek P (2015) Problems and normative evaluation of bond-strength tests for coated reinforcment and concrete. Mater Technol 49(6):847–856
16. Dai J, Ueda T, Sato Y (2005) Development of the nonlinear bond stress-slip model of fiber reinforced plastics sheet-concrete interfaces with a simple method. J Compos Constr 9(1):52–62

Ergodesign Aspects of the Development of Unmanned Aerial Vehicle System

Andrii Bieliatynskyi, Oleh Kardash, Anatolii Rubtsov, and Svirko Volodymyr

Abstract Based on the analysis of the current state and directions of UAVS development, the authors identify and substantiate the main types of ergodesign support for the development of unmanned aerial vehicle systems (UAVS), and consider ergodesign requirements for the main UAVS components. A detailed nomenclature of ergodesign indicators for UAVS assessment is presented; the determination of the optimal compliance level with such indicators underlies and forms the core of UAVS pre-design analysis and ergodesign development.

Keywords Ergodesign · Unmanned aerial vehicle systems · System analysis · Automated workplaces · Ergodesign criteria · Requirements · UAVS quality indicators

1 Analysis of Literature Sources

One of the main current trends in aviation development is the active formation of unmanned aerial vehicle systems (UAVS). Thus, at the beginning of this century, about 150 UAVS types were developed and produced in the world, currently there are more than 1200 of such systems [1]. This growth and diversity is attributed to the wide range of UAVS applications. For example, for the Ukrainian start-up AVISION Robotics created in 2014, there are no restrictions on the number of controlled UAVs and their routes. A large number of drones can operate both in the city and in the field: all routes can be calculated and adjusted automatically [2].

A. Bieliatynskyi (✉)
North Minzu University, Wenchang North St. 204, Yinchuan 750001,
People's Republic of China
e-mail: beljatynskij@ukr.net

O. Kardash
Kyiv Borys Grinchenko University, Boulevard I. Shamo, 18/2, Kyiv 02154, Ukraine

A. Rubtsov · S. Volodymyr
National Aviation University, Liubomyra Huzara Ave., 1 Building 9, Kyiv 903058, Ukraine

A. Bieliatynskyi and V. Breskich (eds.), *Safety in Aviation and Space Technologies*,
Lecture Notes in Mechanical Engineering,
https://doi.org/10.1007/978-3-030-85057-9_11

Currently this Ukrainian team together with Amazon, Verizon and other companies is working on the NASA team to develop common standards, as well as protocols for UAV control and monitoring. These trends were confirmed by the 6th International Aerospace Industry Exhibition “Singapore Airshow” which took place in 2018 in Singapore [3].

The International Civil Aviation Organization (ICAO) considering the UAVS issue highlights the key points characterizing these systems. In particular, in order to distinguish the job title of unmanned aerial vehicle (UAV) pilots who perform their duties related to UAV piloting outside the aircraft, it is proposed to use the term “external pilot” (instead of the term “system operator pilot” proposed in ICAO Circular 328-AN/190 “Unmanned Aircraft Systems” [4]).

In our opinion, a more precise definition is given in the Air Code of Ukraine [5]. It is a “remote pilot”. We will use it further implying that other UAVS operation functions during flight are performed by “operators”.

In science, unmanned aerial vehicle systems are used to obtain a variety of new knowledge. The UAVS application scope has two main areas—military and civil. Military UAVSs in accordance with the functional purpose can be used as surveillance (in particular, to regulate fire on the battlefield), reconnaissance, attack ones [4], etc.

The UAV civil application is much broader and extremely diversified: from agriculture and construction to oil and gas and security sectors. From the ergodesign point of view, it is important that the vast majority of crashes (or “flight accidents” according to the terminology used in civil aviation) at such sites occur due to insufficient adaptation of UAVS elements to operating personnel’s capabilities and needs. Therefore, taking into account the human factor is crucial in creating such systems, and the role of ergodesign as a scientific and practical discipline evaluating and studying the impact of the human factor is dominant in modern UAVS development.

The consideration of UAVSs—complex multilevel systems—as an object of ergodesign determines the use of the system approach as the most effective tool of their formation. Therefore, the ergodesign studies of UAVSs and their individual subsystems (unmanned aerial vehicles, ground control stations, launching and landing aids, etc.) should be based on the anthropometric, functional, psychophysiological, macro- and microanalysis of real UAVSs.

Since in this case the system is the highest form of organization, in general, a UAVS should be considered as a complex of necessary and sufficient elements that are in a stable correlation and interaction within relatively defined limits and form a single whole.

The main ergodesign objective in this regard is to ensure the efficiency, safety and comfort of UAVS management, repair and maintenance. Therefore, the relevance of such studies is out of question and is growing every year, because their subject is the development and justification of ergodesign principles, the requirements and methods of creating and operating effective UAVSs.

The principles of the system approach are quite thoroughly addressed in literature sources [6, 7]. Most authors agree that a characteristic feature of the system design process is a sequence of approximations to the optimal solution. Obviously, the reduction in the number of sequential actions also characterizes the optimization degree of the UAVS ergodesign development process in the context of their creation.

We focus only on the main aspects of ergodesign development of such objects, namely, on operators and UAVS remote pilots' activities. Numerous operational stressors (unsatisfactory ergonomics of pilot stations, unacceptable temperature conditions, low level of information flow organization, etc.) have been established during the study of their working environment [8, 9].

In [10], it has been proved that the main factors of UAVS operators' professional stress are also the lack of specialists, variable work organization, too long work shifts, difficulties in perceiving the interface of used computers, etc. All this indicates the urgent need to determine the ergodesign quality indicators of UAVS components and the formation of ergodesign requirements for UAVS development and operation.

2 Problem Statement

In modern socio-economic conditions, design approaches to UAVS creation are constantly evolving, are being formed, and require focused ergodesign research centered around further development of the relevant regulatory framework based on their results. It should be borne in mind that new UAVS models should be created on the functional principle of designing processes, not products.

After all, the subjective criterion for the UAVS quality from the standpoint of ergodesign is a sense of functional comfort, when their components are perceived as a system of functional and objective-spatial means that create comfortable and safe conditions for operators, and UAVS has sufficient technical means to carry out this activity.

The article proposes the approach to the UAVS creation according to which ergodesign is considered not only as a progressive technology for the creation and production of high quality science-intensive products but also as a research and development activity different from traditional product design. In a broad sense, the purpose of this approach is to ensure UAVS effectiveness in all application areas. It is achieved through synergy at all the stages of UAVS life cycle and, above all, of such aspects of the system as efficiency, safety, and comfort.

The first step in this direction—the identification of ergodesign quality indicators of the main UAVS components, primarily, of operators' automated workstations—has already been taken.

3 Formulation of Research Aims

The aim of the paper is to study the role of the human factor in the UAVS creation and operation as well as to develop its ergodesign evaluation indicators.

4 Research Methods

The applied research methods include systematic ergodesign analysis of UAVS structural and functional components as well as the computer simulation of their operators' activity.

5 Statement of Basic Materials

From the standpoint of the ergodesign support of UAVS design and operation, in their composition typically composed of at least five components—unmanned aerial vehicle (UAV), ground control station (GCS), launching device, landing aids, antenna-rotary device—it is expedient to single out the following main components:

- UAV: a glider, an airborne control system, power plants and other life-critical systems;
- GCS: remote pilots (RP) and payload operators' workstations; control, telemetry, and radio communication systems, etc.

The ergodesign analysis of these components conducted by the authors revealed the following typical weak points inherent in the vast majority of UAVSs.

1. Extremely insufficient UAVS equipment with display devices adequate for operators' the activities.
2. Complicating the operating personnel's work through the use in UAVS design of control effectors operating based on different principles.
3. Insufficiency (in some cases the absence) of the adjustable parameters of UAVS operators' main workstations.
4. Insufficiency and inconsistency (according to modality, the nature of information provision) of the status indication of life-critical UAVS systems, etc.
   Even this far from being complete list of shortcomings indicates the need for the in-depth ergodesign research of each of UAVS components and further ergodesign support of their development and operation.

It should be borne in mind that the main feature of systematic ergodesign development is not only a comprehensive consideration of a wide range of specific problems but also the anthropocentric orientation of design actions. In particular, it is ensuring the ability to meet the requirements for the functional comfort of all participants in UAVS control and service. And the activity approach is a necessary feature of ergodesign development. In this way, such a postulate as "taking into account the human factor" can be most fully realized, in particular, in "remote pilot—UAV" subsystems.

As mentioned before, an integral measure of UAVS ergodesign perfection is the achievement of functional comfort as a criterion for operators' optimal intellectual and psychophysiological state in the process of their activities and as a criterion for the adequacy of UAVS components and elements for individual operators' capabilities. It is well-known that the general criteria and indicators of the quality of socio-technical systems are accuracy, reliability, productivity and operators' fatigue in a system.

Regarding UAVS ergodesign criteria, it is worth noting the following. In the ergodesign development, depending on priority given to the problems of utility, beauty or convenience, it is respectively determined their predominantly utilitarian, aesthetic or ergonomic aspect. The determining factor here is the level of activity, which significantly objectifies the specific UAVS design process. Their subsystems are considered as a means of ensuring human activity and are designed taking into account the human factor. In this case, the focus should be on the structure of operators' activities with all UAVS elements, and actions themselves are multi-component acts determining the operation of a system as a whole.

It should be borne in mind that ergodesign development is the initial stage of the UAVS life cycle, the results of which are crucial for the further functioning of a system and the implementation of project ergodesign activities results in a project of a technical and procedural complex containing psycho-physiological connections, UAVS personnel's anthropometric features, etc. Their structure contains the actual project of a system and a description of its control procedures.

Taken together, they make up the UAVS ergodesign model. The functional content of such a model is a hierarchy of connections "functional area—an operator's activity role (work algorithms)—workstations-equipment systems—equipment units". From an anthropocentric point of view, the ergodesign model determines, in particular, the activity structure and organization (its rationale, algorithms, etc.).

In general, the ergodesign of the UAVS technical-procedural system determines activities carried out in an ergodesign model. Such an approach requires a hierarchical construction of ergodesign support for UAVS development, when each of its "lower" levels is included as a subsystem in a "higher" level: an individual element is included in a complex, the complex becomes a part of a block, the block is integrated into a UAVS subsystem, etc.

In the UAVS ergodesign development process, one of the important principles is compliance with the complexity of individual indicators. For example, the quality of display devices is determined by a set of ergodesign indicators such as accuracy and the operator's visual fatigue when analyzing readings. The set of individual ergodesign indicators determines the complex ergodesign indicator, which can almost completely show the ergodesign level of UAVS components.

The nature of UAVS operators' interaction necessitates the formation of the special indicators and criteria for the quality or efficiency of a system as a whole and its individual components. These criteria should evaluate both each element of a system and the system as a whole. It is also considered proven [6, 7] that the correctness of the general system efficiency determination depends on a number of individual (or internal) indicators characterizing the functioning of system elements.

But in case of UAVSs, the main attention should be focused not on individual indicators, but on their structure, interconnections and finding common formal properties—isomorphism reflecting the content of the interconnections of indicators. After all, ergodesign development should be carried out according to a single efficiency criterion to which other indicators are related and by which they are all directly or indirectly governed. Thus, all indicators should be grouped into a single system based on the efficiency criterion, and their role should be considered and evaluated according to this criterion.

In practice, based on specific tasks, the dominant ergodesign criterion is usually determined, such as functionality, reliability, speed, fatigue, etc. An important requirement for such a criterion is for it to have specific ergodesign content. The following should be taken into account. The nature of the criterion that affects the UAVS ergodesign level is not as important as its significance and weight in a number of other criteria and indicators. This weight should relate to the type of functional dependence of the UAVS ergodesign level on changes in this factor, although the dependence itself is determined by the nature of the current factor.

It should also be borne in mind that the typical ergodesign assumption about the approximation of indicators to the reference (ideal) ergodesign characteristics of the product is valid only to some extent, which is further defined as the level of ergodesign indicators and is detected during UAVS ergodesign development and analysis.

The basis for an objective assessment of a system and the determination of the the ergodesign level of UAVS quality is formed by the results of its functional analysis. At this stage, it is possible to make a comprehensive definition of quality but by one selected criterion only. Thus, one criterion can be used to assess, for example, an operator's seat, or an automated workstation as a whole based on the rate of human fatigue as the main ergonomics characteristic of an operator's seat.

It should be taken into account that among the ergodesign characteristics of objects of such complexity as an UAVS, the most important ones in terms of the "human factor" are its ergonomic characteristics. They underlie other UAVS ergodesign quality indicators determining the main technical parameters of this system. It should be clearly understood that indicators derived from the requirements of the "human factor" have to be "embedded" in the design of the product. Therefore, it is advisable to constantly correlate them with the reference or ideal ergodesign characteristics of the product. It makes it possible to determine the ergonomic level of quality manifested in the UAVS ergodesign development.

The creation of new unmanned aerial vehicle systems should also be based, first of all, on the results of the functional-algorithmic analysis of remote pilots' activities. The authors' experience shows that such an analysis is an effective way of identifying weak points in a particular operating procedure, for example, in the optimization of information flows in an UAVS, etc.

In general, the procedure for determining the required ergodesign level of UAVS quality as the initial stage of its design can be presented in the following sequence:

- based on specific tasks, it is chosen (usually by the expert method) the dominant ergodesign indicators of a developed UAVS and its components;
- following the analysis of reference samples, it is determined the ergodesign parameters of UAVS components (in particular, the parameters and location of workstations, controls, display devices, etc.);
- using instrumental methods and taking into account ergonomic recommendations and expert assessments, it is determined functional relationships between the ergodesign criteria and ergodesign indicators of system components, and it is evaluated the ergonomic level of UAVS components (based on individual parameters);
- establishment (usually according to the analytical method) of the interconnections of ergodesign indicators with each other and their relationships as a whole, primarily with ergonomic criteria;
- on the basis of the received data, it is carried out the ergodesign estimation of the developed UAVS according to each chosen criterion (for example, the rationality criterion—the arrangement of components or the remote pilot's fatigue criterion).

We emphasize once again that the UAVS ergodesign level is determined by several criteria. Their combination is determined by a complex ergodesign criterion which corresponds to the UAVS ergodesign level. To establish it, the values of individual criteria are determined according to a number of parameters and their interdependence is found as well as the equivalent values of individual criteria within the framework of the complex criterion. After all, the complex criterion must combine the values of individual ergodesign criteria taking into account the importance of each of them. This is done according to specially developed ergodesign procedures. The complex ergodesign criterion obtained in this way will express the maximum achievable ergodesign level of quality for the given UAVS type (or its individual component), and its ergodesign indicators will be close to optimal.

We emphasize that this is only the initial, basic stage of UAVS ergodesign development.

- Further ergodesign support of UAVS development according to its stages should be based on the purpose and sequence of its main stages and be centered around the relevant ergodesign requirements and indicators.
- Given the limited volume of the article, we present in tabular form a system of the ergodesign quality indicators of UAVS operator's automated workstations (AW) (Table 1) and a system of the unified ergodesign indicators of UAVS ground control stations (GCS) (Table 2), as the main components of unmanned aerial vehicle systems with regard to the human factor.

Table 1 Ergodesign quality indicators of UAVS operator's automated workstations (AW)

Complex indicator of the 1st level ease of AW use	
Complex indicator of the 2nd level	Individual indicator
Rationality of the function allocation between the operator and technical equipment	The level of the optimal operator's workload
	Rationality of the operator's activity organization and algorithm
	The validity level of functions (actions) performed exclusively with operator's approval
	The human resource reserve level (operator's functions in case of a system failure)
Design operational comfort	The general adequacy of the AW sizes for the performed operating task: seated activity
	Standing activity
	Seated and standing activity
	The level of the provision of adjusting worksurfaces in the AW design
	The level of provision in the AW design of the opportunity to control viewing distances
	Provision of the opportunity in the AW design to change look-out angles
	Provision of the visual markings of specific device groups (color, layout, etc.) in the AW design
	Availability in the AW design of an alarm of the control effector (CE) failure
	Availability in the AW design of an alarm of the visual information means (VIM) failure
	Compliance of work surfaces with ergonomic requirements for firmness, color, and stiffness
	Availability in front of a display of a work surface for keeping records, keyboard and communication equipment placement, etc.
	Compliance with ergonomic requirements for the lighting levels (natural, general, and spotlight) of the AW and its elements
Rationality of layout	Adequacy of the AW layout for the performed task
	Appropriateness of the AW operator's main working posture for the activity structure
	Compliance of mutual arrangement of AW components with the activity requirements
	Correspondence between the CE and VIM mutual arrangement and the structure of the executed tasks

(continued)

Table 1 (continued)

	Compliance with the ergonomic requirements for legroom organization when performing work in a sitting position
	Compliance of the visibility of AW components with ergonomic requirements
	Compliance of the CE and VIM mutual arrangement with ergonomic requirements
	Provision of distances recommended for simultaneous interaction with CEs in the layout of AW components
	Ensuring the functionality of CE and VIM clustering
	Ensuring the sufficient operator's freedom of movement within an AW while performing work in a standing position
Correspondence between design and the operator's anthropometric characteristics	Taking into account in the AW design of the size of the human body and its parts
	Taking into account in the AW design of the shape of the human body and its parts
	Correspondence between distances to control effectors and the operator's anthropometric characteristics
	Correspondence between the form of EC gripping parts and the anthropometric features of the operator's hands (legs) and character of gripping
The operator's psycho-physiological load	Compliance of the principles behind the operator's interaction with CEs and VIM applied in an AW with ergonomic requirements
	Correspondence between the level of accuracy of devices and the required accuracy of indicators
	The monotony level of the operator's activities
	The level of the operator's information load
	The level of the operator's neuropsychological and emotional tension
	The level of ensuring the ability to perform operations with a coordinated system of movements
	The deviation level of the AW operator's optimal movements from physiologically rational characteristics

(continued)

Table 1 (continued)

Compliance with ergonomic requirements for the operator's fatigue level	The level of the AW operator's static stress
	The level of a decrease in the operator's emotional state while working at an AW
	The level of the deterioration of the operator's functional state while working at an AW
	Provision of support for the operator's elbows, arms, forearms, and back
Complex indicator of the 1st level *Ease of AW management and control (controllability)*	
Compliance with ergonomic requirements for control effectors (CEs)	Compliance of the principles behind the operator's interaction with CEs with ergonomic requirements
	Compliance with ergonomic requirements for CEs used in an AW
	The unification level of the organization types and methods of CEs used in an AW
	Compliance with ergonomic requirements for CE coding means used in an AW
	Compliance with the ergonomic requirements for the effort required to activate CEs
	Adequacy of the CE form, sizes, and material for the actions they perform
	The level of ability to determine the state ofthe controlled object (process) according to the CE position (on–off)
	Compliance with ergonomic requirements for contrasts between CEs and AW surfaces
Compliance with ergonomic requirements for the CE placement	Adequacy of the CE placement for performed tasks
	Correspondence between the CE placement and the limits of reach of the operator's motor field in a certain working posture
	Compliance with ergonomic requirements for ways to cluster several CEs into groups (panels)
	Correspondence between the mutual arrangement of panels and control effectors and the operator's activity algorithm
Compliance with ergonomic requirements for visual information means (VIM)	Compliance with ergonomic requirements for visual information means used in an AW
	Adequacy of the VIM form and sizes for the actions they perform
	Compliance of AW information elements with ergonomic requirements for contrast, color, and surface texture

(continued)

Table 1 (continued)

	Compliance with ergonomic requirements for VIM visibility angles
	Compliance with ergonomic requirements for symbolic and graphic information used in an AW (texts, numbers, diagrams, tables, markings, signs, etc.)
	Compliance with ergonomic requirements for working with visual display units
	Adequacy of displays used in an AW for performed tasks
Compliance with ergonomic requirements for the VIM placement	Correspondence between the VIM placement and recommended visual observation zones
	Compliance with ergonomic requirements for the visibility of panels and an individual VIM
	Compliance with ergonomic requirements for VIM placement and their functional interaction with an operator
	Compliance of the brightness levels of VIM information items with ergonomic requirement
Compliance with ergonomic requirements for acoustic and tactile information	Correspondence between the nature of acoustic and tactile messages (simple, complex, periodic, continuous, etc.) and the algorithm of the operator's activity
	Correspondence between acoustic and tactile information media (tone, vibration, configuration, temperature, etc.) and the algorithm of the operator's activity
	Compliance with ergonomic requirements for coding of acoustic and tactile signals
	Compliance with ergonomic requirements for voice message indicator units
AW assimilation	Adequacy, stereotypy, and structural ordering of the information model
	Adequacy of information on AWs and controlled processes
	Redundancy of information on AWs and controlled processes
Compliance with ergonomic requirements for operational documentation	Compliance with ergonomic requirements for operational documentation
	Clarity of information presented in operational documentation
	Quality (clarity, scale, layout logic) of illustrations, schemes, and graphic elements in operational documentation

(continued)

Table 1 (continued)

	Comprehensibility of operation manual for AW components
	Structuring and availability of additional information on AW operation
Ease of AW management and control (controllability)	
Compliance with ergonomic requirements for equipment and tools required for AW operation	The complexity of the algorithm for the maintenance and repair of AW components
	Ease of access to adjustable and replaceable elements
	Availability of technical failure diagnostics means
	Ease of use of instruments and controls and diagnostic equipment
	Provision of comfortable conditions for the repair and maintenance of AW components
	Accessibility of AW components during maintenance and cleaning
Complex indicator of the 1st level *Hygiene of the AW location environment*	
Compliance with ergonomic requirements for physical factors of the AW environment	Compliance with AW noise level standards
	Compliance with the standards of natural and artificial lighting of the GCS room, AW, CE and VIM work surfaces
	Compliance with the standards of AW mechanical vibrations
	Compliance with the standards of electromagnetic fields and static electricity in the AW environment
	Compliance with the standards of the AW thermal environment
	The level of AW protection against draught and dust
AW sefety	
Compliance with ergonomic requirements for chemical and biological AW environment factors	Compliance with ergonomic requirements for the concentration levels of hazardous substances in the air of the GCS room
	Compliance with ergonomic requirements for the presence of hazardous components in AW materials and coatings Compliance with ergonomic requirements for the concentration of microorganism (including fungi) in the air or on the surface of AW elements
	Adequacy of protective means against hazardous events in an AW

(continued)

Table 1 (continued)

	The safety level of the factors of a mechanical, chemical, and electrical origin
	The level of protection against harmful radiation
	The level of protection against extreme temperatures
	Compliance with ergonomic requirements for audio alarm signals
	Compliance with the requirements for visual alarm signals
	The level of compliance with the conditions for the prevention of hazardous events when working with VIM
	The availability of tools and materials needed in hazardous situations at an AW
	The adequacy level of CE protection means
AW artistic expression	
Decorative expression of the AW form	Correspondence between the AW artistic image and its purpose Correspondence between the AW artistic image and modern ideas about the products of a certain type
AW form originality	Originality of the used AW formation principles: plastic, compositional, layout Peculiarity of applied decorative and color-graphic elements Adequacy of originality methods for the expediency requirements
AW form fashionableness	Correspondence of the color-graphic solution, AW finishing to "fashionable" decorating methods Correspondence of AW compositional and plastic characteristics (or its packaging)to "fashionable" methods of form making
Decorative expression of the AW form	Decorative expression of the used materials and coverings Correspondence between the AW decorative expression methods and the expediency requirements
Stylistic unity of the AW form	Compatibility of AW design features with each other within the limits of a chosen style (level of eclecticism) Compatibility of AW design features with other components of a system within the limits of a chosen style

(continued)

Table 1 (continued)

Rationality of the AW form	
Functional and constructive conditionality of the AW form	Adequacy of the AW form for its the purpose and operating conditions (for example, manual and portable AW) Adequacy of the AW form for its composition and layout scheme Suitability of the use of constructive methods of organizing the AW form elements
Technological conditionality of the AW form	Compliance of the AW form with the requirements of its manufacturing technology
Integrity of the awcompositional-plastic form solution	
Harmony of the AW three-dimensional structure	Interdependence of primary and secondary elements of the AW form in size, proportions and scale The degree of AW scale and its elements (visual matching to the size of the human body)
AW architectonic form	Manifestation in the form of the nature of AW structural loads Visual balance of the AW three-dimensional, compositional and plastic structure
Integrity of the awcompositional-plastic form solution	
Harmony of the AW three-dimensional structure	Interdependence of primary and secondary elements of the AW form in size, proportions and scale The degree of AW scale and its elements (visual matching to the size of the human body)
AW architectonic form	Manifestation in the form of the nature of AW structural loads Visual balance of the AW three-dimensional, compositional and plastic structure
Plasticity of the AW form	Integrity of three-dimensional and plastic solution of the AW form Correspondence between the volumetric and plastic solution and applied materials, and manufacturing technology
Artistic and graphic expression	Compositional validity of the arrangement of graphic elements on the AW parts The degree of conformity of the nature of the fonts to the semantic value of the inscriptions Expression of AW functional graphics
Color-graphic compatibility of elements	Interdependence between color and graphic elements Subjection of color and graphic elements to the general AW compositional and color-graphic solution

(continued)

Table 1 (continued)

Color and texture compatibility of elements	Compatibility of different types of materials, composition, textures, coatings used in the AW with each other Consistency of different types of materials, composition, textures, coatings with the AW shape, purpose, and operating conditions
Perfection of production and the preservation of the AW marketable condition	
Fineness of contours	Fineness of contours, fillets, and joints of the elements of the AW form
Quality of the AW surface treatment	Careful treatment of AW surfaces Careful application of decorative and protective coatings
Clarity of signs and accompanying documentation	Quality of AW graphic elements and promotional materials

Table 2 A system of the unified ergodesign indicators of UAVS ground control stations (GCS)

Group of indicators	Complex indicator of the 1st level	
	Complex indicator of the 2nd level	Individual indicator
Ergonomic indicators	Ease of GCS use for its intended purpose	
	Correspondence of GCS design and its Elements to human anthropometric characteristics	Taking into account in the GCS design the size of the human body and its parts Taking into account in the GCS design the form of the human body and its parts
	The operator's physical load (severity of work performed	Dynamic physical activity: the amount of work performed during the transportation of GCS, preparation for use, configuration, adjustment, assembly (disassembly); the mass of the GCS when it is moved Static physical activity (holding effort) Deviation of working posture and movements from physiologically rational characteristics
	The operator's psycho-physiological load (work intensity)	The monotony level of the operator's activity The operator's information load Intellectual intensity of the operator's activity Nervous and mental and emotional tension of the operator's activities

(continued)

Table 2 (continued)

Group of indicators	Complex indicator of the 1st level	
	Complex indicator of the 2nd level	Individual indicator
	Development of fatigue and a reduction in the operator's functional state for a given time	The operator's energyconsumption level The level of changes in the operator's functional state The level of a decrease in the operator's emotional state The level of reduction in motivation for work
	Ergonomics of the form, sizes and the arrangement of GCS control panels	Correspondence between the form of control panels and the GCS service algorithm Correspondence between the sizes of control panels and the GCS service algorithm Correspondence between the mutual arrangement of control panels and the GCS service algorithm Correspondence between the look-out angles of GCS control panels and the operator's anthropometric and psycho-physiological characteristics (taking into account the degree of importance and their application frequency)
Ergonomic indicators	Ease of perception of the displayed information	The levels of direct and inverse contrasts The coefficient of uneven brightness of information units The unevenness of the screen field brightness Linear values of image distortion in the screen area
	Ergonomics of visual information display devices	The levels of direct and inverse contrasts The coefficient of uneven brightness of information units The unevenness of the screen field brightness Linear values of image distortion in the screen area
	Ergonomics of acoustic information devices	Correspondence of message types to the GCS operation algorithm (a bell, buzzer, siren, musical tone or voice) Correspondence of the nature of messages to the GCS operation algorithm (simple, complex, periodic, and continuous with disconnection at response time)

(continued)

Table 2 (continued)

Group of indicators	Complex indicator of the 1st level	
	Complex indicator of the 2nd level	Individual indicator
	Ergonomics of tactile information devices	Conformity of the information provision means to the GCS operation algorithm (vibration, configuration, temperature, and amperage) Compliance of levels of electrical, chemical, and thermal signal levels with the ergonomic requirements
	Convenience of control effectors design	Conformity of the form and the constructive execution of control effectors to ergonomic requirements Conformity of the sizes of control effectors to the ergonomic requirements Correspondence between the effort required to bring the controls in action and ergonomic requirements
	Ergonomic placement of control effectors	Correspondence of the nature of the operator's control movements to the functional state of the controlled system Compliance of the combination methods of several control effectors with the ergonomic requirements Correspondence between distance to controls (taking into account the degree of importance and frequency of their application)with the operator's anthropometric characteristics Availability and adequacy of the protection means for control effectors
	Rationality of GSC layout	Compliance of GCS sizes with ergonomic requirements Optimal placement of display devices and control effectors
	GCS assimilation	
	Information Model Quality	Adequacy of the information model Stereotypy of the information model Adequacy of information on the product and process Redundancy of information on the product and process Structural ordering of the information model
	Completeness and convenience of GCS operation manual	The level of operation manual completeness Clarity of the instructions Quality of material design

(continued)

Table 2 (continued)

Group of indicators	Complex indicator of the 1st level	
	Complex indicator of the 2nd level	Individual indicator
Ergonomic indicators	GCS maintenance	
	–	Comfort and the rate of maintenance, repair, preparation for operation The complexity of the maintenance and repair algorithm Ease of access to adjustable and replaceable elements Availability of technical failure diagnostics means
	Ergonomics of operational documentation	Completeness of operational documentation Convenience of the material presentation structure, the levels of information decoding and re-coding Quality of illustrations, schemes, graphic elements, and documentation format Documentation storage capability
	GCS HYGIENE	
	Physical factors	Indicators of the illumination level of work surfaces and controls
	Chemical factors	Presence of hazardous components in materials and coatings
	GCS safety	
		Safety level of the factors of a mechanical origin Safety level of the factors of a chemical origin Safety level of the influence of an electric current Safety level due to the completeness of taking into account the consumer's psycho-physiological characteristics Safety level conditioned by the algorithm of the GCS operation
Aesthetic indicators	GCS artistic expression	
	Decorative expression of the GCS form	Correspondence between the GCS artistic image and its purpose Correspondence between the GCS artistic image and modern ideas about the products of a certain type

(continued)

Table 2 (continued)

Group of indicators	Complex indicator of the 1st level	
	Complex indicator of the 2nd level	Individual indicator
	GCS form originality	Originality of the used GCS formation principles: plastic, compositional, layout Peculiarity of applied decorative and color-graphic elements Adequacy of GCS originality methods for the expediency requirements
	GCS form fashionableness	Correspondence of the color-graphic solution, GCS finishing to "fashionable" decorating methods Correspondence of GCS compositional and plastic characteristics (or its packaging)to "fashionable" methods of form making
	Decorative expression of the GCS form	Decorative expression of the used materials and coverings Correspondence between the GCS decorative expression methods and the requirements of expediency
	Stylistic unity of the GCS form	Compatibility of GCS design features with each other within the limits of a chosen style (level of eclecticism) Compatibility of GCS design features with other components of a system within the limits of a chosen style
	Rationality of the GCS form	
	Functional and constructive conditionality of the GCS form	Adequacy of the GCS form for its the purpose and operating conditions (for example, manual and portable GCS) Adequacy of the GCS form for its composition and layout scheme Suitability of the use of constructive methods of organizing GCS form elements
	Technological conditionality of the GCS form	Compliance of the GCS form with the requirements of its manufacturing technology

(continued)

Table 2 (continued)

Group of indicators	Complex indicator of the 1st level	
	Complex indicator of the 2nd level	Individual indicator
Integrity of the GCS compositional-plastic form solution		
Functional indicators	Harmony of the GCS three-dimensional structure	Interdependence of primary and secondary elements of the GCS form in size, proportions and scale The degree of GCS scale and its elements (visual matching to the size of the human body)
	GCS architectonic form	Manifestation in the form of the nature of GCS structural loads Visual balance of the GCS three-dimensional, compositional and plastic structure
	Plasticity of the GCS form	Integrity of three-dimensional and plastic solution of the GCS form Correspondence between the volumetric and plastic solution and applied materials, and manufacturing technology
	Artistic and graphic expression	Compositional validity of the arrangement of graphic elements on the GCS parts The degree of conformity of the nature of the fonts to the semantic value of the inscriptions Expression of GCS functional graphics
	Color-graphic compatibility of elements	Interdependence between colour and graphic elements Subjection of colour and graphic elements to the general GC compositional and colour-graphic solution
	Color and texture compatibility of elements	Compatibility of different types of materials, composition, textures, coatings used in the GCS with each other Consistency of different types of materials, composition, textures, coatings with the GCS shape, purpose, and operating conditions
	Perfection of production and the preservation of the GCS marketable condition	
	Fineness of contours	Fineness of contours, fillets, and joints of the elements of the GCS form
	Quality of the GCS surface treatment	Careful treatment of GCS surfaces Careful application of decorative and protective coatings

(continued)

Table 2 (continued)

Group of indicators	Complex indicator of the 1st level / Complex indicator of the 2nd level	Individual indicator
	Clarity of signs and accompanying documentation	Quality of GCS graphic elements, PDT and promotional materials
	Resistance to damage	Protection of the GCS form elements and surfaces against damage, attrition, and decorative covering quality changes
	Perfection of the main GCS function performance	
	Efficiency of GCS use	The degree of satisfaction with the control function in the UAV flight
	Versatility of GCS use	
	The range of GCS use for its intended purpose	The range of conditions and capabilities for the given GCS use for various UAVs
	Perfection of auxiliary operations	
	Perfection of preparatory operations	Suitability of the GCS to perform auxiliary transportation operations and preparation for operation
Functional indicators	Perfection of finishing	Suitability of the GCS to perform auxiliary operations of disassembly, cleaning, packaging and transportation
	Perfection of GCS auxiliary operations	Perfection of operations on watching videos (for example, search of the necessary record)
	Ease of GCS operation	
Operational indicators	–	Perfection of the GCS use during service operations accompanying implementation of the main and auxiliary functions
	Ease of GCS maintenance	
	–	Perfection of preparatory and final operations, and also GCS regulation in the course of operation GCS suitability to perform auxiliary operations of maintenance, storage, and disposal
	GCS Reliability	
	GCS failure-free operation	Preservation of the basic parameters of GCS operation in time and within the limits corresponding to the set operating conditions
	GCS durability	Preservation of the basic parameters of GCS operation before the limit state is achieved at which their

(continued)

Table 2 (continued)

Group of indicators	Complex indicator of the 1st level	
	Complex indicator of the 2nd level	Individual indicator
		fulfillment becomes impossible. When calculating durability, it is determined the GCS service life or resource in conditions as close as possible to its specific operational process
	GCS maintainability	Possibility of GCS urgent repair in field conditions The average duration and complexity of the current GCS repair in stationary conditions
Socio-cultural indicators	GCS Social address and consumer class	
	–	Correspondence of the GCS to the structure of needs of a certain target audience
	Compliance with the optimal GCS nomenclature	
	–	Efficiency of GCS use in the operational or projected GCS system of a certain type
	GCS moral aging	
	–	The GCS service life is limited by the introduction of th new GCS types of higher quality
Design and marketing indicators	The degree of GCS compliance with the world level	
	–	The level of GCS design and ergonomic characteristics as compared to the products of the leading manufacturers of similar products
	Compliance with the requirements of the potential target market	
	–	The degree of market demand for a particular GCS
Design and environmental indicators	The nature and extent of the GCS impact on the environment	
	–	The impact of GCS on the environment during its life cycle
	Utilization degree of GCS materials	
	–	The output of recycled materials

6 Conclusions

The expansion of the UAV application causes a sharp complication of unmanned aerial vehicle systems and the need to adapt their design and functionality to operators' psycho-physiological, intellectual, anthropometric, and other features necessitating the development of the new approaches and methods of UAVS design.

Nowadays, the need for in-depth pre-design UAVS ergodesign research based on experimentally and analytically developed requirements for UAVS components and their ergodesign quality indicators formulated on this basis is becoming more and more urgent.

Currently, based on the results of UAVS ergodesign studies, we can consider the need for the in-depth study of the peculiarities of operators' interaction with the UAVS functional elements in different operational situations. In order take into account the human factor in all UAVS functional structures when creating new UAVS types to the maximum extent and ensuring the greatest cost-effectiveness, it is extremely important to involve specialists in ergodesign at least at the stage of the formation of a technical task for the product.

References

1. Osipov YuV (2011) Requirements for tools and programs for UAV operators' training in unmanned aircraft systems. Armament Polit Convers 3:99
2. Svirko VO, Rubtsov AL, Chemakina OV, Bieliatynskyi AO, Kuzmin OV, Kuzmin AO (2018) ErgoDesign of visual information systems. Int Sci J "Internauka" 22. https://doi.org/10.25313/2520-2057-2018-22-4532
3. Ugnenko EB, Uzhvieva OM, Bieliatynskyi AO (2011) Designing detours of localities based on traffic flow modes. Bull Nat Aviat Univ 4:106–110
4. Circular 328. AN/190 (2011) Unmanned aircraft, p 12
5. Air Code of Ukraine (2011) Vidomosti Verkhovnoyi Rady Ukrayiny, pp 48–49
6. Ergonomics: principles and guidelines. Instructional guidelines (1983)
7. Balashov EP (1985) Evolutionary synthesis of systems
8. Ouma JA, Chappelle WL, Salinas A (2011) Facets of occupational burnout among U.S. Air Force active duty and national guard/reserve MQ-1 predator and MQ-9 reaper operators, OH, pp 45433–7913, 6–7
9. Thompson WT, Lopez N, Hickey P, DaLuz C, Caldwell JL, Tvaryanas AP (2006) Effects of shift work and sustained operations: operator performance in remotely piloted aircraft (OPREPAIR). Technical Report HSW-PE-BR-TR-2006-0001, p 311
10. Chappelle WL, Salinas A, McDonald K (2011) Psychological health screening of usaf remotely piloted aircraft (RPA) operators and supporting units. In: Paper presented at the symposium on mental health and well-being across the military spectrum, Norway
11. Nagy J, Muse K, Eaton G (2006) U.S. Air Force unmanned aircraft systems performance analyses: predator sensor operator front end analysis (FEA Report), p 6
12. URL: http://www.dtic.mil/cgi-bin/GetTRDoc?Location=U2&doc=GetTRDoc.pdf&AD=ADA582856

Structural-Kinematic Analysis and Synthesis of Measuring Systems Using a Scanning Probe Microscope

Andrii Bieliatynskyi, Mariia Kataieva, Volodymyr Kvasnikov, and Dmytro Ornatskyi

Abstract The modern stage in the development of nanotechnology is characterized by a fundamentally new approach to the design of high-precision information-measuring systems (MIS) of mechanical quantities. In this regard, the system approach to solving the problems of structural-kinematic analysis and the synthesis of measuring systems based on scanning probe microscopes (SPM) for various purposes and developing the classification of the basic mechanical part of SPM is topical. Based on the analysis of functional problems of scanning probe microscope, an object-oriented structure of the basic measuring system for working in the nanometer range. An algorithm is developed for combining the control contours of a mathematical model, the measured nano-object, and the parametrization of the trajectory of the measuring sensor along with the object. In the article, the problems of the choice of the velocity of the measuring tip moving along a given trajectory with allowance for a limited number of degrees of freedom are solved, as well as the stabilization of the mechanical part of the system with small errors in the positioning of the measuring tip.

Keywords High-precision information-measuring systems · Measurement in the nanometer range · Structural-kinematic analysis · Synthesis of a scanning probe microscopes

A. Bieliatynskyi (✉)
North Minzu University, Wenchang North St. 204, Yinchuan 750001, People's Republic of China
e-mail: beljatynskij@ukr.net

M. Kataieva · V. Kvasnikov · D. Ornatskyi
National Aviation University, Liubomyra Huzara Ave., 1, Building 9, Kyiv 903058, Ukraine

A. Bieliatynskyi and V. Breskich (eds.), *Safety in Aviation and Space Technologies*, Lecture Notes in Mechanical Engineering,
https://doi.org/10.1007/978-3-030-85057-9_12

1 Introduction

One of the urgent tasks of modern measuring technology in the field of the nano industry, instrumentation, and tool production is to improve the accuracy of measuring in micro- and nanometer range. To solve this problem, optical devices are widely used, which differ in a wide range of measured dimensions, the ability to conduct non-contact remote measurements of dimensions and angles, shape deviations, location, and surface roughness.

Measuring microscopes constitute a special class of optical measuring instruments (SI). They are widely used for high-precision measurements of parts of complex shapes and structures, elements of technological and control devices, cutting tools.

However, along with increasing the precision characteristics of such measuring instruments especially in the micro- and nanometer range, it is often required to solve another urgent problem—increasing their versatility and functionality. This is since in the course of performing complex measuring work, as a rule, when measuring parts of complex shapes and precise, small-sized parts, it becomes necessary to use several different sensors per measuring cycle. Thus, it is required to create a new type of measuring machines that combine the functionality of both universal measuring microscopes.

On all modern microscopes, projectors, and other instruments for coordinate measurements, mainly optoelectronic encoders are used. The rapid development of computer technology allowed SPM to provide developed and sophisticated software and fully automate the process of measuring, calculating geometric parameters, shape, and position, enabling the statistical processing of the results.

This harshly increased the productivity and convenience of measurements on the SPM and expanded the range of tasks to be solved. Improved measurement accuracy and automation capabilities of the SPM contributed to the invention by Renishaw (UK) of a touch sensor on three prismatic supports, which is used on most modern SPMs. The use of the aerostatic guides, which eliminated friction, and linear adjustable electric drives promoted the development of the SPM capabilities and increased their accuracy.

2 Most of the Research

To control the distance between the probe and the sample in SPM methods, a system of their mutual movement with a constant assessment of the force of the resulting interaction is organized. The general principle of organizing negative feedback of a scanning probe microscope is schematically shown in Fig. 1.

If there is an unambiguous dependence of the interaction force on the controlled distance, then the relative magnitude of this force can be used to maintain a constant small distance between the probe and the sample in the information-measuring

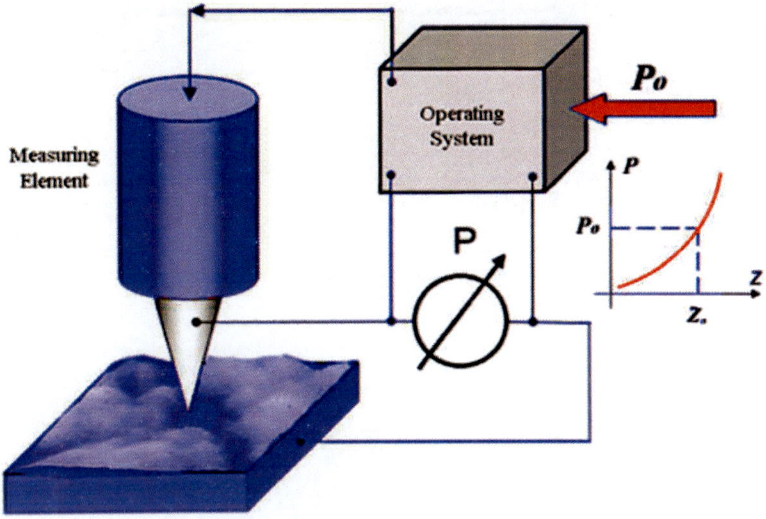

Fig. 1 Organization diagram of the SPM feedback system

system. As a result of changing the distance between the surface and the probe, a difference signal is formed in the measuring system, which is proportional to the magnitude of the change in the force acting between them.

The actuator processes this difference signal by bringing the probe closer to the surface or moving it away until the difference signal becomes equal to zero. Thus, it is possible to maintain a constant probe-sample distance with high accuracy. The set of data on the difference signal obtained at different parts of the sample surface during its line-by-line scanning by the probe is a numerical matrix of the height values equivalent to the surface relief, which can be used to construct a three-dimensional image of this relief.

Modern SPM of different firms is largely unified, they use the same components —touch sensors, aerostatic guides, optoelectronic linear incremental converters, adjustable linear electric drives, and advanced software. And although the accuracy of the modern SPM is very high, a further increase is necessary, which requires fundamentally new approaches and solutions.

The measurement objects installed on the SPM are not subjected to a thorough mechanical reconciliation. The values of the measured coordinates of the nano-object points in the coordinate system of the microscope with the help of transformation programs are given to the coordinate system of the nano-object associated with the nano-object bases (Fig. 2).

For the coordinate measurements, the number of measurement points for an individual component of the nano-object is determined to depend on the type of element, the deviation of its shape, the measurement task (geometric parameters to be determined), the permissible measurement error.

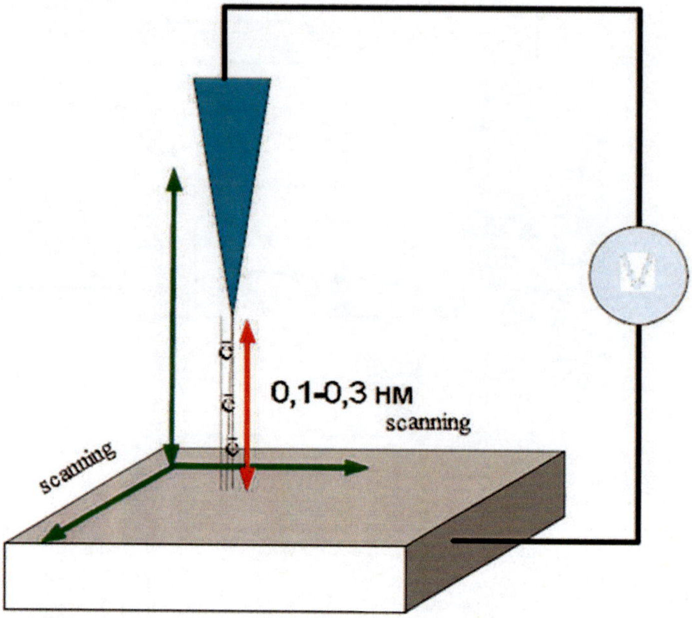

Fig. 2 The relative position of the probe and substrate in SPM

When the surface of a measured nano-object is mechanically felt by a spherical measuring tip with a radius RTK, the coordinates of the centers of the sensing sphere are initially obtained. If we assume that the measurement errors caused by the deflection of the measuring tip are excluded by introducing an appropriate correction for the calibrator, then the measured values of the coordinates of the centers belong to an element equidistant to the measured and spaced from it at a distance RTK.

Many papers have been devoted to the research of the basic kinematic structure of the coordinate SPM [1–10]. However, they do not address issues related to structural-kinematic analysis and synthesis of the SPM. Modern information objects are multidimensional systems, the state of which is characterized by a large number of different parameters.

For the management of such objects, monitoring, and testing, information on their condition is required. For the multidimensional systems, a probabilistic relationship between the parameters is typical, as well as between the values of each parameter at different instants of time.

The accuracy of measuring the geometric dimensions of the nanoobjects with the help of an SPM is limited by constantly acting random noise, the nature of which essentially depends on the operating conditions of the measuring system and the location of its placement [11]. For IMS built into TIMS, random vibrations are primarily caused by vibration and noise.

External random perturbations, causing fluctuations in the sensing element of the measuring system, form random errors, the value of which can be determined by a statistical description of the deviation from the geometric dimensions of the object as a dynamic system. In the statistical description of an SPM with a dynamic type sensitive element, it is assumed that external random perturbations can be represented as—correlated Gaussian noise, whose effect on the dynamical system is adequately described by the Fokker–Planck equation. Real spectra of random perturbations for measuring systems in both shop and laboratory execution often differ from the white noise spectrum, while remaining a Gaussian random process.

Therefore, for an adequate statistical description of IMS for nano measurements based on SPM, it is necessary to develop a procedure that would allow its random errors to be determined with allowance for the real spectrum of external disturbances. Also, this statistical description should allow us to find analytical expressions for the variance of random oscillations of the sensing element of the system, taking into account the actual form of the spectrum of external disturbances, which in turn ensures the possibility of optimizing the parameters of the primary converter to reduce the level of its random errors.

At the design stage, it is necessary to take into account the influence of all possible types of error on the accuracy of measurements and speed. This makes it possible to compile a vector balance of the accuracy of the TIMS, on the whole, and on its basis to formulate the requirements for the accuracy of all links of the TIMS for nano measurements involved in the process of shaping.

To take into account the deformations of the elements of the mechanical part, it is necessary to introduce the effect of factors—the weight of the elements of the system—into the model of accuracy and to take into account other destabilizing factors, use the appropriate dependencies (for example, thermal deformations, oscillations, etc.).

Since the IMS for nano measurements, as a rule, is designed to measure several complex surfaces, calculations are performed for each surface, after which from the whole set of the obtained variants of the structure of the mechanical part, those that are characterized by the minimum ductility of the links are selected. A characteristic feature of the proposed method of designing the TIMS based on SPM is that the underlying mathematical model of accuracy makes it possible to consider such methods of ensuring accuracy and speed as the regulation of force factors, deformation of the geometric parameters of the mechanical part, including adaptive control.

The modern stage in the development of instrument making is characterized by a fundamentally new approach to the design of high-precision information-measuring systems of mechanical quantities. The concept of the new approach is that the three-coordinate IMS based on SPM for nano measurements should be designed as a single whole.

The need for such an approach is dictated by two main reasons: the first—the possibility of realizing at the present stage of instrumentation development the idea of deep integration, primarily due to the advent of computers of a new generation with high productivity, and the second—the fight against destabilizing factors of various nature.

Widely used in instrument engineering methods of structural, algorithmic, or functional design allow obtaining models that meet all the characteristics of well-organized systems. However, design methods are based on algorithmic decomposition do not have mechanisms that reflect the generality of the behavior and structures of abstractions that are allocated. When designing such complex systems as three-coordinate IMS for nano measurements of mechanical quantities, this limitation becomes a significant drawback of this method.

In the design of the modern automated measuring instruments, a systematic analysis of the problems solved by IMS based on SPM is performed and the decomposition of these tasks is carried out following certain characteristics, followed by a generalization of the components of the functional components into various multilevel associations of tasks based on systematization, ordering, and unification, organization of rational links, hierarchical methodology.

In the system analysis of problems, first of all, we single out the tasks of special-purpose designation, which are the main content in the mechanical measurements of nano-objects with complex spatial surfaces.

Based on the analysis of the functional problems of IMS based on SPM and the use of the object-oriented approach 12, 13, an object-oriented structure of the basic three-coordinate IMS developed in Fig. 3.

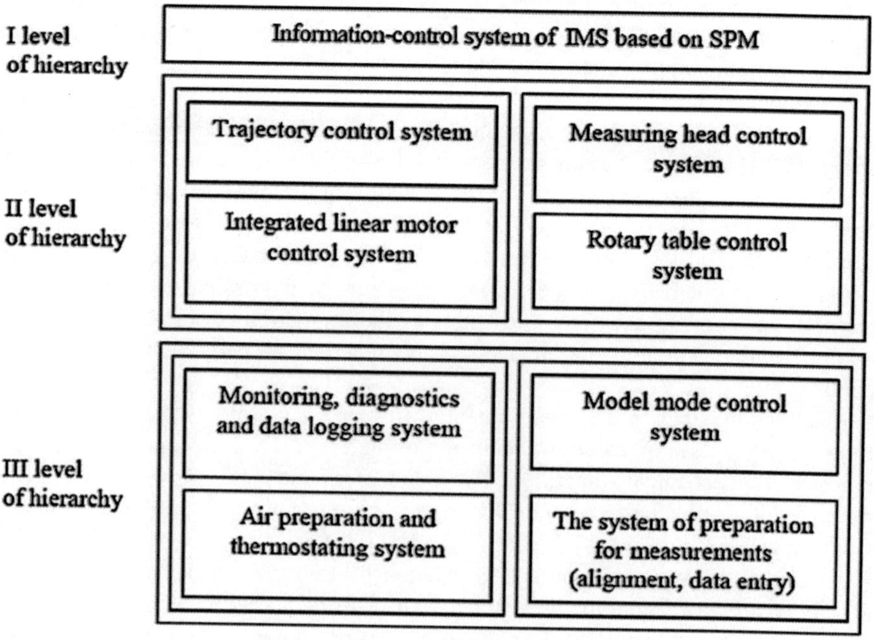

Fig. 3 The object-oriented hierarchical structure of the TIMS of mechanical quantities

The first level of the hierarchy of the object-oriented hierarchical structure of the TIIS of mechanical quantities is a complex object of the IMS as a whole.

The second level of the hierarchy combines the following objects: the information-control system of the TIMS; the system of trajectory control; a measuring head control system; a system for monitoring, diagnosing, and recording data.

The third level of the hierarchy includes a model mode management system, a system for preparing for measurements, including adjustment and input means.

The information and control system includes an information and control field that includes a reference model system, a control system for operating modes, and an integrated expert system for intellectual support of the measurement process.

The initial data for the design of the IMS are the specified nomenclature of the measured objects and the requirements for the accuracy of their measurement. After synthesizing several variants of the TIMS structure for measuring a multitude of surfaces corresponding to a given set of details, one should select from them those which provide the necessary accuracy and speed with the best economic indicators.

The problem of the structural synthesis of the mechanisms of the al system at the stage of conceptual design now has a special significance. The specified structure of the mechanical part does not always ensure the exact reproduction of the required trajectory and therefore the problems approximating to it are solved by various methods.

The variety of measured surfaces and trajectories of working bodies depends on the form-building capabilities of the executive part of the IMS. Analyzing measured SPEs, we can draw the following conclusion about the further design and development of a multivariant solution of the inverse problem, namely: for a given surface or trajectory, to determine the structure of the mechanical part of the IMS.

In most small and medium-sized SPMs, basic details (bases, portals, vertical pinoles) are made of fine granite. This material is suitable for obtaining and maintaining a stable geometric accuracy of the guide surfaces on which the aerostatic supports of the mobile units are mounted. As reference systems, photoelectric pulse converters with glass or metallic raster scales are used. Laser interferometers are used for periodic SPM attestation, revealing systematic errors and inputting corrections into the memory of computers that are part of UVK SPM. SPMs are equipped with measuring zero heads of the firm “Renishow”, three-coordinate measuring deviation heads, which greatly expand the functionality of SPM.

When synthesizing the schemes of the basic part of the IMS based on SPM, the output parameters are the shape of the measuring zone, the coordinate system in which the transporting movements are described, and the number of degrees of freedom of the measuring tip.

The internal parameters of synthesis are the number of links, the sizes of links, the maximum displacement in kinematic pairs. At the same time, it is necessary to satisfy the quality requirements: the maximum volume of the measuring zone, the minimum of the motor redundancy of the displacements by the actuators. The shape of the measuring space of the IMS based on SPM depends on the actuating movements of the measuring tip, which is possible in one of the four coordinate systems.

Three-coordinate IMS based on SPM has six movements of the executive bodies, which are controlled in general by nonlinear laws by modern drives, which provide the required relative movements of the measuring tip and the turntable. These six movements will be conditionally divided into two categories: rotations B and rectilinear translational displacements P. In total, more than 1000 variants of the arrangements of the basic part of SPM [2, 12].

Given the influence of these options on the design and ease of maintenance of the CIIS, as well as on the ranges and laws of movement of the executive bodies, they were identified by their primary classification characteristics.

We introduce the notation: T—is the Z-axis, changing its direction; R—is the Z-axis, which does not change its direction; G—rotation axis of the turntable changing its direction; J—axis of rotation of the turntable, which does not change its direction.

The first group includes the layouts for which R = J, T = G; and to the second those for which: T = J, R = G (Figs. 4 and 5).

One of the most important characteristics of the base part is the orientation in the space of the Z-axes, which depends largely on the convenience of a manual or automated change of the measuring tip. According to the orientation of the quill, each of the types is divided into three classes:

- Th (axis T—horizontal);
- Tv (axis T—vertical);
- Tv/h (axis T—oblique).

Denote by D the axis of the angular displacement T between J and G. It is possible to construct the basic part of the TIIS, in which the T axis, when moving around the D axis, describes the P-plane, the cone K, or the hyperboloid of one sheet.

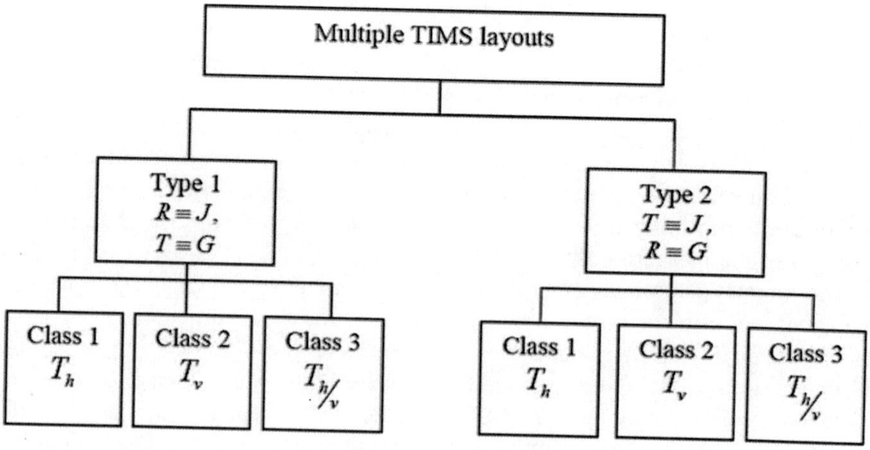

Fig. 4 Classification scheme of the IMS based on SPM layouts: many layouts

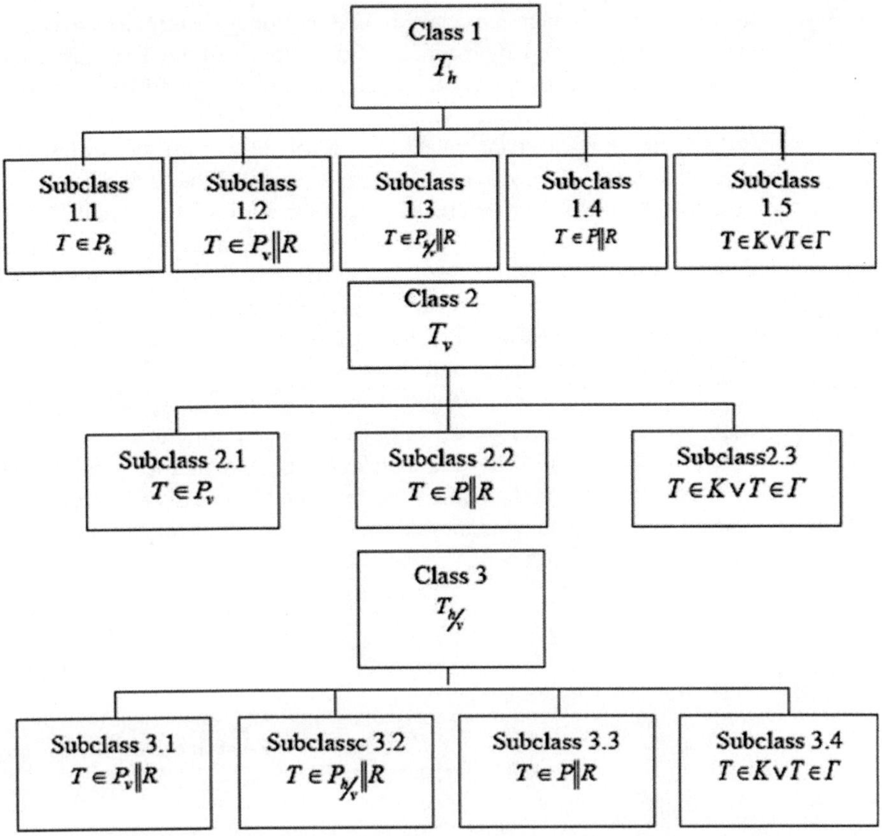

Fig. 5 Classification scheme of the IMS layouts: subclasses of layouts

Depending on the type of surface that the T-axis forms, and also on the orientation of this surface in space, the classes are divided into subclasses (Fig. 5).

Subclasses are divided into three groups according to the method of implementing three movements along the axes OX, OY, OZ. The executive body along the OZ оси axis performs 7 motions: one rotational and 6 translational, and the actuator along the OY axis—two: one translational and one rotational.

3 Conclusion

As a result, questions of the kinematics of the executive mechanisms of SPMs common in the nano-measurements have been considered. A structural-kinematic analysis of SPM for various purposes was carried out and the basic mechanical part of the SPM was classified.

Based on the structural-kinematic analysis of traditional coordinate-measuring machines, general principles, and approaches to the synthesis of the mechanical part of the three-coordinate IMS based on SPM of mechanical quantities with laser measuring subsystems are formulated. The problems of choosing the speed of the measuring tip moving along a given trajectory are solved taking into account the limited number of degrees of freedom and the capabilities of the actuators.

The problem of stabilizing the mechanical part of the system with small errors in the positioning of the measuring tip has been solved, and an adaptive algorithm for controlling three-coordinate IMS has been proposed. It is shown that the system approach to solving the problems of structural-kinematic analysis and the synthesis of a three-coordinate IMS based on SPM taking into account the constraints and requirements common to the entire system is relevant.

The following new results were obtained: a theory of measuring objects with a complex spatial surface was developed, an object-oriented structure of the basic three-coordinate IMS based on SPM was developed, an algorithm for combining the control contours of the mathematical model and the measured object was developed, and an algorithm for parametrizing the motion path of the measuring head along the object was developed.

References

1. Gapshis AA, Kasparaitis AYu, Modestov MB, Ramanauskas ZA, Serkov NA, Chudov VA Coordinate measuring machines and their application, p 328
2. Gapshis AA, Kasparaitis AYu (1986) Coordinate measuring machines. Machine-tool construction of Lithuania 14:5–11
3. Kobrinsky AE, Kochenov MI, Podlazov SS (1979) Coordinate measuring machine with CNC for adaptive machine systems. Machines Tools 3:24–25
4. Bekintis ZN, Kasparaitis AYu, Kumetaitis YuP (1981) New automated coordinate measuring machines. Machines Tools 4:12–15
5. Muller E (1994) 3D Koordinen-Messtechnik der Zukunst. Masehinenbau 23, 12, 21
6. Kobrinsky AA, Kobrinsky AE (1985) Manipulation systems of robots, p 343
7. Kasparaitis AYu (1976) Theoretical analysis of the spindle axis of precision instruments. Machine Tool Constr Lithuania 8:163–170
8. Kayashev AI, Mitrofanov VI (1995) Methods of adaptation in the management of automated machine tools. Mech Eng 240
9. Pavlenko II (1996) Synthesis of structural promises of robots and analysis of mechanical systems for versatile from CNC, p 230
10. Zubarev IM, Kosarevsky SV, Revin NN (2011) Automation of coordinate measurements, p 160
11. Naboka EV (2010) Analysis of measurement errors on coordinate measuring machine. Technologies in mechanical engineering: a collection of scientific papers, pp 84–88
12. Kvasnikov VP (2002) Mathematical model of the process of measurements that of the impulses of folding spacious vertexes with a three-coordinate informational-measurement system. Vimyryuvalna Enumerated Technol Proc 3:51–53

Designing High-Rise Residential Building, Given the Progressive Collapsing

Andrii Bieliatynskyi, Valerii Pershakov, Oleksandra Akmaldinova, and Victoria Akmaldinova

Abstract The article presents the features of designing a high-rise residential building, taking into account the progressive collapsing. On the basis of numerical modeling of high-rise frame-monolithic residential building stress and strain state by means of the "LIRA CAD" PC, the feasibility of recommendations stipulated in construction standards and constructive measures proposed in literature sources are analyzed. And the effectiveness of high-rise residential buildings protection against progressive destruction is assessed.

Keywords Reliability · Progressive destruction · Progressive collapse · High-rise residential building

1 Introduction

Modern multi-storey buildings are complex systems for various functional purposes (dwelling houses, offices, shopping and entertainment centers, etc.) having special requirements for safe operation and evacuation of people in case of emergencies and accidents of natural or technical nature (earthquakes, fires, terrorist acts, etc.). Tragic destruction of Ronan Point building in London, Skyline Plaza, Virginia (USA), L'Ambiance Plaza, Connecticut (USA), World Trade Center (New York) due to the progressive collapse led to the death of a large number of people and great material damage to the countries as a whole [1–3].

The word combination "progressive collapse" means the spread of the initial local destruction in the form of a chain reaction from structure to structure, resulting in the destruction of the building or a part of it two or more floors high that has lost support due to local destruction on any floor. The destruction may be caused by any

A. Bieliatynskyi
North Minzu University, Wenchang North St. 204, Yinchuan 750001, People's Republic of China

V. Pershakov · O. Akmaldinova · V. Akmaldinova (✉)
National Aviation University, Liubomyra Huzara Ave.,1, Building 9, Kyiv 903058, Ukraine

A. Bieliatynskyi and V. Breskich (eds.), *Safety in Aviation and Space Technologies*, Lecture Notes in Mechanical Engineering,
https://doi.org/10.1007/978-3-030-85057-9_13

of a great number of emergencies or events that are not considered in usual designing, such as: flooding by groundwater; erosion of the foundation due to accidents on the water lines; destruction of structural members through their overloading or as a result of explosion, collision; weakening of the material structure because of corrosion; design errors in calculating fasteners and load-bearing elements; gas explosion or fire. In accordance with the construction standards, buildings and structures are calculated for earthquakes, fires and strong winds, which should not lead to progressive collapse. In view of this, problems of ensuring seismic and fire resistance of building structures are particularly relevant, as well as the high-rise residential buildings resistance to progressive destruction [2, 3]. Preventing progressive destruction is one of important conditions of safe operation for both new and existing buildings. Thus, under the requirements of the State Construction Standards, the structural system of the building must provide resistance to progressive destruction in case of emergency [1–4].

A review of recent publications has shown that the research issue is extremely relevant. A number of domestic and foreign researchers are engaged in studying progressive destructions: Perelmuter A.V., Eremeiev P.G., Shapiro G.I., Rzhanitsyn A.V., Shpete G., Veryuzhsky Y.V., Barabash M.S., Gorodetsky A.S. State standards on designing high-rise buildings prescribe measures against progressive collapse [5–8].

2 Most of the Research

Architectural and constructive decisions. The twenty-six-storey, one-section residential building with built-in rooms has a rectangular plan shape with dimensions: 32.1 × 16.2 m in the axes. Building class—II. Degree of durability—II. Degree of fire resistance—II. Difficulty category—II. Class of responsibility—SS-1. The building is of a frame-monolithic type, 79.8 m high. It has 26 floors, 175 apartments, including: single room apartments—75; two-room—75; three-room—25 (Figs. 1 and 2).

The structural scheme of the building is frame-monolithic. Brick walls have effective warming. External walls are 380 mm thick, internal walls—250 mm thick. Brick partitions are 120 mm thick. The floor is monolithic. Flat archs and stairs are of prefabricated reinforced concrete. Roof is of two-layer roofing material. The building is 26-storey with a floor 2.7 m high. The total height of the building is 79.8 m. The foundations are reinforced concrete piles, 300 × 300 mm in cross section and 9 m long.

External walls are of ceramic brick. Laying is carried out by a multi-row system of seam dressing with M50 cement-sandy solution. Partitions are made of brick 120 mm and 250 mm thick; plaster partitions are 200 mm thick. The floor is a monolithic reinforced concrete slab, 200 mm thick. Flat archs are also monolithic. The roof is combined with rolled roofing.

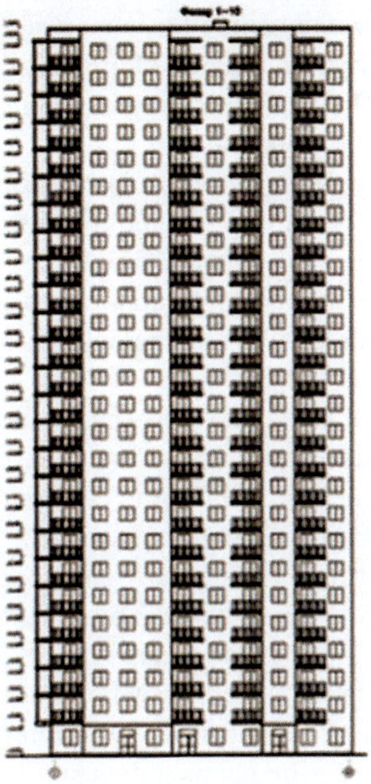

Fig. 1 Facade

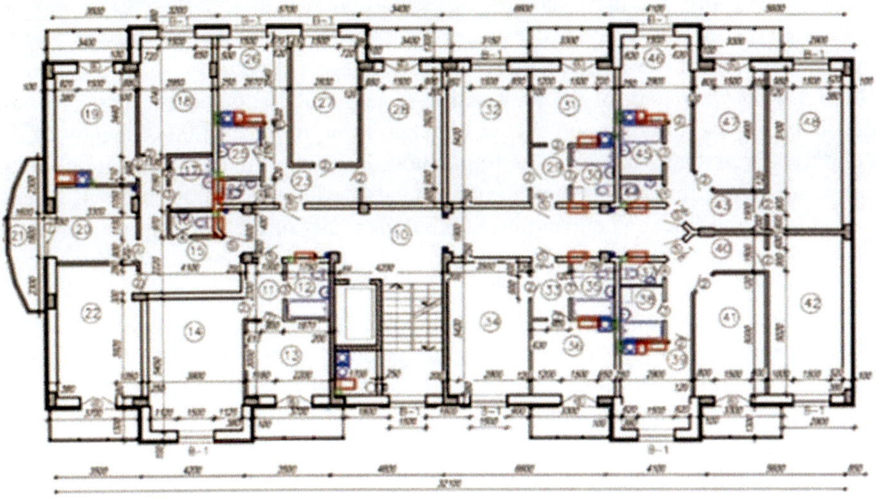

Fig. 2 Typical floor plan

Calculation of the monolithic frame by means of "LIRA CAD" PC [8, 9]. The 26-storey frame-monolithic residential building has a combined, frame-binding, sceleton-trunk structural scheme. The strength, stability and spatial rigidity of the frame are provided by the joint operation of the floor and vertical structures (pylons and vertical rigidity diaphragms). The floor is considered to be rigidly clamped at the joints with vertical elements. The sceleton scheme of the building is characterized by incomplete coincidence of the frame axes of the lower walls and the walls of the upper floors. The building supports are pylons, walls and a stair-lift units. The pylons are designed with rectangular cross-section to be easily concreted in the same formwork set as the walls.

The walls of the stair-lift units and two spatial closed shape elements, located near them, extend to the full height and are the core of the building rigidity. In the combined calculation scheme, the building is considered as a plate-rod system. To calculate such a scheme, the finite element method is used as the basis of the "LIRA CAD" software package. In this calculation scheme, rod elements of 10 types are taken as a universal type of the element. The core elements of the building are pylons and columns of the 1st and 2nd floors, all the other elements being referred to as lamellar. In the calculation scheme, such elements are modeled by 41 types, taking into account that these elements can withstand both bending and compression, i.e. they work as a shell. The characteristics of the calculation scheme are the number of nodes in the scheme—67,863, the number of elements—69,122, the number of equations—407,174 and duration of the calculation—12.8 min.

Loads and impacts taken into account when calculating the building structural schemes. According to the standards, residential buildings more than 40 m high are calculated taking into account the pulsation of the high-speed wind pressure, while the acceleration of the building structures oscillations should not exceed 0.1 m/s^2. Buildings are designed for the joint action of horizontal and vertical loads. So, the deflections of the building top should be taken as no more than 0.001 height of the building, given the pliability of the foundation. Buildings are calculated for climatic and temperature impacts, if their dimensions exceed the temperature block values. In this case, the forces occurring in different elements of the building as a result of these impacts are determined. To take into account concrete shrinkage, the concrete elastic modulus is corrected. It is reduced by 15% for monolithic buildings.

Horizontal loads are received by bonds—pylons, the structural purpose of which in the skeleton is to redistribute the frame loads. Stiffness diaphragms change the epures of Q and M struts (Q and M are higher at the struts top and lower at the base). In the frame system, the epures of M struts and crossbars increase downwards under the action of the horizontal load. Therefore, there is a need to increase the support reinforcement of the columns on the lower floors and in cross-bars intersection. In the frame-binding systems—on the contrary: the bending moment in framework elements decreases downwards, which allows to keep constant cross section of pylons along the height of the high-rise building. The main load-bearing structures of a multi-storey skeleton building are reinforced concrete frames with vertical bond diaphragms and intermediate floors binding them. Vertical permanent and temporary load and horizontal wind load were applied to the nodes and elements of the spatial scheme. 13 downloads were considered [1].

Determination of load values. Wind load. For Zhytomyr, the characteristic wind load is W_0 = 0.46 kN/m^2 [2]. The maximum calculated value of wind load is determined by the formula: $W_m = \gamma_{fm} W_0 C$, where $\gamma_{fm} = 1.14$ is the reliability factor for the limit value of wind load

$$C = C_{aer} C_n C_{alt} C_{rel} C_{dir} C_d$$

where C_{aer} is the aerodynamic coefficient on the wind side λ= 1/b = 59/25 = 2.36; λ_e = 2 = 4.72 ⇒ k = 0.6; λ_e = 5

Reynolds number:

$$Re = 0.88\ 103 = 0.88 \cdot 103 == 1.2 \cdot 105$$

C_{aer} = 1.16; C_h, the coefficient of the structure height,
C_{h5} = 0.2; C_{h10} = 0.4; C_{h20} = 0.65; C_{h40} = 1.0; C_{h60} = 1.1;
C_{alt} = 1—coefficient of geographical altitude; C_{rel} = 1—relief coefficient;
C_{dir} = 1—coefficient of wind direction; C_d = 0.95—dynamism coefficient.
Appropriate wind load limits at 5, 10, 20, 40 and 60 KN/m:

$$W_{fm}^{5} = 1.14 \cdot 0.46 \cdot 1.16 \cdot 0.2 \cdot 1 \cdot 1 \cdot 1 \cdot 0.95 = 0.116\ (\text{kN/m});$$

$$W_{fm}^{10} = 1.14 \cdot 0.46 \cdot 1.16 \cdot 0.4 \cdot 1 \cdot 1 \cdot 1 \cdot 0.95 = 0.231\ (\text{kN/m});$$

$$W_{fm}^{20} = 1.14 \cdot 0.46 \cdot 1.16 \cdot 0.65 \cdot 1 \cdot 1 \cdot 1 \cdot 0.95 = 0.376\ (\text{kN/m});$$

$$W_{fm}^{40} = 1.14 \cdot 0.46 \cdot 1.16 \cdot 1.0 \cdot 1 \cdot 1 \cdot 1 \cdot 0.95 = 0.578\ (\text{kN/m});$$

$$W_{fm}^{60} = 1.14 \cdot 0.46 \cdot 1.16 \cdot 1.1 \cdot 1 \cdot 1 \cdot 1 \cdot 0.95 = 0.636\ (\text{kN/m});$$

Snow load. For Zhytomyr, the characteristic snow load is S_0 = 1.46 kN/m^2. The maximum calculated value of the snow load on the pavement horizontal projection is determined by the formula:

$$S_m = \gamma_{fm} S_0 C \tag{1}$$

where $\gamma_{fm} = 1.14$.

The coefficient C is calculated by the formula:

$$C = C_e C_{alt}, \tag{2}$$

where C_e = 1, and C_{alt} = 1; S_e = S_0C, $\gamma_{fm} = 1.46\ (\text{kN/m})$.

Loads from walls, partitions and floors. Load from partitions −2.0 kN/m^2 ($\gamma_{fm} = 1.1$); load from the floor −2.0 kN/m^2 ($\gamma_{fm} = 1.3$); payload on the floor

($\gamma_{fm} = 1.2$); for apartments of residential buildings -1.5 kN/m^2; corridors, stairs -3.0 kN/m^2; beauty salon -4.0 kN/m^2. The effect of concrete creep was taken into account by introducing a stiffness correction: $B = 0.85 \cdot E_b \cdot I/\varphi_{b2}$.

Comparison of options. The comparison of options was made on the basis of the reinforcement consumption for reinforcing the 26th floor slab located at 79.8 m mark. The comparison was performed by reinforcing oilcloths for two options: the first option—vertical elements (pylons) and floor slabs made of C20/25 concrete; the second option—vertical elements (pylons) made of C25/30 concrete, and horizontal slabs of C20/25 concrete. When comparing the oilcloths of the reinforcement options, the difference was determined in the calculated area of the longitudinal reinforcement in the places of contact of the vertical bearing elements contact with the floor slab. The comparative analysis of the options for the floor slab reinforcement is given in Table 1.

The result of the analysis shows that for option № 2 the reinforcement consumption is, on average, 9.72% less, so for further development option № 2—vertical elements (pylons) made of C25/30 concrete and horizontal slabs of C20/25 concrete was taken in the project.

Table 1 Comparative analysis of options for the floor slab reinforcement

Name	1 variant A_s	2 variant A_s	Difference (%)
Along the X axis near the lower face of the plate (upper part of the plate)	11.5	10.4	10.6
Along the X axis near the upper face of the plate (upper part of the plate)	24.9	23.7	5.06
Along the Y axis near the lower face of the plate (upper part of the plate)	11.5	9.4	22.3
Along the Y axis near the upper face of the plate (upper part of the plate)	24.9	23.8	4.7
Transverse along the X axis near (upper part of the plate)	2.01	1.93	4.15
Transverse along the Y axis near (upper part of the plate)	4.31	3.2	4.2
Along the X axis near the lower face of the plate (lower part of the plate)	4.23	3.87	9.3
Along the X axis near the upper face of the plate (lower part of the plate)	7.36	7.06	4.24
Along the Y axis near the lower face of the plate (lower part of the plate)	4.78	4.02	18.9
Along the Y axis near the upper face of the plate (lower part of the plate)	12.2	10.19	19.72
Transverse along the X axis near (lower part of the plate)	0.65	0.6	8.3
Transverse along the Y axis near (lower part of the plate)	1.12	1.07	4.67
			16.64/12 = 9.72%

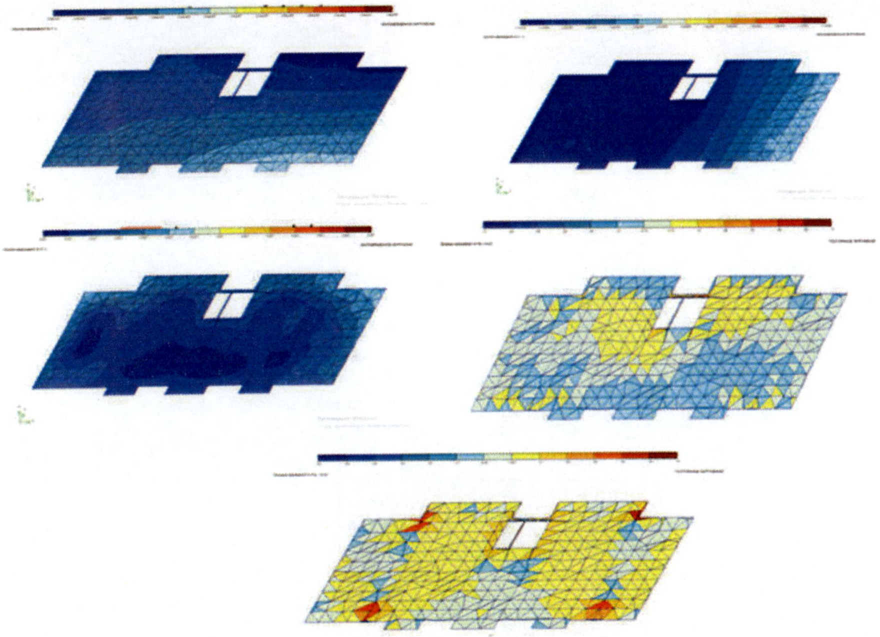

Fig. 3 Results of ITU calculation

Reinforcement of the sixteenth floor slab. For the 26th floor slab reinforcement "oilcloths" were obtained by means of "LIRA CAD" PC, on the basis of which the design drawing of armature arrangement was developed. The results are shown in Fig. 3. The slab is reinforced by a solid knitted net consisting of rods Ø8 mm arranged with a pitch of 200 mm. In places of significant internal loading, additional reinforcing rods are arranged with 200 mm pitch. Additional rods are accepted Ø10 .... 25 mm. The skeleton reinforcement is Ø8, 10 mm. Frames are installed at places of slab adjoining vertical structures. The frame rods are connected by means of a binding wire.

Calculation of the foundation grille. Initial data: bored pile with a diameter of 0.25 m (U = 0.785 m; A = 0.049 m^2). Dry concreted; working length of the pile Lp = 11.65 – dp = 11.65 – 1.65 = 10.0 m. The soil characteristics are given in Table 2.

Calculation of the allowable pile load under indenting strain.

Permissible pile design load taking into account the negative friction forces P_n [9]:

$$N_p = \frac{F_d}{\gamma_k} - \gamma_c P \tag{3}$$

$$F_d = \gamma_c \left(\gamma_{cR} RA + U \sum \gamma_{cf} f_i h_i \right) \tag{4}$$

Table 2 Physical and mechanical properties of soils

Lay number	Specific weight of particles (kN/m^3)	Humidity at the swing limit, Wp	Humidity at fluidity limit, WL	The porosity factor, e	Specific weight of soil γ* (kN/m^3)	Specific adhesion, c*	Internal friction angle φ* (deg)	The modulus of total deformation E* (MPa)	Poisson's ratio υ*
I	6.8	0.21	0.36	0.815	8.3	2.0	5.0	2.0	0.35
II	6.7	0.20	0.33	0.91	7.1	2.0	2.0	2.0	0.35
III	27.0	0.23	0.39	0.71	19.3	25.0	20.0	12.0	0.35

Note *—characteristics of the soil in a water-saturated state

Tabular characteristics for calculations by formulae are determined at the IL fluidity indicators corresponding to total water saturation of the soil:

$$I_L = \frac{\frac{0.9e\gamma_w}{\gamma_s} - W_p}{W_L - W_p} \quad (5)$$

For the first layer $I_L = \frac{\frac{0.9 \cdot 0.81 \cdot 10}{26.8} - 021}{0.15} = 0.42$

For the second layer $I_L = \frac{\frac{0.9 \cdot 0.91 \cdot 10}{26.7} - 0.2}{0.13} = 0.8$

For the third layer $I_L = \frac{\frac{0.9 \cdot 0.71 \cdot 10}{27.0} - 0.23}{0.16} = 0.04$

According to the calculation scheme, we find the calculated soil supports on the side surface f_i and under the lower end R of the pile, given that the side surface resistance is taken below the h_{sl} limit (lowest limit up to which the negative friction forces act): $Z_4 = 6.7$ m – $f_i = 8$ kPa; $Z_5 = 8.1$ m – $f_i = 8$ kPa; $Z_6 = 9.8$ m – $f_i = 64.7$ kPa; $Z_7 = 11.225$ m – $f_i = 66.7$ kPa; $Z_R = 11.65$ m – R = 1455 kPa. Then the pile bearing capacity, taking no account of the negative friction forces, will be equal to:

$$F_d = 0.8 \cdot (1 \cdot 1455 \cdot 0.049 + 0.7 \cdot 0.785 \cdot (8 \cdot 1.4 + 8 \cdot 1.4 + 64.7 \cdot 2 + 66.7 \cdot 0.85)$$

$$F_d = 148.7 \text{ kH}$$

Calculation of the negative friction force acting on the h_{sl}:

$$\text{P}_\text{n} = \text{U} \sum_0^{h_{sl}} \tau_i \cdot h_i \quad (6)$$

The side surface designed resistance τ_i to the depth of 6 m is determined by the formula:

$$\tau_i = \xi \cdot \sigma_{zg} \cdot tg\varphi + c \quad (7)$$

According to the calculation scheme:
for Z_1 = 2.45 m

$$\tau_i = 0.7 \cdot \gamma \cdot z_1 \cdot tg15 + c = 0.7 \cdot 18.3 \cdot 2.45 \cdot 0.268 + 12 = 20.4 \text{ kPa}$$

for Z_2 = 4.07 m

$$\tau_i = 0.7 \cdot \gamma \cdot z_1 \cdot tg15 + c = 0.7 \cdot 18.3 \cdot 4.07 \cdot 0.268 + 12 = 25.9 \text{ kPa}$$

for Z_3 = 5.45 m

$$\tau_i = 0.7 \cdot \gamma \cdot z_1 \cdot tg15 + c = 0.7 \cdot 18.3 \cdot 5.45 \cdot 0.212 + 8 = 22.5\ \text{kPa}$$

Then the force of negative friction will be:

$$\mathrm{P_n} = 0.785 \cdot (20.41 \cdot 1.6 + 25.9 \cdot 1.65 + 22.5 \cdot 1.1) = 78.61\ \text{kN}$$

As a result, according to the formula, the allowable pile design load will be equal to:

$$N_p = \frac{148.7}{1.4} - 0.8 \cdot 78.61 = 43.4\ \text{kN}$$

Parameters required to model different variants of the pile foundation are given in Table 3.

Physical and mechanical characteristics of soils are shown in Table 4.

The results of the pile foundation calculation are shown in Figs. 4, 5, 6, 7, 8, 9 and 10.

Table 3 Pile foundation parameters

Piles number	16 piles				9 piles				5 piles			
Foundation grille size (m)	3,3*3,3 m		6,0*6,0 m		2,4*2,4 m		4,2*4,2 m		2,1*2,1 m		3,3*3,3 m	
Distance between piles	3d	6d	3d	6d	3d	6d	3d	6d	3d	6d	3d	6d
Foundation grille thicknes (mm)	500	1000	1500	500	1000	1500	500	1000	1500	500	1000	1500

Table 4 Physical and mechanical characteristics of soils required for modeling

	γ_{II}, kN/m^3	$\mathrm{I_L}$	e	ν	E (MPa)	c (kPa)	φ (°)
Clay	19	0.4	0.85	0.42	15	43	16
Loam	18	0.7	0.7	0.35	14	44	18
Sand	17	–	0.55	0.3	40	2	38

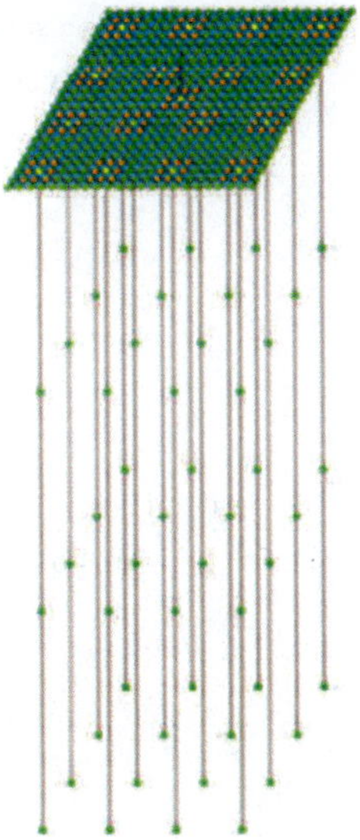

Fig. 4 Initial model of the pile foundation

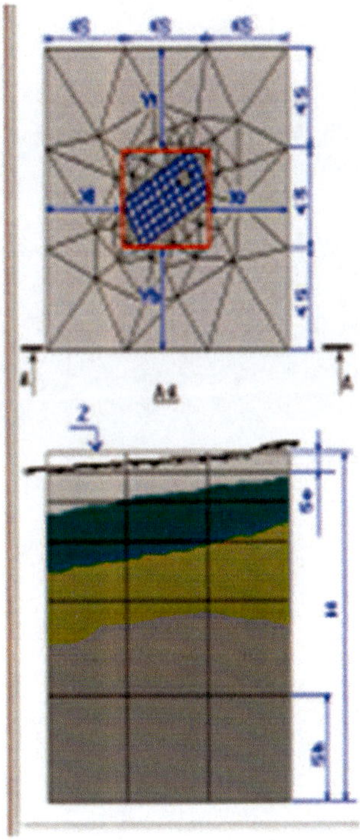

Fig. 5 Parameters required to create a three-dimensional soil mass

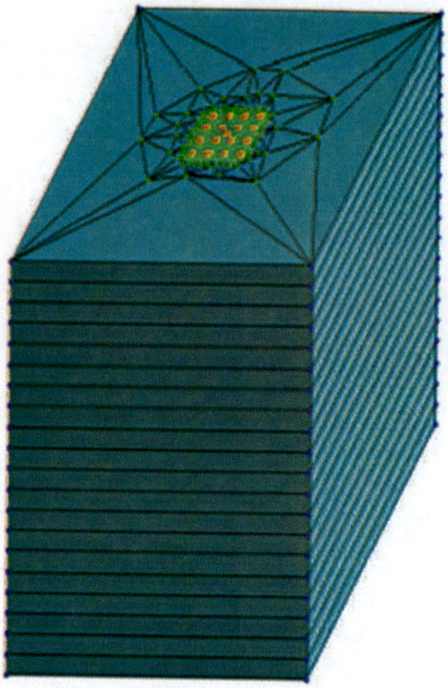

Fig. 6 Three-dimensional pile foundation model

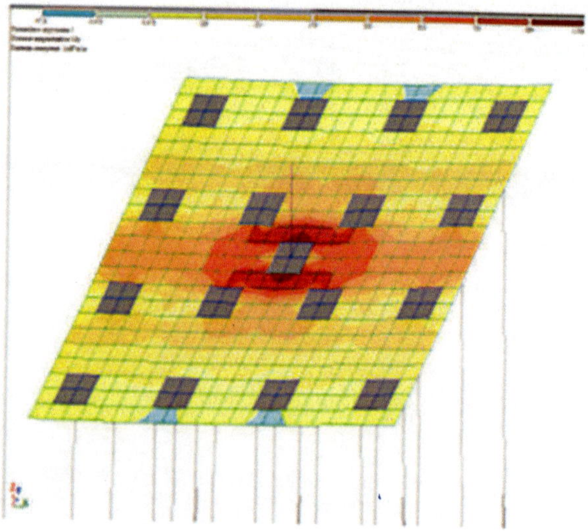

Fig. 7 The moment along the OY axis arising in the foundation grille

Fig. 8 Longitudinal forces arising in the piles

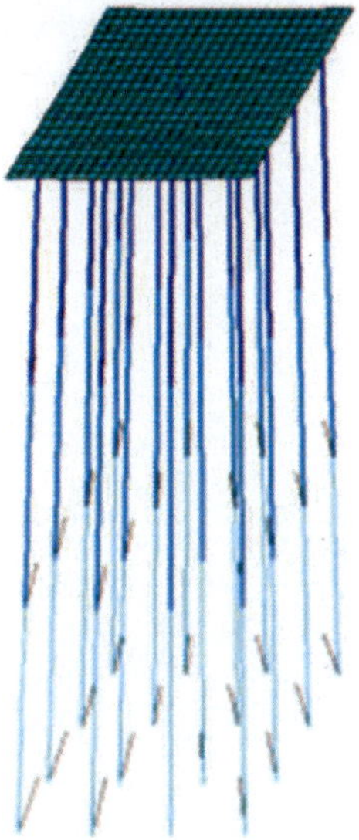

Fig. 9 Joint work of piles with grille after calculation

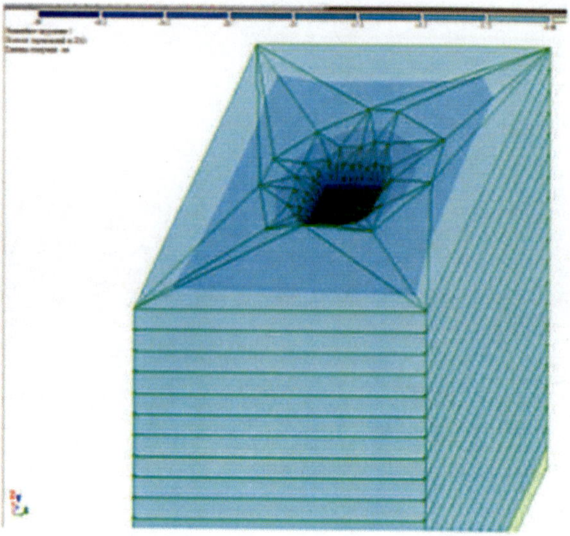

Fig. 10 Sedimentation of the foundation after calculation

3 Conclusion

Investigation of the rigid foundation grille impact on the operation of the columnar pile foundation, taking into account the progressive collapse has shown that the main parameters affected by the grille rigidity are: the stress–strain state of the grille as a component of the pile foundation and load redistribution between the piles of the group.

In addition, the operation of the pile foundation is influenced by the following factors:

- the number of piles in the group;
- mutual location of piles and distance between them;
- soil condition in which the pile foundation is located;
- foundation grille thickness.

Based on the analysis of forces redistribution between the piles, it was found that with increasing grille thickness, the middle piles are unloaded, the end piles are almost unaffected by the grille stiffness, while the corner piles are loaded more intensively. As the thickness of the grille increases, the grille itself takes on a larger share of the load. This value ranges from 9% to 65% depending on the soil medium. In this case, the greater the rigidity of the grille, the greater the moment arising in it.

When laying foundations of the same size, technology and methods of installation, the economic efficiency may be achieved by saving the material. Thus, when installing the grille, the main materials are concrete and reinforcement. The volume of concrete does not change, and the reinforcement amount is taken by determining the maximum moment arising in the foundation grille. With the proposed numerical modeling of the pile foundation, the actual moment is lower (from 4% to 27% depending on the soil condition between the piles) than that calculated analytically according to the regulatory documents. This allows to model and choose a more economical version of the pile foundation, involving the piles in more intensive operation.

References

1. Shapiro G, Eisman Y, Zalesov A (2005) Recommendations for the protection of monolithic residential buildings from progressive collapse. Moscomarchitecture, Moscow, p 40. www.complexdoc.ru
2. Bielostotskii A (2007) Analysis of the causes of collapse of the structures of the coating of SOC "Transvaal Park". ANSYS Solutions, Russian edn, Winter, p 8
3. Bielostotskii A et al (2008) Calculations of buildings for resistance to progressive collapse, taking into account physical and geometric nonlinearities. Theory and practice of calculation of buildings, structures and structural elements. In: Analytical and numerical methods: Sat. labor interd. scientific-practical conference. MGSU, Moscow, pp 183–193
4. Shapiro G, Eisman Y, Travush V (2006) Recommendations for the protection of high-rise buildings from progressive collapse. Moscomarchitecture. Moscow, p 34. www.complexdoc.ru
5. Dudar I, Prilipko T, Potapova T (2001) Directory of regulatory and technical data for projects of a complex of works of a zero cycle in construction. Manual. VSTU. Vynnytsia, p 133
6. Almazov V, Plotnikov A, Rastorguiev B (2011) Problems of resistance of buildings to progressive destruction. Sci Tech J "Herald" 2:15–20
7. Davydov I (2010) Features of calculations and protection of building structures from progressive collapse. Bull Dnieper State Acad Civil Eng Archit PDABA. Dnipropetrovsk 11:38–46
8. Gorodetsky A, Barabash M, Vodopyanov R, Titok V, Artamonova A (2013) LIRA-CAD® software package Textbook. Kyiv, Moscow, p 394
9. Nazarov Y, Zhuk Y, Simbirkin V, Egorov M (2007) Basmanny market: analysis of constructive decisions and possible mechanisms of building destruction. Constr Mech Calculation Constr 2:49–55

Tribocorrosive Behavior of Aluminum Alloy AA2024 Under Electrochemical Polarization

Myroslav Khoma, Vasyl Vynar, Aleksandr Dykha, and Chrystyna Vasyliv

Abstract The purpose of the work—to study the frictional interaction of tribotechnical pairs: alloy AA2024—co-rundum indenter, interacting in an acidic environment under conditions of cathodic and anodic polarization. The influence of external polarization on the tribocorrosive behavior of AA2024 alloy paired with a ceramic ball in acid rain was studied. The rate of destruction of the oxide film on the surface, material losses and friction coefficient depending on the applied external potential is established. The dual influence of cathodic polarization on the tribocorrosion of AA2024 alloy is established. The destruction of metal is inhibited at the potentials below the potentials of the freshly renewed surface. It is due to the reduction of the corrosion rate. Above this potential, corrosion-mechanical wear intensifies due to alkalization of the near-electrode layer of the electrolyte as a result of hydrogen depolarization. Changes in the coefficient of friction are insignificant. It is shown that the anodic polarization intensifies the formation of secondary structures. As a result, there is an increase in losses of materials and the decrease of friction coefficient. It indicates the lubrication properties of products that are intensively removed from the friction zone.

Keywords Tribocorrosion · Polarization · Fracture · Secondary structures · Coefficient of friction · Potential

1 Introduction

Tribocorrosion is the irreversible destruction of the surface layers of metals from the action of chemical and mechanical processes during frictional contact. This type of destruction is common in friction units operating in various aggressive environ-

M. Khoma · V. Vynar · C. Vasyliv
Karpenko Physico-Mechanical Institute of the NAS of Ukraine,
Naukova St., 5, Lviv 79060, Ukraine

A. Dykha (✉)
Khmelnitsky National University, Institutska St., 11, Khmelnitsky 29016, Ukraine

A. Bieliatynskyi and V. Breskich (eds.), *Safety in Aviation and Space Technologies*,
Lecture Notes in Mechanical Engineering,
https://doi.org/10.1007/978-3-030-85057-9_14

ments. It is difficult to predict the consequences of the simultaneous action of friction and corrosion due to the synergy of the processes. In this case, friction changes the sensitivity of the material to corrosion, and the corrosive factor influences the friction conditions. Electrochemical research methods are widely used to study the processes of tribocorrosion: changes in the electrode potential of metals, polarization currents during friction, and the like. However, the interpretation of these parameters is not always correct, since the data obtained depend on the relative position of the electrode and the friction track. Therefore, it is rational to study the electrochemical state of the surface during frictional interaction using the method of microprobing the friction zone. Friction units operating in active media are the subjects of destruction. In such systems, wear can occur from the flow of liquids, collisions of particles or contact of solids. Therefore, the study of tribocorrosion processes and the influence of various factors on them is an urgent scientific and technical problem. There are two interrelated areas in the field of research of electrochemical processes at the friction of metals: the use of electrochemical parameters to obtain information about contact processes and an attempt to control friction and wear of materials using electrochemical methods [1]. The second direction—active intervention in the electrochemical processes on the friction contact—has received much less development than the first, although it seems quite promising. One of the control methods is the polarization of the friction system from an external source. The method consists in the fact that the systems with the help of an external source of polarization is shifted to a certain area for the processes of friction and wear, and is maintained at a given level during the operation of the friction pair. Polarization of the friction system pursues various goals: to reduce surface energy, to improve the wettability and adsorption of environmental components, and others. Some cases of this method are anodic and cathodic protections, which change the corrosion-mechanical processes and the formation and destruction of secondary structures. The purpose of the work is to study the frictional interaction of tribotechnical pairs: alloy AA2024—corundum indenter, interacting in an acidic environment under conditions of cathodic and anodic polarization.

2 Literature Review

Much attention is currently being paid to the study of the processes of tribocorrosion of metals and alloys. This is due to the problems of increasing the durability of modern friction units of machines that are operated in difficult external conditions. The destruction of materials in a result of tribocorrosion occurs in many industries: processing, ener-gy, chemical and others [1]. The results are presented on the behavior of metals associated with wear and corrosion through the development of simulation models of tribocorrosion experiments [2]. These results demonstrate that mathematical models of tribo-electrochemical contacts can complement a physical experiment and contribute to an understanding of the tribological behavior of metals and alloys in an electrochemically active medium. The work [3] shows that the action of Sn in a five-component system leads to a significant increase in hardness and resistance to tri-bocorrosion. The presence of a hard intermetallic compound and a viscous phase helps to resist crack propagation and subsequent material removal during wear. The high polarization resistance of the surface film and the high corrosion po-tential help to prevent the corrosive effects of chlorides. In [4] the influence of external polarization with different poten-tials on the tribocorrosion behavior of steel in contact with a corundum ball in were studied. It is shown that the mechanical factor shifts the potential of pitting formation to the negative side as compared with the surface without friction. The aim of the work [5] was to study the tribocorrosive behavior of single-layer thin films of titanium oxycarbide TiCxOy on a reciprocating tribometer. The films were obtained by DC reactive sputtering using carbon. Corrosion studies have shown high corrosion resistance of the films, as evidenced by constant values of polarization resistance before and after the sliding process. The effect of hardness, thickness and structure of films on their tribocorrosion properties was studied depending on the oxygen content. In [6] a systematic study was carried out to study the electrochemical and tribo-corrosive behavior of the Monel K500 alloy sliding on an Al_2O_3 alloy in seawater. It is shown that the total volumetric losses from tribocorrosion increase with an increase in the applied potential. It is noted that the total loss of material at an applied potential is more than three times higher than the loss at mechanical wear, which confirms the synergy between wear and corrosion. In studies [7–10], attention is paid to the effect of the surface structure of materials and coatings on corrosion and bonds with tribochemical processes. The results of the practical application of research methods for tribocorrosion processes are considered in works [11–13]. Attention is paid to the establishment of correlations between wear and corrosion processes on the friction surfaces of machine parts. The influence of different types of technical and biological fluids on the processes of tribocorrosion are reflected in studies [14, 15].

3 Materials and Methods

Tribocorrosion studies of aluminum alloy AA2024 (size of samples 50 × 40 × 5 mm) were performed on a reversible friction test plant according to the scheme "ball—plane". The load on the ceramic ball was 1 N, the length of the friction track —24 mm, the speed of movement—1.6 mm/s, friction time 20 min. During the tribocorrosion tests, the change in the friction moment was recorded and the polarization was set using the PI-50-1.1 potentiostat. A saturated silver chloride reference electrode and an auxiliary platinum electrode were used. The kinetics of changes in the measuring parameters were recorded by an analog–digital device by using a personal computer with a measurement step of 0.25 s. The working medium was acid rain with a pH = 4.5, reduced by the addition of Na_2CO_3.

A scanning electron microscope EVO-40XVP (Carl Zeiss) with a system of micro-X-ray spectral analysis using an energy dispersion spectrometer INCA ENERGY 350 was used to study the surface of the aluminum alloy after corrosion mechanical wear.

4 Research Results and Their Discussion

It is established (Fig. 1a) that the character of cathode processes changes during the frictional interaction of the AA2024 alloy with the corundum ball in the acid rain during the reciprocating motion of the indenter. There is an oscillation of corrosion currents in the polarization curve from the corrosion potential to −1.2 V (potential of freshly renewed surface) because there is a local change of cathode–anode processes on the deformed surface. It indicates the formation and destruction of secondary structures. At a potential below −1.2 V, hydrogen depolarization blocks

Fig. 1 Polarization curves of AA2024 alloy (**a**) and its damage during tribocorrosion at a potential of −1.6 V in acid rain: 1—without load; 2—during friction (P = 1 N, counter body: ball Al_2O_3)

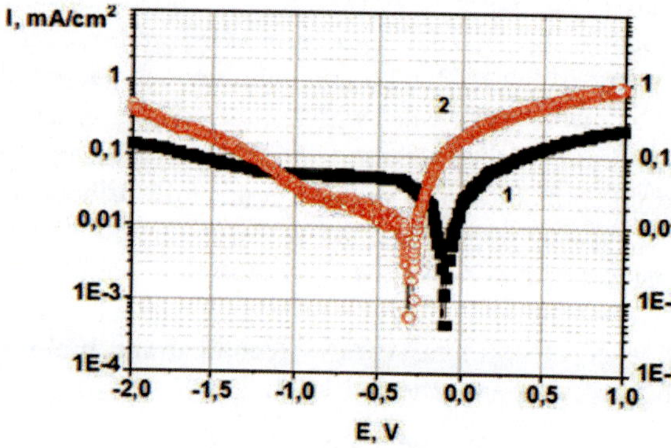

a

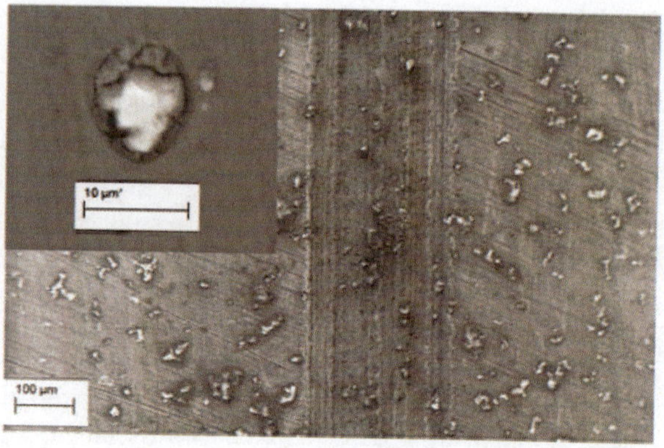

b

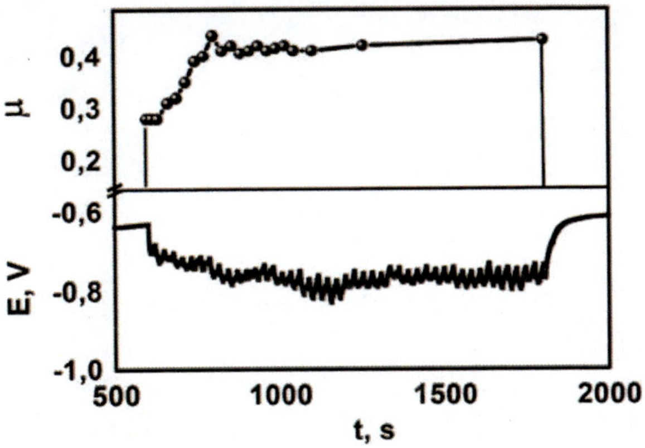

Fig. 2 Time dependences of friction coefficient and corrosion potential of aluminum alloy AA2024 during tribocorrosion studies in acid rain solution

the formation of films due to near-surface alkalization of the interaction zone, which intensifies the rate of chemical corrosion. This is evidenced by the dissolution in the contact zone (Fig. 2b), especially intense local corrosion at the sites of intermetallic compounds.

Under anodic polarization, both in stationary conditions and under frictional interaction, corrosion processes are activated, and their speed during friction increases 5 times.

Tribocorrosion studies have shown that potential on the surface of the alloy AA2024 in acid rain is −0.520 V without applying a load. Its potential shifts sharply to the negative side and becomes −0.770 V with the beginning of mechanical activation of the surface by a moving indenter. The average value of the friction coefficient gradually increases with the beginning of research and its value was established at the level of ~0.43 after a stage of running-in. Changes of the tribopotential and the coefficient of friction are characterized by a gradual shift of values at the initial stage of research. The oxide film is formed on the metal surface in neutral and acidic environments [4]. This film contacts with the indenter, destructs and is removed from friction zones during the friction.

It was found (Fig. 3a) that, this film is destroyed in 6 passes of the indenter without the potential applying. The change of the polarization potential from stationary potential in the cathode and anode sides significantly changes the lifetime of the film. The lifetime of the film increases due to electrochemical protection at cathodic polarization in the range of −0.700... −1.200 V. It is most evident at the electrode potential of pure alloy AA2024 without oxide film (−1.200 V). The destruction of the film intensifies as the potential increases in the negative direction from the potential −1.200 V. The reason for this is the near surface alkalinization of the environment due to the increase in the rate of hydrogen depolarization [10]. When level of polarization increases, the chemical corrosion intensifies, and the destruction increases. Friction during anodic polarization significantly accelerates processes of electrochemical corrosion and intensifies the destruction of the film. Damage of film occurs without the frictional interaction at potentials of 0.750... 1.000 V and the indenter already removes its remnants from the contact zone.

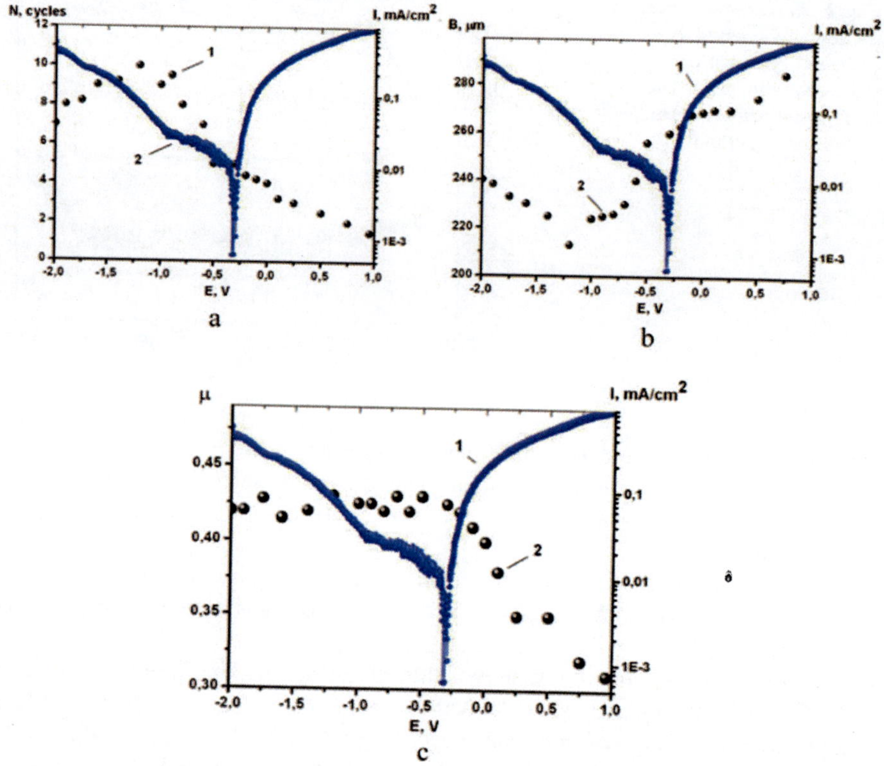

Fig. 3 Dependence of the rate of destruction of the oxide film (**a**), material loss (**b**) and friction coefficient (**c**) on the polarization potential of the alloy AA2024 by friction in acid rain in contact with a ceramic ball (1 N): 1—polarization curve of AA2024 alloy for tribocorrosion; 2—parameters: **a**—destruction of the film; **b**—the width of the friction track; **c**—coefficient of friction

Polarization changes the lifetime of the film in the initial stages of tribocorrosion tests. Also it changes the material loss and friction coefficient for the entire period of tribocorrosion tests (Fig. 3b, c). Cathodic polarization reduces material loss. So the width of the track on the surface without polarization is 230 μm and it is reduced to 225 μm with a potential shift in the negative direction. The smallest material losses are observed at a potential of −1.200 V, due to the maximum inhibition of corrosion processes, the track width is 212 μm. Shifting the potential to more negative values, as in the case of natural oxide film (Fig. 3a), leads to alkalinization of the environment and intensification of corrosion processes. Material losses increased due to corrosion-mechanical wear. Anode polarization, which is only 0.100 V relative to the steady-state potential, increases the width of the wear track by 20 μm. With a shift of the polarization potential to a more positive side, the loss increases the material loss and at a polarization potential of +1.000 V is 45% compared to an alloy without polarization.

It is established that the dependence of the friction coefficient on the polarization potential for tribocorrosion has the opposite character to the material loss. It is to some extent abnormal for most tribological processes. Cathodic polarization does not change the value of the coefficient of friction, the increase occurs only at the potentials −0.120 V (maximal inhibition of corrosion processes). Anodic polarization intensifies corrosion processes, accelerates the formation of secondary structures, which consist mainly of Al (63.2 mass. %) and O (32.19 mass. %) (Fig. 4). It reduces adhesion interaction of friction surfaces. So, with increasing polarization potential, the coefficient of friction decreases from 0.43 to 0.32 at a potential of 1.000 V.

The nature of the surface destruction also changes in a result of the application of the polarization potential during friction (Figs. 1b and 5). Thus, the greatest surface damage is observed during anodic polarization. For example, there are etches and products of the combined action of corrosion and friction—especially active local corrosion—at a potential of 0.500 V on the surface (Fig. 5a). The appearance of microcutting elements is found on the material without the application of polarization (Fig. 5b). Under these conditions the rate of secondary structures formation decreases compared to anodic polarization and the adhesive interaction between the contact surfaces increases. The surface damage decreases in the case of the polarization potentials shift in the cathode direction, as electrochemical corrosion is inhibited, but the role of chemical dissolution increases. There is a change in the mechanism of corrosion, and therefore the mechanism of friction at the potential of the freshly renewed surface (−1.200 V). At this potential, electrochemical corrosion is practically blocked, but chemical corrosion begins to intensify due to near-surface alkalinization of the environment.

As the rate of hydrogen depolarization increases, the concentration of hydrogen in the near-surface layer increases and the pH of the solution changes intensively. It is manifested in the local etching of intermetallic inclusions in the contact zone (Fig. 1b). Thus, depending on the level of cathodic and anodic polarization, the intensity of destruction of the AA2024 alloy during tribocorrosion in pair with corundum counter body changes.

Element	Mass. %	At.%
O K	32.19	45.27
Mg K	1.17	1.08
Al K	62.30	51.95
S K	0.24	0.17
Mn K	0.37	0.15
Fe K	1.29	0.52
Cu K	2.44	0.86

Fig. 4 Chemical composition of secondary structures, which are formed by anodic polarization (0.100 V) of AA2024 alloy during tribocorrosion

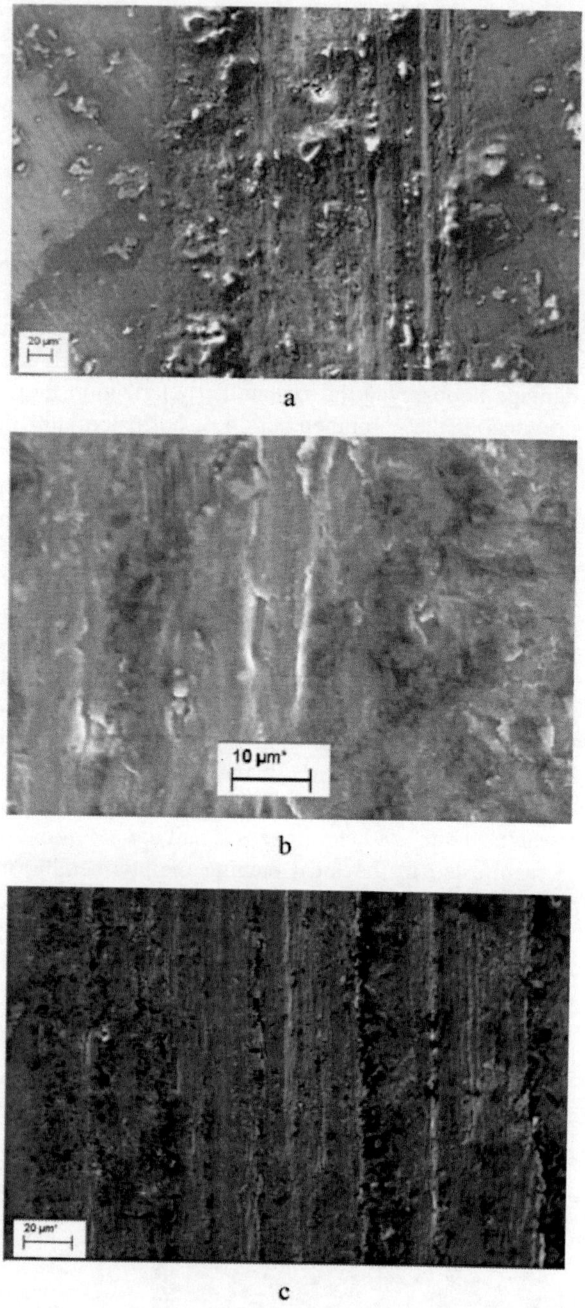

Fig. 5 Topography of the surface of AA2024 alloy after tribocorrosion studies in acid rain at different potentials of polarization: **a**—polarization at +0.500 V; **b**—without polarization; **c**—polarization at −1.200 V

5 Conclusions

The influence of external cathodic and anodic polarization on tribocorrosion of AA2024 alloy paired with a ceramic ball in acid rain has been studied.

The dual influence of cathodic polarization on the tribocorrosion of AA2024 alloy is established. The destruction of metal is inhibited at the potentials below the potentials of the freshly renewed surface. It is due to the reduction of the corrosion rate. Above this potential, corrosion-mechanical wear intensifies due to alkalization of the near-electrode layer of the electrolyte in a result of hydrogen depolarization. Changes in the coefficient of friction are insignificant.

It is shown that the anodic polarization intensifies the formation of secondary structures. As a result, there is an increase in losses of materials and the decrease of friction coefficient. It indicates the lubrication properties of products that are intensively removed from the friction zone.

References

1. Landolt D, Mischler S (2011) Tribocorrosion of passive metals and alloys. Woodhead Publishing Cambridge UK, p 584. ISBN: 978-1-84569-966-6
2. Papageorgiou N, Mischler S (2012) Electrochemical simulation of the current and potential response in sliding tribocorrosion. Tribol Lett 48:271–283. https://doi.org/10.1007/s11249-012-0022-9
3. Muangtong P, Namus R, Goodall R (2021) Improved tribocorrosion resistance by addition of Sn to CrFeCoNi high entropy alloy. Metals 11(1):13. https://doi.org/10.3390/met11010013
4. Pokhmurs'kyi VI, Khoma MS, Vynar VA et al (2018) Influence of the external polarization potential on the tribocorrosion behavior of 08Kh18N10T steel. Mater Sci 54:279–285. https://doi.org/10.1007/s11003-018-0184-1
5. Mathew M, Ariza E, Rocha L, Fernandes A, Vaz F (2008) TiCxOy thin films for decorative applications: tribocorrosion mechanisms and synergism. Tribol Int 41(7):603–615. https://doi.org/10.1016/j.triboint.2007.11.011
6. Chen J, Wang J, Yan F, Zhang Q, Li Q (2015) Effect of applied potential on the tribocorrosion behaviors of Monel K500 alloy in artificial seawater. Tribol Int 81:1–8. https://doi.org/10.1016/j.triboint.2014.07.014
7. Merl D, Panjan P, Panjan M, Čekada M (2007) The role of surface defects density on corrosion resistance of PVD hard coatings. Plasma Process Polym 4:613–617. https://doi.org/10.1002/ppap.200731416
8. Yang M, Allen A, Nguyen M, Ralston W, MacLeod M, DiSalvo F (2013) Corrosion behavior of mesoporous transition metal nitrides. J Solid State Chem 205:49–56. https://doi.org/10.1016/j.jssc.2013.06.026
9. Mischler S (2008) Triboelectrochemical techniques and interpretation methods in tribocorrosion: a comparative evaluation. Tribol Int 41:573–583. https://doi.org/10.1016/j.triboint.2007.11.003
10. Zhang L, Chen Y, Feng Y, Chen S et al (2015) Electrochemical characterization of AlTiN, AlCrN and AlCrSiWN coatings. Int J Refract Met Hard Mater 53:68–73. https://doi.org/10.1016/j.ijrmhm.2015.03.018

11. Dykha A, Marchenko D, Artiukh V, Zubiekhina-Khaiat O, Kurepin V (2018) Study and development of the technology for hardening rope blocks by reeling. East Eur J Enterp Technol 2(1(92)):22–32. https://doi.org/10.15587/1729-4061.2018.126196
12. Dykha A, Zaspa Y, Slashchuk V (2018) Triboacoustic control of fretting. J Friction Wear 39 (2):169–172. https://doi.org/10.3103/S1068366618020046
13. Dykha A, Makovkin O (2019) Physical basis of contact mechanics of surfaces. J Phys Conf Ser 1172(1):012003. https://doi.org/10.1088/1742-6596/1172/1/012003
14. Stachowiak G, Salasi M, Stachowiak G (2015) Three-body abrasion corrosion studies of high-Cr cast irons: benefits and limitations of tribo-electrochemical methods. J Bio Tribo Corros 1:6. https://doi.org/10.1007/s40735-015-0006-z
15. Hacisalihoglu A, Samancioglu A, Yildiz F, Purcek G, Alsaran A (2016) Tribocorrosion properties of different type titanium alloys in simulated body fluid. Wear 332–333. https://doi.org/10.1016/j.wear.2014.12.017

Influence of Misalignment of Poles on Arcing During Current Collection

Ildar Bayanov, Timur Badretdinov, Saidaziz Saydivaliev, Elyor Saliyev, and Sardor Muminov

Abstract One of the important systems in the traction power supply of railways is the contact lines, which is responsible for the reliable transmission of electricity to the electric rolling stock (ERS) through the pantograph, regardless of weather conditions and the intensity of train traffic. Increased reliability requirements are imposed on all overhead line devices. The process of transferring electrical energy from the overhead wire to the pantograph takes place in a dynamic mode. Large currents are transmitted and even the slightest separation of the sliding contact causes the formation of an arc. Separation can occur during high-speed movement of the rolling stock when the slope of the contact wire changes. The arc creates a strong electromagnetic disturbance that spreads in the surrounding space. This disturbance interferes with communication channels, facility management systems and telecommunications located in the zone of its influence. Determining the magnitude of the potential of the electric field at the point of contact will determine the degree of the harmful effect of arcing on adjacent objects and develop a technology to reduce the interfering effect. This article examines the influence of the pole tilt on a stable contact and, consequently, on the effective operation of the overhead contact lines, and proposes a method for controlling the misalignment of the poles to exclude its influence on arc formation.

Keywords Contact lines · Reinforced concrete poles · Misalignment arc · Sliding pair · Pantograph · Electrical system · Field potential

1 Introduction

The supporting structures of the overhead catenary of electrified railway transport are responsible for the safety and correct positioning of the overhead catenary in the span, while at the same time in severe climatic and operating conditions without a

I. Bayanov (✉) · T. Badretdinov · S. Saydivaliev · E. Saliyev · S. Muminov
Tashkent State Transport University, Temiryo'lchilar street, 1, Tashkent, Uzbekistan100167
e-mail: ildar-lili@yandex.ru

A. Bieliatynskyi and V. Breskich (eds.), *Safety in Aviation and Space Technologies*, Lecture Notes in Mechanical Engineering,
https://doi.org/10.1007/978-3-030-85057-9_15

possible reserve. Deviations of the pole structure from the standard values lead to a violation of the geometric parameters of the overhead catenary, deviation of the stagger from the permissible values, a sharp change in the slope of the overhead wire, which increases the likelihood of the pantograph separating from the overhead wire and thereby leading to the formation of an arc with subsequent burnout of the overhead wire and stopping the train, violation of the train schedule due to the creation of an emergency. Even with a short-term separation of the current collector from the contact wire in the air gap, as a result of its breakdown, a high-temperature electric arc occurs. The resulting electric arc creates its own magnetic influence on the objects adjacent to it, operating at high frequencies. The time for the formation of an electric arc is instantaneous, is of the order of a thousandth of a second and always depends on many factors accompanying its formation—the atmospheric pressure of the air surrounding the electric arc, the temperature of the environment surrounding the electric arc, the distance between the current collector and the contact wire, i.e. anode and cathode of the electric arc [1–4].

The temperature of the electric arc at normal atmospheric pressure and current strength at the time of the formation of an electric arc of several hundred amperes reaches a value of about 10,000° K. The distribution of high temperature in different parts of the electric arc between the overhead wire of the overhead suspension and the ski of the electric locomotive pantograph and the current strength of 250 A [5].

2 Materials and Method

The current–voltage characteristic of an electric arc at low currents has a falling shape, while an increase in current is accompanied by a decrease in the voltage between the electrodes. At currents over tens of amperes, the arc voltage does not depend on the current, but depends on the arc length. In this case, the voltage Ud on the arc is defined as [4, 6].

In Fig. 1b shows a diagram for calculating the vertical distribution of the charge field of the overhead wire and the skis of the electric locomotive current collector.

In the presence of an electric arc in the air gap between the wire of the contact lines and the ski of the current collector of an electric locomotive, for calculating the electric potential, they can be represented as two mutually perpendicular straight lines located in two planes parallel to the xoy plane.

The appearance of an electric arc has practically no effect on the distribution of the electric field potential of the sliding contact charge. However, the appearance of an electric arc leads to accelerated wear of the copper contact wire and graphite lining of the electric locomotive pantograph, to environmental problems associated with the generation of electromagnetic radiation by an electric arc and the influence of generated electromagnetic disturbances on the objects of train and household communications, telecontrol of power supply facilities and operational control, etc.

The electric field in the junction area of the pantograph of an electric rolling stock with a high-voltage overhead wire has a great influence on the ignition of the

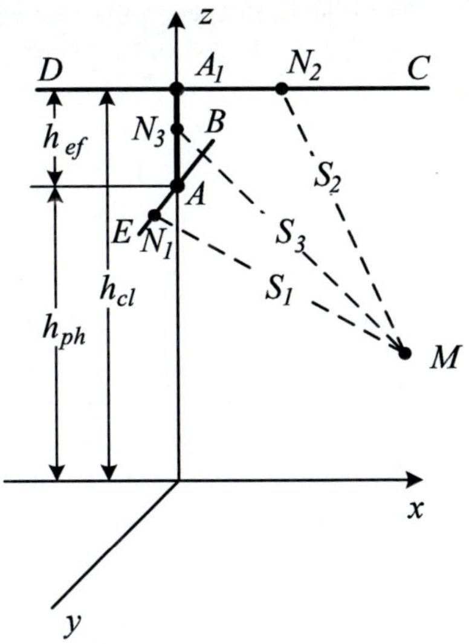

Fig. 1 The diagram of the distribution of the potential of the electric field of the contact lines

arc between the pantograph and the overhead wire [7, 8]. Knowing the magnitude of the electric field potential in this zone will make it possible to develop a technology to reduce the wear of the overhead contact wire and the contact insert of the pantograph of the electric rolling stock [9].

In the first approximation, the sliding joint A between the overhead wire, which is under a voltage of 27.5 kV, and the electric rolling stock pantograph contcat strips, can be represented as two straight lines DA1C and EAB, located in a plane parallel to the xoy plane (Fig. 1b). The line DA1C, located at the height z = how along the vertical axis oz, corresponds to the wire of the electric railway overhead line. The contcat strips of the electric rolling stock pantograph corresponds to the EAB located at the height z = hpcs line having a length of 2R. The oz coordinate axis passes through point A.

In space, the points of these lines have the following coordinates: A (0, 0, hpcs); E (0, R, hpcs); B (0, −R, how); D (−A1, 0, how); C (A1, 0, how). Let us assume that lines DA1C and EAB have a distributed electric charge of a positive sign with linear density τ [4, 5].

In space, we choose an arbitrary point M (x, y, z), in which we define the potential of the electric field of the charge as the sum of two potentials φ = pcs + φow + φef, where φow is the potential of the charge field of the DA1C line (overhead wire), and φpcs is the potential of the field of the charge of the EAB line (pantograph contcat strips) and φef, is the potential of the charge field of the AA1 line. The value of the electric charge of the element of the DA1C line with length dlx will be equal to dl1 = τdx1, and the element of the line EAB dl2 = τdy2 and AA1 line dl2 = τef·dz3. First, we determine the distribution of the potential of the charge field in the space between the wire of the contact lines and the level of the rail head [10, 11].

The potential φpcs of the electric field of the charge at the point N1 (0, y1, hpcs) on the EAB line at an arbitrary point M (x, y, z) is determined by the general solution of the Poisson equation [5]:

$$\varphi_{pcs} = \int_{-1}^{+1} \frac{\tau R dl_{x1}}{2\pi\varepsilon\varepsilon_0 S_1} = C_{pcs} \int_{-1}^{+1} \frac{dy_1}{\sqrt{x^2 + (y - y_1)^2 - (z - h_{pcs})^2}}, \tag{1}$$

where ε0 = 8.85 • 10 −12—electrical constant; ε is the relative dielectric constant of the medium; S1—distance from an arbitrary point M (x, y, z) of space to point N1 (0, y1, hpcs) on the EAB line;

Cpcs = Rτ/(2π ε0εU). We take the voltage of the overhead lines U = 27.5 kV.

The potential φow at an arbitrary point M (x, y, z) of the space from the charge at the point N2 (×2, 0, how) on the line DA1C has the form:

$$\varphi_{ow} = \int_{-A}^{+A} \frac{\tau R dl_2}{2\pi\varepsilon\varepsilon_0 S_2} = C_{ow} \int_{-A}^{+A} \frac{dx_2}{\sqrt{(x - x_2)^2 + y^2 - (z - h_{ow})^2}}, \tag{2}$$

S2—distance from an arbitrary point M (x, y, z) of space to point N2 (×2, 0, how) on the EAB line;

The potential φef of the electric field of the charge at an arbitrary point M (x, y, z) of space is also determined by solving the Poisson equation:

$$\varphi_{ef} = \int_{h_{pcs}}^{h_{ow}} \frac{\tau_{ef} R dl_2}{2\pi\varepsilon\varepsilon_0 S_2} = C_{ef} \int_{h_{pcs}}^{h_{ow}} \frac{dx_2}{\sqrt{x^2 + y^2 + (z - h_{ef})^2}}, \tag{3}$$

where, Cef = Rτef/(2π ε0εUef).

Integrals (1), (2) and (3) are tabular, after integration we get:

$$\varphi_{pcs} = C_{pcs} ln \left| \frac{(y + 1) + \sqrt{x^2 + (y + 1)^2 + (z - h_{pcs})^2}}{(y - 1) + \sqrt{x^2 + (y - 1)^2 + (z - h_{pcs})^2}} \right|, \tag{4}$$

$$\varphi_{ow} = C_{ow} ln \left| \frac{(x + A) + \sqrt{(x + A)^2 + y^2 (z - h_{ow})^2}}{(x - A) + \sqrt{(x - A)^2 + y^2 + (z - h_{ow})^2}} \right|, \tag{5}$$

$$\varphi_{ow} = C_{ef} ln \left| \frac{(z + h_{ow}) + \sqrt{x^2 + y^2 (z - h_{ef})^2}}{(z + h_{pcs}) + \sqrt{x^2 + y^2 + (z - h_{ef})^2}} \right|, \tag{6}$$

We obtain the desired particular solution of the Poisson equation if we take into account the following boundary conditions of the problem: under the boundary conditions z and z1 formulas (4), (5) and (6), taking into account the difference of logarithms.

At z1 = 0, the distribution of the potential of the charge field in the space between the CL wire and the level of the rail head φpcs = 0, φow = 0 and φef = 0. Taking into account formulas (1–6), and transformations φpcs, φow and φef take the form:

$$\varphi_{pcs} = C_{pcs} ln \left| \frac{\dfrac{(y+1)+\sqrt{x^2+(y+1)^2+\left(z-h_{pcs}\right)^2}}{(y-1)+\sqrt{x^2+(y-1)^2+\left(z-h_{pcs}\right)^2}}}{\dfrac{(y+1)+\sqrt{x^2+(y+1)^2+h_{pcs}^2}}{(y-1)+\sqrt{x^2+(y-1)^2+h_{pcs}^2}}} \right|, \tag{7}$$

$$\varphi_{ow} = C_{ow} ln \left| \frac{\dfrac{(x+A)+\sqrt{(x+A)^2+y^2(z-h_{ow})^2}}{(x-A)+\sqrt{(x-A)^2+y^2+(z-h_{ow})^2}}}{\dfrac{(x+A)+\sqrt{(x+A)^2+y^2+h_{ow}^2}}{(x-A)+\sqrt{(x-A)^2+y^2+h_{ow}^2}}} \right|, \tag{8}$$

$$\varphi_{ow} = C_{ef} ln \left| \frac{\dfrac{(z+h_{ow})+\sqrt{x^2+y^2\left(z-h_{ef}\right)^2}}{\left(z+h_{pcs}\right)+\sqrt{x^2+y^2+\left(z-h_{ef}\right)^2}}}{\dfrac{h_{ow}+\sqrt{x^2+y^2+h_{ef}^2}}{h_{pcs}+\sqrt{x^2+y^2+h_{ef}^2}}} \right|, \tag{9}$$

3 Research Results

The numerical calculation of the potential φ of the electric field of the charge in the form of the sum of the potential of the overhead wire (8) and the potential of the pantograph ski of the electric rolling stock (7) was performed in the MathCAD program, at A = 5; hpcs = 5.9 m.; how = 6.0 m. hef = 0.1 m.

Thc calculated curves of the distribution of potentials φpcs and φow the electric field of the charge in the presence of an electric arc are shown in Fig. 2. The vertical distribution along the oz axis of the unnormalized potential φow = f1 (z) wire-rail and potential distribution of the electric field of the contact strips pantograph φpcs = f 3(z).

The unnormalized distribution of the potential φ, defined as the sum of the potential of the wire of the wire-rail contact lines φow and the contact strips of the

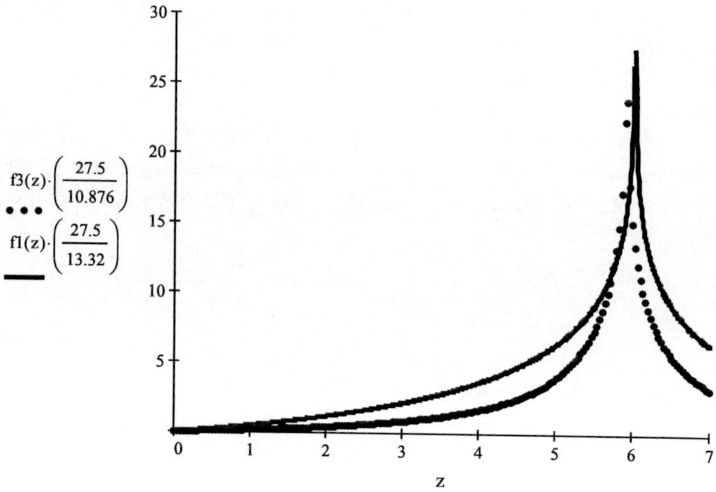

Fig. 2 The distribution of the potential of the electric field of the overhead wire and contact strips of pantograph

pantograph φpcs, is shown in Fig. 3a and the required normalized distribution of the potential φ is shown in Fig. 3b.

The vertical distribution of the electric field strength φ of the charge is determined by differentiating the potential function:

$$E_{ow} = \frac{d(f_1)}{dz}; E_{pcs} = \frac{d(f_3)}{dz}; E_{ow+pcs} = \frac{d(f_1 + f_3)}{dz}. \tag{10}$$

The results of calculating the intensity E (kV) are shown in Fig. 4.

An analysis of the graphs of the potential distribution of the non-standardized and normalized by the parameters of the suspension height of the overhead wire and the voltage of the contact lines of the wire of the contact lines and the contcat strips

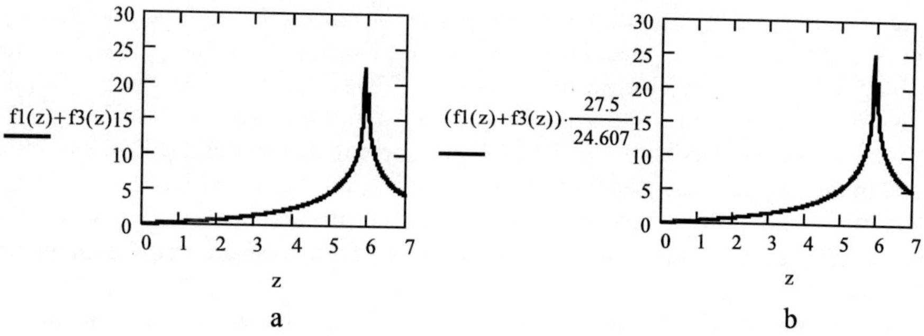

Fig. 3 Unnormalized potential distribution (**a**) and normalized potential distribution (**b**)

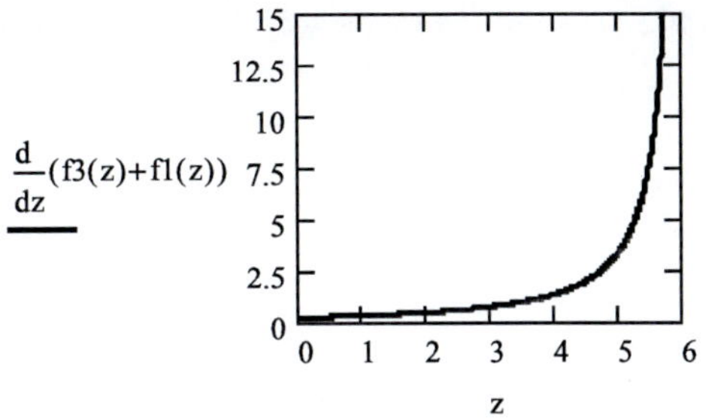

Fig. 4 Results of calculating the tension E

pantograph of the electric rolling stock in the space between the wire of the contact lines and the rail, between the wire of the contact lines and the upper surface of the car shows that the potential distribution at an arbitrary point M the same form irrespective of whether standardized parameters or non-standardized parameters are accepted. The vertical distribution of the electric field strength E of the charge increases with an increase in the height parameter of the contact wire suspension [12].

A system is proposed for determining the angle of deviation of the supports in all possible directions relative to the path and increasing the speed of the process of measuring the angle of inclination of the poles, by installing the system on a self-propelled unit, and making the measurement in motion. This system consists of 3 pairs of laser rangefinders, a gyroscope sensor with an accelerometer to filter the accuracy of measurements on the curve (when the outer rail rises) and possible vibration from the interaction of the wheel pair with the rail, GPS/GLONASS sensors to determine the location of the measured support, as well as a mobile device for data collection and processing based on its own algorithm. The process of measuring in one of the planes is shown in Fig. 5. The angle of inclination in each plane is measured using a pair of laser distance sensors, the first sensor measures the distance to the support perpendicular, and the second at an angle of 30° up with respect to the first sensor.

This method is convenient when measuring the angle of inclination of reinforced concrete support stoics and their taper does not affect the measurement process, since the taper angle for all supports of the SS type is the same β0 ≈ 0.425° (0°25′ 30″) and it is taken into account in the calculations. In the formed triangle ABC, the angle of inclination φ is determined from the following expression:

$$tg\phi = \frac{l \cdot \cos 30^\circ - d}{l \cdot \sin 30^\circ} = ctg30^\circ - \frac{d}{l \cdot \sin 30^\circ}$$

Let

$$\omega = \frac{d}{l}.$$

Fig. 5 Process measuring the angle of inclination of the poles in one of the planes

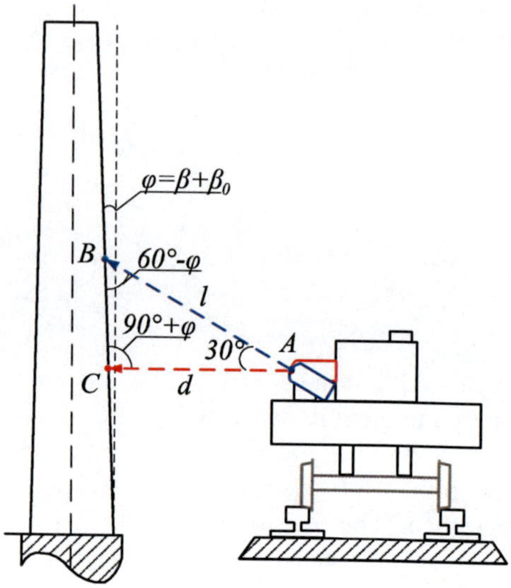

Then

$$\phi = arctg\left(\sqrt{3} - 2\omega\right),$$

and the angle of inclination is determined by the expression:

$$\beta = \phi - \beta_0$$

From the study, it was revealed that $\omega = 0.8623$ the pole is in the correct position. If $\omega < 0.8623$, then pole has a slope towards the path, if $\omega > 0.8623$—towards the field. Measurements in other planes are defined in a similar way.

Processing all the obtained values of the angles in all the measured planes, we get the true deviation of the pole and its angle of inclination.

This technique is intended to control the correct installation of the pole during installation and to accelerate the control of misalignment during the life cycle.

4 Conclusion

At the moment, on the electrified railways of the Republic of Uzbekistan, little attention is paid to detailed monitoring of diagnosing the state of traction power supply facilities, and the available diagnostic tools for the overhead lines do not fully meet the requirements for operating power supply devices due to the insufficient number of determined parameters of the overhead lines [13, 14].

Timely detection of deviations of the parameters of the overhead catenary from the norm would make it possible to bring the speed of high-speed trains closer to the optimal mode with a stable and high-quality current collection, to exclude the occurrence of an arc when the slope of the overhead wire changes.

Thus, we have developed a method for calculating the spatial distribution of the potential and electric field strength of a sliding contact pair consisting of an overhead wire and a pantograph ski, based on the analytical solution of the Poisson equation in integral form. The technique makes it possible to determine the emerging potential on adjacent communication lines of the electrified railway and provide methods for protection against interfering and dangerous effects of the electromagnetic field, and also allows using this approach to numerically estimate the amplitude of radio interference of various interfering factors.

References

1. Li V, Demina L, Vlasenko S (2020) Assessment of the concrete part of the contact system supports in the field. E3S web of conferences, vol 164, p 03028. https://doi.org/10.1051/e3sconf/202016403028
2. Galkin A, Kovalev A, Okunev A (2019) Assessing remained service time of contact-line support under the constant load. In: International scientific Siberian transport forum, pp 693–702. Springer, Cham. https://doi.org/10.1007/978-3-030-37916-2_68
3. Li VN, Shnurova NK (2016) On the method of determining optimal electricity consumption from an electric traction network. Russ Electr Eng 87(2):97–99. https://doi.org/10.3103/S1068371216020127
4. Amirov S, Bayanov I, Turdibekov K, Abdullayev R (2020) Assessment and prediction of the quality of the contact network under high-speed motion of electric transport. J Adv Res Dyn Control Syst 12(2):160–165. https://doi.org/10.5373/JARDCS/V12SP2/SP20201056
5. Bayanov I, Badretdino, T, Muminov S, Karimov I, Saydivaliev S, Saliyev E (2021) The electric field of the sliding contact during the interaction of the pantograph and the contact wire. E3S web of conferences, vol 264, p 04029. https://doi.org/10.1051/e3sconf/202126404029
6. Rauter F, Pombo J, Ambrosio J (2007) Contact model for the pantograph-catenary interaction. J Syst Des Dyn 1(3):447–457. https://doi.org/10.1299/jsdd.1.447
7. Liu Z, Song Y, Han Y, Wang H, Zhang J, Han Z (2018) Advances of research on high-speed railway catenary. J Mod Transp 26(1):1–23. https://doi.org/10.1007/s40534-017-0148-4
8. Wu G, Wei W, Gao G et al (2016) Evolution of the electrical contact of dynamic pantograph–catenary system. J Mod Transp 24(2):132–138. https://doi.org/10.1007/s40534-016-0099-1
9. Massat JP, Laine JP, Bobillot A (2006) Pantograph–catenary dynamics simulation. Int J Veh Mech Mob 44:551–559. https://doi.org/10.1080/00423110600875443
10. Midya S, Bormann D, Larsson A, Schutte T, Thottappillil R (2008) Understanding pantograph arcing in electrified railways—influence of various parameters. In: 2008 IEEE international symposium on electromagnetic compatibility, vol 1–3, pp592–597. https://doi.org/10.1109/ISEMC.2008.4652112
11. Collina A, Bruni S (2002) Numerical simulation of pantograph-overhead equipment interaction. Veh Syst Dyn 38(4):261–291. https://doi.org/10.1076/vesd.38.4.261.8286
12. Chater E, Ghani D, Giri F, Haloua M (2015) Output feedback control of pantograph–catenary system with adaptive estimation of catenary parameters. J Mod Transp 23(4):252–261. https://doi.org/10.1007/s40534-015-0085-z

13. Pastucha E (2016) Catenary system detection, localization and classification using mobile scanning data. Remote Sens 8(10):801. https://doi.org/10.3390/rs8100801
14. Geng Y, Pan F, Jia L, Wang Z, Qin Y, Tong L, Li S (2021) UAV-LiDAR-based measuring framework for height and stagger of high-speed railway contact wire. IEEE Trans Intell Transp Syst. https://doi.org/10.1109/TITS.2021.3071445

Determination of Thermal Conductivity Coefficient in the Turbulent Boundary Layers

R. S. Selim

Abstract The heat transfer process in a developed turbulent boundary layer is addressed. The fluid flow of study is assumed to be incompressible flowing over a flat plate with turbulent mean velocity profile which is based on the eddy viscosity concept. Theoretically, based on a waveguide model of turbulent boundary layer, the expression of thermal conductivity coefficient is formulated. This formula is dependent on the behavior of least damping modes eigenvalues, and eigenfunctions, that are deduced from solution Orr-Sommerfeld equation by spectral Chebyshev collocation Method. Numerical code, based on a Chebyshev spectral collocation method is used to solve the Orr-Sommerfeld eigenvalue problem, at fixed values of the wave number, and Reynolds number. Spectral characteristics of the dispersion relation are contributed to study the three wave resonance of the least damping mode Tollmien-Schlichting. As a result, a system of differential equations for multiple 3-wave resonance is considered to define the coherent structure. The weight factors values are deduced to capture the condition of system solutions. Theory of Birkhoff-Khinchinis applied to obtain the time average of the square modulus amplitudes of the main harmonic, and sub harmonics.

Keywords Coefficient of thermal conductivity · Incompressible fluid flow · Orr-Sommerfeld eigenfunctions

R. S. Selim (✉)
Moscow Institute of Physics and Technology (MIPT), 9 Institutsky per., Dolgoprudny, Moscow reg. 141701, Russia
e-mail: selim.rs@phystech.edu

R. S. Selim
Mathematics Department, Faculty of Science, Tanta University, Tanta 31527, Egypt

A. Bieliatynskyi and V. Breskich (eds.), *Safety in Aviation and Space Technologies*, Lecture Notes in Mechanical Engineering,
https://doi.org/10.1007/978-3-030-85057-9_16

1 Introduction

Last investigations in works [1–3] introduce the theory of a turbulent boundary layer as well as heat transfer in a developed turbulent boundary layer. The missing knowledge, in addition to the conservation rules, is taken from the experiment in these monographs, which, from a practical standpoint, provides detailed information about momentum, energy, and heat transfer processes. Based on the qualitative study of the physics of the processes discussed in monographs [1, 4], which allowed to determine the scales of quantities [5]. Next, we look at the implications of the developed turbulent boundary layer's waveguide model [6, 7], in which asymptotic features emerge naturally from physical representations of the existence of the pulsations of the developed turbulent boundary layer. Consider the problem of heat transfer in a boundary layer on a plate that is flowed by an incompressible fluid as an example. In this paper, we consider a model composed of developed turbulent flow of an incompressible fluid over a flat plate in a boundary layer. The objective of the present work is to formulate the equation of temperature fluctuation in an approximation form. Moreover, give an expression for the turbulent thermal conductivity coefficient in the discrete representation of a coherent structure. The plan of this work is as follows: In the next section, we will give a description of the statement problem. In the third section, the expression of the turbulent thermal conductivity coefficient is formulated to describe its behavior. Finally, the salient results of our analysis are discussed in the conclusion.

2 Statement of the Problem

We consider a model of fluid flow over a flat plate in the turbulent boundary layer. The fluid is assumed to be incompressible with zero pressure gradients. Here, the Reynolds number is large and the Mach number is small, the density ρ can be assumed to be constant in this case. The Prandtl molecular is of order unity. The governing equation that describe the heat transfer of the model [1]:

$$\frac{\partial T}{\partial t} + \mathbf{U} \cdot \nabla T = \frac{\lambda}{\rho c_p} \nabla^2 T + \frac{\mu}{\rho c_p} \Phi, \quad \nabla \cdot \mathbf{U} = 0 \tag{1}$$

Here Φ is the dissipative function, and λ denotes the coefficient of molecular thermal conductivity [1]. Using the traditional Reynolds decomposition of the velocity, and temperature

$$T = T_0 + T', \quad \mathbf{U} = \mathbf{U}_0 + \mathbf{u}' \tag{2}$$

After the substitution of the above decomposition into the heat transfer, and continuity equations and further manipulations, we arrive to the following equation:

$$\frac{\partial(T_0+T')}{\partial t}+(\mathbf{U}_0+\mathbf{u}')\cdot\nabla(T_0+T')=\frac{\lambda}{\rho c_p}\nabla^2(T_0+T')+\frac{\mu}{\rho c_p}(\Phi_0+\Phi') \tag{3}$$

We introduce the averaging operation [6],

$$\overline{a(t_0,t_1)}=\left\langle \lim_{t_1\to\infty}\frac{1}{t_1}\int_0^{t_1}\lim_{t_0\to\infty}\frac{1}{t_0}\int_0^{t_0}a(t_0,t_1)dt_0dt_1\right\rangle \tag{4}$$

Here $\langle\ldots\rangle$, is ensemble average, and L is the characteristic longitudinal scale determined by the decrement of the least damped Tollmien-Schlichting wave on the profile of the developed turbulent boundary layer, and δ^{**} is the momentum thickness.

$$t_0=\delta^{**}/U_\infty,\ t_1=\delta^{**}/\varepsilon U_\infty,\ \varepsilon^2=L/\delta^{**} \tag{5}$$

Then the equation for the average values will take the form

$$\begin{gathered}\frac{\partial T_0}{\partial t}+\mathbf{U}_0\cdot\nabla T_0+\overline{\mathbf{u}'\cdot\nabla T'}=\frac{\lambda}{\rho c_p}\nabla^2 T_0+\frac{\mu}{\rho c_p}\Phi_0,\ \overline{\mathbf{u}'\cdot\nabla T'}=\nabla\cdot\overline{\mathbf{u}'T'},\\ \nabla\cdot\mathbf{U}_0=0,\ \mathbf{U}_0=\{U_0,V_0\},\ \nabla\cdot\mathbf{u}'=0,\ \mathbf{u}'=(u',v',w')\end{gathered} \tag{6}$$

Subtracting the averages from the original equation gives the expression for temperature fluctuations

$$\frac{\partial T'}{\partial t}+\mathbf{U}_0\cdot\nabla T'+\mathbf{u}'\cdot\nabla T_0+\mathbf{u}'\cdot\nabla T'-\overline{\mathbf{u}'\cdot\nabla T'}=\frac{\nu}{Pr^2\frac{\nu}{c_p}(\Phi-\Phi_0)}$$

$$\left(\frac{\rho c_p\nu}{\lambda}\right)=\mathrm{Pr},\quad \nabla=\left(\frac{\partial}{\partial x},\frac{\partial}{\partial y}\right) \tag{7}$$

The governing Eqs. (6 and 7), are subject to boundary conditions, at the boundaries, we assume that zero values of temperature fluctuations. Consider the case in which, the average flow is two-dimensional, while the fluctuations are three-dimensional.

In this case, we get

$$\frac{\partial T_0}{\partial t}+\mathbf{U}_0\cdot\nabla T_0+\overline{u'\cdot\frac{\partial}{\partial x}T'}+\overline{v'\cdot\frac{\partial}{\partial y}T'}+\overline{w'\cdot\frac{\partial}{\partial z}T'}=\frac{\nu}{Pr_0^2\frac{\nu}{c_{p0}}} \tag{8}$$

$$\frac{\partial T'}{\partial t} + \mathbf{U}_0 \cdot \nabla T' + u' \cdot \frac{\partial}{\partial x} T_0 + v' \cdot \frac{\partial}{\partial y} T_0 + \mathbf{u}' \cdot \nabla T' - \overline{\mathbf{u}' \cdot \nabla T'} = \frac{\nu}{Pr^2 \frac{\nu}{c_p} (\Phi - \Phi_0)} \tag{9}$$

$$\Phi' = \Phi - \Phi_0 = \Phi = 4 \frac{\partial U_0}{\partial x} \frac{\partial u}{\partial x} + 4 \frac{\partial V_0}{\partial y} \frac{\partial v}{\partial y} + 2 \left(\frac{\partial v}{\partial z} + \frac{\partial u}{\partial y} \right) \frac{\partial U_0}{\partial y}.$$

2.1 Dimensionless of the Problem

It is possible, and sometimes preferable, to write governing equations in non dimensional form. To do this, we need to select characteristic quantities that describe the flow problem, such as a characteristic scales.

Length L, characteristic velocity, and temperature are denoted by U_∞, T_∞ respectively. Now we need to enter dimensionless quantities into the Eqs. (8 and 9), for the averaged, and pulsation fields respectively.

For the characteristic scales of the values for the mean field:

$$x - L,\ y - \delta^{**},\ U_0 - U_\infty,\ V_0 - \frac{\delta^{**}}{L} U_\infty,\ t - \frac{L}{U_\infty},\ Re_{\delta^{**}} = \frac{U_\infty \delta^{**}}{\nu}$$

For the fluctuation values, the derivatives are

$$(x, y, z) - \delta^{**},\ (u', v\prime, w') = \sqrt{\delta^{**}/L}(u, v, w),\ T' = \sqrt{\delta^{**}/L}\, T_\infty,\ \varepsilon^2 = \delta^{**}/L.$$

Using the last scales, we have

$$\begin{aligned} \frac{\partial T_0}{\partial t} + \boldsymbol{U}_0 \cdot \nabla T_0 &= -\nabla \cdot \overline{\{u', v', w'\} T'} + \frac{1}{Pr \varepsilon^2 Re_{\delta^{**}}} \frac{\partial^2}{\partial y^2} T_0, \\ \nabla \cdot \overline{\{u', v', w'\} \cdot T'} &= \frac{\partial}{\partial x} \overline{u'T'} + \frac{\partial}{\partial y} \overline{v'T'} + \frac{\partial}{\partial z} \overline{w'T'} \end{aligned} \tag{10}$$

Here, The Prandtl molecular number is often believed to be on the order of one. We can deduce from the estimates of the characteristic values (assuming the Mach number is small) that dissipation can be ignored in relation to thermal conductivity terms, leaving only the derivative with respect to y in the heat flow. In this case, the problem for average temperatures becomes analogous to the problem of average velocity, where the terms of thermal conductivity are given by the value $1/Pr\varepsilon^2 Re_{\delta^{**}}$, with respect to which, in the statement about the developed turbulent boundary layer, it is assumed that this value is small together with the value $1/Re_{\delta^{**}}$, i.e. $\mathrm{Re}_{\delta^{**}} \gg 1$, $\varepsilon^2 \mathrm{Re}_{\delta^{**}} \gg 1$.

Let us now consider the estimates of the terms of the equation for temperature pulsation (9). If we follow the above estimates, then in the equation, first of all, we must leave the terms

$$\frac{\partial T'}{\partial t} + U_0 \cdot \frac{\partial}{\partial x} T' + v' \cdot \frac{\partial}{\partial y} T_0 = \frac{1}{Pr\varepsilon^2 Re_{\delta^{**}}} \nabla^2 T' \tag{11}$$

In this form, this equation resembles the Squire equation in the theory of hydrodynamic stability. It also has Tollmien-Schlichting mod-induced members. In Eq. (11), the induced term creates a turbulent heat transfer with a certain turbulent thermal conductivity coefficient. There are no derivatives of temperature in z, since it is assumed that the average values are homogeneous in this direction.

3 Formulation of Thermal Conductivity Coefficient

Following the spectral representation of the fluctuations [5], then

$$T'(t, \boldsymbol{r}) = \frac{1}{(2\pi)^3} \int\int\int T'_{\omega \boldsymbol{k}} e^{i\omega t_0 - i\boldsymbol{k}\cdot \boldsymbol{r}_0} d\omega\, d\alpha\, d\beta,$$

$$v(t, \boldsymbol{r}) = \frac{1}{(2\pi)^3} \int\int\int v_{\omega \boldsymbol{k}} e^{i\omega t_0 - i\boldsymbol{k}\cdot \boldsymbol{r}_0} d\omega\, d\alpha\, d\beta$$

$$= \frac{1}{(2\pi)^3} \int\int\int A_{\omega \boldsymbol{k}} \delta(\omega - \omega_{T-S}(\boldsymbol{k})) e^{i\omega t_0 - i\boldsymbol{k}\cdot \boldsymbol{r}_0} d\omega\, d\alpha\, d\beta,$$

$$T'_{\omega \boldsymbol{k}} = \iiint T'(t, \boldsymbol{r}) e^{i\omega t_0 - i\boldsymbol{k}\cdot \boldsymbol{r}_0} dt\, dx\, dz, \ \boldsymbol{r} = (x, z), \ \boldsymbol{k} = (\alpha, \beta), \ i^2 = -1$$

Here $\omega_{T-S}(\boldsymbol{k})$ is the frequency of the Tollmien-Schlichting wave with the wave vector $\boldsymbol{k}$.

The equation for temperature fluctuation T' is formulated by using Fourier transform, then we have

$$(-i\omega + i\alpha U_0)\, T'_{\omega k} + \left(v' \cdot \frac{\partial T_0}{\partial y} \right)_{\omega k} = \frac{1}{Pr\varepsilon^2 Re_{\delta^{**}}} \left(\frac{\partial^2}{\partial y^2} - k^2 \right) T'_{\omega k} \tag{12}$$

In the first approximation, assuming that there are no resonances with temperature modes, we obtain

$$T'_{\omega k} \cong - \frac{v_{\omega k} \frac{\partial T_0}{\partial y}}{(-i\omega + i\alpha U_0)} \tag{13}$$

According to Eqs. (12 and 13), we calculate the term of the heat equation responsible for turbulent heat transfer.

$$\overline{-v'.\partial T'/\partial y} = \nabla.\frac{1}{(2\pi)^6}\iiint\iiint \overline{\Delta_{\omega_1 k_1}.\Delta_{\omega_2 k_2}}\frac{\partial T_0}{\partial y}d\omega_1 d\boldsymbol{k}_1 d\omega_2 d\boldsymbol{k}_2$$

$$\Delta_{\omega_1 k_1} = \{0, v, 0\}_{\omega_1 k_1},\ \Delta_{\omega_2 k_2} = \frac{v_{\omega_2 k_2} e^{-i(\omega_1+\omega_2)t - i(\boldsymbol{k}_1+\boldsymbol{k}_2).\boldsymbol{r}}}{(-i\omega_2 + i\alpha_2 U_0)}$$

$$\lambda_t(y) = \frac{\varepsilon^2\phi^2}{(2\pi)^6}\left(h_1(\boldsymbol{k}_1, y) + 2\sum_{s=1}^{s=5}\sum_{l=2}^{l=3} h_2\left(\boldsymbol{k}_l^s, q_{1,l}^s, y\right)\right),$$

$$\langle\ \rangle_{t_1} = \frac{1}{t_1}\int_0^{t_1} dt_1,\ q_{1,l}^s = q_{1l}^s \qquad (14)$$

$$h_1(\boldsymbol{k}_1, y) = \frac{\omega_i(\boldsymbol{k}_1)\left\langle|\tilde{a}_1|^2\right\rangle_{t_1}\alpha_1|\varphi_0(\boldsymbol{k}_1)|^2}{\omega_i^2(\boldsymbol{k}_1) + (\omega_r(\boldsymbol{k}_1) - \alpha_1 U(y))^2}$$

$$h_2\left(\boldsymbol{k}_l^s, q_{1,l}^s, y\right) = \frac{\omega_i\left(\boldsymbol{k}_l^s\right)\left\langle|\tilde{a}_l^s|^2\right\rangle_{t_1}\alpha_l^s|\varphi_0\left(\boldsymbol{k}_l^s\right)|^2}{q_{1,l}^s\left(\omega_i^2\left(\boldsymbol{k}_l^s\right) + \left(\omega_r\left(\boldsymbol{k}_l^s\right) - \alpha_l^s U(y)\right)^2\right)}$$

where $v_{\omega_1 k_1}, v_{\omega_2 k_2}$ are defined by eigenfunctions of Orr-Sommerfeld equation for the least damping mode which are corresponding to two wave vectors $\boldsymbol{k}_1$, $\boldsymbol{k}_2$ respectively. Since the near-wall region can generally be subdivided into viscous sub layer, buffer layer and fully turbulent layer for turbulent flow over a flat plate, as displayed in Fig. 1 [7]. Therefore, the meshed elements need to be carefully arranged to capture the velocity and turbulence in the different layers.

As outlined in formula (14), the left side represent the coefficient of turbulent thermal conductivity. The terms on the right hand side depend on parameter values

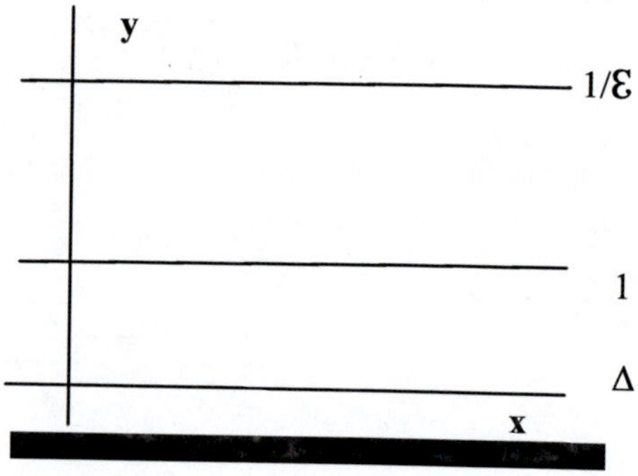

Fig. 1 Structure of the flow field in the inner, intermediate, and outer regions by asymptotic expansion [7]

$(\varepsilon, \phi, q_{12}^s, q_{13}^s)$, time averages $\left\langle |\widetilde{a_l^s}|^2 \right\rangle_{t_1}$ for the main harmonic, and sub harmonics. In this study, the obtained results are formulated on the basis of finite model, which consist of the main harmonic, and five sub harmonics, as a dynamical system for describing the coherent structure [6], in the case of two-dimensional incompressible fluid flow, with zero pressure gradient, flowing over a flat plate.

$$\frac{da_1}{dt_1} = \sum_{s=1}^{n} \Lambda_1^s a_2^s a_3^s, \; \frac{da_2^s}{dt_1} = \Lambda_2^s a_1 a_3^{s*}, \; \frac{da_3^s}{dt_1} = \Lambda_3^s a_1 a_2^{s*}, \; s = 1, 2, \cdots n \tag{15}$$

The boundedness of the solution of the system (15) of differential equations will then follow from the positivity of the quantities q_2^s, q_3^s which should be checked from determination the coefficients $(\Lambda_1^s, \Lambda_2^s, \Lambda_3^s)$. The solution of system (15) is formulated with statistical initial conditions, which are distributed on the surface 6-dimensional ellipsoid. But a change of variables [8], in the model version, by

$$a_i(t_1, t_2) = \phi(t_2)\tilde{a}_i(\phi(t_2)\tilde{t_1}, t_2), \phi(t_2) = \sqrt{\frac{\lambda}{\Phi}}$$

Those are introduced here to capture solutions of the dynamical system (15), on the unit sphere, as formulated below in Eq. (16)

$$\begin{gathered} I^{(n)}(t_1) = |\tilde{a_1}|^2 + \sum_s q_{12}^s \left|\widetilde{a_2^s}\right|^2 + q_{13}^s \left|\widetilde{a_3^s}\right|^2 = 1, \\ q_{12}^s = \frac{q_2^s}{q_1}, q_{13}^s = \frac{q_3^s}{q_1}, s = 1, 2, \cdots n \end{gathered} \tag{16}$$

Based on the above explanation, and applying the theory of Birkhoff-Khinchin [9], the time averages of amplitude solutions of the dynamical system (15) are deduced.

In addition, the mean velocity profile Mucker $U(y)$ [10], eigenvalues, and eigenfunctions for the least damping mode of different wave vectors $\boldsymbol{k}_l^s = (\alpha_l^s, \beta_l^s)$ are important factors to define the coefficient of turbulent thermal conductivity.

The value of parameter $\varepsilon = \varepsilon(Re_{\delta^{**}}) = 0.07$, that deduced from Reynolds number, $\mathrm{Re}_{\delta^{**}} = 10^4$ is provided in [7], and also $\phi = 0.417133$, in a stationary case is defined in [8].

The real and imaginary parts of leading mode eigenvalues are outlined by $\omega_r = Re[\omega], \omega_i = Im[\omega]$ respectively. The asterisk, $\varphi_0(\boldsymbol{k})^*$, denotes a complex conjugate of the least damping mode eigenfunctions $\varphi_0(\boldsymbol{k})$. The index sum l is considered for two wave vectors $\boldsymbol{k}_2, \boldsymbol{k}_3$, which are satisfied the triad resonance with the main wave vector. The first part in the right, concerns only the leading mode eigenfunctions, for the main wave harmonic $\boldsymbol{k}_1 = (1, 0)$, that shown in Fig. 2b with index number 1. The second and third terms are represented leading mode eigenfunctions for five sub harmonics amplitudes, which are corresponding to five wave vectors $\boldsymbol{k}_2^s, \boldsymbol{k}_3^s$

respectively. In addition, the first, and the second terms are multiplied by 2, due to the symmetry of the wave vector component β. The wave vector, $\boldsymbol{k}_2^s, s = 1, \ldots, 5$ are indicated in Fig. 2b by index numbers, where the power is referred to the number of sub harmonic $(2^1, 2^2, 2^3, 2^4, 2^5)$, which are corresponding to five multipliers values $(q_{12}^1, q_{12}^2, q_{12}^3, q_{12}^4, q_{12}^5)$, respectively as outlined below in Table 1, and the similar situation is also defined for $\boldsymbol{k}_3^s, s = 1, \ldots, 5$, and their multipliers $(q_{13}^1, q_{13}^2, q_{13}^3, q_{13}^4, q_{13}^5)$. Noting that the sum extends to all sub harmonics in Eq. (14), and since the imaginary part of the eigenvalues are negative [6], then the value of the turbulent thermal conductivity coefficient is positive. Figure 3a describes properties eigenfunctions of the main wave vector $\boldsymbol{k}_1$ and some different combination of the wave vectors $\boldsymbol{k}_2^s, \boldsymbol{k}_3^s$ which are defined from solving the Orr-Sommerfeld equation by the spectral Chebyshev Collocation method, at $Re_{\delta^{**}} = 10^4$.

It is clear that $\lambda_t(y) = O(y^4), y \to 0$, as shown in Fig. 3b, the dotted curve is represented by $\zeta_1 y^4$, where $\zeta_1 = 1.585 \times 10^6$, while the dotted-dashed curve $\zeta_2 + \zeta_3 y$, $\zeta_2 = -6.36$, $\zeta_3 = 251.0769$. This behavior is due to the fact that the

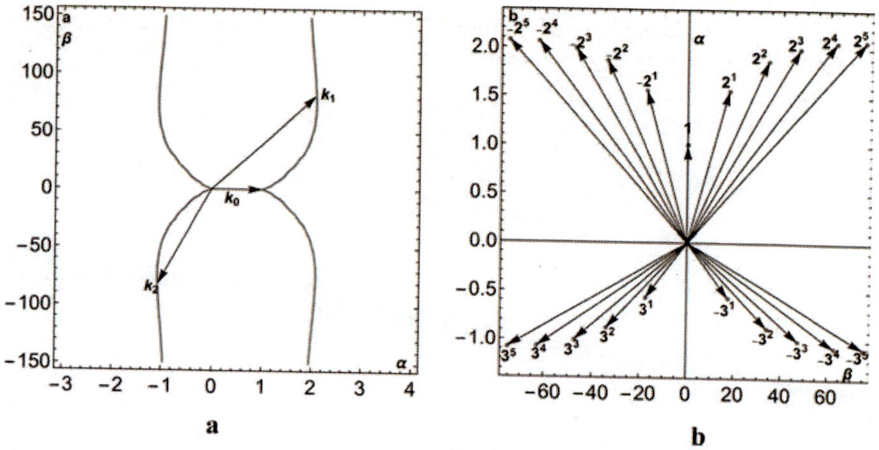

Fig. 2 **a** Curve 3—wave resonance, for the last damping mode Tollmien-Schlichting, *at* $k_0 = (1, 0), Re_{\delta^{**}} = 10^4$. **b** Structure of wave vectors which are located on three wave resonance curve, and formulated in Eq. (14), [6]

Table 1 Multiplier values, for five sub harmonics, which are corresponding to five wave vectors [6]

$\boldsymbol{k}_l^s = (\alpha_l^s, \beta_l^s)$	q_{12}^s	q_{13}^s
$s = 1,\ \alpha_1 = 1.566,\ \beta_1 = 17.38$	0.166531	0.122806
$s = 2,\ \alpha_2 = 1.874,\ \beta_2 = 33.99$	0.318746	0.29306
$s = 3,\ \alpha_3 = 1.988,\ \beta_3 = 47.2$	0.226102	0.180738
$s = 4,\ \alpha_4 = 2.068,\ \beta_4 = 62.68$	0.116389	0.0526377
$s = 5,\ \alpha_5 = 2.077,\ \beta_5 = 75.13$	0.0609696	0.0376014

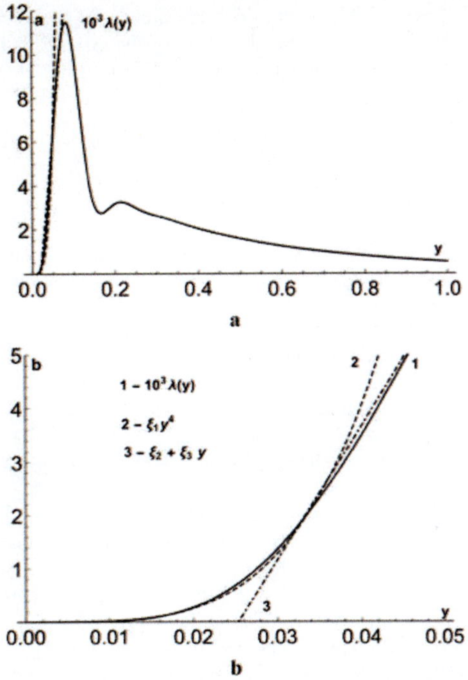

Fig. 3 **a** The behavior of the turbulent thermal conductivity coefficient $\lambda_t(y)$ *when the main wave vector* $\boldsymbol{k}_1 = (1,0), Re_{\delta^{**}} = 10^4$. **b** The behavior of the turbulent thermal conductivity coefficient $\lambda_t(y)$ *near zero, 1-solid line:* $10^3\lambda_t(y)$, *2-dotted:* $\zeta_1 y^4$, *and dotted-dashed:* $\zeta_2 + \zeta_3 y$

eigenfunctions $\varphi_0(\boldsymbol{k},\boldsymbol{y})$ of the Orr-Sommerfeld equation near the wall have asymptotic $\varphi_0(\boldsymbol{k},\boldsymbol{y}) = O(y^2)$, which coincides with more general conclusions based on the theory of dimension and similarity [4].

4 Conclusions

In this paper, we determine the expression of turbulent thermal conductivity in terms of the waveguide model in the turbulent boundary layer, for given mean velocity profile Mucker. The behavior of coefficient thermal conductivity is mainly dependent on the properties of eigenvalues, and eigenfunctions Orr-Sommerfeld equation, that obtained by using the spectral Chebyshev collocation method. Generation of eigenfunctions, for different combination of the wave vectors $\mathbf{k}_2^s, \mathbf{k}_3^s$ in the formula of thermal conductivity coefficient are related to the structure of the main harmonic, and a set of sub harmonic amplitudes, in the case of multiple three wave resonance. It is found that, the behavior of thermal conductivity coefficient in the turbulent boundary layer that formulated in Fig. 3a, b, has approximate form close to $y = 0$ are noticed in forms of $\zeta_1 y^4$, $\zeta_2 + \zeta_3 y$, due to the behavior of eigenfunction Orr-Sommerfeld equation near the wall.

Acknowledgements The researcher R. S. Selim is funded by a partial scholarship from the Ministry of Higher Education of the Arab Republic of Egypt, and by scholarship from Russian government scholarship with application no. EGY – 6177/17.

References

1. Schlichting G (1974) Theory of the boundary layer. Nauka, Moscow
2. Kutateladze SS (1979) Fundamentals of the theory of heat transfer. Atomizdat, Moscow
3. Kutateladze SS, Leontiev AI (1985) Heat and mass transfer and friction in a turbulent boundary layer. Energoatomizdat, Moscow
4. Landau LD (1986) Theoretical Physics, vol. 6. Hydrodynamics. Nauka, Moscow
5. Dixit SA, Gupta A, Choudhary H, Singh A, Prabhakaran T (2020) Asymptotic scaling of drag in flat-plate turbulent boundary layers. Phys Fluids 32:041702
6. Zharov VA, Lipatov II, Selim RS (2020) Wave model of organized structures in a turbulent boundary layer on a plate with zero longitudinal pressure gradient. TsAGI Sci J 51(6):631–640
7. Zharov VA, Van Ha L (2016) Construction of a longitudinal velocity profile based on a given tangential stress in a developed turbulent boundary layer on a plate. Scientific notes of TsAGI 8, XLVII, pp 50–60
8. Zharov VA (2014) Model representation of a coherent structure in a developed turbulent boundary layer, Scientific Notes of TsAGI, XLV 5, pp 33–46
9. Kolmogorov AN (1985) Selected works of mathematics and mechanics. Nauka, Moscow
10. Musker AJ (1979) Explicit expression for the smooth wall velocity distribution in turbulent boundary layer. AIAA J 17(6):655–657

Pressure Amplification Coefficient of the "Nozzle-Flap" Transducer

Vitaly Golubovsky, Vladimir Konovalov, and Marina Dontsova

Abstract The principle of operation of automatic control system of hydraulic actuator based on hydraulic measuring transducer of angular velocity of a "nozzle-flap" type is considered. The aim of the research was mathematical modelling of the influence of geometrical parameters of the measuring converter on its pressure intensification coefficient in the MathCAD program. The research methodology included a theoretical substantiation of the proposed expressions describing the transducer, as well as an analysis of numerical results obtained by numerical simulation in the MathCAD program. Numerical researches of the known pressure intensification coefficient in the interval of diameter changes of nozzle and constant choke of 0.5…1.5 mm at a pressure in the system of 0.5…2.0 MPa showed the presence of extreme character of pressure intensification coefficient changing depending on a flap stroke from the nozzle end. An adequate linear dependence for the pressure intensification coefficient was observed in the section of the flap stroke up to (0.5 h). The highest value of the coefficient of amplification corresponded to a flap stroke of (1 h). An active growth of the area under the pressure intensification coefficient graph occured up to (1.7 h). It should be kept in mind, that when the flap stroke increases more than 0.25 of nozzle orifice diameter, the annular gap no longer functions as an adjustable hydraulic resistance. Thus, for nozzle and constant choke orifice diameters (0.0011; 0.0010 m) the recommended initial clearance is 0.5 h or, based on the purpose of the measuring transducer, can be increased to a value of 1.7 h/2.

Keywords Angular velocity measuring transducer of the "Flap-Nozzle" type · Pressure gain · Mathematical modeling

---

V. Golubovsky · V. Konovalov (✉) · M. Dontsova
Penza State Technological University, 1A/11, Pr. Baydukova/Gagarina St., Penza, Russia
e-mail: konovalov-penza@rambler.ru

A. Bieliatynskyi and V. Breskich (eds.), *Safety in Aviation and Space Technologies*, Lecture Notes in Mechanical Engineering,
https://doi.org/10.1007/978-3-030-85057-9_17

1 Introduction

The development of industrial facilities leads to the emergence of new processes which require automation [1–3], or which cannot be operated without the use of automatic control [4–6]. Creation of fast accurate and highly sensitive automatic control systems requires a more careful approach to the design of measuring transducers. The design and calculation of transducers, which are a complex of different elements, is an important task in control system design [7–9], since the properties of the primary transducers affect the quality of the control process [10–12]. For this reason, the calculation of basic design and hydraulic parameters of measuring transducers is of great importance [13, 14].

These systems are often based on hydraulic nozzle-flap type amplifiers in the first amplification stage and a spool-type throttle valve in the second amplification stage. The design and principle of operation of the angular acceleration transducer are discussed in [15, 16].

Figure 1 shows a schematic diagram of the system of automatic control of angular speed of the working body of the machine on the basis of the nozzle-flap type transducer.

The angular velocity transducer is attached to the shaft of the working body of the machine, which is rotated by the hydraulic motor. When rotating uniformly, the flap is in equilibrium, occupying a symmetrical position between the nozzles, resulting in equal pressures P_1 and P_2 and no output signal.

When the speed of rotation changes, the resulting acceleration shifts the flap from its neutral position, resulting in a pressure difference $\Delta p_{mt} = P_1 - P_2$ in the working chambers, which is used to operate the actuator (spool) of the automatic control system regulator included in the diagonal of the bridge. The actuator is a spool throttle valve that controls the fluid inlet and outlet flow (lines A and B) of the hydraulic motor of the machine drive, while the line T is connected to the drain. Due to the pressure difference $\Delta p_{mt} = P_1 - P_2$ in the lines under the spool ends, it shifts, changing the flow-through sections of the respective slots, thus generating a signal that is used to automatically control the drive of the industrial machinery.

Stabilization of rotation speed is achieved by generation of control action by automatic regulator of the system, which is aimed at early transition from accelerated or decelerated motion of the working body to steady motion—at a constant speed.

The aim of the research is to obtain expressions describing the influence of geometrical parameters of the nozzle-flap type transducer on its pressure amplification coefficient, and on their basis to conduct a numerical study to determine the effect of the parameters of the orifices of the constant choke and nozzle on the pressure amplification coefficient, with the determination of rational design parameters.

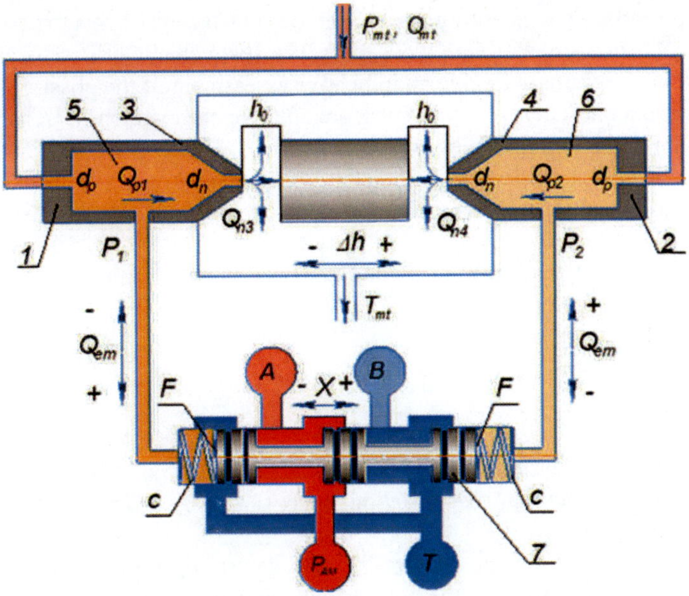

Fig. 1 The schematic diagram of the automatic control system: 1, 2—constant chokes; 3, 4—nozzles; 5, 6—working chambers; 7—actuator; P_{mt}—the supply pressure of the measuring transducer, MPa; Q_{mt}—the flow rate of the power fluid through the measuring transducer, m^3/s; Q_{p1}, Q_{p2}—flow rates through constant chokes, m3/s; Q_{n3}, Q_{n4}—flow rates through the nozzles, m^3/s; T_{mt}—the power-fluid drain line; P_1, P_2—the pressure in the working chambers, MPa; Q_{em}—the flow rate determining the spool travel speed, m^3/s; P_{AM}—the actuator supply pressure, MPa; A, B—the control lines of industrial equipment drive; T—the power fluid drain line; d_p—the diameter of the constant choke orifice, m; d_n—the diameter of the nozzle orifice, m; h_0—the initial clearance between the nozzle and the flap, m; Δh—the displacement of the flap from the middle position, m; X—the spool displacement, m; F—the area of the spool plunger ends, m^2; c—the total stiffness of the centring springs, N/m

2 Research Methods

The method of researches provides a theoretical substantiation of the offered expressions describing geometrical and tuning parameters of the measuring transducer, the analysis of the numerical results received by numerical modelling in the MathCAD program.

The development of method of calculation of geometrical and adjusting parameters influence on the measuring transducer amplification coefficient is caused by necessity of reception of analytical dependences, allowing to determine parameters and characteristics of transducers and their elements at a design stage. The calculations help to determine the operating ranges of the parameters or to establish their optimum values when using transducers in automatic control systems. Such parameters can include: characteristics of elements included in transducers (the diameter and length of a constant choke orifice, the diameter and length of a nozzle orifice) and setting parameter of transducers themselves (the initial clearance between a nozzle end and a flap). The static characteristics of the angular velocity transducer allow to give an indicative assessment of the possibility of using this device in a particular scheme. For a quantitative comparison of transducers, the pressure amplification coefficient is generally used.

The pressure amplification coefficient of the measuring transducer is the derivative of the pressure difference in the inter-choke chambers over the movement of the spool valve.

From the spool's equilibrium condition, its displacement from the neutral position is found (see Fig. 1):

$$X = \frac{F}{c}\Delta p_{mt} - \frac{N}{c}, \tag{1}$$

where F—the area of spool plug ends; c—the total stiffness of centring springs; N—the total force including dry friction force, obliteration force and axial hydrodynamic force.

Thus, the offset of the spool valve from its neutral position is a function of the differential pressure Δp_{mt} in the inter-choke chambers. The value Δp_{mt} in turn depends on the value of the damper displacement. In terms of speed and sensitivity, the best measuring transducer parameters are those that give the greatest value of differential pressure Δp_{mt} for the smallest possible control signal (gate offset).

It is advisable to have amplifiers with a pressure amplification coefficient $(K_p)_{\Delta h=0} = \left(\frac{d\Delta p_{mt}}{dh}\right)_{\Delta h=0}$ that maintains its maximum value over the largest possible area of change Δh.

The pressure difference in the inter-choke chambers is described by the expression:

$$\Delta p_{mt} = P_1 - P_2 = \left[\frac{\sigma_2^2}{\sigma_2^2 + a_4^2(h_0 - \Delta h)^2} - \frac{\sigma_1^2}{\sigma_1^2 + a_3^2(h_0 + \Delta h)^2}\right] P_{mt} \tag{2}$$

where $\sigma_{1,2}$—the hydraulic conductivity of constant bore chokes; $a_{3,4}$—the relative conductivity of variable bore chokes (nozzles); h_0—the initial clearance between the nozzles and the flap; Δh—the movement of the flap; P_{mt}—the power supply pressure of the measuring transducer.

The hydraulic conductivity of constant bore chokes is determined by the formula:

$$\sigma_{1,2} = \mu_p \frac{\pi d_p^2}{4}\sqrt{\frac{2}{\rho}} \tag{3}$$

where μ_n—the coefficient of flow through a constant bore choke; d_p—the diameter of the orifice of the constant bore choke; ρ—power fluid density.

The relative conductivity of variable bore chokes:

$$a_{3,4} = \mu_n \pi d_n \sqrt{\frac{2}{\rho}} \quad (4)$$

where μ_n—the coefficient of flow through nozzles with variable cross-section d_n—nozzle orifice diameter.

To determine the pressure amplification coefficient, take the partial derivative of h from the right-hand side of the Eq. (2). We obtain:

$$\frac{\partial(\Delta p_{mt})}{\partial h} = \left[\frac{2\sigma_2^2 a_4^2 (h_0 - \Delta h)}{(\sigma_2^2 + a_4^2 (h_0 - \Delta h))^2} + \frac{2\sigma_1^2 a_3^2 (h_0 + \Delta h)}{(\sigma_1^2 + a_3^2 (h_0 + \Delta h))^2}\right] P_{mt} \quad (5)$$

Since with $\Delta h = 0$; $\sigma_1 = \sigma_2 = \sigma_p$; $a_3 = a_4 = a_0$, then

$$(K_p)_{\Delta h=0} = \left(\frac{\partial(\Delta p_{mt})}{\partial h}\right)_{\Delta h=0} = \frac{4\sigma_p^2 a_0^2 h_0}{(\sigma_p^2 + a_0^2 h_0^2)^2} P_{mt} \quad (6)$$

and the pressure in the inter-inter-choke chambers will be:

$$P_1 = P_2 = \frac{\sigma_p^2}{\sigma_p^2 + a_0^2 h_0^2} P_{mt} \quad (7)$$

Transform the Eq. (6) by using the expression (7) and introducing the notation $\lambda = \frac{\sigma_p}{a_0}$. The result is an expression for determining the pressure amplification coefficient:

$$(K_p)_{\Delta h=0} = \left(\frac{\partial(\Delta p_{mt})}{\partial h}\right)_{\Delta h=0} = \frac{4P_1}{\lambda P_{mt}} \sqrt{P_1 (P_{mt} - P_{1)}} \quad (8)$$

—for bridge connection of constant chokes and nozzles;

To perform a numerical investigation of the effect of amplifier design parameters on the pressure amplification coefficient, use the expression in the f.6 version:

$$K_{P_{s,i,j}} = 4 \cdot P_s \cdot \frac{\sigma_i^2 \cdot a_i^2 \cdot h_j^2}{\left(\sigma_i^2 + a_i^2 \cdot h_j^2\right)^2} \quad (9)$$

For this purpose add a specified formula for calculation of pressure amplification coefficient to the previously approved models [15, 16] for flow rate and force values of the flow stream. During calculations the following was changed: the diameter of orifice of the choke with constant flow cross-section and the diameter of the nozzle orifice (in the range of values 0.5…1.5 mm). The calculation was carried out at the transducer supply pressure of ($P_0 = 0.5; P_1 = 1.0;\ P_2 = 1.5; P_3 = 2.0$ MPa). The flap stroke h_j (m) varied from the extreme flap position (pressed against the nozzle end, $h_j = 0$) to a certain set point.

3 The Results of the Studies

When examining the effect on the pressure amplification coefficient K_P of the flap stroke h_j, an extreme character of this dependence was found (Fig. 2).

The nature of the variation of the pressure amplification coefficient $K_{P_{i,j},p}$ shows that there is an extremum at some value of the flap stroke h_j (m). At the same time, the value of pressure in the control network influences the numerical values of the specified coefficient. The coordinate of the extremum is independent of the supply pressure. Since the specified pressure amplification coefficient $K_{P_{i,j},p}$ is an indicator of the control system performance, it is therefore necessary to find the flap stroke h_j corresponding to the extremum of the pressure P_p amplification coefficient K_P. Taking into account that to compare the value of the flap stroke and nozzle orifice diameter, we used the flap stroke coefficient K_x ($K_x = d_s/h_{max}$ as a ratio of nozzle orifice diameter to flap stroke h_{max}, corresponding to the maximum of the pressure amplification coefficient K_P), it is possible to determine the effect of nozzle diameter d_s and restrictor diameter d_m on the specified indicators. In this case, for

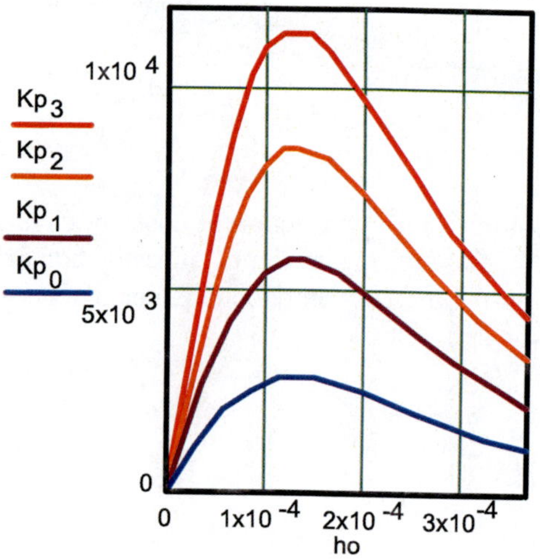

Fig. 2 The influence of the flap stroke hj (m) on the value of pressure amplification coefficient KP for supply pressure indexes: 0—0.5 MPa; 1—1.0 MPa; 2—1.5 MPa; 3—2.0 MPa; at different range of hj for diameters (m) of choke orifices (dm = 0.001) and nozzles (ds = 0.0011)

comparability of the results, the lengths of the choke and nozzle orifices are ten times the value of their diameters.

As a result of numerical simulation of the processes based on a computer programme that implements the theoretical statements of the flow and force characteristics of the system [15, 16], regression equations for the flap stroke coefficient were obtained:

$$K_x = \frac{d_s}{0.2339 \cdot (0.1838 \cdot d_s)^{-1.0227} \cdot 0.3292 \cdot d_s^{1.99}} \tag{10}$$

The correlation coefficient for the estimated values from the regression model of the initial values is R = 0.99994. The result of F-test = 0.998058, which is sufficient to describe the model adequately. Accordingly, we can use the obtained model.

The lowest values of the flap stroke coefficient correspond to the smallest nozzle diameters and the largest choke orifices, therefore the flap stroke coefficient K_x will be lower in this section. And on the contrary, the highest values of the flap stroke coefficient K_x will be with the largest diameter of the nozzle orifice and the smallest orifices of the choke. It is logical to assume that increasing the orifice diameter increases the cross-sectional area of the flow and its force effect. The dependence has a character close to hyperbolic, according to the exponent in the equation. Increasing the diameter of the constant choke decreases the fluid flow rate through the nozzle with a dependence that is close to quadratic (Fig. 3).

The regression model of the flap stroke h (mm) providing extreme (maximum) values of the pressure amplification coefficient K_P:

$$h = 0.65866 \cdot (0.8989 \cdot d_s)^{-0.986} \cdot (0.44456 \cdot d_m)^{1.9957} \tag{11}$$

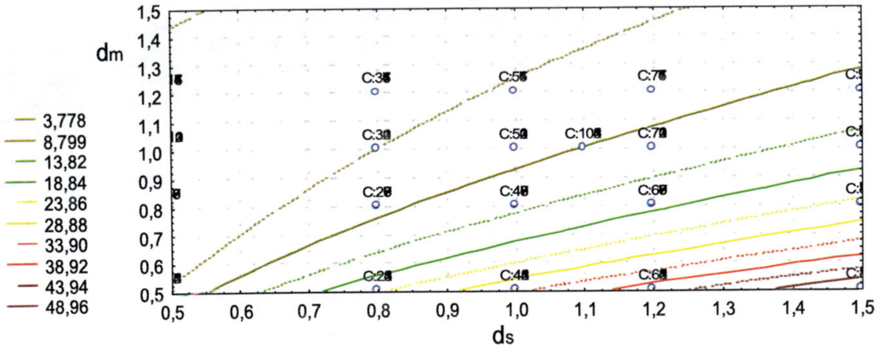

Fig. 3 The influence of the diameters (m) of the choke orifices dm and nozzle orifices ds on the flap stroke coefficient Kx

The correlation coefficient of calculated values on initial values of regression model is R = 0.99991. At the same time, the result of F-test = 0.991012, which indicates that the model adequately describes the initial data.

Taking into account that the flap stroke coefficient Kx is a relative indicator, and the flap stroke itself h is an absolute indicator, and they have an inverse reciprocal relationship, the conclusions for these indicators are generally the same. Increasing the nozzle diameter increases the flap stroke to the extremum of the pressure amplification coefficient KP. At the same time, an increase in the diameter of the nozzle orifice contributes to an increase in the flap stroke, both absolute and relative. Increasing the orifice size of the choke reduces these values.

However, the investigated model is applicable only in the section where the area of the gap between the flap and the nozzle does not exceed the area of the nozzle orifice. The specified zone is located in Fig. 4 the red line ($0.25d_s$). Accordingly, only the indicated zone should be used in the analysis.

The regression model of the pressure amplification coefficient K_P:

$$K_P = 108.9 \cdot P \cdot \left(47.73 \cdot d_s \cdot d_m^2 - 0.0031\right) \quad (12)$$

The correlation coefficient for the estimated values from the initial values of the regression model is R = 0.999999. The result of F-test = 0.999984, which adequately describes the model. The convergence plot of the calculated and initial values (Fig. 5) shows good convergence of the results.

The pressure amplification coefficient K_P (Fig. 6) increases with an increase in applied pressure. Reducing the diameter of the orifice of the choke contributes to an

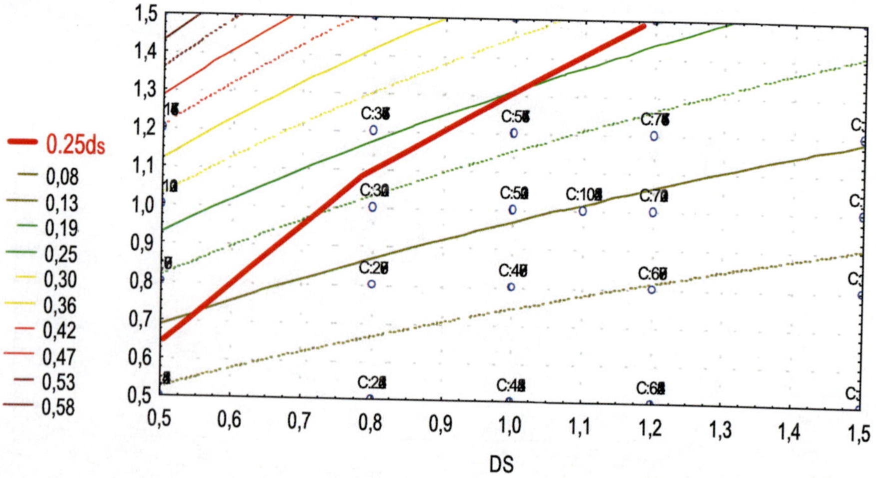

Fig. 4 The influence of the diameters (m) of the choke orifice channel dm and nozzle ds on the flap stroke h (mm), providing the maximum values of the pressure amplification coefficient:—two-dimensional section

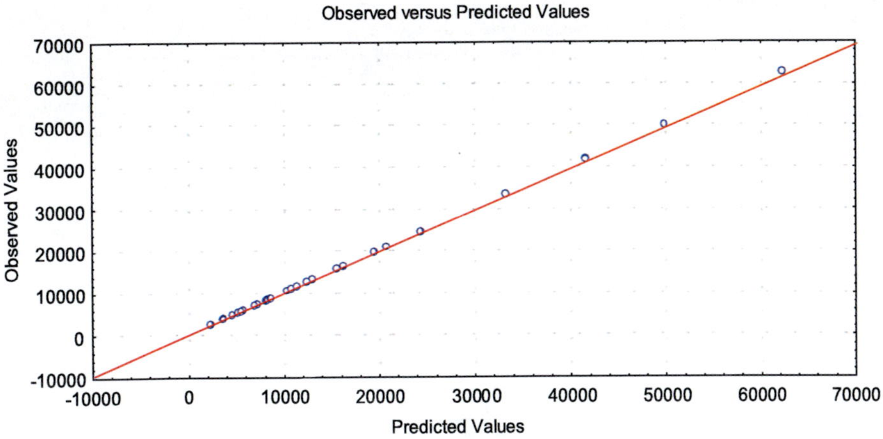

Fig. 5 The diagram of correspondence of the calculated and initial values for the pressure amplification coefficient model

increase in the values of this coefficient, especially with a reduction of the orifice of the choke less than 1 mm. Increasing the nozzle orifice slightly increases the pressure amplification coefficient K_P.

For experimentally tested parameters at the sizes of nozzle and constant choke orifice diameters: ($d_s = 0.0011$ and $d_m = 0.001$) we can find the stroke length when the pressure amplification coefficient K_P is adequately described by a linear model. The resulting graphs are different (Fig. 7, Table 1, f.13), but based on the same data samples, when the flap stroke is reduced (in fractions) from the flap stroke h_{max}, which corresponds to the maximum value of the pressure amplification coefficient.

The regression equations of the linear model of the pressure amplification coefficient as a function of the investigated flap stroke fraction $\mathrm{d_h}$ (Y_1—at $\mathrm{d_h} = 1\mathrm{h}$; Y_{08}—at $\mathrm{d_h} = 0.8\mathrm{h}$; Y_{06}—at $\mathrm{d_h} = 0.6\mathrm{h}$; Y_{05}—at $\mathrm{d_h} = 0.5\mathrm{h}$; Y_{04}—at $\mathrm{d_h} = 0.4\mathrm{h}$):

$$Y_1 = 481.245 + 2859 \cdot h; \; Y_{08} = 288.285 + 3481 \cdot h;$$

$$Y_{06} = 140.795 + 4086 \cdot h; \; Y_{05} = 87.506 + 4362 \cdot h;$$

$$Y_{04} = 48.265 + 4606 \cdot h. \tag{13}$$

Table 1 summarises the results of the Fisher tests (F-test) and Pearson correlation coefficients R. The regression equations become inadequate when the significance level of the error is greater than 10%, i.e. the F-test must be at least 0.9. If the fraction of the flap stroke is less than 50% (inclusive), the adequacy of the linear model is observed. If the stroke (greater than ~55%), the adequacy of the linear model of the pressure gain is not ensured.

The linear dependence of the amplification coefficient is valid for a section of no more than half of the flap stroke up to the maximum value of the pressure

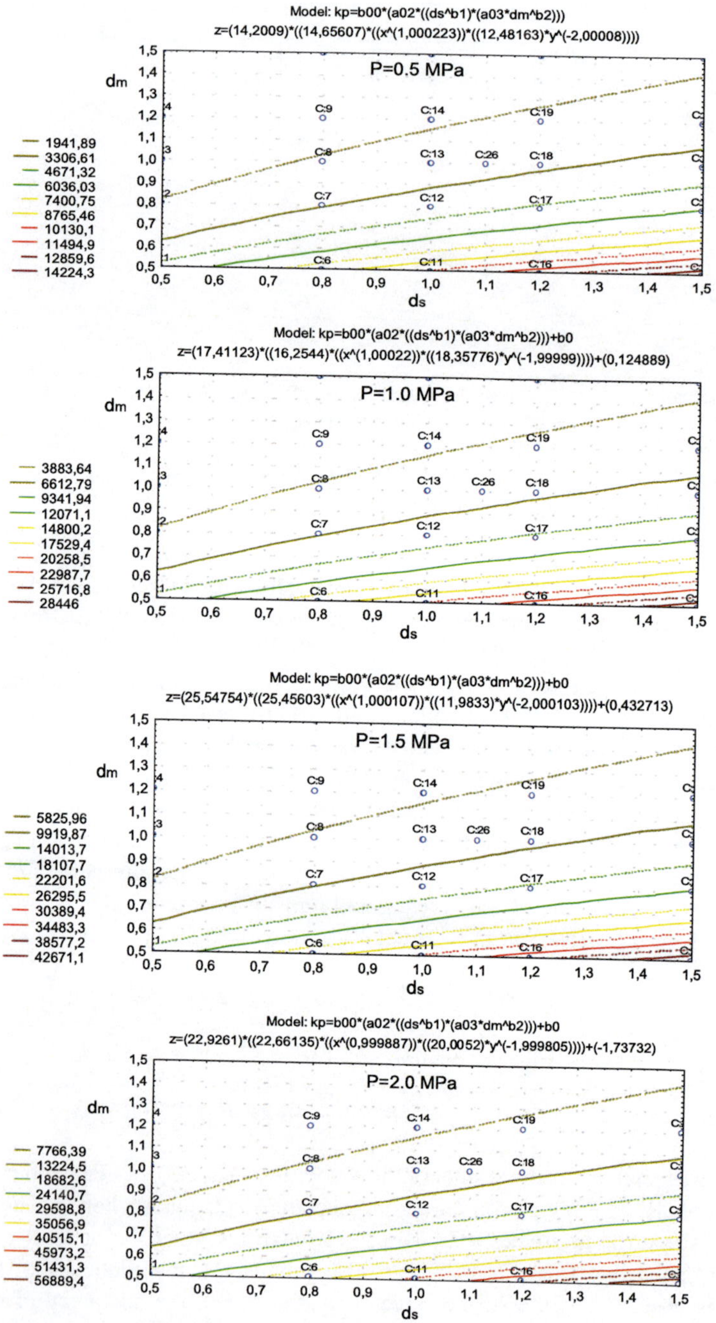

Fig. 6 The influence of the diameters (m) of the choke d_m and nozzle d_s orifice channels on the value of the pressure amplification coefficient K_P for different supply pressures (0.5 MPa; 1.0 MPa; 1.5 MPa; 2.0 MPa, respectively)

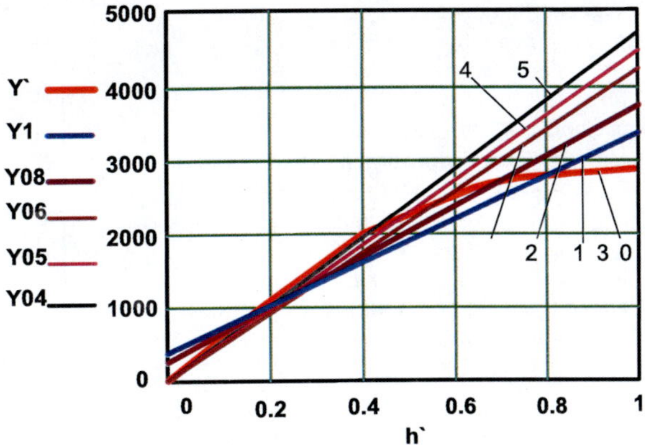

Fig. 7 The graphs of the linear functions (*Y*) of the pressure amplification coefficient in relation to the flap movement coordinate (*h*) depending on the investigated fraction of the flap stroke (d_h): 0—initial values; 1—at $d_h = 1h$; 2—at $d_h = 0.8h$; 3—at $d_h = 0.6h$; 4—at $d_h = 0.5h$; 5—at $d_h = 0.4h$

Table 1 The results of the analysis of linear pressure amplification coefficient models

The fraction d_h of the flap stroke (h) up to K_P = max	1.0	0.8	0.6	0.5	0.4
F-test	0.612059	0.779059	0.880581	0.915395	0.978799
R-corr	0.967293	0.986647	0.99582	0.99801	0.999205

amplification coefficient. In this case the F-test is over 90%, which indicates the adequacy of the linear model.

For an experimental nozzle of diameter ($d_s = 0.0011$) the flap stroke to match the areas of the nozzle orifice and the area of the gap between the nozzle and the flap corresponds to $h_0 = \pi d_s^2/(4\pi d_s) = d_s/4 = 0.000275$ m. That is, the critical value of the flap stroke coefficient $K_x = d_s/h_0 = 4$. The coordinate of maximum of pressure amplification coefficient K_P corresponds (see Fig. 2) to the stroke $h_{max} = 0.00013$. The section of adequate linear dependence of pressure amplification coefficient K_P corresponds to the stroke $h = 0...0.00013 \cdot 0.5 = 0...0.000075$ m. An analysis of changes in the cumulative sum (Fig. 8) of pressure amplification coefficient K_P shows an increase in the area under the graphs up to a flap stroke of 0.00022 m. This value corresponds to $1.7h_{max}$ and falls within the interval of up to $h_0 = 0.000275$ m.

Thus, the maximum is within the area of the flap stroke survey and the results obtained do not fall out of the process under investigation. The actual value of the flap stroke coefficient $K_x = d_s/h_{max} = 8.47$, and the pressure amplification coefficients K_P correspond to 2860, 5710, 8570 and 11,430 for the supply pressures: 0.5; 1.0; 1.5; 2.0 MPa (see Fig. 2).

Increasing the stroke of the flap increases the range of measured acceleration, but reduces the sensitivity. Decreasing the stroke increases the sensitivity, but requires a significant increase in the complexity of the oscillation compensation system.

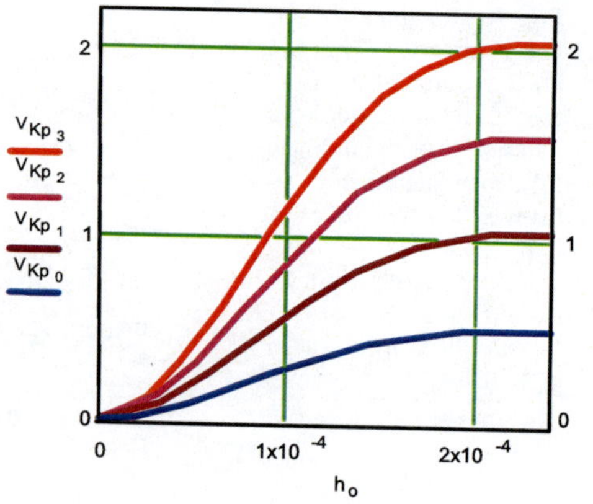

Fig. 8 The influence of possible flap movement h_j (m) on cumulative sum of pressure amplification coefficient K_P for the pressure indices in the control system: 0—0.5 MPa; 1—1.0 MPa; 2—1.5 MPa; 3—2.0 MPa; at different interval of change h_j for diameters (m) of choke orifice channel ($d_m = 0.001$) and nozzle channel ($d_s = 0.0011$)

4 Conclusion

Numerical investigations of the known pressure amplification coefficient in the interval of changing nozzle and the constant choke orifices diameters of 0.5… 1.5 mm at a pressure in the system of 0.5…2.0 MPa showed the presence of an extreme character of pressure amplification coefficient change depending on the flap stroke from the nozzle end. Adequate regression expressions linking nozzle and constant choke orifice diameters with the flap stroke (h) at the maximum pressure gain, as well as with the value of the pressure gain itself, were established. An adequate linear dependence for the pressure amplification coefficient is observed in the section of the flap stroke up to (0.5 h). The highest value of the amplification coefficient corresponds to the flap stroke of (1 h). The active growth of the area under the pressure amplification coefficient occurs up to the value of (1.7 h). It should be kept in mind, that when the flap stroke increases more than 0.25 of the nozzle orifice diameter, the annular gap ceases to work as an adjustable hydraulic resistance. Thus, for nozzle and constant choke orifice diameters ($d_s = 0.0011$, $d_m = 0.001$), the recommended initial clearance is 0.5 h or, based on the purpose of the measuring transducer, it can be increased to the value of 1.7 h/2.

References

1. Stein E, Valdiero C, Hubert MA, Goergen R, Souza M, Rasia LA, Valdiero AC (2019) Development of an automatic machine for sensor manufacturing by the gop technique Smart Innovation. Syst Technol 198:239–247. https://doi.org/10.1007/978-3-030-55374-6_24
2. Chen XS, Huang ZJ, Ma X, Qi L, Fang GJ (2021) Design and test of control system for rice mechanical weeding and seedling-avoiding control Jilin Daxue Xuebao (Gongxueban). J Jilin Univ (Eng Technol Ed) 51(1):386–396. https://doi.org/10.13229/j.cnki.jdxbgxb20190848

3. Kafiev I, Romanov P, Romanova I (2020) Fuzzy logic based control system for automated guided vehicle. In: International multi-conference on industrial engineering and modern technologies 2020, FarEastCon 9271513. https://doi.org/10.1109/FarEastCon50210.2020.9271513
4. Wei B, Liang H, Feng J, Liu J, Xie S (2020) Innovative and optimized design of automatic adjustment system for beaded rope of new diamond wire sawing machine. IOP Conf Ser Mater Sci Eng 892(1):012078. https://doi.org/10.1088/1757-899X/892/1/012078
5. Shoshiashvili ME, Shoshiashvili IS, Kartashova TP (2020) Analysis of drive of mechatron control system of rotary trencher 2020. Lect Notes Electr Eng 641:50–57. https://doi.org/10.1007/978-3-030-39225-3_7
6. Karabegović E, Šemić E, Isić S (2019) Automatic control of tube hydroforming process in experimental conditions. Lect Notes Netw Syst 42:101–106. https://doi.org/10.1007/978-3-319-90893-9_11
7. Shilin AN, Shilin AA, Sedov MN, Mustafa MN (2019) Intelligent low-frequency electromechanical vibration transducers. Stud Syst Decis Control 174:327–337. https://doi.org/10.1007/978-3-319-99759-9_27
8. D'Emilia G, Natale E (2018) Instrument comparison for integrated tuning and diagnostics in high performance automated systems. J Eng Appl Sci 13(3):795–803. https://doi.org/10.36478/jeasci.2018.795.803
9. Satybaldiyeva FA, Beysembekova RN, Sarybaev AS, Esenbekova GJ (2017) The development of intelligent systems for solar panels station and methods for determining the precision of the software tracking systems, the use of wireless communication device. J Theor Appl Inf Technol 95(21):5848–5858
10. Malarkodi A, Sridhar PSSR, Latha G (2016) Development of automated measurement setup for characterizing underwater acoustic transducers. Mar Technol Soc J 50(6):69–75. https://doi.org/10.4031/MTSJ.50.6.9
11. Rybarczyk D (2021) Application of the MEMS accelerometer as the position sensor in linear electrohydraulic drive. Sensors 21(4):1479. https://doi.org/10.3390/s21041479
12. Francia M, Milani M, Montorsi L (2016) An Innovative approach to kinematic analysis of multibody hydraulic actuation systems. SAE Int J Commer Veh 9(2):313–320. https://doi.org/10.4271/2016-01-8120
13. Stefanowski A (2015) Increased accuracy of ship's steering gear with constant delivery pump using the rudder angular velocity signal. Zeszyty Naukowe Akademii Morskiej w Szczecinie 44(116):29–33. https://doi.org/10.17402/053
14. Sokolov V, Krol O, Stepanova O (2018) Automatic control system for electrohydraulic drive of production equipment. In: International Russian automation conference, RusAutoCon-2018, 8501609. https://doi.org/10.1109/RUSAUTOCON.2018.8501609
15. Golubovsky V, Konovalov V, Doncova M (2020) Modeling static characteristics of angular velocity measuring transducer of the "nozzle-damper" type. J Phys Conf Ser 1614(1):012084. https://doi.org/10.1088/1742-6596/1614/1/012084
16. Golubovsky V, Konovalov V, Doncova M (2020) Modelling the force action of a liquid on the shutter of a measuring transducer. E3S web of conferences vol 157, p 02007. https://doi.org/10.1051/e3sconf/202015702007

Dependence Between the Mechanical Characteristics of the Material and the FDM Sample Made From This Material

Vera Raspopina, Alexandra Perelygina, Lev Shemetov, and Pavel Grigorov

Abstract The possibilities of FDM printing make it attractive for the transport engineering industry. Due to its specificity, FDM printing technology is becoming in demand for almost all areas of this industry: automotive, aerospace, shipbuilding, and the railway industry. Along with modeling and prototyping, it is of interest to use FDM technology to create products with a given bearing capacity. The specificity of the technology allows to create products of almost any geometry and size with a high accuracy of compliance with the 3D model. However, the description of the physical and mechanical characteristics of these products is currently very difficult. The layering and discreteness of the medium filling the volume of the product do not allow using the tools of continuum mechanics. Hence, it is difficult to analyze the strength and stiffness of such structural elements. To eliminate these difficulties, it is necessary to establish the dependence between the mechanical characteristics of the product and the filament material with which the product is manufactured. For this, the samples were tested for central tension with unloading under a static method of force action. The samples were made using FDM printing. The purpose of the tests was a comparative analysis of the behavior of FDM sample deformation and the filament material from which this sample was made. This analysis made it possible to establish that their behavior at the initial stage of loading is qualitatively the same.

Keywords Transport engineering industry · Transport · Reliability · FDM printing · Filament · Strength · Sample tensile diagram · Material tensile diagram · Unloading · Proportionality limit · Elastic limit · Forced elastic limit · PetG thermoplastic · Polymer · Filling pattern · Maximum tensile force

V. Raspopina (✉) · A. Perelygina · L. Shemetov · P. Grigorov
Irkutsk National Research Technical University, 83, Lermontova str., 664074 Irkutsk, Russia

A. Bieliatynskyi and V. Breskich (eds.), *Safety in Aviation and Space Technologies*, Lecture Notes in Mechanical Engineering,
https://doi.org/10.1007/978-3-030-85057-9_18

1 Introduction

The relevance of 3D printing technology is constantly growing, since the prospects for its application in various industries are the most optimistic [1]. In particular, these technologies are promising for transport engineering industry: automotive, aerospace, shipbuilding, and the railway industry. One of the factors determining this trend is the ability to improve the manufacturability of a product both at the manufacturing stage and at the stage of operation, without conflicting with the requirements of economy. In particular, the FDM-technology (method of layer-by-layer deposition), which is based on the extrusion of the material, is of interest [2]. Thermoplastics and composites are currently one of the most popular materials for FDM printing.

The most important requirement for any type of transport is to ensure reliability in operation, in particular, to ensure the strength of the product while minimizing its weight. This is especially actual for the structures of airplanes, cars, and water transport. FDM printing technology allows to create parts with optimal geometry with specified strength and stiffness characteristics. One of the limiting factors of the widespread use of this technology in transport engineering industry is the lack of a coherent theory of strength analysis of products made with the use of FDM printing (FDM product).

Currently, there is a wide range of specialized software products for engineering analysis that allow to obtain a reliable assessment of the strength reliability of structures of any complexity. As a rule, to build a computational model in these software products, the physical and mechanical characteristics of products are set from the standpoint of continual mechanics [3–7]. Such an approach to the analysis of an FDM product will not allow to correctly take into account the effect of the physical and mechanical characteristics of the layered structure filling its volume on the final product model. Therefore, it is impossible to obtain reliable information about the strength reliability of an FDM product.

However, the engineering automation industry is always sensitive to new trends. The intensive development of 3D technologies and their widespread demand in high-tech production did not go unnoticed by the leading manufacturers of software products. In particular, ANSYS Mechanical 2020 R1 contains the Workbench Additive toolkit [8], the functionality of which allows modeling the SLS technology (laser sintering of powder components) from the moment the geometric model is formed to the analysis of the thermal and strength properties of the product. The inclusion of the Inherent strain method allows to take into account such an important factor of SLS technology as material shrinkage during cooling. This significantly increases the degree of reliability of the results of the strength calculation.

Another giant of digitalization of engineering calculations, MSC Software, is working in the same direction. Its software package Digimat offers the industry a solution for the implementation of additive technologies for products made of plastics or reinforced plastics [9, 10]. The Digimat platform currently supports SLS technology and FDM technology. The toolkit of the complex covers all stages of product formation: design, modeling of the manufacturing process by 3D printing, obtaining parameters of strength and rigidity of the product, taking into account the residual phenomena of the technical process, namely, warpage, residual stresses and "porosity". Manufacturers claim to simulate the characteristics of thermoplastics, thermosetting plastics and composite materials with different matrices (polymer, metal, rubber). In this case, the construction of the model is carried out using a multilevel approach: micro-, meso- and macrolevels.

2 Description of the Problem

However, the correct description of the parameters of the strength and stiffness state of an FDM product requires the construction of a model that takes into account a wide range of influencing factors, both individually and in combination. For example, these factors include the multiphase behavior of the filament material at a specific point in the structure at all stages of a complex temperature regime due to the printing process; the behavior of the filament material due to the inhomogeneity of the residual stress fields in the volume of the product; interstitial and interlayer discontinuous adhesion; multidirectional structure of filament material; the presence of "voids" and "non-voids" within the volume of the structure, etc. The multi-component characteristics that affect the physical and mechanical properties of the product determine the complexity of its design model. In other words, the model is based on a combination of the taken into account the ambiguous characteristics of the material of the structure filaments filling the volume of the product, and the declared properties of the geometry of the structure in particular, and the product as a whole. At the same time, there is no complete information about which characteristics of the product were not taken into account. The guarantee of the correct result will be the degree of adequacy of the multilevel model of the FDM product.

The practice of analyzing the parameters of strength and rigidity of a part created by the traditional "subtractive" technology [1] shows that it is impossible to obtain a model that would reliably describe the physical and mechanical characteristics of a material at the micro level. In particular, the strength of real bodies is much less than the theoretical values, which are based on taking into account the binding energies between atoms and ions. This difference can reach hundreds of times [11]. This is so, since not all factors can be taken into account by the design model, for example, defects in the material structure. To assess the physical and mechanical characteristics of the material, we switched to the macrolevel, namely, a phenomenological approach was used in combination with continual mechanics [12]. That is, the output data after testing the corresponding samples is processed taking into account the theory of solid material [13]. The result was the use in engineering calculations of the averaged mechanical characteristics of materials—proportionality limit, yield limit, tensile strength.

A similar approach can be applied when analyzing the parameters of the state of strength and stiffness of FDM-products at the meso-level. Namely, to switch from the volume of a part filled with a filament structure and, as a consequence, having "voids" and "non-voids", inhomogeneity of the stress state, unequal strength of structural elements, to a volume filled with a continuous medium of the printing filament material equivalent in terms of mechanical characteristics. The implementation of this approach requires a significant experimental component.

The theory of engineering calculations of continuous mechanics is based on the postulate that there is an unambiguous predictable relationship between the mechanical characteristics of a part and a material, based on theoretical and experimental foundations [14–17]. The transition from material characteristics to the characteristics of a specific FDM product is not so unambiguous. It can be described as two-level. This can be clearly seen if we take into account that an FDM product is a volume of a given geometry, filled in a certain way with a printing filament, while the volume of a filament, in turn, is filled with a specific material [2]. Obvious logic dictates that in order to take into account the physical and mechanical characteristics in the product model, it is necessary to take into account the distribution structure of the printing filament that fills the volume of the product and the characteristics of the material in the volume of this filament. Even at this level of description, the physical and mechanical model is complex and ambiguous in the context of constructing a finite element model. Judging by this, this model can be interpreted as a "structure within a structure".

A different approach to the description of the physical and mechanical properties of an FDM product is based on the experimentally established relationship between the mechanical characteristics of the material of the filament of the structure and the characteristics of the structure itself, which forms the volume of the product.

Currently, thermoplastics are actively used for FDM printing. These materials include PetG thermoplastic.

3 Material Characteristics

PetG thermoplastic (polyethylene terephthalate glycol) is a modification of the thermoplastic polymer polyethylene terephthalate (Pet). The modification has improved a number of performance characteristics of products made from this material. At the same time, the modified form retained the properties of the thermoplastic polymer. Thermoplastics are characterized by the ability to retain their properties upon repeated heating to temperatures not exceeding the depolymerization temperature. In other words, the material of the internal structure of the sample prepared using the FDM technology must have the initial characteristics of the material. In the solid state, thermoplastics are highly elastic, resilient and strength. In this case, elasticity refers to the ability of a material to undergo significant reversible deformations under the influence of small external forces. Such deformations can reach values of the order of hundreds of percent.

The tensile diagram of a typical amorphous thermoplastic is shown in Fig. 1 [18]. Polyethylene terephthalate glycol is a representative of this category of materials. The diagram clearly demonstrates the behavior of a solid material at different stages of deformation. Section 0A is characterized by a linear dependence between deformations and stresses, that is, formally, this section is described by Hooke's law. At this stage of deformation of the material, the deformations are elastic. The shape of the sample remains practically unchanged. After point A, the tangent of the slope of the curve begins to change in the direction of decrease. At this stage, forced elasticity begins to develop in the material. It should be clarified that forced elasticity is usually understood as the ability of polymers in the solid state to undergo large reversible deformations under conditions of temperatures exceeding the glass transition temperature of the polymer. The ordinate of the maximum point B in this region corresponds to the mechanical characteristic of the thermoplastic, called the "forced elasticity limit" [11]. In the vicinity of the maximum and in the zone of stress decay, the sample neck begins to form. The formation of the neck is completed in the vicinity of point C. The section of the curve parallel to the abscissa corresponds to the growth of the neck. The term "plateau" is used for this section of the diagram in specialized sources [19]. Point D corresponds to the completion of the growth of the neck. The neck extends to the entire working part of the sample. Then the sample is destroyed.

Since the growth process of forced elastic deformation is analogous to plastic deformation in a metal [20], for a thermoplastic the analogue of the yield limit is the forced elasticity limit, σf (see Fig. 1). In fact, for products made of such materials, the bearing capacity will correspond to the operating stresses, below the limit of forced elasticity. In other words, of interest are the stresses of the thermoplastic corresponding to the section of the curve bounded by point A. This section corresponds to the elastic model of material behavior and is determined by Hooke's law. According to this model, upon unloading, the sample returns to its original size and shape. That is, the descending branch of the diagram falls into 0 (see Fig. 1).

The strength characteristics of polymers largely depend on both external conditions (temperature, loading rate, loading time, environment) and internal (structural defects, molecular weight, molecular orientation, etc.). For solid

Fig. 1 Amorphous polymer tensile diagram

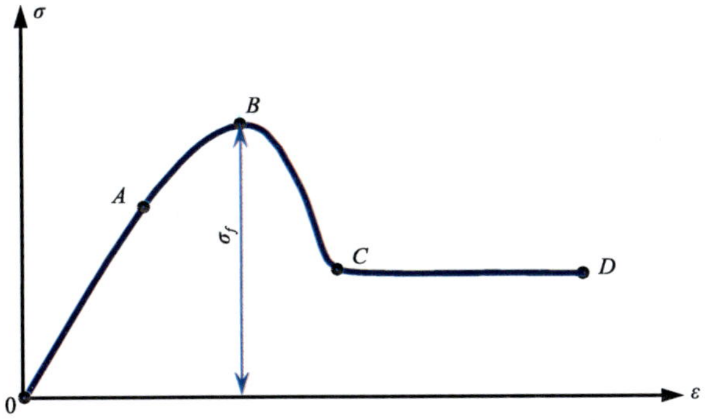

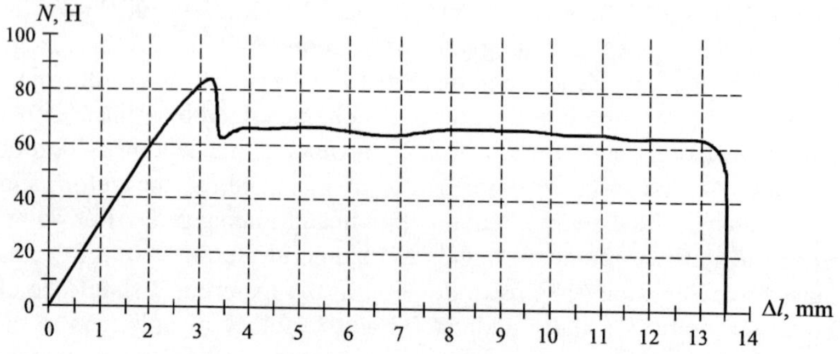

Fig. 2 Tensile diagram of the PetG thermoplastic filament

thermoplastics under static loading at room temperature, the qualitative picture of the tensile diagram of the material in the region of the 0-A-B-C curve is stable [11, 12, 18–20], that is, the region of elastic deformations determined by Hooke's law is stable.

In order to establish the relationship between the behavior of the deformation of the material and the filament made from this material, static tensile tests were carried out on a PetG thermoplastic yarn intended for the manufacture of a sample. Filament diameter 1.75 mm. Based on the test results, a diagram was obtained, shown in Fig. 2. The ordinate is the value of the tensile load in newtons, N, N; abscissa is the value of the absolute longitudinal deformation in millimeters, Δl, mm.

Analysis of the manufacturing technology of a product [2], made using FDM printing, from thermoplastic filament, taking into account the structure filling the sample volume, suggests that the qualitative picture of the behavior of the sample subjected to central tension deformation should not differ from the qualitative picture of behavior of filament material to a certain section of the thermoplastic tensile curve (see Fig. 1).

4 The Essence of the Experiment

To confirm or refute this assumption, a complex experiment was performed, the purpose of which was to determine:

- whether the mechanical behavior of the filament material and the mechanical behavior of the FDM sample coincide;
- if they coincide, then to what characteristic point of the tension diagram this coincidence will be observed.

The complex experiment consists of three series of tests of samples made using FDM-printing for central tension under static loading. Samples were prepared using

PetG thermoplastic filament. Each series of tests was carried out within the framework of the established protocol.

The first and second protocols determine the loading of the samples to certain values of the tensile force, then their unloading, followed by loading and bringing to failure. The third protocol of the experiment determines the loading of the samples without unloading with bringing to failure.

5 Description of Samples

The dimensions of the samples for testing for central tension were taken in accordance with GOST 11262-2017. Figure 3 shows a standard sample of type A.

Samples were printed with a printing speed of 50 mm/sec, a working table temperature of 75 °C, and a nozzle temperature of 220 °C. A template was applied to print samples with lines, at an angle of 90° to each other, with 60% coverage (Fig. 4). In this case, the direction of the printing filaments periodically changes from layer to layer. Accordingly, the layers are "attached" to each other "point-wise", namely, in the limited areas of contact of the printing filaments.

Within the framework of this model, it can be argued that the areas of the printing filament between their intersection points are free from the tight fit of the material of the adjacent parallel filament. As a result, in this area the filament

Fig. 3 Sample of type *A*

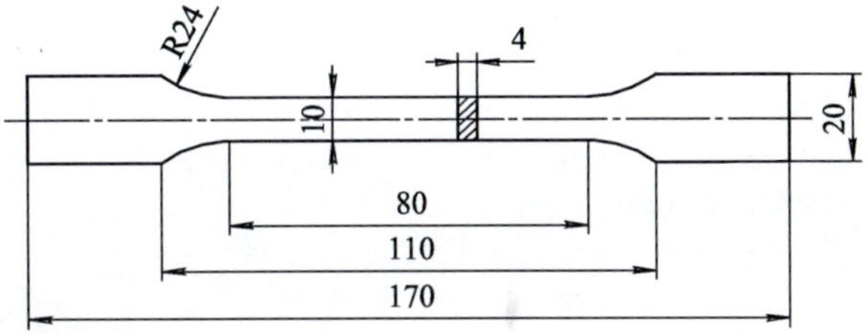

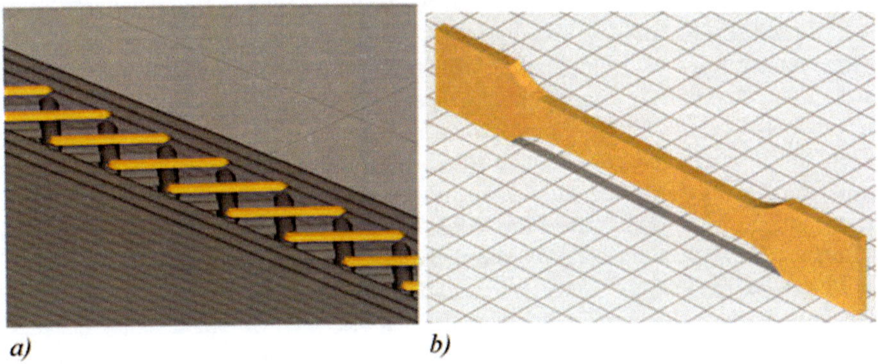

Fig. 4 Sample model for FDM printing: **a**—printing template; **b**—sample geometry

Fig. 5 Inner structure of printed sample

practically "independently" perceives the corresponding part of the load with minimal influence of adjacent layers (Fig. 5).

The good adhesion strength of PetG makes it possible to assume that in the vicinity of the filament bonding, the load is perceived by all layers as a monolith. In other words, in the vicinity of the bonding surfaces of the layers, it is possible, as a first approximation, to consider the model of the behavior of the macro-volume of the material, which is continuously distributed over the height of the layers.

6 Test Results

The first protocol of the experiment assumed that the tensile force of the samples should be brought to a certain level of 700 N and then the samples should be unloaded, followed by loading and bringing them to failure. The second experimental protocol is identical to the first with the only difference that the tensile force of the sample before unloading reaches 900 N.

Two series of experiments were carried out in accordance with these protocols.

During the experiments, some of the tested samples from each series were rejected due to their destruction outside the working area. As a result, after tests with a tensile force before unloading of 700 N, six samples remained, after tests with a tensile force before unloading of 900 N, four samples remained. In both the first (Fig. 6) and the second (Fig. 7) cases, the unloading of the samples is accompanied by an almost complete disappearance of deformations. More precisely, the unload line does not quite coincide with the load line. However, it can be seen from the diagrams that the discrepancy is very insignificant relative to the scale of the diagram along the abscissa. And this fact makes it possible to level the difference between the ascending and descending lines of the diagrams.

Moreover, when the tensile-unloading diagrams are combined with the tensile-unloading diagram of an identical sample, their complete coincidence is observed (Fig. 8). A series of experiments with tensile testing without unloading such samples was described in [21]. As part of these experiments, a third complex

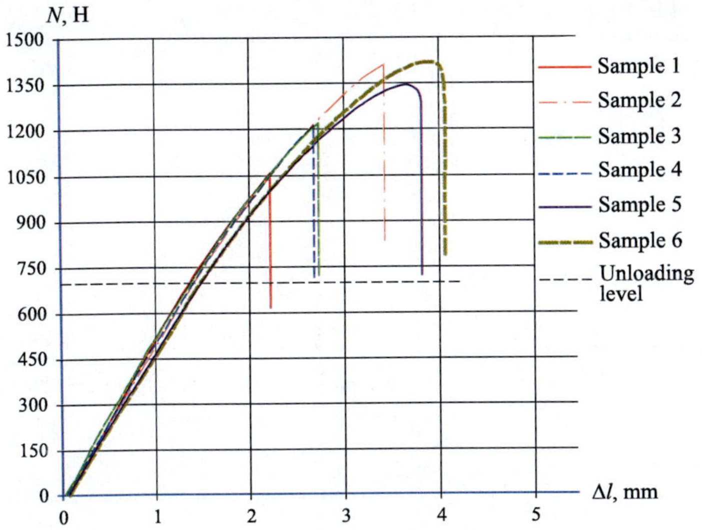

Fig. 6 Tensile diagram of sample with unloading after 700 N

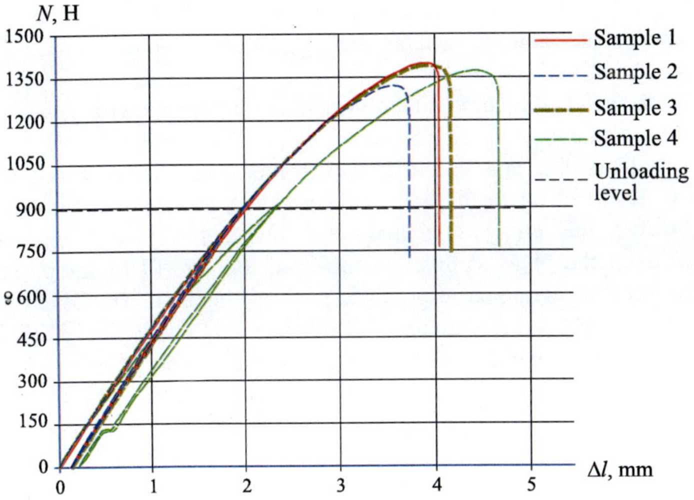

Fig. 7 Tensile diagram of sample with unloading after 900 N

experiment protocol was implemented to establish a correspondence between the characteristics of the filament material and the characteristics of the sample formed by this filament.

Analysis of the tensile diagrams of the samples obtained in accordance with three test protocols (see Fig. 8), the tensile diagrams of the filament (see Fig. 2) and

Fig. 8 Tensile diagram of sample with loading at different conditions

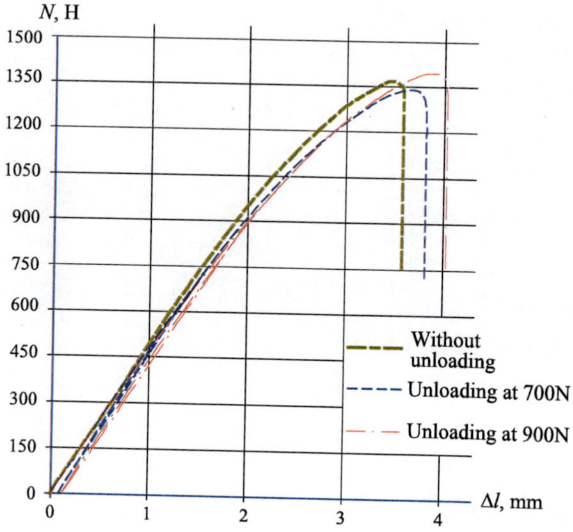

the canonical tensile diagram of amorphous thermoplastic (see Fig. 1) confirms the assumption that the mechanical the behavior of a sample made with the use of FDM printing, with its static central tension, should not differ from the mechanical behavior of the material under similar loading conditions up to a certain limit of the material tensile curve.

At the initial stage of loading, both the samples and the material have only elastic deformations. Sample and material diagrams contain the Hooke's law area. The coincidence of the diagrams is not limited to this area. Visual coincidence of the diagrams is observed up to the area corresponding to the maximum tensile load value. However, this area of the diagrams corresponds to different mechanical characteristics of the FDM sample and material. For the FDM sample, the maximum of the tensile diagram corresponds to the breaking load. This load for standard metal samples is called the ultimate tensile load.

For the filament material, this maximum of the tensile diagram is the forced elasticity limit (see Fig. 1), that is, a kind of analogue of the yield limit for metals. The material tensile diagram does not stop there. It continues with the formation of forced elastic deformations, and only then does destruction occur.

7 Conclusion

The following facts became the result of theoretical and experimental research.

First, the diagrams of the static central tension of the studied FDM samples begin with a section of direct proportionality between the force and deformation characteristics. That is, within these limits, Hooke's law operates and elastic deformations take place. The basis for the statement about the presence of only elastic deformations at this stage of loading the sample are the results of experiments, the diagrams of which are shown in Figs. 6 and 7. The same pattern is observed in the behavior of the filament material.

Second, the phenomenon of sample unloading at the elastic loading stage does not affect the behavior of the sample under repeated loading. The sample under repeated loading behaves as if it were loaded without unloading. This fact follows from a comparative analysis of the tensile diagrams of the samples obtained in accordance with the three loading protocols described above, Fig. 8. This behavior correlates with the behavior of the material.

Third, in the vicinity of the diagram point corresponding to the maximum tensile load, and after it, the behavior of the FDM sample and the filament material becomes qualitatively different. Further theoretical and experimental studies are needed to explain this phenomenon.

Within the framework of this work, it was found that there is a coincidence of the mechanical behavior of the filament material and the FDM sample at the initial stage of loading. Namely, at this stage, in both cases, Hooke's law is in effect. That is, their behavior at this stage of loading is qualitatively the same. And this coincidence is a consequence of the obvious dependence between the mechanical characteristics of a material and a sample made from a filament of this material. However, despite the fact that this dependence is obvious, it is not unambiguous as in the case of a similar solid sample. To formalize the pattern underlying this dependence, further experimental studies are required taking into account the parameters that affect the mechanical characteristics of both the samples and the material of the filament. In particular, it is necessary to add the sample print template, sample filling percentage, sample print direction, sample loading rate, material loading rate, temperature regime of the experiment to these influence parameters [22–26]. Along with the above-mentioned influence parameters, there are undoubtedly parameters, the identification of which will be carried out in the course of further experimental research in combination with an analysis of the fundamental theory of mechanics of a rigid deformed body. Such work is being carried out with plans for the future.

References

1. Dresvyannikov VA, Strakhov EP (2018) Additive technologies as technological innovation: concept, content, analysis of development. Economics and management of innovative technologies, 1 [Electronic resource]. http://ekonomika.snauka.ru/2018/01/15666 (date of access: 26.03.2021)
2. ISO/ASTM 52900: 2018—06 Additive manufacturing—General principles—Terminology (2018) [Electronic resource]. https://www.iso.org/obp/ui/#iso:std:74514:en. (date of access: 23.06.2021)
3. Izuka RC, Odokwo VE (2020) Design and stress analysis of marine diesel engine piston using solid works and finite element method (FEM). Int J Sci Eng Res 11(1):6

4. Protasov AV, Nikolaychuk OA (2011) Applying the finite-element method for evaluating the reliability of mechanical systems. J Mach Manuf Reliab 40(1):27–30. https://doi.org/10.3103/S105261881101016X
5. Klochkov YV et al (2016) Stress-strain analysis of a thin-shell part of fuselage using a triangular finite element with Lagrange multipliers. Russ Aeronaut. 59(3):316–323. https://doi.org/10.3103/S1068799816030041
6. Govorkov AS, Lavrenteva MV, Fokin IV (2018) Mathematical modeling of making mechanical engineering products based on an information model. MATEC Web of Conf. 224:02022. https://doi.org/10.1051/matecconf/201822402022
7. Chernykh EV, Zaidès CA, Kudryavtsev AA (2019) Peculiarities of making bolt and rivet connections with interference when assembling the aircraft MS-21. IOP Conf Ser Mater Sci Eng 632:012090. https://doi.org/10.1088/1757-899X/632/1/012090
8. Avedyan A (2020) We open new possibilities of ANSYS discovery 2020 R1. CAD Graph 2:50–56
9. Gnotiuk AP (2018) Digimat integrated solution for additive technologies: composite material, manufacturing process, construction characteristics. CADmaster 1(87):84–89
10. Gumenyuk AV, Gontyuk AP (2020) MSC Software solutions for additive manufacturing. Design and topological optimization of parts, modeling of processes. Addit Technol 1:22–25
11. Tager AA (2007) Physicochemistry of polymers. 4th edition, revised and enlarged. M. Scientific world
12. Makarov PV (1999) Modeling the processes of deformation and fracture at the mesoscale. Rigid body Mech 5:109–130
13. Novikov AD, Reznik SV, Denisov OV (2020) An experimental study to determine mechanical and thermophysical characteristics of thin-walled carbon plastic antenna reflectors. BMSTU J Mech Eng 3:84–91. https://doi.org/10.18698/0536-1044-2020-3-84-91
14. Leonovich DS, Zhuravlev DA, Karlina YuI, Govorkov AS, Karlina AI (2020) Automated assessment of the low-rigid composite parts influence on the product assemblability in the GePARD system. IOP Conf Ser Mater Sci Eng 760(1):0120382019
15. Osipok TV, Zaides SA (2020) Evaluation of the anisotropy of the mechanical properties of carbon steel sheet. Bull Irkutsk State Tech Univ 24(5):1007–1018. https://doi.org/10.21285/1814-3520-2020-5-1007-1018
16. Larchenko AG (2019) Assessment of the quality of products made of polymer materials for machine-building purposes. Bull Irkutsk State Tech Univ 23(3):463–471. https://doi.org/10.21285/1814-3520-2019-3-463-471
17. Mironenko VV, Govorkov AS (2021) Identification of problem parts made by longitudinal stretching when compensating for springing. IOP Conf Ser Mater Sci Eng 1064(1):012051. https://doi.org/10.1088/1757-899x/1064/1/012051
18. Kryzhanovsky VK, Burlov VV, Panimatchenko AD, Kryzhanovskaya YuV (2003) Technical properties of polymeric materials: Study guide. SPb., Publishing house “Profession”
19. Arzhakova OV et al (2017) The role of the scale factor in the structural and mechanical behavior of glassy polymers. Colloid J 79(6):679–699
20. Batenev GM, Zelenev YuV (1983) Physics and mechanics of polymers: Study guide for technical colleges. Higher school
21. Perelygina AYu, Raspopina VB, Rudykh AV, Grigorov PA (2020) Assessment of strength of elements produced using 3D printing technology. In: IOP conference series: materials science and engineering, vol 1064, International conference on mechanical engineering, automation and control systems (MEACS 2020) 17th September 2020, Novosibirsk, Russian Federation. https://doi.org/10.1088/1757-899X/1064/1/012052
22. Wu W et al (2015) Influence of layer thickness and raster angle on the mechanical properties of 3D-printed PEEK and a comparative mechanical study between PEEK and ABS. Materials 8:5834–5846. https://doi.org/10.3390/ma8095271
23. Kovan V et al (2018) Effect of printing parameters on mechanical properties of 3D printed PLA/carbon fibre compos. Int Sci J Mater Sci Non-Equil Phase Transf 4:126–128

24. Cantrell J et al (2017) Experimental characterization of the mechanical properties of 3D-printed ABS and polycarbonate parts. Adv Opt Methods Exp Mech 3:89–105. https://doi.org/10.1007/978-3-319-41600-7_11
25. Coogan TJ et al (2020) Prediction of interlayer strength in material extrusion additive manufacturing. Addit Manuf 35:1–16. https://doi.org/10.1016/j.addma.2020.101368
26. Popescu D, Baciu F, Vlăsceanu D, Cotruţ C, Marinescu R (2020) Effects of multiple sterilizations and natural aging on the mechanical behavior of 3D-printed ABS. Mech Mater 148:1–12. https://doi.org/10.1016/j.mechmat.2020.103423

Impact of Self-Vacuuming of Lithoid-Pumice Aggregate on Concrete Properties

Ashot Antonyan

Abstract Development of modern construction material science follows the way of cardinal increase of durability of erected constructions. The durability of concrete itself can be evaluated in different ways, but in practice it is convenient to use and test its water resistance, as this property is a qualitative indicator of capillary-porous microstructure of the material. Waterproofing is also closely related to other properties that predetermine the durability of concrete, such as frost resistance, corrosion resistance, etc. Increasing the water resistance of concrete is achieved by increasing the density of all components: cement stone and aggregates. Basically, waterproof concrete uses aggregate from dense rocks such as basalt, granite, diabase, etc. The idea of using porous rocks as aggregates for waterproof concrete at first glance may not be appropriate, but a deeper approach to the issue of application shows the obvious advantage of some of the lightweight aggregates compared with dense. The article provides information about the effect of self-vacuuming of litho-aluminous aggregate in hardening concrete. It is shown that the introduction of these aggregates leads to a decrease in the permeability of concrete, and in some cases to an increase in its strength. The test results are given in comparison with concrete on dense aggregates of basalt rubble and river sand.

Keywords Lithoid pumice · Permeability · Concrete · Strength · Self-vacuuming

1 Introduction

At the moment, heavyweight concrete is the most produced type of concrete in the world. The volume of lightweight concrete accounts up to 20–30% of the total amount of concretes produced. However, the constant search for solutions to improve the strength of erected structures forces us to consider the use of light porous aggregates in concrete in a qualitatively new way, since in some cases the

A. Antonyan (✉)
The National University of Architecture and Construction of Armenia,
Teryan street 105, index 0009, Yerevan, Armenia

A. Bieliatynskyi and V. Breskich (eds.), *Safety in Aviation and Space Technologies*,
Lecture Notes in Mechanical Engineering,
https://doi.org/10.1007/978-3-030-85057-9_19

light aggregates provide high values of properties responsible for the strength of concrete (water resistance, frost resistance, etc.).

Typically, modern publications related to this topic cover the development and application of high-functional lightweight concretes based on artificial porous aggregates such as expanded clay and agglоporite [1, 2]. The structure formation of such concrete is significantly different from the structure formation of concrete made on natural lightweight aggregates. Continuous study of various natural porous rocks is quite reasonable due to the wide geography of their deposits.

We studied lightweight concretes based on lithoid-pumice aggregates widespread in Armenia, Italy, and the northern Caucasus in publications [3–5]. Various tests of the water resistance of such concretes under equal conditions (the same mobility of concrete mixtures and the content of cement) showed an obvious advantage of lightweight aggregates over traditional heavy ones.

Pumice is a product of volcanic activity, which is a porous, loose and lumpy rock in the form of erupted solidified acid volcanic glass. The porosity of pumice is determined by the content of dissolved and trapped gas in the molten glass released during the eruption. Depending on the viscosity of molten volcanic glass and the amount of gas dissolved in it, the character and number of pores in pumice vary widely, from spongy to fibrous, from coarse to finely porous [6].

There are two types of pumice depending on petrographic and physical and mechanical data—normal type (or Ani type) and lithoid.

Unlike the normal type, lithoid pumice is characterized by lower porosity and, accordingly, higher strength. They also differ in geological age: the Ani type pumice is Quaternary, and the lithoid pumice is Tertiary.

Pumices are found in many areas of volcanism occurrence. Large deposits of pumice are found in Armenia, Italy (on the Lipari Islands) and in the northern Caucasus (in the Nalchik region). There are more than 70 pumice deposits in Armenia. Material of the above is successfully used in construction, in the chemical industry, etc.

Naturally, lithoid pumice occurs in the bedrock and in the form of secondary deposits. The first case shows solid deckes, the second—layers of lumpy pumice, chestnut and sand [6]. Generally, fragments of obsidian are found in deposits of lithoid pumice. Obsidian is identical to lithoid pumice in terms of chemical composition with the only difference that it does not have time to bubble before cooling.

Depending on the content and valence of iron, the color of lithoid pumice is white with a pink tint, light gray, yellow, brown, and also black.

The chemical composition of lithoid pumice (in comparison with normal) is shown in Table 1, and the main physical and mechanical properties of lumpy pumice—in Table 2.

Lithoid pumice in concrete can be used as a coarse and fine aggregate after crushing lumpy rock and enriching the grain-size distribution, as an active mineral additive in the production of cement, as well as in concrete [3].

The increased water resistance of lithoid-pumice concrete is explained by the effect of self-vacuuming of the porous aggregate. After the water is mixed, it is sucked out and accumulated in the pores and microcapillaries of the pumice. It

Table 1 Elemental composition of pumices

Type of pumice	SiO_2	Al_2O_3	Fe_2O_3	CaO	MgO	$Na_2O + K_2O$	Loss-on-ignition
Lithoid	69.82–70.20	16.90–17.68	1.91–2.14	1.63–1.70	0.38–0.45	4.35–5.30	3.80–3.97
Normal	68.84–70.08	13.33–15.19	1.44–3.77	1.48–2.35	0.21–0.71	6.83–7.84	3.99–4.54

Table 2 Basic physical and mechanical properties of lumpy lithoid pumice

Properties	Range of values/Average
Mass specific gravity, kg/m^3	1310–1473/1409
True density, g/cm^3	2.36–2.39/2.38
Porosity, %	37.9–44.6/40.6
Water absorption by mass, %	8.1–18.1/12.5
Compression strength, MPa	18.9–20.8/20.0
Softening coefficient	0.91

should be noted that the water is not sucked off as is, but in the form of a saturated solution of cement. The real W/C ratio near the aggregate decreases and the compaction of the entire system occurs.

The opposite process takes place later, during the hardening of concrete. The water accumulated in the pores of the aggregate starts to feed the hardening cement stone. The rate of water release depends on the radius of the aggregate capillaries. For comparison, volcanic slag and some tuff rocks have much larger average capillary diameter (>0.5 mm) than lithoid pumice (0.0002–0.5 mm), therefore, the latter will give water more slowly, which is preferable from the point of view of normal hardening of cement stone (Fig. 1).

Self-vacuumining of lithoid pumice was covered in publications [3, 7, 8]. The increase in the water resistance of concrete in comparison with the heavy one was explained in [7] by self-vacuuming and pozzolanic effects. However, the fact of self-vacuuming was stated in these publications without its quantitative and qualitative assessment. In addition, the information found in publications provides a comparison of lithoid-pumice concretes with heavy ones at different W/C ratios due to the high water demand of light aggregates, which brings with it some uncertainty. The following studies were organized to identify these dependencies.

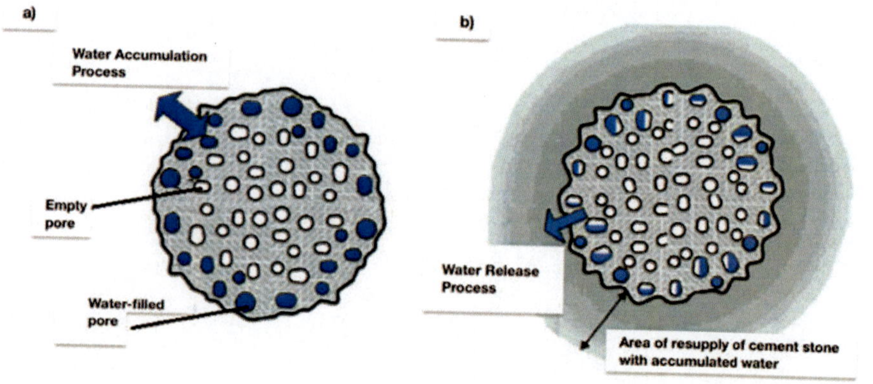

Fig. 1 Outline of self-vacuuming of a porous aggregate. **a** Freshly prepared concrete mix, **b** hardening concrete

2 Methods

The composition of heavy concrete on crushed trap rock (5–20 m) and gravel sand was considered as the basic composition. In subsequent mixes one of its aggregates was changed to the corresponding one of light variety, until they were completely replaced with lithoid-pumice. That is, there were cases when the lithoid-pumice rock was featured in the matrix, and in another one—in the concrete frame. At the same time, in all cases, the W/C ratio and the flowability of the concrete mixture remained unchanged. This was achieved by varying the dosage of the superplasticizing admix, and, in the case of a complete replacement of aggregates with lithoid pumice—by changing the type of aggregate. Table 3 provides the properties of the aggregates used, and Table 4 includes compositions of the concretes.

In accordance with the above compositions, the following concrete samples were formed: cylinders of 150 × 150 mm for testing for water resistance by the method of determining the penetration depth (according to EN 12,390–8) [10] and cubes of

Table 3 Physical and mechanical properties of the aggregates used

Property	Lithoid pumice		Crushed trap rock	Gravel sand
	Crushed rock	Sand		
Dry bulk density, kg/m^3	874	1156	1301	1680
True density, g/cm^3	2.31	2.31	2.70	2.72
Mass specific gravity, kg/m^3	1400	–	2551	–
Porousness, %	38	50	50	38
Obsidian amount, %	0	7	–	–
Amount of pulverulent particles, %	1.0	27.0	1.10	1.26
Amount of particles smaller than 0.16 mm, %	–	25.2	–	1.41
Fineness modulus	–	2.5	–	

Table 4 Compositions of the concretes under consideration

Aggregates variations	Content of components, kg/m^3				
	CR	S	C	C	SP
Crushed trap rock + gravel sand	923	872	410	226	1.64 (0.4% of cement mass)
Crushed trap rock + lithoid-pumice sand	962	640	427	238	8.54 (2% of cement mass)
Lithoid-pumice crushed rock	669	831	391	215	5.9 (1.5% of cement mass)
Lithoid-pumice crushed rock + lithoid-pumice sand	684	599	419	231	8.38 (2% of cement mass)

Notes CR—crushed rock, S—sand, C—cement, W—water, SP—superplastificating admixture
- Portland cement was used as a cement CEM II/A-P 42.5 N (Ararat plant),
- Sikament 3235 superplastificator (based on a mixture of polycarboxylate with lignosulfonate) in all cases except the last one in the table where Sika Viscocrete 20HE (based on polycarboxylate) was used.

100 × 100 × 100 mm to determine the strength. The tests were carried out after 28 days of normal hardening. The results of these tests are shown in Table 5 and Figs. 2 and 3.

3 Results and Discussions

The analysis of the test results shows that the properties of concrete change significantly regardless of the type of the appended lithoid-pumice aggregate (crushed rock or sand). Its density decreases by 186 and 322 kg/m^3 where pumice is placed in a matrix or in a frame, respectively. Technically, such concrete still belongs to the heavyweight type, since the density of lightweight concrete should not exceed 2000 kg/m^3 according to GOST 26,633. However, according to the publication [8], even a partial introduction of a porous aggregate gives concrete properties attributed to lightweight type.

In the case of replacing crushed trap rock with lithoid-pumice one, concrete permeability decreases by 323% (from 110 to 34 mm), however, there is also a decrease in concrete strength by 6.9%: from 38.7 to 35.7 MPa. In this case, the decrease in the permeability of concrete can be explained only by the effect of self-vacuuming, since a possible pozzolanic reaction, conditioned to the chemical composition of lithoid pumice, is applicable only to dust-like fractions of sand. Pozzolanic effect is negligible on the surface of a large aggregate, especially on the 28th day of hardening.

The decrease in the strength of concrete is explained by the processes occurring during the testing of a concrete sample. In this case, there is a redistribution of internal stresses caused by an external load through the matrix onto the frame (lithoid crushed rock), the strength of which is much lower (about 20 MPa) than the strength of crushed trap rock. However, due to the effect of self-vacuuming and surface hardening of the porous aggregate, concrete has a strength exceeding that of a coarse aggregate [11–13].

In contrast to the determination of the strength of lightweight concrete, where the frame part of concrete is tested over, it is mainly the matrix part of it that is tested for water resistance [14–16].

All properties improve in the case of the appending of pumice-lithoid sand instead of gravel sand. Thus, the permeability of concrete decreases by 354% (from 110 to 31 mm), and the strength increases by 17.5% (from 38.7 to 45.5 MPa). Obviously, such a jump is attributed to the synergism of two processes: self-vacuuming of large sand particles (1.25–5 mm) and pozzolanic activity of small particles [9], so that the hydrolysis product of clinker minerals, calcium hydroxide ($Ca(OH)_2$), binds into a strong and stable low-basic calcium hydrosilicate of the C-S–H(I) type [17–20].

In addition, mechanical blockage of small pores and capillaries occurs when using lithoid-pumice sand. The increasing water demand of the concrete mixture in

Table 5 Test results for water resistance and strength of concrete with different types of aggregates

Aggregate type	Concrete mix			Concrete				
	W/C	Cone slump, cm	Density, kg/m^3	Density, kg/m^3	Water penetration depth, mm	Strength, MPa	Property change compared to the test specimen mixture	
							Water penetration depth (%)	Strength (%)
Crushed trap rock + gravel sand	0.56	22	2433	2401	110	38.7	–	–
Crushed trap rock + lithoid-pumice sand	0.56	21	2277	2215	31	45.5	−354	+17.5
Lithoid-pumice crushed rock + gravel sand	0.56	23	2112	2079	34	35.7	−323	−6.9
Lithoid-pumice crushed rock + lithoid-pumice sand	0.56	24	1940	1906	20	39.8	−550	+2.8

Note—all the results of water penetration depth and strength are the average value of the test of 3 samples, "–" —decreasing property change, "+"—increasing property change.

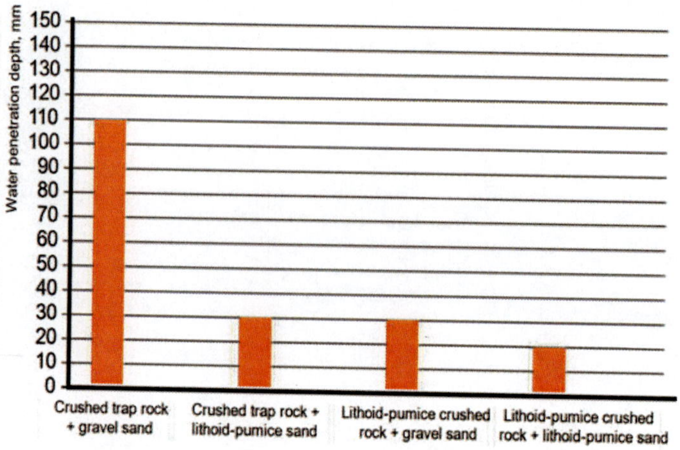

Fig. 2 Changes of water penetration depth depending on the aggregate type

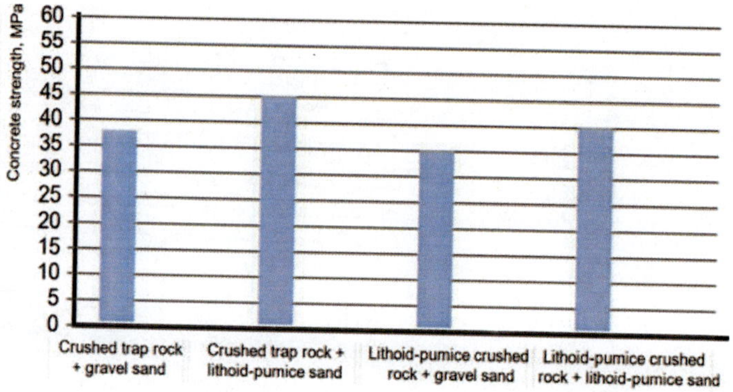

Fig. 3 Changes in concrete strength depending on the type of aggregates

this case was compensated by an increase in the dosage of the superplasticizing admix from 0.4 to 2% of the cement mass.

In the case of a complete replacement of aggregates with a lithoid-pumice type, the water demand of the concrete mixture increases to a greater extent. A superplasticizing admix based on polycarboxylate esters in an amount of 2% of the cement mass was used to maintain a constant W/C ratio and flowability of the concrete mix.

The concrete permeability in this case decreased by 550% compared to the test specimen (from 110 to 20 mm), and the strength increased by 2.8% (from 38.7 to 39.8 MPa).

The results are quite logical. Each of the lithoid-pumice aggregates contributes to the reduction of permeability according to the outlines described above. The change in strength by such a small amount (+2.8%) is explained by the fact that the positive contribution of the lithoid-pumice sand compensates for the negative contribution of crushed rock to the strength of concrete.

4 Conclusions

The following conclusions can be drawn:

Self-vacuuming of lithoid-pumice crushed rock and sand induces a decrease in the permeability of concrete by 5.5 times and an increase in its strength by 17.5% compared to concrete on dense aggregates.
The contribution of lithoid-pumice sand to the improvement of the set of properties of interest exceeds the contribution of crushed rock to the same properties.

References

1. Shuguan H, Chzhou VF (2016) Lightweight concretes. ASB publishing house, Moscow
2. Rinker ME (2013) Determination of acceptance permeability characteristics for performance—related specifications for portland cement concrete. Report Submitted to Florida Department of Transportation, University of Florida
3. Antonyan AA (2018) Lithoid pumice, a material for water-resisting concretes. Technol Concr 3–4:18–23
4. Antonyan AA (2019) Change in water resistance of highly functional concrete during operation in a dry hot climate. Concr Technol 7–8:4–8
5. Antonyan A (2020) Water resistance of lightweight concrete on porous aggregates of volcanic origin. Int J Adv Sci Technol 29(9s):6454–6464
6. Acagorcyan ZA (1967) Natural stone materials of Armenia. Publishing of construction literature, Moscow
7. Simonov MZ, Khudaverdyan VM (1958) Hydraulic concrete based on lithoid pumice. In: Yerevan (ed). Publishing house of Science Academy
8. Simonov MZ (1973) Fundamentals of lightweight concrete technology. Stroyizdat, Moscow
9. Antonyan AA (2019) Water resistance of concrete with activated pumice-lithoid microfiller. In: Building materials, equipment, technologies of the XXI century vol 5–6, pp 44–49
10. Homepage https://www.researchgate.net/figure/a-Water-penetration-apparatus-b-Water-penetration-testing-and-c-Depth-of-water_fig5_323945345
11. Cantero B, Saez de Bosque IF (2019) Water transport mechanisms in concrete bearing mixed recycled aggregates. Cem Concr Compos. https://doi.org/10.1016/j.cemconcomp.2019.103486
12. Nataadmadja AD, Runtuwene JAP (2018) Analysis of concrete permeability with additional waterproofing admixture. IOP Conf Ser Earth Environ Sci 195:012002
13. Lee SH, Kim HJ, Sakai E, Daimon M (2003) Effect of particle size distribution of fly ash-cement system on the fluidity of cement pastes. Cem Concr Res 33:763–768
14. Sisomphon K, Copuroglu O (2011) Self healing mortars by using different cementitious materials. In: Proceedings of the international conference on advances in construction materials through science and engineering, Hong Kong, China, 5–7 September
15. Wang H, Sun X, Wang J, Monteiro PJM (2016) Permeability of concrete with recycled concrete aggregate and pozzolanic materials under stress. Materials 9(4):252

16. Sandhu AR, Lakhiar MT, Jhatial AA, Karira H, Jamali QB (2019) Effect of river indus sand and recycled concrete aggregates as fine and coarse replacement on properties of concrete. Eng Technol Appl Sci Res 9(1):3832–3835
17. Adnan SH, Lee YL, Abdul Rahman I, Mohd H, Wimala M (2018) Water permeability of recycled aggregate concrete, presented at the technology and innovation for sustainable development conference, Khon Kaen, Thailand
18. Shin KJ, Bae W, Choi SW, Son MW, Lee KM (2017) Parameters influencing water permeability coefficient of cracked concrete specimens. Constr Build Mater 151:907–915
19. Yi ST, Hyun TY, Kim JK (2011) Effects of hydraulic pressure and crack width on water permeability of penetration crack-induced concrete. Constr Build Mater 25(5):2576–2583
20. Rani BD, Rao BK (2019) Service life prediction of high-performance concrete with respect to chloride ion penetration by incorporated with fly ash and silica fume. In: International conference on advances in civil engineering (ICACE-2019)

Selection Criteria of Aggregates for Waterproofed Concrete Production

Ashot Antonyan

Abstract The selection and use of aggregates for concrete are based on the test results of any characteristics given in the standards. However, as the article shows, not all aggregates meeting the requirements of these standards are equally good for the concrete waterproofing qualities. For more detailed consideration of this issue, both the model aggregates produced in the laboratory by means of screening and repeated mixing in the required proportions and the aggregates taken from different concrete plants were used in the study.

Keywords Permeability · Concrete · Crushed stone · Sand · Fineness modulus · Standard · Strength

1 Introduction

The studies on the increase of permeability of the used concretes are now extremely relevant. This is explained by the necessary drastic prolongation of the life of the constructed buildings and facilities in the transition to the sustainable development concept of the society.

The concrete is a composite material consisting of aggregates and cement stone. Consequently, by studying the issue of permeability, it is necessary to consider each component of the system. Then, the study of the impact of aggregates on the change of permeability of concrete is given.

Aggregates occupy up to 3/4 of total concrete volume. By creating a rigid frame, they reduce the shrinking deformations and the total amount of the cement stone.

As a rule, change in quality of aggregates, in contrast to the quality of the cement stone, has less impact on the permeability of concrete [1–3]. Nevertheless, the impact of variation of some characteristics of aggregates on hydrophysical properties of concrete is given below.

A. Antonyan (✉)
The National University of Architecture and Construction of Armenia, Teryan street 105, index 0009, Yerevan, Armenia

A. Bieliatynskyi and V. Breskich (eds.), *Safety in Aviation and Space Technologies*, Lecture Notes in Mechanical Engineering,
https://doi.org/10.1007/978-3-030-85057-9_20

2 Coarse Aggregate

Aggregate is considered coarse when its particle size is more than 5 mm. It is divided into gravel and crushed stone depending on the surface type, however, to produce waterproofed concrete, the crushed stone is mainly used. This is explained by that when the latter is used, the contact area between the cement stone is less defective and porous than with the gravel. That is why, the impact of change of the crushed stone quality on the permeability of concrete is only considered below.

Coarse aggregate is characterized by grain-size composition, porosity/density, grain shape, impurity with dust and clay particles, alkali activity [4]. However, not all listed characteristics impact the permeability of concrete. The main influential factors are given below.

Grain shape. The grain shape is the most important quality indicator of the crushed stone. It can be cubic, ball-shaped, octahedral, icosahedral, and elongated (acicular and nut-like). From the point of view of the permeability of concrete, the elongated shape is the most unfavourable. Firstly, because with this shape, the void ratio of aggregate increases, and secondly, the elongated shape facilitates the water accumulation under aggregate after concrete mix application and the formation of macroscopic cavities that then serve as water filtration channels [5–9].

The work [10] shows that depending on the type of the crushed stone (light or heavy) with the nut-like grains, the permeability of heavy concrete increases by 3.1 times, and of light lithoid-pumice concrete—by 1.2 times as compared to concretes with ball-shaped aggregates.

Grain-size composition. By improving the grain-size composition of the crushed stone and thus reducing the void ratio of the mix, the permeability can be increased by times [11]. The compact-grained crushed stone creates a dense frame, thus reducing the cement layer. In practice, it is desirable that the grain-size composition of the crushed stone is close to the Fuller's curve or within the standard recommended area [12].

Impurity of the crushed stone by dust and slay particles impacts the permeability of concrete in two ways. On the one hand, as the number of the dust particles increases, the water demand of concrete mix increases that leads to the negative impact on the permeability of concrete. On the other hand, the dust particles serve as mud grouting and physically clog the ducts and channels that, in its turn, increases the permeability of concrete. However, taking into account the fact that the dust particles have negative impact on the concrete strength and prevent rigid adherence of aggregate to the cement stone, the washed crushed stone should be preferred to ensure the complex of the required properties of concrete.

Impact of the maximum crushed stone size. The maximum size of the used coarse aggregate depends on the thickness of the waterproofed structure. In the thin-walled structures, the maximum size of aggregate does not exceed 10 mm, and in massive hydraulic engineering structures, it can reach 70–150 mm.

The impact of this factor on the permeability of concrete was considered in experimental way.

3 Methods

The base composition of concrete is as follows: CS = 1055 kg/m^3, S = 800 kg/m^3, C = 380 kg/m^3, W = 170 kg/m^3, superplasticizer − 1% of cement mass = 3.8 kg/m^3. For concrete in the framework of this study, the following materials were used: basalt crushed stone of Ararat origin (Armenia) 15–25 mm, 15–30 mm, 25–40 mm, river sand washed ($S_{кр}$ = 2.9), portland cement CEMII/A-P 42.5 N (Ararat plant), superplasticizer MasterPozzolith 519 (Basf) based on naphthalene-sulfoformaldehyde.

3 cases were considered. In the first case, the amount of the crushed stone specified in the base composition was only made up by 15–25 mm size (D_{max} = 25 mm).

In the second case, 60% of the crushed stone was made up by 15–30 mm size (D_{max} = 30 mm), and remaining 40%—by 15–25 mm size.

In the third case, 60% of the crushed stone was made up by 25–40 mm size (D_{max} = 40 mm), and remaining 40%—by 15–25 mm size.

3 cylinder samples 150 × 150 mm were produced from each concrete composition to determine the permeability and 3 cylinder samples 150 × 300 mm to determine the strength. Hardening occurred in the normal conditions. After 28 days, the cylinder samples were tested for permeability according to EN 12,390–8 (Determination of the depth of penetration of water under pressure) and for strength. The test results are given in Table 1 and Fig. 1.

4 Results and Discussion

As it can be seen from the given data, the permeability of concrete increases as the maximum crushed stone size increases. This is explained by the following. The contact area coarse aggregate-cement stone is a place for accumulation of portlandite mineral (CH) formed from hydrolysis of alite and belite. Also, water accumulates under aggregate when separated after concrete mix application. This

Table 1 Test results of concrete with different maximum crushed stone size

Maximum crushed stone size, mm	Concrete mix		Concrete		
	W/C	CS, cm	Strength, MPa	Depth of penetration of water, mm	Concrete water-tightness grade*
25	0.45	12	38.5	73	W12
30	0.45	14	37.8	84	W8
40	0.45	16	36.2	110	W6

Note—Each result is average for 3 samples
*according to the work [13]

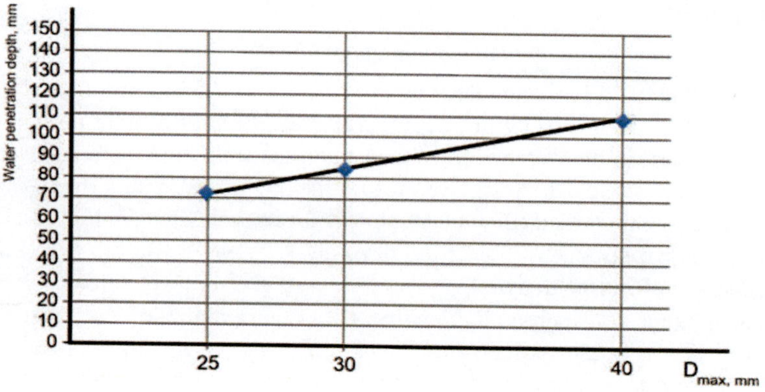

Fig. 1 Dependence of depth of penetration of water on maximum crushed stone size

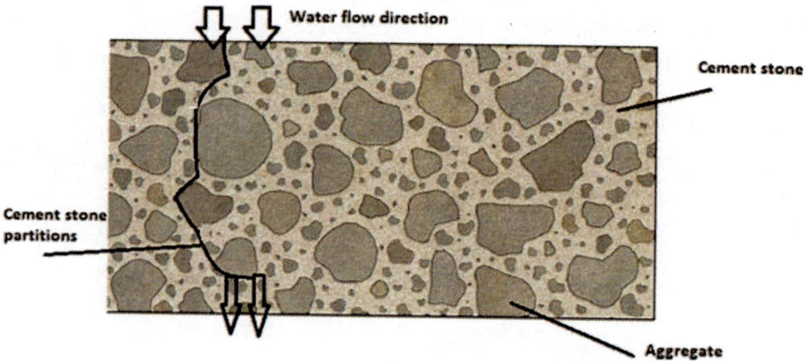

Fig. 2 Diagram of water flow in the contact area aggregate-cement stone [14]

leads to that the contact area 5–10 mcm thick is characterized by more porous structure than the cement stone and conducts water when filtered.

Despite that as the maximum crushed stone size increases, its specific surface decreases, but the total length of the contact area increases, consequently, the permeability of concrete increases (Fig. 2).

It can be concluded that to increase the permeability of concrete, it is necessary to reduce the maximum crushed stone size to Dmax = 15–20 mm, if possible.

5 Fine Aggregate (Sand)

Aggregate is considered fine when its particle size is within 0.14–5 mm.

The quality of fine aggregate (sand) used for waterproofed concrete is characterized by grain composition and/or fineness modulus, clay and dust impurities, alkali activity.

The impact of impurities affecting the sand on the qualities of concrete mixes and hardened concrete is very negative. With the large amount of the latter, the water demand of concrete mix increases, all physical–mechanical and hydrophysical properties of concrete are impaired, including freeze–thaw resistance and permeability, that is why we used washed river sand with dust particle content of no more than 0.5–1.3% in these and further tests.

The aggregate reactivity with the cement alkali impacts in no way the permeability of concrete but leads to further internal corrosion, that's why prior to use in waterproofed concrete, aggregate should be also tested for alkali activity.

The impact of the grain composition and integral fineness characteristics, the fineness modulus of the used sand [15], on the permeability of concrete is of certain interest. It is considered in the general concrete technology that the fineness modulus increase, i.e. the use of coarser sands, promotes for decreased water demand of concrete, improved physical–mechanical properties. This makes the basis for using sands with the fineness modulus of more than 2.8, and in some cases of more than 3 in high-functioning and high-strength concrete [16–22].

The another situation with waterproofed concrete. The increased fineness modulus of sand for decreased water demand does not always ensure a notable gain in permeability.

As the fineness modulus does not generally characterize the grain composition of sand, it is good to consider the impact of the amount of each size.

By analysing the recommended grain composition ranges for sands (Table 2, Fig. 3) for GOST, ASTM (US) and BS (UK), it can be concluded that the GOST values have the narrowest range, i.e. the grain composition requirements are the highest. That's why the experiments were conducted by studying grain compositions of this standard.

To study the impact of grain composition of sand on the permeability of concrete, washed river sand was screened by standard sizes, and then mixed again in the corresponding amount to produce sands for the upper (Sand 1), lower (Sand 2) and medium limit (Sand 3) of the GOST recommended ranges (Table 3, Fig. 4).

Along with this, sands used in production (Sand 4 and Sand 5) were studied that, generally, showed different grain composition but had approximately similar amount of fine particles 0.14–0.315 mm.

Table 2 Recommended grain composition ranges for sands

Sieve size, mm	Fill sieve residue, %, according to the standard		
	GOST 8736	ASTM D448-08	BS 882
5	0	5–18	0–11
2.5	0–20	15–58	0–40
1.25	15–45	20–60	0–70
0.63	35–70	40–78	0–85
0.315	70–90	65–95	30–95
0.14	90–100	80–95	85–100
<0.14	0–10	5–20	0–15

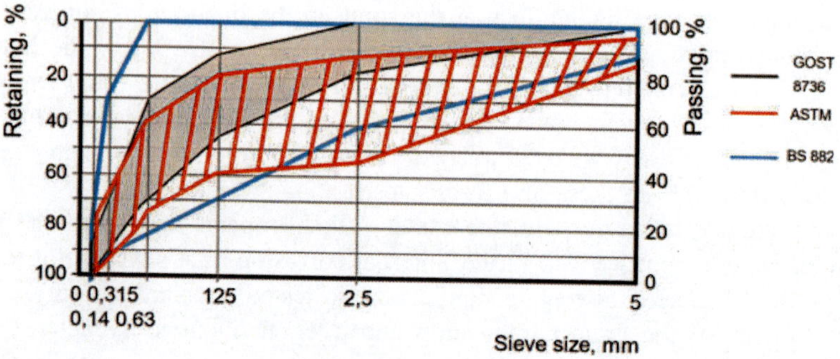

Fig. 3 Recommended grain composition ranges for sands according to different standards

Table 3 Grain composition of studied sands

	Fill sieve residue, %						Fineness modulus
	5	2.5	1.25	0.63	0.315	0.14	
Sand 1	0	20	45	70	90	100	3.25
Sand 2	0	0	15	35	70	90	2.1
Sand 3	0	10	30	52.5	80	95	2.68
Sand 4	8.25	25.08	36.85	66.69	87.52	99.01	3.15
Sand 5	13.61	12.61	20.26	51.2	86.23	98.66	2.68

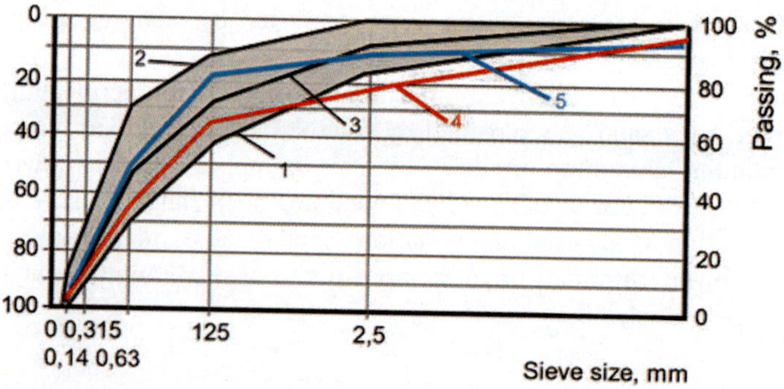

Fig. 4 Grain composition of studied sands

The studies were conducted on the same concrete composition with the unchanged W/C ratio by measuring the type of sand every time. The base composition of concrete is as follows: CS = 1100 kg/m^3 (basalt 5–20 mm of Aegean origin, Armenia), S = 740 kg/m^3, C = 320 kg/m^3 (CEM IIA/42.5 N Ararat plant), W/C = 0.75.

From each batch wit different sands, 6 cylinder samples 150 × 150 mm and 9 cube samples 100 × 100 × 100 mm were formed. Hardening of 6 cylinders and 3

Table 4 Test results for impact of grain composition of sand on properties of concrete and concrete mix

	Concrete mix			Concrete			
				Strength, MPa		Hydrophysical properties	
	Particle composition 0.14–0.315 mm, %	Cone slump, cm	W/C	7 days	28 days*	Water-tightness grade	Penetration rate, mm/s
Sand 1	10	23	0.75	17	25/23	W10	2.37×10^{-3}
Sand 2	20	20	0.75	18	27/27	W12	4.56×10^{-4}
Sand 3	15	21	0.75	21	28/24	W14	3.46×10^{-4}
Sand 4	12	21	0.75	20	30/29	W12	4.43×10^{-4}
Sand 5	13	22	0.75	20	28/27	W12	4.36×10^{-4}

Note above the line is the concrete strength, Day 28, hardening in water, under the line—in the normal conditions. The samples tests on Day 7 also got hardened in water

cubes occurred in the normal conditions, hardening of the remaining cubes occurred in water (to determine the strength accuracy—in 7 and 28 days). The cylinders were tested for permeability on Day 28 after production according to GOST 12,730.5 using the wet spot method. The test results are given in Table 4.

As the W/C ratio was the same in all cases, the capillary porosity should be the same, consequently, the permeability of concrete. But the table shows that for coarse sand concrete (Sand 1) with fine particle content of 10%, the ducts are interconnected that leads to the impaired permeability. The concrete water-tightness grade with this sand is the lowest—W10, and the penetration rate of water determined on Germann GWT-4000 is the highest—2.37×10^{-3} cm/s. It should be noted that for the batch with Sand 1, water separation of concrete mix was observed after formation of samples.

When testing concretes with other sands with the fine particle content of 12–20%, increased permeability by a grade and decreased penetration rate of water are observed.

If, for example, to compare Sand 1 and Sand 4 having the close fineness modulus, concretes based on such sands show different permeability, consequently, the one fineness modulus, without specification of the amount of fine particles, cannot be a factor determining the permeability.

6 Conclusions

- When choosing the crushed stone for waterproofed concrete, it is better to prefer aggregates with maximum grain diameter up to 20–22 mm.

- To improve the permeability of concrete, it is reasonable to use sands with large amount of fine particles (0.14–0.315 mm) or sands which grain composition is within the upper limits recommended by the standards.
- Firstly, fine particles clog the ducts making them discrete, and secondly, prevent water separation of concrete mix, thus improving the concrete macrostructure as a whole.

References

1. Lei B, Li W, Tang Z, Tam VWY, Sun Z (2018) Durability of recycled aggregate concrete under coupling mechanical loading and freeze-thaw cycle in salt-solution. Constr Build Mater 163:840–849
2. Wittmann FH, Zhang P, Zhao T (2006) Influence of combined environmental loads on durability of reinforced concrete structures. Int J Restor Build Monum. 12(4):349–362
3. Park SS, Kwon SJ, Jung SH, Lee SW (2012) Modeling of water permeability in early aged concrete with cracks based on micro pore structure. Constr Build Mater 27(1):597–604
4. Homepage https://www.cement.org/learn/concrete-technology/durability/alkali-aggregate-reaction
5. Smyl D, Ghasemzadeh F, Pour-Ghaz M (2016) Modeling water absorption in concrete and mortar with distributed damage. Constr Build Mater 125:438–449
6. Claisse PA, Elsayad HI, Shaaban IG (1997) Absorption and sorptivity of cover concrete. J Mater Civ Eng 9:105–110
7. Özalp F, Yılmaz HD, Kara M, Kaya O, Sahin A (2016) Effects of recycled aggregates from construction and demolition wastes on mechanical and permeability properties of paving stone, curb and concrete pipes. Constr Build Mater 110:17–23. https://doi.org/10.1016/j.conbuildmat.2016.01.030
8. Bravo M, de Brito J, Evangelista L, Pacheco J (2018) Durability and shrinkage of concrete with CDW as recycled aggregates: Benefits from superplasticizer's incorporation and influence of CDW composition. Constr Build Mater 168:818–830. https://doi.org/10.1016/j.conbuildmat.2018.02.176
9. Bravo M, de Brito J, Pontes J, Evangelista L (2015) Mechanical performance of concrete made with aggregates from construction and demolition waste recycling plants. J Clean Prod 99:59–74. https://doi.org/10.1016/j.jclepro.2015.03.012
10. Simonov MZ, Khudaverdyan VM (1958) Lithoid pumice based hydroengineering concrete. Publishing House of the Academy of Sciences, Yerevan
11. Cachim PB (2009) Mechanical properties of brick aggregate concrete. Constr Build Mater 23:1292–1297. https://doi.org/10.1016/j.conbuildmat.2008.07.023
12. ASTM C33/C33M-11a Standard Specification for Concrete Aggregates, 2011 (n.d.)
13. Anronyan AA (2017) About some peculiarities of the modern methods for determining the permeability of concretes. Concr Technol 9–10:29–33
14. Stark J, Wicht B (2004) Durability of concrete. Oranta, Kiev
15. Omary S, Ghorbel E, Wardeh G (2016) Relationships between recycled concrete aggregates characteristics and recycled aggregates concretes properties. Constr Build Mater 108:163–174. https://doi.org/10.1016/j.conbuildmat.2016.01.042
16. Villagrán Zaccardi YA, Alderete NM, De Belie N (2017) Improved model for capillary absorption in cementitious materials: progress over the fourth root of time. Cem Concr Res 100:153–165. https://doi.org/10.1016/j.cemconres.2017.07.003

17. Evangelista L, de Brito J (2010) Durability performance of concrete made with fine recycled concrete aggregates. Cem Concr Compos 32:9–14. https://doi.org/10.1016/j.cemconcomp.2009.09.005
18. Liu X, Chia KS, Zhang MH (2011) Water absorption, permeability, and resistance to chloride-ion penetration of lightweight aggregate concrete. Constr Build Mater 25(1):335–343
19. Cark AI, Sumer M (1996) The effect of curing temperature on compressive strength and water permeability of concrete, presented at Concrete Technology for Developing Countries, Forth International Conference, East Mediterranean University, Gazimagusa, North Cyprus
20. Demirboğa R, Örüng I, Gül R (2001) Effects of expanded perlite aggregate and mineral admixtures on the compressive strength of low density concretes. Cem Concr Res 31 (11):1627–1632
21. Standard B (1992) Specification for aggregates from natural sources for concrete. BSI, London
22. Adnan SH, Lee YL, Abdul Rahman I, Mohd H, Wimala M (2008) Water permeability of recycled aggregate concrete, presented at the Technology and Innovation for Sustainable Development Conference, Khon Kaen, Thailand, Jan

Process Mining Subsystem in Automotive Enterprise Quality Management System

Elena Malysheva, Sergey Bobrovskiy, Sergey Krapivka, and Svetlana Pivneva

Abstract The paper considers the advantages of process mining methods in the study of automotive enterprise production processes. The general processes structure in the automotive enterprises quality management systems and the use of process mining methods in the automotive enterprises are considered. It is proposed to supplement the quality management system of automotive industry enterprises with a processes mining subsystem. This subsystem should include modules: process selection and data preparation, process model creation and analysis, visualization tools, and process mining algorithms library. The interaction of process mining subsystem with enterprise quality management system elements and enterprise information system is shown. The sequence of actions when creating process models in the process mining subsystem and the criteria for process mining algorithms selection are considered. The possibility of using both commercial and free software products for the subsystem implementation is noted. It is concluded that the intelligent analysis of processes within the quality management system leads to a more effective automotive enterprise quality management, helps to improve business processes, increase productivity and competitiveness.

Keywords Quality management system · Automotive enterprise processes · Process mining

1 Introduction

The widespread introduction of quality management systems [1] has led to the emergence of new quality management methods related to process improvement, such as Six Sigma, Total Quality Management (TQM), reengineering and business process management, as well as the emergence and improvement of relevant standards, including the ISO 9000 quality management standards family [2, 3]. The quality specialist tasks include the interrelated system development and the pro-

E. Malysheva (✉) · S. Bobrovskiy · S. Krapivka · S. Pivneva
Russian State Social University, Wilhelm Pieck street, 4, build.1, Moscow 129226, Russia

A. Bieliatynskyi and V. Breskich (eds.), *Safety in Aviation and Space Technologies*, Lecture Notes in Mechanical Engineering,
https://doi.org/10.1007/978-3-030-85057-9_21

cesses indicators determination, the QMS processes implementation, the collection, calculation and monitoring of performance indicators. Business processes analysis using process mining can significantly help quality management specialists. Process mining often leads to new discoveries and understanding of real processes taking place in the organization [4]. The results of the process study can explain the problems faced by enterprises, identify the reasons for their inefficient work or explain the lack of progress in their activities [5].

Process mining offers a set of techniques to obtain information about business processes based on data from event logs. Event logs are sets of data about events or processes with timestamps indicating when a process starts and ends [6]. These logs are a special database used to identify and verify how the company's processes work and to search unused or unnecessary processes [7]. Process mining technology make it possible to analyze business processes based on the enterprise information system data, to get a clear picture of how these processes are actually performed.

The purpose of this study is to develop the process mining subsystem structure for the automotive industry quality management systems. This subsystem will be used to create quality management system process models and improve processes through audit and process reengineering.

2 Materials and Methods

2.1 The Role of Processes in the Automotive Company Quality Management System

Since the first versions of the ISO 9000 series of standards, the fundamental principle of the quality management system is the process approach. ISO 9001:2015 "Quality management systems—Requirements" [3] introduces a direct requirement to apply a process approach when creating an enterprise quality management system. For each process, it is necessary to define the required inputs and expected outputs, formulate risks and opportunities, including opportunities for improving processes. The process approach in quality management systems implies the need to define the methods used to manage the processes, the resources to support the processes, and the responsibility and authority for these processes.

In accordance with the requirements of the ISO 9001 standard, the organization's quality system must be documented. The structure of quality management system documentation, built according to the ISO 9001 standard, is a hierarchical system of interrelated documents, which has a "constant" component, defined by the standard and a "variable" component, depending on the specific organization. Quality system documentation is usually carried out in the form of quality system procedures.

The company identifies processes that are important for the QMS, studies how they function and interact with each other. QMS specialists monitor the state of these business processes and manage their work. This allows them, together with other specialists, to plan and implement the "continuous improvement" that is part of ISO 9001, to identify business processes elements in which improvements can be made, to assess the risks from changes, to verify that the impacts actually lead to improvements.

Figure 1 shows the ISO 9001:2015 standard structure in the PDCA cycle (Plan-Do-Check-Act) [3]. The PDCA cycle can be applied to all processes and to the quality management system as a whole.

For the automotive industry, the IATF 16949 quality management system standard has been developed [8, 9]. IATF 16949 applies to the design, development, production and, when relevant, installation and servicing of automotive-related products. The IATF 16949:2016 standard shares the same section headings and clause structure as ISO 9001:2015. It should be understood as a supplement to ISO 9001:2015 and used in conjunction with ISO 9001:2015.

The IATF 16949:2016 standard, in comparison with the ISO 9001:2015 standard, expands and complements the requirements for the management system processes. In particular, the requirements for sources and methods of collecting processes information are expanded and supplemented. Let's look at some of them.

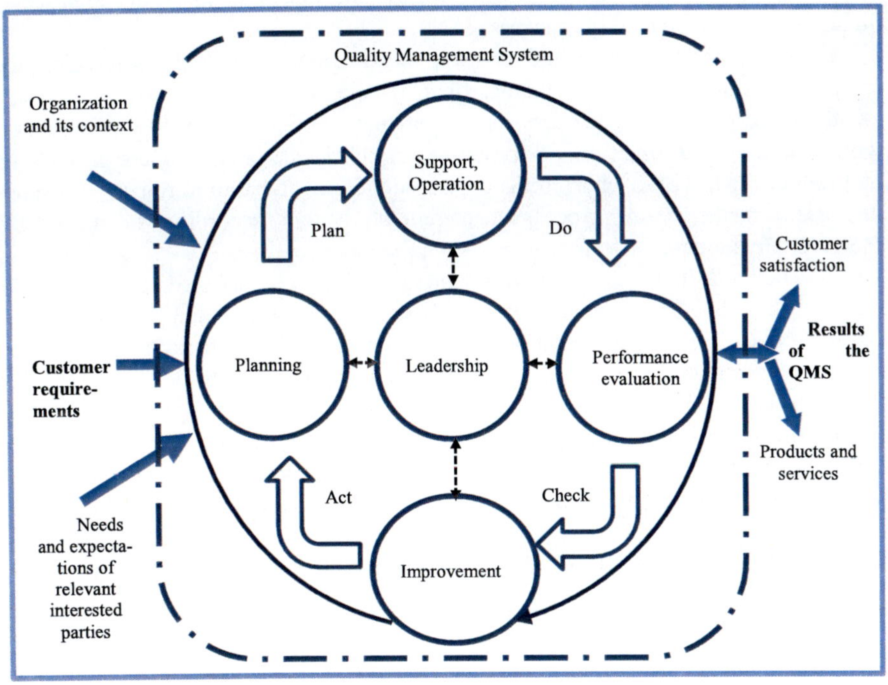

Fig. 1 Representation of ISO 9001 International Standard structure in the PDCA cycle

A multi-disciplinary approach is defined as one of the main collecting information methods for analysis [8]. A multi-disciplinary approach is a method of collecting information from all parties that may affect the management of the management system processes. Stakeholders are organization personnel, customer representatives, supplier representatives. In a multi-disciplinary approach, team members can be internal or external to the organization. The team that manages the process can include existing groups and ad-hoc groups. A multi-disciplinary approach extends to all major organizational processes, including design, manufacturing, engineering, quality, manufacturing, purchasing, suppliers, and other relevant organizational processes.

The organization should use a multidisciplinary approach to analyze information about the management system processes. Based on this analysis of information, the organization determines the processes ability to meet all requirements. These can be technical requirements and requirements set by the consumer. This feasibility study should be carried out for any production or product technology. The analysis is carried out for new processes and products of the organization, and for all changes to existing production processes or product design.

The main process requirements include requirements for monitoring and measuring manufacturing processes. For example, for all new manufacturing processes, the organization must conduct research on those processes. These studies are conducted to test the process's ability to meet specified requirements. As a result of these studies, additional data is obtained for the analysis and control of the process, including data on special characteristics.

When analyzing product and process information, special characteristics are entered separately. A special characteristic is "classification of a product characteristic or manufacturing process parameter that can affect safety or compliance with regulations, fit, function, performance, requirements, or subsequent processing of product" [8]. These characteristics are identified and monitored as the most important in the entire process management system according to the IATF 16949:2016 standard.

The standard includes 13 recommended process characteristics and process information related to product safety. These include, for example: information about products FMEA; information about processes FMEA; identification of product safety related characteristics and the management of these characteristics; information of control plans, etc.

One of the important tools in IATF 16949:2016 is the reaction plan. This plan is based on the results of processes information analysis and is included in the control plan. A reaction plan is an action or series of steps. Reaction plans are developed for unsatisfactory or unstable processes.

These IATF 16949:2016 requirements examples show the importance of developing and applying modern methods for analyzing information about automotive industry processes.

2.2 Process Mining in Automotive Company

The main idea of process mining is to identify, track and improve real processes by extracting knowledge from event logs that are widely available in modern information systems. Process mining includes identifying processes in the form of creating a process model from the event log, verifying compliance by comparing the known process model and log, organizational analytics, and a variety of operations with process models (creating simulation models, adding and correcting models). Petri nets predominate as a modeling language, but recently models are often created in BPMN, which is more practice-oriented than Petri nets. The modeling languages used also include causal networks, finite automata, and simple directed acyclic graphs. The main problems with automated process detection methods when applied to real logs [10] are that they create large and spaghetti-like models, or create models that do not correspond well to the event log, or are overly generalized [11] presents an overview and comparative evaluation of automatic process detection methods using an open source reference test. More than 30 methods that create procedural and declarative models are considered. More than half of the methods provide implementation as a plugin for the open source ProM platform.

Process mining is often used in areas such as sales, purchasing, banking, insurance, and only recently it has been used in manufacturing, including in the automotive industry. An example of such a company is the BMW Group, which has been using Process Mining since 2017 [12]. First of all, this technology is used in the analysis of the main processes: processing and warehousing of raw materials, assembly, testing of products, warehousing and distribution of finished products, marketing and sales, advertising, promotion, pricing, working with sales channels. At the same time, the BMW Group is exploring the possibilities of using Process Mining in support processes, which include general management, finance, accounting, strategic planning, personnel management, technology development, procurement of raw materials, machinery, consumables, and service.

The article [13] discusses the use of Process Mining in the automotive industry, taking into account the basic car production principles in Toyota [14]. Various types of scenarios have been generated in the event logs covering standard and non-standard vehicle manufacturing procedures for a single production line, including scenarios containing production problems. Based on the event logs, several experiments and tests were carried out to automate the car manufacturing processes models creation and to verify that the actual car manufacturing processes were consistent with standard car manufacturing processes.

The ability to apply Process Mining methods is provided by a large amount of data in the semi-automatic production processes event logs of modern automotive companies. An example of such a company is e. GO, founded as electric vehicles semi-automatic manufacturer. E.GO Mobile AG uses a flexible manufacturing model with Autonomous Guided Vehicles (AGVs) that autonomously control each vehicle during its production, and which records every step of the manufacturing process [15].

Process mining methods in combination with other data analysis mechanisms may be of great interest to the automotive industry. In [16], the approach of applying process mining methods consistent with the DMAIC (Define-Measure-Analyze-Improve-Control) model of the Six Sigma model is presented.

3 Process Mining Subsystem Structure

The proposed process mining subsystem is closely related to the information system of the enterprise, its main processes and other subsystems of the enterprise quality management system. It includes the following modules: process selection and data preparation, process model creation and analysis, visualization tools, and process mining algorithms library. The process mining subsystem structure and its interaction with the enterprise quality management system elements and the enterprise information system is shown in Fig. 2. Working with a process mining subsystem begins with process selection and data preparation. The investigated processes, as a rule, are selected from the group of the main enterprise processes and the QMS processes.

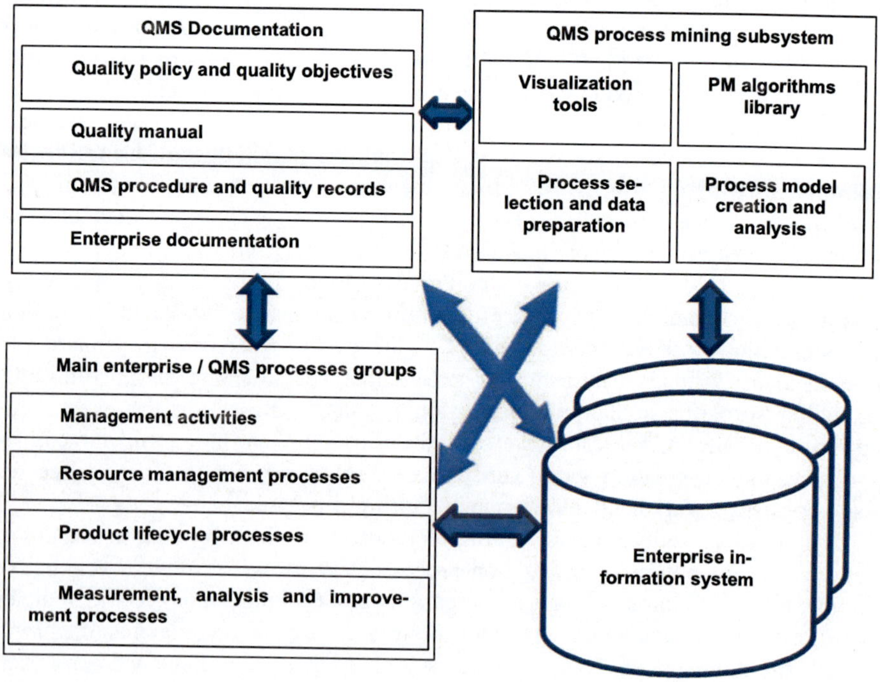

Fig. 2 Process mining subsystem structure in the enterprise quality management system

First of all, these are the main production processes and measurement processes, the data about which are recorded in the enterprise information system. Event log data is used as input to create a business process model. Examples of processes that generate information for event logs in the automotive industry are casting, stamping, painting, car assembly, and quality control. The choice of processes is determined by the tasks solved by the QMS, the company characteristics, the analyst qualifications and the process mining technology capabilities.

Since the data storage format in the enterprise information system may differ from the presentation required for restoring processes, the module should include functions for converting and preparing data from the enterprise information system into the required set of event logs. It should be noted that the log quality can be very different, and it may be necessary to adapt the enterprise information system to be able to obtain reliable data of the required volume and quality.

In the process model creation and analysis module, the prepared event logs are used as input data for the algorithms that create the process model. We propose to use the Petri net as the main tool for representing the process model, since it is more universal and is able to visualize the transition of events into each other based on event logs. However, the subsystem should be able to represent the process in the form of a BPMN model, which is widely used in process modeling. Recently, many process mining tools have appeared that work with BPMN models. The process model creation and analysis module works together with the PM algorithms library and the visualization module.

The sequence of actions for creating process models is standard for Process mining. At the first stage, a process model is extracted from the logs using process discovery methods. Then, using conformance checking mechanisms, discrepancies between the initial model and the event protocol are determined. At this stage, the resulting model can be compared with the process model from the quality management system documentation and the QMS model can be analyzed using real data from the event log. The third stage is the model enhancement stage, in which the model is corrected or supplemented. The resulting model can be used to further analyze and improve the investigated process.

Since the quality management system modeling language and the subsystem algorithms modeling language may differ, it requires tools for converting the model from one modeling language to another, as well as highly qualified specialists working with the subsystem.

The same event logs can be used by different Process mining algorithms. Creating different models based on the same data and presenting them in different ways allows analysts to compare the simulation results and look at the process under study from different perspectives.

The visualization module is used at all stages of working with processes. Includes the ability to use the dotted chart to display and pre-evaluate events, shows the identified model and its behavior. The peculiarity of this module in the subsystem is that it should not only show the restored and analyzed process, but also make it possible to compare it with the representation of the process in the quality management system documentation.

The PM methods library allows the use of various PM algorithms, such as Alpha algorithms, Inductive Miner, Heuristics Miner, Genetic Miner, ILP Miner, Fuzzy Miner. The choice of algorithms depends on many criteria: the specifics of the tasks to be solved, the quantity and quality of the event log data, the capabilities and limitations of the algorithms used, and the execution time.

The process mining subsystem implementation environment depends primarily on the means of enterprise information systems implementing and the qualifications of analysts. For the process mining subsystem, you can use both commercial software products—such as Celonis, Minit, ARIS Cloud, LANA Process Mining, and such reliable open source tools as ProM and RapidProM. The ProM platform also offers a selection of plug-ins for integration with popular information systems.

4 Discussion

The process mining subsystems capabilities development requires the development of approaches in several directions. The first direction is the selection of quality management system processes optimal set that generates data for the subsystem. This requires additional preliminary research to analyze the degree of various processes influence on the efficiency and effectiveness of the QMS and the enterprise as a whole.

The second direction involves the study of the universality of the solutions obtained. To what extent can the structural and software solutions of the automotive industry process mining subsystem be applied in other industries?

The third direction is related to the further development and optimization of the subsystem structure itself. For example, how can you implement feedbacks from the subsystem operation analytical results to the inputs of the analyzed processes? A separate direction can be formulated to solve the problem of optimal selection and expansion of the composition of the analytical tools.

5 Conclusion

In connection with the study of the process mining subsystem capabilities in the automotive enterprises quality management systems and the development of the subsystem structure, the following conclusions can be drawn:

- the subsystem should be able to obtain information about the enterprise's processes and the quality management system processes from the enterprise information system;
- since the event logs at automotive enterprises are of very different quality, it is necessary to adapt the enterprise information system in order to obtain reliable data of the required volume and quality;

- the subsystem should be able to store and display process models from the quality management system documentation;
- to compare models of different types, including QMS models and models created by process mining algorithms, the subsystem must have tools for converting the model from one modeling language to another;
- both commercial software products and free PM tools can be used to create a subsystem. The process mining subsystem implementation environment depends on the qualifications of analysts and the means of implementing enterprise information systems.

Identification, analysis, modification and simulation of business processes by PM tools will help to implement "continuous improvement" of business processes in accordance with the ISO 9001 standard requirements. Thus, the use of the process mining subsystem within the automobile manufacturing enterprise quality management system allows solving problems related to automobile production processes, leads to more effective quality management, increased productivity and competitiveness.

References

1. Peris-Ortiz M, Alvarez-Garcia J (2014) Action-based quality management: strategy and tools for continuous improvement. Springer International Publishing. https://doi.org/10.1007/978-3-319-06453-6
2. ISO 9000 family quality management. International Organization for Standardization. https://www.iso.org/iso-9001-quality-management.html. Last accessed 2021/04/10
3. ISO 9001:2015 Quality management systems—requirements. International Organization for Standardization. https://www.iso.org/standard/62085.html. Last accessed 2021/04/10
4. van der Aalst WM, Reijers HA, Weijters AJ, van Dongen BF, De Medeiros AA, Song M et al (2007) Business process mining: an industrial application. Inf Syst 32(5):713–732. https://doi.org/10.1016/j.is.2006.05.003
5. van der Aalst WM (2011) Process mining: discovery, conformance and enhancement of business processes. Springer, Berlin. https://doi.org/10.1007/978-3-642-19345-3
6. van der Aalst WM (2016) Process mining. Data science in action. Springer, Berlin, Heidelberg. https://doi.org/10.1007/978-3-662-49851-4
7. González López de Murillas E, Reijers HA, van der Aalst WMP (2016) Connecting databases with process mining: a meta model and toolset. In: Schmidt R, Guédria W, Bider I, Guerreiro S (eds) Enterprise, business-process and information systems modeling. BPMDS 2016, EMMSAD 2016. Lecture notes in business information processing, vol 248. Springer, Cham. https://doi.org/10.1007/978-3-319-39429-9_15
8. IATF 16949:2016. International Automotive Task Force. https://www.iatfglobaloversight.org/iatf-169492016. Last accessed 2021/04/10
9. Gruszka J, Misztal A (2017) The new IATF 16949:2016 standard in the automotive supply chain. Research in logistics and production, vol 7, No 4, pp 311–318. https://doi.org/10.21008/j.2083-4950.2017.7.4.3
10. De Weerdt J, De Backer M, Vanthienen J, Baesens B (2012) A multi-dimensional quality assessment of state-of-the-art process discovery algorithms using real-life event logs. Inf Syst 37(7):654–676. https://doi.org/10.1016/j.is.2012.02.004

11. Augusto A, Conforti R, Dumas M, La Rosa M, Maggi FM, Marrella A, Mecella M (2019) A.r Soo: automated discovery of process models from event logs: review and benchmark. IEEE transactions on knowledge and data engineering, vol 31, No 4, pp 686–705. https://doi.org/10.1109/TKDE.2018.2841877
12. Lechner P (2020) BMW: process mining @ Production. In: Reinkemeyer L (eds) Process mining in action. Springer, Cham. https://doi.org/10.1007/978-3-030-40172-6_11
13. Siek M, Mukti RMG (2020) Process mining with applications to automotive industry. 2020 IOP Conf Ser Mater Sci Eng 924, 012033. https://doi.org/10.1088/1757-899X/924/1/012033
14. Toyota. How we manufacture. http://www.toyotaindonesiamanufacturing.co.id/manufacturing/how-we-manufacture?lang=id
15. Bauernhansl T et al (2018) The digital shadow of production—a concept for the effective and efficient information supply in dynamic industrial environments. Pr. CIRP 72:69–74. https://doi.org/10.1016/j.procir.2018.03.188
16. Graafmans T, Turetken O, Poppelaars H, Fahland D (2020) Process mining for six sigma: a guideline and tool support. Bus Inf Syst Eng. https://doi.org/10.1007/s12599-020-00649-w

Flexible Maintenance Systems to Prevention Road Failures of Vehicles

Sergey Moroz

Abstract Flexible systems of maintenance of modern vehicles, integrated at the same time into the on-board local network and information systems of service enterprises, adjust the maintenance schedule individually for each vehicle. Flexible maintenance systems replaced the fixed frequency of service books, automated accounting of operating time, operating intensity and displaying an adjustable maintenance schedule. But auto manufactures put in flexible maintenance systems adjusting only the minimum service intervals in accordance with the conditions and intensity of operation. This does not ensure the prevention of road failures or the timeliness of repairs. To counteract road failures and unjustified downtime in repairs, it is proposed to reduce maintenance intervals of larger labor intensity individually for each vehicle in accordance with a relative increase in downtime in repair and maintenance or a decrease in operating time as the vehicle is worn out and aging. As a result, the known correction will be supplemented by a new function of increasing the frequency of maintenance of greater labor intensity as the vehicle's resource is depleted. At the same time, maintenance intervals for vehicle maintenance of larger labor intensity will need to be adjusted to a much greater extent than for maintenance of less labor intensity. The proposed development of flexible vehicle maintenance systems will be a further individualization of the vehicle service, ensuring timely repairs.

Keywords Vehicle · Maintenance · Flexible maintenance systems · Interservice interval · Failure · Running time · Repair

---

S. Moroz (✉)
Moscow Automobile and Road Construction State Technical University (MADI), 64, Leningradsky Prosp., Moscow 125319, Russia
e-mail: sm-moroz@mail.ru

A. Bieliatynskyi and V. Breskich (eds.), *Safety in Aviation and Space Technologies*, Lecture Notes in Mechanical Engineering,
https://doi.org/10.1007/978-3-030-85057-9_22

1 Introduction

Wear and aging increase the rate of deterioration of the technical condition of vehicles, the frequency of failures and exits from the line. The preventive capabilities of maintenance decrease as the vehicle resource is exhausted, as a result, the frequency of failures, labor intensity and downtime in maintenance and repair increase by 4…5 times or more (Figs. 1, 2).

The Russian scheduled preventive maintenance system since the time of the planned economy was not able to counteract this. With constant periodicities of all

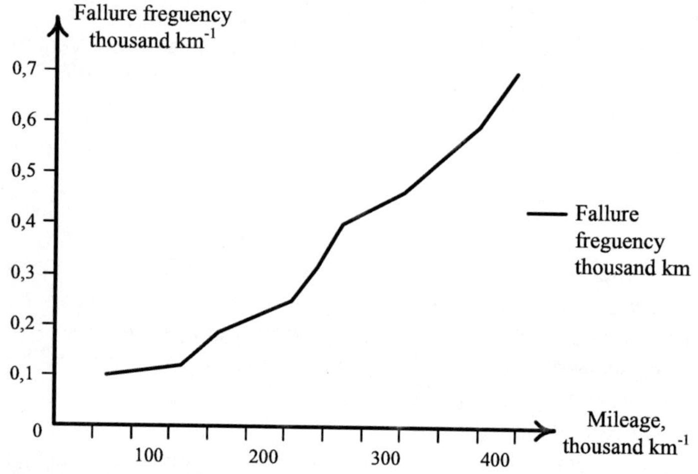

Fig. 1 Change failure flow from the beginning of operation up to the vehicle disposal

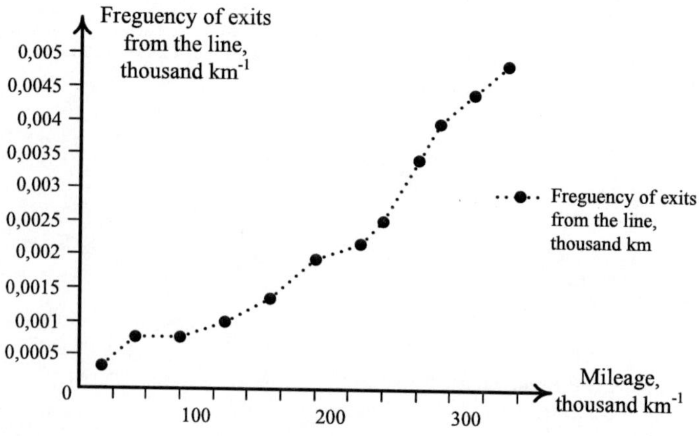

Fig. 2 Frequency dynamics of exits from the line as vehicle resource is exhausted

types of maintenance, it only recorded a three-fold increase in the standards of labor-intensive maintenance from the beginning of the vehicle operation to the resource expiration, and with further continuation of the operation, by 5 times. Thus, the convenience of planning and managing the installation of vehicle for maintenance with manual accounting of mileage and maintenance planning was achieved, and variable periodicities would significantly complicate the work of technical services of motor transport enterprises (MTE). Moreover, MTEs with a different brand fleet equalized the maintenance intervals of different vehicles, regardless of manufacturers' recommendations.

Service book maintenance systems also kept the service intervals unchanged. To counteract the depreciation and aging of the vehicle, such measures were proposed as diagnosing with a frequency greater than the frequency of maintenance; scheduled preventive repair and two-stage regulation of the maintenance frequency as the vehicle wears out. However, these proposals were not implemented.

Scheduled repair or diagnostic are not economically justified: the capital costs of creating the required production capacities and the current costs of the frequent implementation of these influences significantly exceed the cost of eliminating failures. A two-stage increase in the frequency of maintenance is not consistent with the separation of maintenance by type, taking into account the resources of motor and transmission oils and lubricants, and a five-fold increase in the flow of failures.

Equipping the vehicle with a local network has led to the creation of flexible maintenance systems that adjust the maintenance schedule individually for each vehicle. Flexible maintenance systems are already crowding out the maintenance system with service books. Flexible maintenance systems created in order to reduce the manufacturer's costs for warranty repair have an increasingly significant impact on the technical operation of cars in general. Initially, they only reduced the fixed base frequency of service books during intensive operation and in the harsh operating conditions of vehicle. Flexible systems automate the accounting of running time, operating intensity and display on the electronic dashboard of an individual adjustable maintenance schedule for vehicle, as well as monitoring its compliance [1].

The first generation of flexible maintenance systems (for example, the Telligent system of Mercedes-Benz trucks) provided for software adjustment of running time only to maintenance less labor intensity and only depending on the intensity indicators and operating conditions, but not on the "age" or degree of resource expiration of the vehicle. At the same time, the adjustment range did not exceed 50% of the basic standards.

Modern flexible maintenance systems are integrated not only into the on-board local area networks of the vehicle, but via GSM or GPS also into the information systems of the technical support center of the branded technical service network of the vehicle manufacturers, as well as into information systems of the motor transport enterprises and service stations (for example, Volkswagen Long Life system). This provides not only convenience to the owners, reduction of downtime in anticipation of repairs or spare parts, but also opportunities for dealers to monitor during the warranty period the intensity of operation and compliance with the requirements for servicing the vehicle. However, the trajectory of the previous development of flexible maintenance systems pursued mainly minimization of downtime and arrivals of vehicle for repairs and maintenance [2, 3].

Such systems are already used by all leading vehicle manufacturers [4–7]. But even in them, the running time to maintenance is corrected only depending on the automatically controlled parameters of the driving style and engine load, and the interservice intervals do not decrease with an increase in the flow of failures as the vehicle resource is exhausted [8]. Only the minimum interservice intervals (similar to first maintenance) are corrected by the signals of the sensors of aging and engine oil level. As a result, timely repair is not achieved, and road failures are not prevented.

A number of studies have worked out methods for adjusting the nomenclature of maintenance operations depending on the projected residual life of the components of heavy trucks [9] and light army trucks [10]. New technologies for remote interaction and technical support for service stations and owners of vehicle for maintenance and repair were developed, ensuring the minimization of downtime and arrivals for repairs and maintenance [11], as well as the reduction of errors during their implementation [12] and remote diagnostics [13, 14].

These are promising areas for further improvement of the maintenance system and repair by means of informatization. But they do not solve the initial problem of counteracting a five-fold increase in the frequency of road failures and unjustified downtime in the repair as the vehicle resource is exhausted.

Further "intellectualization" of the maintenance performance gives new opportunities to counteract road failures and unjustified downtime of vehicle in repair.

2 Material and Methods

However, there are not many possibilities for regulating maintenance intervals.

The number of types (steps) of maintenance and their frequency are appointed by manufactures expertly, depending on only three groups of indicators:

1. Resources of oils and lubricants on the chemotological map of vehicle.
2. Resources of the most wearing parts of vehicle (filters, brake linings, belts, seal covers, etc.).
3. Average statistical resources of more durable components, dependent mainly on the operating time, and not the conditions or intensity of operation (catalytic converters, spark plugs, seals of wheel brake cylinders, steering mechanisms, etc.).

The frequency and number of types of maintenance of less labor intensity (similar to first maintenance) are connected with the resources of oils and the most wearing components. Individual adjustment of shorter interservice intervals for each vehicle is productive only if automation of control of oils, filters, brake linings, etc. is implemented. It was implemented by flexible maintenance systems depending on operating conditions and intensity and became their first function. But its effectiveness is limited by the manipulation of interservice maintenance intervals of less labor intensiveness. This adjustment does not take into account changes in the frequency of vehicle failures as it deteriorates and ages and is not able to reflect them.

It is advisable to supplement such correction with a new, no less significant function. As a counter to road failures and unjustified repair downtimes, it is proposed to increase the frequency of maintenance of greater labor intensity (similar to second maintenance and performed less frequently) as the vehicle resource is exhausted, and to carry out the timeliness repair with it. More frequent maintenance will make it possible to more frequently monitor and diagnose the vehicle, and therefore, proactively identify the development of a larger number of emerging faults and failures, and prevent the development of chain failures.

For this, the interservice maintenance intervals of greater labor intensity will need to be adjusted by flexible systems in a much larger range than for maintenance of lower labor intensity. This second correction mechanism will complement the correction already implemented by flexible systems within 100% of the minimum maintenance intervals (similar to first Russian maintenance) depending on the intensity and operating conditions.

The purpose of this supplement is to minimize line exits due to road vehicle failures in commercial transport and the costs associated with them by transferring the elimination of the majority of failures from the claimed to the related and preventive repair. Such a transfer is quite real for gradually developing failures, accompanied by proactive signs. At the same time, the claimed repair will be reduced by more than half to the volume of sudden unpredictable failures (such as chip failures, spring breaks, clamping of pipelines, etc.).

The dynamics of preventable failures (let's call them gradual failures, accompanied by proactive signs) that interrupt the operation of vehicle, is usually characterized by regression dependences on the mileage of an exponential form. In particular, this was experimentally confirmed for small-class buses Mercedes-Benz Sprinter 411cdi and Iveco Daily 50C15 operated by Sever-Avto LLC. As a result of processing the statistics of their failures for four years of passenger transportation, a regression dependence of the frequency F(L) of failures on the mileage L from the beginning of operation was obtained taking into account the dispersion of the mean time between failures and the frequency of failures. By the least squares method, it is approximated by an exponential dependence of the form:

$$F(L) = 0.5524e^{0.001856L} \tag{1}$$

However, preventing exits from the line or absenteeism on the line of vehicle is only possible with respect to failures accompanied by proactive signs. Identification of such signs by organoleptic methods or diagnostics during planned maintenance prior to failure will prevent returning from the line and absenteeism to the line for claimed repair [10]. At the same time, a more frequent maintenance "replaces" not the failure itself, but its unplanned "claimed" elimination and the associated costs that are transferred from the claimed to a related or preventive repair.

To assess the proportion of preventable failures, an expert survey of specialists from the technical service of Sever-Avto LLC was carried out. According to its results, a classifier of signs of preventable failures was formed. It included 26 signs of failures, confidently detected organoleptically, and 50 signs detected by diagnostics within the 20…50% running time until the maintenance of a higher periodicity (for Mercedes-Benz Sprinter 411cdi—5 thousand km, and for Iveco Daily—6.6 thousand km). The developed classifier made it possible to single out the nomenclature of failures, accompanied by proactive signs, from the general list of failures. It included 70% of failures that required the performance of claimed repair on a mileage of 100 to 300 thousand km. An expert analysis showed that at least 85…90% of them could have been prevented by means of concomitant repair during the next maintenance, or preventive (before failure) repair.

The regression dependence of the change in the frequency of "predictable" failures F1(L) (which cannot be identified with the predicted ones) of vehicle, on the mileage L from the beginning of operation, is approximated by a power-law function (Fig. 3):

$$F_1(L) = 0.303e^{0.001856L} \tag{2}$$

An analysis of the change in the ratio of the number of preventable failures F1 (L) to their total number F(L) according to data collected over a mileage of 100 to 300 thousand km, which were approximated by the given power functions (2) and (1), respectively, showed that it does not go beyond the limits of 0.595 … 0.638.

This confirms the possibility of detecting during maintenance and proactive elimination of at least 60% of failures. These are not yet expressed vehicle failures requiring immediate elimination, but malfunctions, accompanied by visible signs of an approaching failure. Therefore, their elimination can be carried out depending on

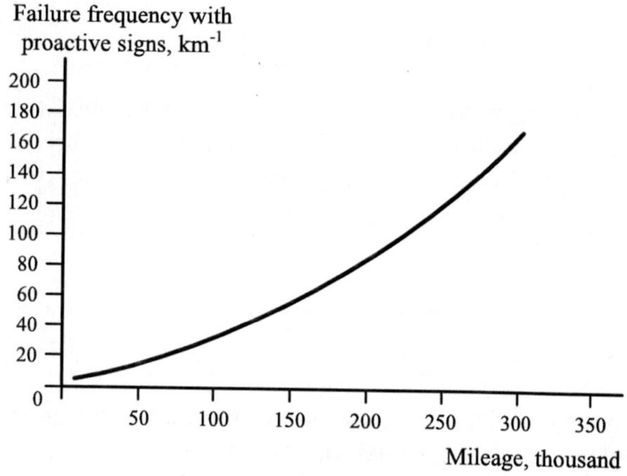

Fig. 3 Approximation by the exponential dependence of the increase in the frequency of failures, accompanied by proactive signs, as the bus resource is exhausted (green—data about failures with proactive signs, red—approximation)

the degree of the malfunction development and the expected consequences of the failure in the convenient period for the vehicle owner by means of the planned preventive repair or the associated repair timed to the completion of the maintenance.

In addition, proactive repair work will reduce the dependent failures of the vehicle components caused by a chain of failures. This will reduce the cost of manufacturers for spare parts during the warranty period of operation.

3 Theory/Calculation

However, the straightforward use of the laws governing the increase in the flow of vehicle failures, accompanied by proactive signs, for regulating the interservice maintenance intervals in practice is not feasible. For vehicles of different families, manufacturers and years of manufacture, the dynamics of these failures will be different and the proactive signs will be different. The accumulation of data to approximate their failures and programming the laws of reduction of running time to maintenance would have to be repeated "manually" for each new vehicle model. Today this is unacceptable, as it does not lend itself to automation and thereby excludes such an approach.

Instead of the dynamics of failures, it would be possible to adjust the interservice maintenance intervals by the frequency of exits from the line and absenteeism to the line. But the accounting of these data is carried out manually, and only at the best motor transport enterprises and it is difficult to automate. Data on the accounting for the labor intensity of repair accumulated by the technical service of enterprises is also unsuitable for this, since they require preliminary manual sorting.

Instead, it is proposed to adjust the interservice maintenance intervals of greater labor intensity depending on the increase in downtime of the vehicle in repair as the resource is exhausted, or the duration the operation on the line is reduced. The accounting of downtime and time in work is easily automated by software according to accounting data at motor transport enterprises. The proposed method eliminates the need for preliminary accumulation of data before the operation of the vehicle. Instead of determining the form of the mathematical dependence of the dynamics of the exits from the line or the labor intensity from the running time of the vehicle, it is sufficient to monitor the relative increase in downtime in repair and maintenance or the reduction of the operating time as the resource of the vehicle is exhausted.

The regularity of the correction of maintenance intervals between maintenance as the vehicle resource is exhausted can be the function of a relative reduction in the duration of work on the line:

$$L_j = L_0 \frac{t_i}{t_1} \quad (3)$$

where

L_j and L_0 the rational value of the j-th interservice interval and its initial value at the beginning of the vehicle operation, respectively;

t1 and ti the duration of work on the line, recorded respectively in the 1st and i-th periods of operation.

Instead of function (3), the relative increase in specific (per week or month) vehicle downtime in maintenance and repair can also be used to correct interservice maintenance intervals:

$$L_j = L_0 \frac{q_i}{q_1} \quad (4)$$

where

q1 and qi the vehicle downtime in maintenance and repair, respectively, for the 1st and i-th periods of operation during a given duration ti.

At the same time, the accounting of downtime in maintenance and repair is automated by summing up the duration of the vehicle stay in the production premises of the motor transport enterprises, recorded, for example, on the facts of entry-exit to the production premises of the maintenance and repair, or the presence of the vehicle in the zone of Wi-Fi points in these rooms on the territory of the motor transport enterprises.

Previously, in both correction options, the value of the i-th operation periods ti should be set smaller than the initial value T0 of the interservice interval (in units of time). For example, ti is convenient to take equal to the month. Initial values L0 of interservice maintenance intervals will be assigned by the vehicle manufacturers, as in modern flexible systems. At the same time, adjustments by flexible systems of the initial interservice maintenance intervals of less labor intensity in terms of driving style and engine load indicators, reflecting the intensity and operating conditions, will not undergo changes. It is only advisable to supplement the nomenclature of these indicators, taking into account the share of mileage in lower gears, engine starting frequencies at low temperatures, the frequencies of acceleration and braking maneuvers. Thus, it is proposed to supplement flexible maintenance with a new additional function for correcting interservice maintenance intervals of greater intensiveness.

The structure of the algorithm of the proposed automatic correction of the interservice maintenance intervals of greater labor intensity is shown in Fig. 4.

Such additional adjustment of interservice intervals will be equally demanded not only at passenger, but also at trucking motor transport enterprises and individual entrepreneurs engaged in the maintenance and repair of their vehicles by third-party contractors.

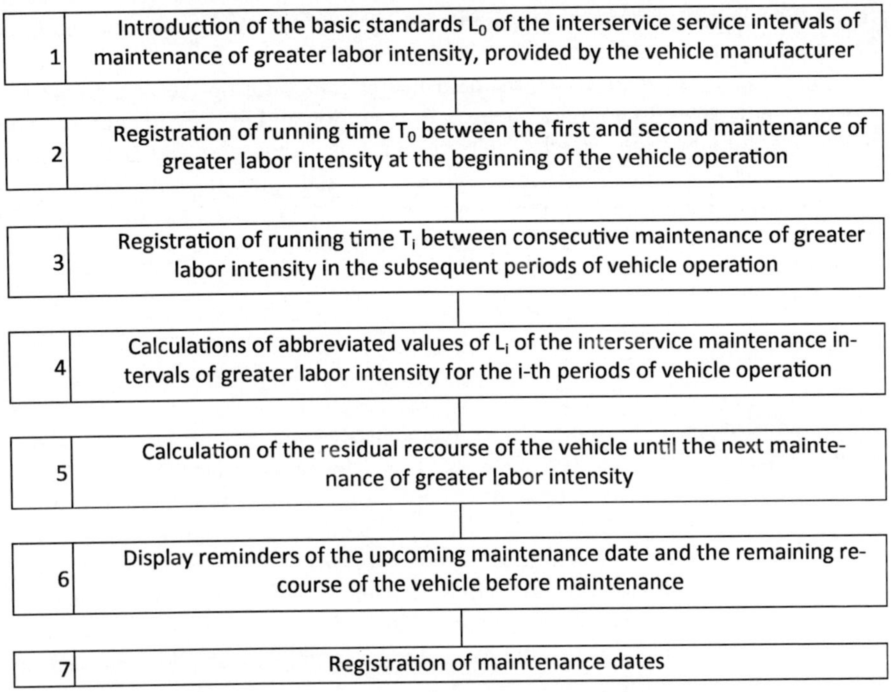

Fig. 4 Structure of the algorithm for automatic correction by a flexible maintenance system of interservice intervals of greater labor intensity

Implementation of the developed algorithm for automatic correction by a flexible maintenance system of interservice intervals of greater labor intensity is proposed for vehicle manufacturers and will not be economically justified on operating vehicles. For the repair and maintenance personnel of motor transport enterprises, it will not be able to do and will require additional hardware costs for the corresponding modernization of the on-board local area network of vehicles.

4 Results

Thus, instead of a single priori regularity of adjustment for each vehicle model not provided with the initial data, before the beginning of operation, it is proposed to reduce the interservice maintenance intervals of the greater labor intensity of each vehicle individually. Such a reduction is automated by software based on the results of monitoring the relative increase in downtime in maintenance and repair or reducing the time in the vehicle operation. The accumulation of the indicated initial data for performing the correction can be automated by software for almost all modern vehicle structures equipped with an onboard local area network and GSM communications.

The reduction of interservice maintenance intervals during the adjustment process will be adapted to both the intensity of operation and the degradation decrease in the reliability of the vehicle as its resource is exhausted, and the degree of reduction will change in accordance with its wear and aging from the beginning of operation to disposal and will be 5…6 times higher than the reduction maintenance intervals of less labor intensity in operating flexible systems.

5 Discussion

Calculations of decreasing interservice maintenance intervals can be automated by means of the on-board local area network of vehicles in accordance with changes in the average daily mileage or downtime during operation. It is proposed to implement the correction algorithm and the display of recommendations on the timeliness of maintenance by the on-board local area network of the vehicle and by means of data transmission via GSM or GLONASS channels using the computing capabilities and the display of the electronic control unit of the electronic dashboard of the vehicle. To do this, the vehicle will not need to be equipped with additional sensors, any additional units or dashboard controls.

All vehicles of modern designs are equipped with an on-board local area network with the functions of a flexible maintenance system and automatic transmission of data on the planned timing of maintenance to the information networks of the dealer and the motor transport company (or service station). Therefore, the organization of the introduction of a new service will be just a buildup of the functions of the already used flexible maintenance systems.

Such an introduction is implemented in the shortest possible time for vehicle manufacturers and will not require additional hardware costs and capital investments, or preliminary design and settlement work. It is almost completely doable by software with the next software updates of the electronic control unit of the intelligent dashboard, the onboard local area network of the vehicle and the proprietary information system of the vehicle manufacturer.

The proposed automatic reduction of interservice intervals will not complicate the planning of vehicles for maintenance [14]. It is acceptable for vehicles equipped with a local network of different designs, purpose and technical level.

6 Conclusions

The proposed correction of interservice maintenance intervals of greater labor intensity will allow transferring most of the troubleshooting work from the claimed to the accompanying and pre-planned preventive repair. Instead of operating the vehicle to failure and the subsequent forced elimination of emergency failures by means of claimed repair, it will be possible to proactively eliminate at least half of the total number of failures during a convenient maintenance period for the vehicle owner. Thus, it will be possible to minimize road failures and the downtime caused by them, associated with the greatest cost of eliminating them, by involving technical assistance teams or tow trucks. With a significant exhaustion of the resource of city buses, this measure is able to reduce the number of road failures and decommissioning cases for unscheduled statement to claimed repair by 2...3 times, and as a result, reduce to a minimum the failures in bus schedules and the number of required reserve buses for passenger motor transport enterprises.

References

1. Robert E, Bérenguer C, Bouvard K, Tedie H, Lesobre R (2018) A predictive approach to jointly schedule missions and maintenances for a deteriorating vehicle. Safety and reliability—safe societies in a changing world—Proceedings of the 28th international European safety and reliability conference, ESREL 2018, pp 529–538
2. Zhang D, Hu H (2014) An optimization on subway vehicle maintenance using a multi-population genetic algorithm. Sustainable development of critical infrastructure—Proceedings of the 2014 international conference on sustainable development of critical infrastructure, pp 316–323
3. Furch J (2014) Advanced maintenance systems of military vehicles. Intelligent technologies in logistics and mechatronics systems, ITELMS 2014—Proceedings of the 9th international conference, pp 96–103
4. VW Free Multi-Point Inspection|Volkswagen Basic Services—URL: https://www.vwserviceandparts.com/service/basic-services/multi-point-inspection/
5. BMW check up & maintenance—URL: https://www.bmw.in/en/topics/owners/service-and-warranty-original/workshop-offers/bmw-check-up-and-maintenance.html
6. Scania is introducing flexible maintenance plans|Scania Group.—URL: https://www.scania.com/group/en/scania-is-introducing-flexible-maintenance-plans-increased-availability-when-actual-usage-governs-truck-maintenance/
7. Service Plans|Service Plans, Warranty & Assistance|Peugeot UK.—URL: https://www.peugeot.co.uk/servicing/peugeot-service-plans/
8. Sirvio KM (2015) Intelligent systems in maintenance planning and management. Intell Syst Ref Libr 87:221–245
9. Biteus J, Lindgren T (2017) Planning flexible maintenance for heavy trucks using machine learning models, constraint programming, and route optimization. SAE Int J Mater Manufact 10(3)
10. Grantner J, Bazuin B, Dong L, Al-shawawreh J, Hathaway R, Fajardo C, Castanier MP, Hussain S (2010) Condition based maintenance for light trucks. Conference proceedings—IEEE international conference on systems, man and cybernetics 5641796:336–342
11. Lesobre R, Bouvard K, Bérenguer C, Grall A (2012) Tailored maintenance optimization within maintenance on demand project. 11th international probabilistic safety assessment and management conference and the annual European safety and reliability conference 2012, PSAM11 ESREL 2012, vol 1, pp 491–499
12. James AT, Gandhi OP, Deshmukh SG (2017) Assessment of failures in automobiles due to maintenance errors. Int J Syst Assur Eng Manag 8(4):719–739

13. Yang X, Gong Z, Wang S, Zhang J (2013) Research on network centric based remote fault diagnosis and maintenance system to vehicle equipment. QR2MSE 2013—Proceedings of 2013 international conference on quality, reliability, risk, maintenance, and safety engineering, статья No 6625936, pp 1843–1845
14. Prasad MRS (2017) Optimum level of preventive maintenance for a fleet of vehicles. J Inst Eng India Part I, V 62 (pt 1–1)

Complex for Research of Power and Speed Losses in Tires of the Car, at Their Rolling on Rollers of the Diagnostic Stand

Oleg Yankov, Anton Chernyshkov, and Roman Sergushko

Abstract The article presents the calculation of the efficiency of an elastic tire of an automobile wheel. Its influence on the errors associated with the development of mathematical models of the car acceleration process in both road and bench conditions is revealed. A detailed description of the design and device of the stand for the study of power losses and speed losses in the tires of the car when they roll along the rollers of the diagnostic stand is given, as well as the principle of its operation. Conducting experimental studies on this stand allows us to obtain characteristics that describe the dependence of power and speed losses on the speed of rotation and the supplied torque to the wheels of passenger cars with elastic tires of various sizes and brands. The dependences obtained as a result of experimental studies on the equipment described in the article are planned to be used in the future in the development and debugging of mathematical models for describing the processes of functioning of cars on stands with running drums while controlling their traction and dynamic qualities.

Keywords Tire · Speed loss · Tire efficiency · Power loss · Wheel radius · Skating resistance · Roller stand · Strain gauge · Optical sensor · Electric current frequency converter

1 Introduction

Diagnostic stands with running drums are now quite widespread. Such stands allow monitoring and diagnostics of aggregates, systems and units of wheeled vehicles by implementing the principle of reversibility of movement when the car is standing, and the road moves under its wheels. Often, comparing the results of analytical studies of the processes of functioning of wheeled vehicles on stands or in road conditions with their experimental counterparts, you can find significant differences of 4–12%, called calculation errors [1–4].

O. Yankov · A. Chernyshkov (✉) · R. Sergushko
Irkutsk National Research Technical University, 83 Lermontov str., 664074 Irkutsk, Russia

A. Bieliatynskyi and V. Breskich (eds.), *Safety in Aviation and Space Technologies*, Lecture Notes in Mechanical Engineering,
https://doi.org/10.1007/978-3-030-85057-9_23

Despite the fact that modern mathematical models take into account all the nuances of the functioning processes of vehicles, their results always contain errors associated with power losses in the tires. The total efficiency of the bus ηK, in the General case, is determined from the following expression:

$$\eta_{\mathrm{k}} = \frac{N_{\mathrm{M}}}{N_{\mathrm{k}}} = \frac{R_{\mathrm{x}} \cdot V}{M_{\mathrm{k}} \cdot \omega_{\mathrm{k}}} = \frac{M_{\mathrm{k}} - M_f}{r_{\mathrm{ko}}} \cdot \frac{V}{M_{\mathrm{k}} \cdot \omega_{\mathrm{k}}} = \frac{M_{\mathrm{k}} - M_f}{M_{\mathrm{k}}} \cdot \frac{r_{\mathrm{k}}}{r_{\mathrm{ko}}} \tag{1}$$

where

N_K и N_M	accordingly, the power supplied to the wheel and allocated from the wheel, W;
ωK	wheel speed, rad/s;
V	car speed, m/s;
Mf	tire rolling torque, Nm;
RX	longitudinal reaction at the contact patch, N;
rk and rko	rolling radius and rolling radius in slave mode, respectively, m;
MK	wheel torque, H [1–4].

Having simplified the expression (1), the efficiency of the tires of the driving wheels can be determined by power and speed losses that occur during their rolling, according to the formula:

$$\eta\mathrm{k} = \frac{M_{\mathrm{k}} - M_f}{M_{\mathrm{k}}} \cdot \frac{r_{\mathrm{k}}}{r_{\mathrm{ko}}} \tag{2}$$

An analysis of expression (2) shows that the first factor reflects the power (hysteresis) power loss due to friction in the tire during its rolling. The parameter on which power losses in the tire of the automobile wheel will depend is M_f—the moment of rolling resistance of the tire. The second factor in expression (2) takes into account that the number of revolutions made by the wheel and the path it has traveled do not always exactly correspond to each other. This is explained by the fact that the rolling wheel slides relative to the supporting surface, which leads to high-speed power losses. The parameter on which the speed loss of power will depend upon rolling a wheel with an elastic tire is its kinematic radius (rolling radius)

$$r_k = \frac{V}{\omega_{\mathrm{k}}} \tag{3}$$

where

V	vehicle speed (linear wheel speed), m/s;
ω_{k}	wheel angular speed, rad/s.

In accordance with this expression, the value of the kinematic radius of the wheel with an elastic tire varies in the range:

$$\infty \geq r_k \geq 0.$$

To take into account the errors associated with power losses in the tires arising from the calculation of mathematical models, earlier analytical [1, 5, 6] and experimental studies were carried out. Unfortunately, these works do not fully describe the relationship of power losses in elastic tires from important factors, such as: wheel angular speed ω_k, tire pressure pt, normal load per wheel GK.

2 Material and Methods

For a comprehensive experimental study of power and speed loss of power in the tire of a car wheel, the bench was modernized to study the process of interaction of an elastic tire with two running drums [1].

The modernized stand design provides for two independent but interconnected systems [2]:

- power system for setting test modes, providing acceleration of the wheel 7 (Pic. 2) and the running drums of the stand 6 and 15 with the specified (steady-state) angular velocities of the wheel ω_K and the running drums $\omega_Б$;
- measuring system registering the parameters of the process of interaction of the tire with the cylindrical surfaces of the running drums 6 and 15;
- the supporting system formed by the assembly of the frame of the stand 1, the wheel frame 3, the running drums 6 and 15. The running drums 6 and 15 correspond in size to the running drums of the diagnostic stands, which is very important when implementing test actions.

In addition to the systems described above, the stand design has a mechanism for fastening and loading the wheel with a given normal load GK, the value of which can be incrementally increased to a value GK ≈ 3500 N. The wheel mounting mechanism is equipped with special dampers—shock absorbers. This solution allowed us to eliminate the free vibrations of the elements of the fastening and loading mechanism, which had a negative impact on the experimental results.

Power system of bench, which drives the running drums and the test wheel into rotation, with specified angular velocities ω_K and $\omega_Б$, consists of two power plants, each of which includes: three-phase asynchronous electric motor 30 and 36, with power $N = 16$ kW, worm gear 32 and 33, type 1Ch-160, with gear ratio $U_p = 10$.

Structure wheel drive power plant, consists of an electric motor 36, which is controlled directly by the operator from the control panel 18. To change the wheel speed in a larger direction, in order to set test modes for studying speed losses, the output shaft of the electric motor 36 is connected by a chain gear 35 to the input shaft of the gearbox 33. The control link to The power supply circuit of the motor is a voltage frequency converter 17, by means of which the frequency of the supply voltage of the electric motor 36 is changed. This solution makes it possible to change the value of slippage S, due to changes in the angular velocity of the wheel ωK. The power unit of the drive of the test wheel is mounted on a special platform, moving depending on the size and pressure in the test wheel, as well as on the position of the wheel 7 on the running drums 6 and 15 of the stand. This eliminates the unevenness of the torque on the shaft of the driveshaft 27, on which a torque sensor with an amplifier 28 is mounted.

Structure power unit drive cross drum, includes: a three-phase asynchronous motor 30, a worm gear 32, an elastic sleeve-finger coupling 31 connecting the motor shaft to the drive shaft of the gearbox, and an elastic sleeve-finger coupling 29 connecting the driven shaft of the gearbox to the shaft of the running drum 15.

Starting and stopping the electric motors 30 and 36 is carried out by turning on the starting relays installed in the electrical box 34. The starting relays are controlled from the control panel 18, the power plant control unit 20 (Figs. 1 and 2).

Wheel mounting and loading mechanism power G_K represents a wheel frame 3, with a wheel 7 mounted on it, pivotally fastened by straps 19 with a loading frame, which is mounted to the frame of the stand 1. The pressing of the wheel 7 to the running drums 6 and 15 occurs due to the installation of the load 2 on the loading frame.

Measuring system, consists of sensors for measuring stresses in the contact patch of the tire with the surface of the running drum 4 and 11; signal amplifiers 5, 13 and 23; the sensor of angular velocity of rotation of the wheel 8; signal converters 10 and 12; a sensor of angular velocity of rotation of the running drums 14; block receiving signals from sensors 21; low voltage source 22; brake/traction force sensor 25; torque sensor and integrated amplifier 28. Measuring system measures and records the parameters of the process of interaction of an elastic tire of a car wheel with two cylindrical supports. The result of the measuring system is the obtained dependencies normal $R_Z = f(t)$ and tangent $R_X = f(t)$ reactions, inhibitory $M_T = f(t)$, traction torque $M_K = f(t)$ or torque of resistance $M_f = f(t)$, brake force

Fig. 1 Appearance of the stand for the study of the processes of interaction of fire with two running drums

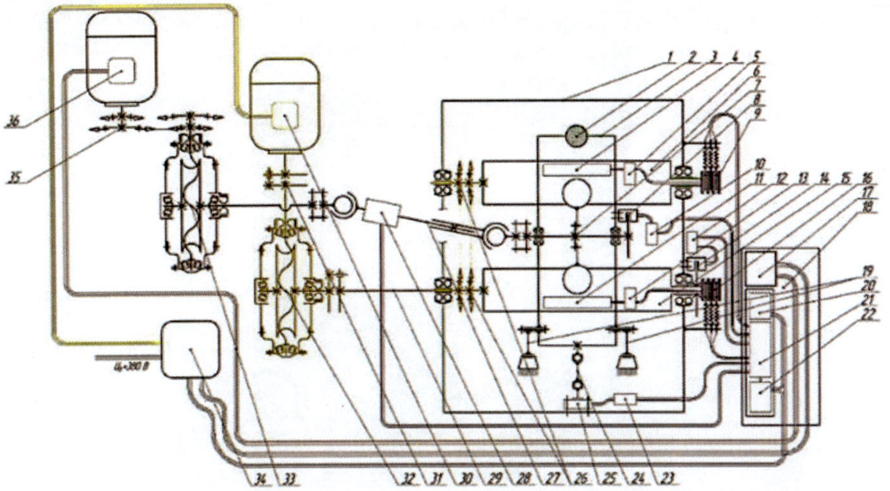

Fig. 2 Functional diagram of the stand for the study of the processes of interaction of the tire with two running drums: 1—stand frame; 2—cargo; 3—wheel frame; 4, 11—a sensor for measuring stresses in the contact patch of the tire with the surface of the running drum; 5, 13, 23—signal amplifier; 6, 15—running drum; 7—wheel; 8—sensor of angular velocity of rotation of the wheel; 9, 16—collector signal transmission; 10, 12—signal converter; 14—sensor angular speed of rotation of the running drums; 17—voltage frequency Converter; 18—control panel; 19—strips connecting the wheel frame with the loading frame; 20—motor control unit; 21—block receiving signals from sensors; 22—source of low voltage; 24—ball thrust; 25—brake/traction force sensor; 26, 35—chain transmission; 27—cardan transmission; 28—torque sensor; 29—clutch drive drums; 30, 36—electric motor; 31—coupling; 32, 33—gear; 34—box with starting relays

$F_T = f(t)$, traction forces $F_K = f(t)$ as well as angular speeds of rotation of the wheel $\omega_K = f(t)$ and running drums $\omega_Б = f(t)$ by time.

To measure the angular velocity of the wheel ω_K and the angular velocity of the running drums $\omega_Б$, the existing measurement systems are used.

To measure the rolling resistance of a tire Mf A strain gauge type sensor designed and manufactured with a signal amplifier in one housing was designed and manufactured (Fig. 3). Sensitive elements of the sensor are strain gauges VDR1 and VDR2 glued to a portion of the cardan shaft and connected to a bridge circuit with resistors R1 and R3 (Fig. 4). Resistors R1 and R3 are also located on the shaft and, in conjunction with the strain gauges VDR1 and VDR2, form the measuring part of the sensor. The electrical connection of the sensor and amplifier is organized using slip rings.

The signal amplifier is based on the DA2 instrumental operational amplifier—AD623AN. Resistor R2 balances the signal relative to the zero point. Resistors R4 and R5 perform a matching function. The RG resistor allows you to adjust the gain of the circuit. Capacitors C4-C6—noise suppressing. The power of the amplifier is implemented by the DA1 chip, which is an integrated voltage regulator of the power supply L7809, with an output voltage of 9 V. Capacitors C1-C3 eliminate

Fig. 3 The appearance of the torque sensor is mounted on the cardan shaft

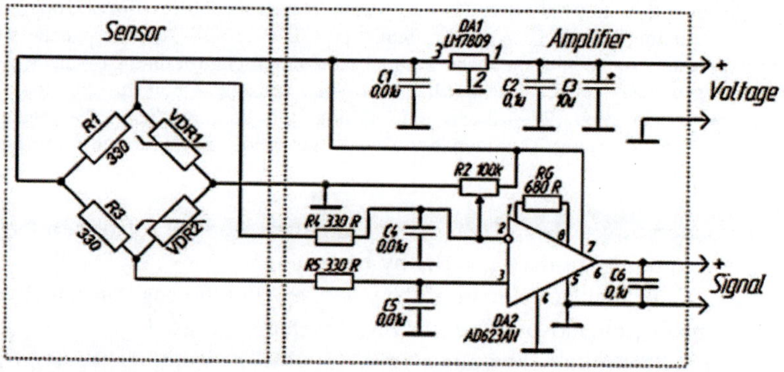

Fig. 4 Schematic diagram of the torque sensor and amplifier

interference in the supply voltage of the DA2 chip. The torque sensor allows you to measure torque or braking torque in the range from 0 to 1500 Nm. The measurement error is 2.5%. The sensors, amplifiers and transducers are supplied with power from block 21 (see Pic 2) connected to power supply 22. All registered parameters are sent to signal receiving block 21, to which the L-Card E-154 analog-to-digital converter is connected. Further processing of the measurement results is performed on a PC at the software level.

3 Results

In the course of the conducted experimental studies, the dependences of changes in the kinematic parameters of the tire rolling process on the cylindrical support surface of rolling drums rKO = f(Gk), rKO = f(psh), rK = f(Gk), rK = f(psh), Mf = f(Gk), Mf = f(Mk), Mf = f(psh). Also, the dependence of the power efficiency on the torque on the wheel, the dependence of the speed efficiency on the torque on the wheel, in addition, the dependence of the total efficiency of the tire on the torque on the wheel is obtained. The regularities of the influence of the load on the car wheel with an elastic tire, the air pressure in the tire, the applied torque, on the main indicators of the studied process are revealed (Figs. 5, 6 and 7).

The result of the conducted experimental studies was the dependence of the influence of power and speed losses in the tire of an automobile wheel on the process of acceleration of a wheeled vehicle on a stand with running drums:

1. The dependence of the rolling resistance moment of the tire M_f on the air pressure in the tire pt, the normal load on the wheel G_K and the linear speed of the wheel *V*. It was found that when the normal load G_K changes from 1765.8 N to 3531.6 N, the air pressure in the tire changes from 0.04 MPa to 0.24 MPa, and the linear speed *V* changes from 3.1 km/h to 14.7 km/h, the rolling resistance moment changes from 32.2 N to 86.81 N, which is 62.8%.
2. The dependences of the power efficiency of the tire on the air pressure in the pt tire, the normal load on the G_K wheel and the torque on the M_K wheel are revealed. It was found that when the normal load G_K changes from 1765.8 N to 3531.6 N, the air pressure in the pt tire changes from 0.04 MPa to 0.24 MPa,

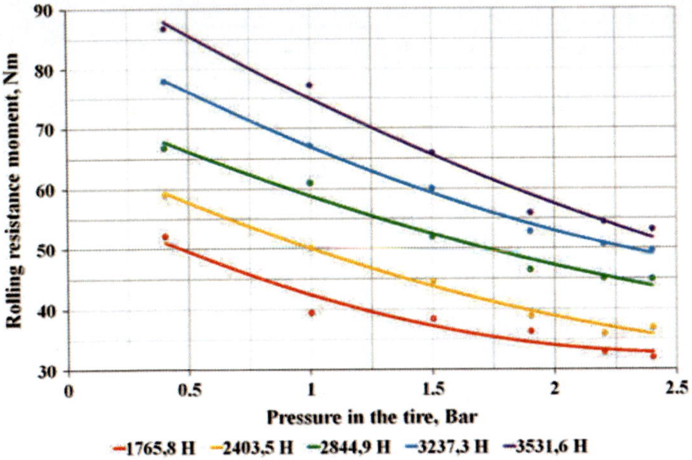

Fig. 5 The dependence of the rolling resistance moment of the tire on the air pressure when changing the normal load on the wheel at a linear wheel speed of 14.7 km/h

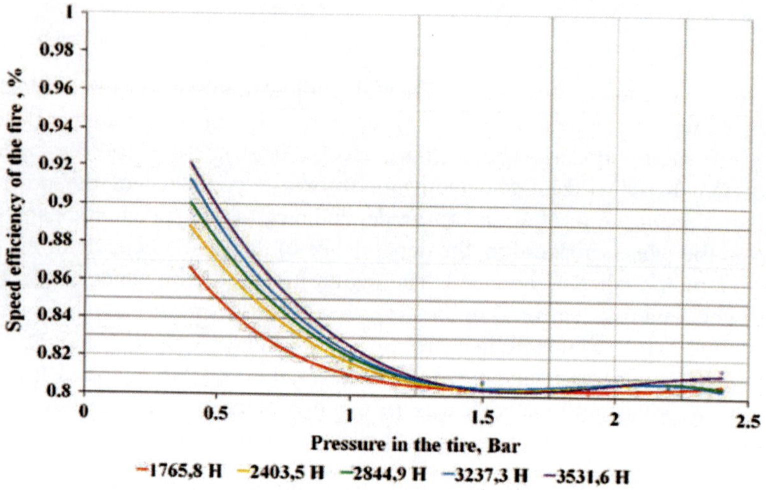

Fig. 6 The dependence of the speed efficiency of the tire on the air pressure when changing the normal load on the wheel at a linear wheel speed of 14.7 km/h

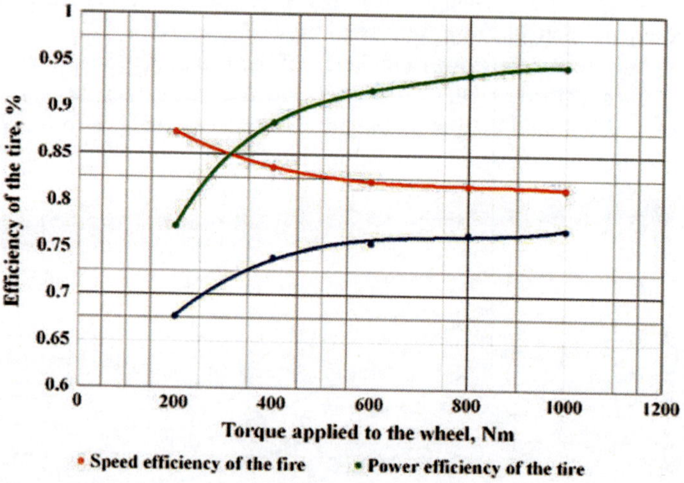

Fig. 7 Tire efficiency at normal load 3531.6 N, tire air pressure 0.24 MPa

and the torque on the M_K wheel changes from 200 to 1000 Nm, the power efficiency changes from 0.84 to 0.91, which is 7.6%.

3. The dependences of the high-speed efficiency of the tire on the air pressure in the pt tire, the normal load on the G_K wheel and the torque on the M_K wheel are revealed. It was found that when the normal load GK changes from 1765.8 N to 3531.6 N, the air pressure in the tire changes from 0.04 MPa to 0.24 MPa, and

the torque on the M_K wheel changes from 200 to 1000 Nm, the speed efficiency of the tire changes from 0.871 to 0.921, which is 5.44%.

4. The dependences of the tire efficiency on the air pressure in the tire, the normal load on the G_K wheel and the torque on the M_K wheel are revealed. It was found that when the normal load G_K changes from 1765.8 N to 3531.6 N, the air pressure in the R_S tire changes from 0.04 MPa to 0.24 MPa, and the torque on the M_K wheel changes from 200 to 1000 Nm, the tire efficiency changes from 0.735 to 0.841, which is 12.6%.

4 Discussion

The dependences obtained as a result of the study will further improve the accuracy of modeling the process of acceleration of a wheeled vehicle on stands with running drums by taking into account the power and speed losses in tires that occur during the rolling of the wheel on the rollers of the diagnostic stand. This data will improve the quality of monitoring the technical condition and diagnostics of vehicles on power stands in operating conditions.

5 Conclusions

The modernization of the test bench and the implementation of experiments on this test bench will allow the study of power and speed power losses in the tire of a car wheel mounted on the running drums of a diagnostic bench. As a result of these studies, it will be possible to improve mathematical models, which will have a beneficial effect on the results of analytical studies of the processes of functioning of wheeled vehicles on stands or in road conditions.

References

1. Fedotov AI, Vlasov VG, Krivtsov SN (2012) Calculation of speed losses in car tires under operating conditions. Automobile Ind 9:16–19
2. Alexandrov IK (2018) Stand method for determining mechanical losses in a pneumatic tire. Automobile Ind 1:34–36
3. Boyko AV, Yankov OS, Markov AS, Kuznetsov NYu (2015) Stand for the study of the interaction of elastic tires with two running drums. Car for Siberia and the Far North. Construction, operation. Economics: proceedings of the 90th conference, Irkutsk: publishing house of the Irkutsk national research technical University, pp 105–110
4. Chernyshkov AS (2021) Improving the quality of control of traction and dynamic properties of electric vehicles on stands with running reels. Problems of technical operation and car service of rolling stock of motor transport. Collection of scientific papers dedicated to the 85th

anniversary of the Department of EATiS MADI, based on the materials of the 79th scientific-methodological and research conference of MADI. MADI. Moscow, pp 196–202
5. Dik AB (1985) Description of the characteristics of the brake wheel slip. SB. nauchn. Tr. “Reliability and active safety of the car”. MAMI, pp 205–216
6. Fedotov AI, Yankov OS, Chernyshkov AS (2019) On speed and power losses in the tires of the driving wheels of the PBX when diagnosing them on stands with running drums. Energy efficiency of motor vehicles: nanotechnology, information and communication systems, alternative energy sources: Mat. Vseros. nauch. - prakt. konf. VOLGSTU. Voronezh, pp 234–241

Influence of Microfiber on Fire Resistance of Reinforced Concrete Tunnel Tubings

Vladimir Golovanov, Andrey Pehotikov, Marina Gravit, Nikolay Novikov, and Vladimir Pavlov

Abstract The fire resistance of reinforced concrete tunnel linings of underground structures depends to a large extent on the explosive destruction of concrete under the influence of high fire temperatures on the surface of the tunnel lining. The use of reinforced-concrete tunnel liners made of fiber concrete with polypropylene fibers for the tunnel enclosing structures will allow to reduce considerably the cost of fire protection device and shorten the construction period. Reinforced concrete tunnel tubes made of heavy concrete with 6% moisture content and 1 kg/m^3 of polypropylene fiber added have been tested. Large-scale fire tests on a specially made stand were conducted at loading of samples with vertical 490.5 kN and horizontal 294.3 kN loads. As a result of experimental studies, it was found, that the fire resistance limit of the structure was REI 120.

Keywords Fire resistance · Reinforced concrete structures · Explosive concrete destruction · Tunnel lining · Polypropylene microfiber

1 Introduction

Reinforced concrete structures made of heavy concrete with high moisture (more than 3.5%) are prone to explosive destruction. This phenomenon can lead to premature onset of the fire-resistance limit of load-bearing reinforced concrete structures and, as a consequence, to partial or complete collapse of buildings and structures. Increased moisture of such structures is usually found in underground structures and newly constructed buildings [1–6].

V. Golovanov · A. Pehotikov · N. Novikov · V. Pavlov
All-Russian Research Institute for Fire Protection of Ministry of Russian Federation for Civil Defense, Emergencies and Elimination of Consequences of Natural Disasters, Balashikha, Russia

M. Gravit (✉)
Peter the Great St. Petersburg Polytechnic University, Saint Petersburg, Russia

A. Bieliatynskyi and V. Breskich (eds.), *Safety in Aviation and Space Technologies*, Lecture Notes in Mechanical Engineering,
https://doi.org/10.1007/978-3-030-85057-9_24

Studies on the protection of reinforced concrete structures against explosive destruction of concrete conducted in a number of countries. The addition of polypropylene fibers to the concrete mixture is the most effective way to protect reinforced concrete structures from explosive fracture [7–21].

However, the addition of fibers to the composition of concrete mixture changes the physical–mechanical and thermal characteristics of concrete, which are very important in assessing the fire resistance of reinforced concrete structures by calculation methods. The values of these characteristics were obtained in earlier studies [7–11].

As polypropylene fiber prevents the explosive destruction of concrete, it is advisable to use this material on such objects as underground structures, the moisture content of which may exceed the threshold values. The main constructions of such constructions are reinforced-concrete tunnel liners which form tunnel lining and are used for underground laying of motor-road and subway tunnels. The fire load carrying capacity of structures is ensured by sufficient fire resistance limits of reinforced concrete structures, which are estimated by both experimental and calculation methods.

Experimental methods allow by means of large-scale fire tests to estimate the fire resistance of a structure and to consider visually its behavior in conditions of fire.

The purpose of the research is:

- determining the effect of adding polypropylene fibers to the concrete mixture of reinforced concrete tunnel liner tunnel tunneling to prevent explosive fracture of concrete with a moisture content of 6%;
- establishment of the fire-resistance limit of reinforced concrete tunnel lining with polypropylene fiber.

2 Materials and Methods

Samples of reinforced concrete tubes for the experiment were made of heavy concrete. Composition of concrete: portland cement, PC I-500-N grade, fine aggregate—quartz sand, coarse aggregate—granite crushed stone (fraction 5–15 mm), plasticizer—«Glenium 51». The sizes of tubings—2984 × 1400 × 300 mm. Polypropylene fiber «ProZASK IGS» with fiber length 6 mm, diameter 18 μ in quantity 1 kg/m^3 was introduced into reinforced concrete tubing. Reinforcement: 28Ø12 mm, steel class B500C. Reinforcement frames were made using spot welding. The thickness of the protective layer from the heated surface to the edge of the reinforcement was 34 mm. The general view of the tested specimens is shown in Fig. 1.

The fire resistance tests performed according to Russian State standards GOST 30247.0 «Elements of building constructions. Fire-resistance test methods. General requirements» and GOST 30247.1 «Elements of building constructions. Fire-resistance test methods. Loadbearing and separating constructions»,

Fig. 1 General view of the tested reinforced concrete tubing

harmonized with series of standards ISO 834-1:1999 «Fire-resistance tests—Elements of building construction—Part 1: General requirements», and where requirements to equipment, placement of thermal converters and registration of required parameters are established. During the tests on fire resistance of tunnel lining it is necessary to reproduce the load, which affects the structure during the operation of the object with the maximum moment and longitudinal force in the tunnel lining cross-section. The experiments were carried out in VNIIPO of EMERCOM of Russia. Since both vertical and horizontal loads act on this structure during its operation, it was necessary to improve the support and loading system of the new installation for testing the fire resistance of floors, coverings and beams with an additional loading system. Figure 2 shows a schematic of the improved heating system, which was outside the heating zone during the fire test.

Before the experiment according to National standard of Russia GOST 30247.1, the maximum deflection was calculated, which was 99.5 mm, and the limiting rate of deformation—0.33 cm/min, the excess of which characterizes the limit state on the bearing capacity. In the course of the experiment, the deflection was fixed with the device «Progibomer MP-3». To establish the limit state in terms of thermal insulating capacity, thermal sensors were placed on the unheated surface to measure temperatures.

Thirty minutes prior to the beginning of fire tests, the structure was loaded with a design load which was 490.5 kN vertically, and 294.3 kN horizontally. After that, the test specimen of the tubing was exposed to the temperature regime of the standard fire.

As a result of experimental studies, graphs of temperature changes in the fire chamber (Fig. 3) and, deflection (Fig. 4), as well as the temperature on the unheated surface (Fig. 5) were plotted.

The normative fire resistance limit of tubing according to the design documentation for this object of underground construction was REI 120, so the fire tests

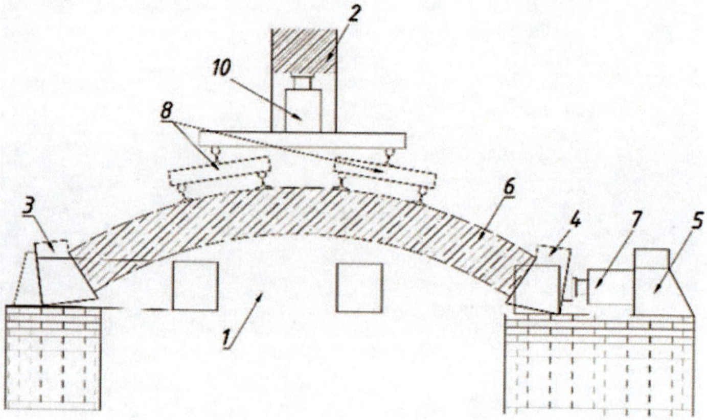

Fig. 2 The advanced loading system for the study of fire resistance of reinforced concrete tubings: 1—horizontal frame; 2—vertical frame; 3—fixed jack support; 4—movable support; 5—hydraulic jack support; 6—test sample; 7—hydraulic jack for horizontal load; 8—vertical load frames; 9—jack rest; 10—hydraulic jack for vertical load

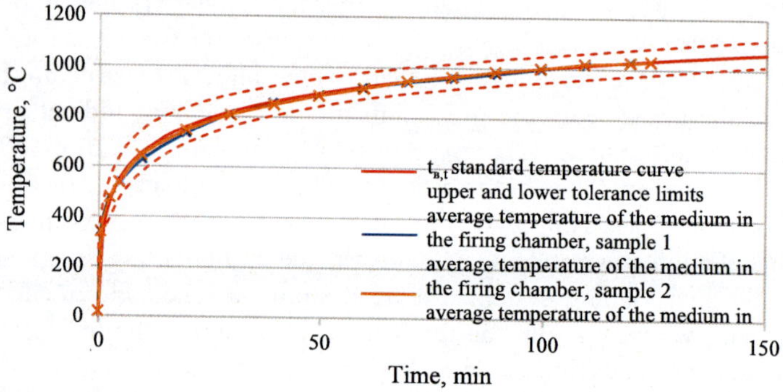

Fig. 3 The temperature curve of the medium in the firing chamber resistance of reinforced concrete tubing with polypropylene fiber

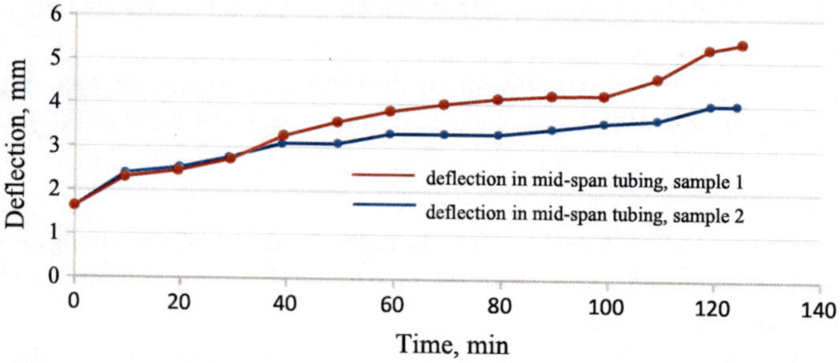

Fig. 4 The test results on the fire resistance of reinforced concrete tubing with polypropylene fiber

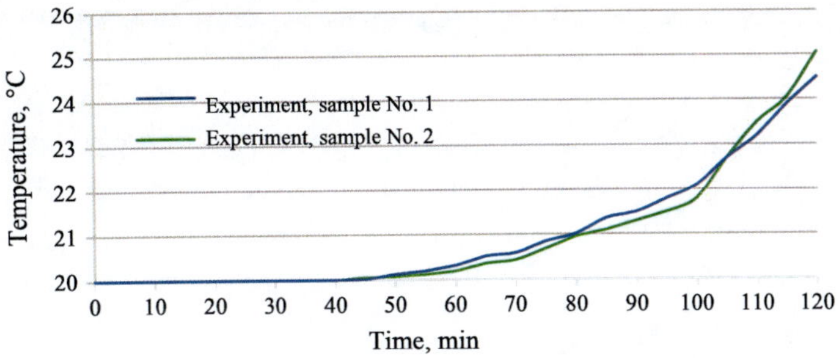

Fig. 5 The time dependence of the temperature on the unheated surface of the tubing during fire tests

Table 1 Properties of concrete with polypropylene fiber and reinforcement

Property	Value
Concrete with polypropylene fiber	
Thermal conductivity λ_t, W/(m K)	$1.3 - 0.0006t$
Heat capacity c_t, J/(kg K)	$481 + 0.92t$
Density, kg/m^3	2400
Humidity of the concrete W, %	6
Reinforcement	
Thermal conductivity λ_t, W/(m K)	$78 - 0.048t$
Heat capacity c_t, J/(kg K)	$310 + 0.48t$
Density, kg/m^3	7800

were carried out for 125 min. Properties of concrete and reinforcement are given in Table 1.

According to the results of the experiment with tubing without the addition of polypropylene fiber, an explosive destruction of concrete under the influence of high temperatures, which was accompanied by strong pops. This led to peeling of the protective layer of concrete and rapid heating of the reinforcement to critical values. The specimen after fire tests with the destroyed protective layer of concrete is shown in Fig. 5.

In the course of fire tests of tubing with polypropylene fiber addition, no explosive destruction of concrete was noted. The maximum deformation was 7 mm, which does not exceed the maximum allowable deflection (99.5 mm). The average temperature of five thermocouples in the interval of 0–120 min was 25 °C, which is much lower than the allowable temperature (160 °C). During the tests it was visually determined that evaporation of moisture on the unheated surface of the samples occurs at the 30–37th minute and continues until the end of the tests.

At the end of tests and removal of the specimen from the installation, two transverse cracks with width 1–3 mm and depth 20–30 mm and numerous filament cracks were found on the heated side (Fig. 6).

Fig. 6 Reinforced concrete tubes after testing without the addition of polypropylene fiber (left) and with the addition of polypropylene fiber (right)

The values obtained from the test were used to simulate the performance of the structure. In preparation for the calculation, the volumetric structure in the form of a reinforced concrete tube was modeled; the concrete and reinforcement were assigned the properties presented in Table 1. The model was divided into finite elements.

To calculate in the software package, the boundary conditions of the 3rd kind were set. For the heated surface, the characteristics shown in Table 2 were specified.

On the unheated surface, the heat transfer coefficient was given in tabular form a ′:

$$\alpha' = 1.5 \cdot \sqrt[3]{t_i - t_{\mathrm{H}}} + 5.77 \cdot \varepsilon_{\pi\mathrm{p}} \cdot \left(\frac{\left(\frac{t_i}{100}\right)^4 - \left(\frac{t_{ref}}{100}\right)^4}{t_i - t_{\mathrm{H}}} \right) \tag{1}$$

where

- a′ heat transfer coefficient from the unheated surface to the medium, W/(m^2 K);
- t_i the temperature of the unheated surface, K;
- t_{ref} initial temperature of construction, K;
- $\varepsilon_{\pi\mathrm{p}}$ The degree of blackness of the system fire chamber—concrete.

Table 2 Boundary conditions for heated surface

Property	Value
Convection heat transfer coefficient α_{conv}, W/(m^2 K)	29
The degree of blackness of the system *fire chamber—concrete surface* $\varepsilon_{\mathrm{ref}}$	$0.67 - 0.0004t$
The temperature dependence of the standard mode of fire (t_{ref}—initial temperature of construction, °C)	$345\lg(0.133t + 1) + t_{ref}$

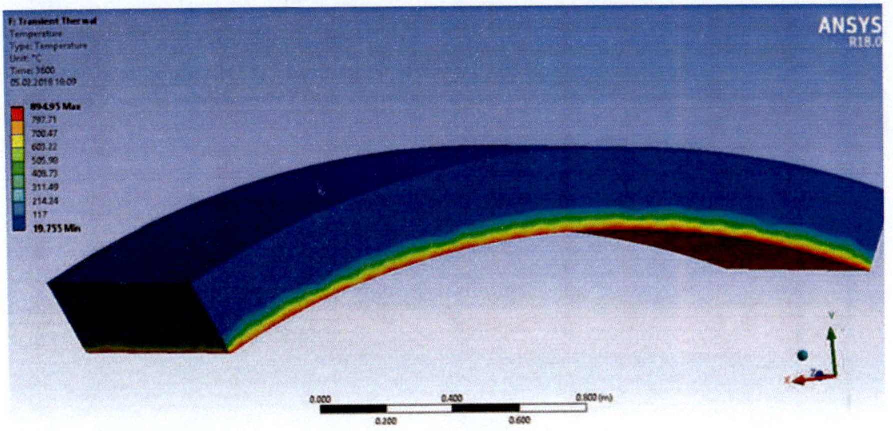

Fig. 7 The results of the calculation of temperature fields for reinforced concrete tubing in the software package

Since concrete is a hygroscopic material, the evaporation of free moisture from the pores of concrete by increasing the coefficient of heat capacity in the temperature range 100–140 °C was taken into account.

The calculation was carried out under the influence of the standard fire temperature regime for 120 min. The calculation results are shown in Fig. 7.

At the end of the thermotechnical calculation, the results of the numerical experiment were processed in order to determine the main parameters required for solving the strength problem.

To determine the reliability of the obtained results, the calculated and experimental curves of temperature growth on the unheated surface of the tubing were compared (Fig. 8). The comparison showed that the difference between the calculated and experimental data does not exceed 10%.

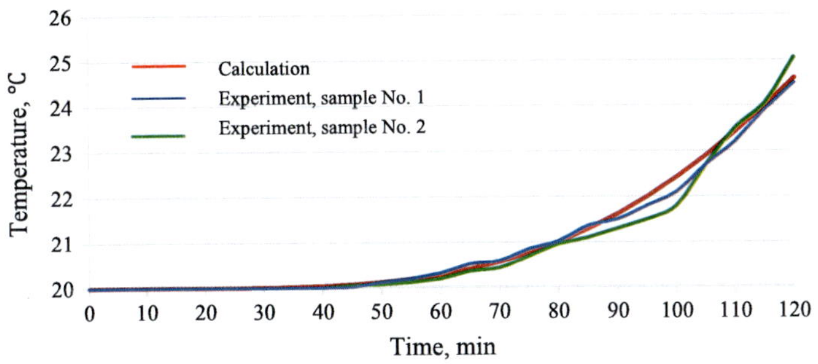

Fig. 8 The time dependence of the temperature on the unheated surface of the tubing during fire tests and thermal engineering calculation

Table 3 Properties of concrete with polypropylene fiber and reinforcement

Properties of concrete	Value
Concrete with polypropylene fiber	
Thermal conductivity λ_t, W/(m K)	$1.3 - 0.0006t$
Heat capacity c_t, J/(kg K)	$481 + 0.92t$
Density, kg/m^3	2400
humidity of the concrete W, %	6
Reinforcement	
Thermal conductivity λ_t, W/(m K)	$78 - 0.048t$
capacity c_t, J/(kg K)	$310 + 0.48t$
Density, kg/m^3	7800

3 Results

Visual inspection after fire tests of the heated surface of the tube with polypropylene fiber additive (see Fig. 5b) showed no damage to the protective layer of concrete. Only insignificant chipping of protective layer of concrete, formed as a result of specimen dismantling from the supports of the installation, was detected. This testifies to the absence of explosive destruction of concrete on the surface of the studied sample.

As a result of experimental studies of reinforced concrete tubing with polypropylene fiber, it was found that the fire-resistance limit of tubing was REI 120, which corresponded to the rated fire-resistance limit. The experiment was terminated at 125th minute. The properties of fiber-reinforced concrete are shown in Table 3.

4 Discussion

As a result of fire tests of reinforced concrete tubing with polypropylene fiber added in an amount of 1 kg/m^3 there was no explosive destruction of concrete, which confirms the positive effect of the additive on the fire resistance of structures made of heavy concrete with increased humidity. Values of the maximum allowable deflection (99.5 mm) and the temperature at the unheated surface of the structure (160 °C) have not been reached, therefore, limit states for the loss of carrying and insulating capacity during 125 min of exposure to temperature fire on the samples did not occur. As a result of experimental studies, it was found that the fire resistance limit of the structure was REI 120.

The thermal engineering problem was solved in the ANSYS software package, as a result of which the data on the heating of the reinforcement and concrete were obtained, and the experimental and calculated temperatures on the unheated surface of the structure were compared. The discrepancy between them was not more than 10%, which indicates the reliability of the numerical experiment.

References

1. Serrano R, Cobo A, Prieto MI, González MdLN (2016) Analysis of fire resistance of concrete with polypropylene or steel fibers. Constr Build Mater 122:302–309
2. Mohammad R, Nasser Al-Nuaimi I, Rabie M (2021) Hybrid effect of carbon nanotubes and polypropylene microfibers on fire resistance, thermal characteristics and microstructure of cementitious composites. Constr Build Mater 266(B):121154. https://doi.org/10.1016/j.conbuildmat.2020.121154
3. Maraveas S, Vrakas A (2014) Design of concrete tunnel linings for fire safety. C. Struct Eng Int 2014 24(3):319–329. https://doi.org/10.2749/101686614X13830790993041
4. Werther N (2006) Brandversuche an Tunnelinnenschalenbetonen für den M 30-Nordtunnel in Madrid. Beton und Stahlbetonbau 101(9):729–731. https://doi.org/10.1002/best.200608187
5. Golovanov V, Pekhotikov A, Novikov N, Pavlov V, Kuznetsova E (2019) Fire resistance of reinforced concrete tubings of underground structures with polypropylene fiber. Fire Explosion Safety 28:60–70. https://doi.org/10.18322/PVB.2019.28.05.60-70
6. Rodrigues J, Laim L, Correia A (2010) Behaviour of fiber reinforced concrete columns in fire. Compos Struct 92(5):1263–1268. https://doi.org/10.1016/j.compstruct.2009.10.029
7. Gravit M, Antonov S, Nedryshkin O (2016) Research features of tunnel linings with innovations fireproof panels. Proc Eng 165:1651–1657. https://doi.org/10.1016/j.proeng.2016.11.906
8. Heo Y, Sanjayan J, Han G, Han M (2016) Synergistic effect of combined fibers for spalling protection of concrete in fire. Cem Concr Res 40(10):1547–1554. https://doi.org/10.1016/j.cemconres.2010.06.011
9. Gravit M, Antonov S, Nedryshkin O, Nedviga E, Pershakov V (2016) Fire resistant panels for the tunnel linings. MATEC Web Conf 73:1–7. https://doi.org/10.1051/matecconf/20167304007
10. Gravit M, Antonov S, Nedviga E, Shkorko M, Kozlova K (2016) Strength test of high-precision tunnel lining blocks. Proc Eng 165:1658–1666. https://doi.org/10.1016/j.proeng.2016.11.907
11. Golovanov VI, Novikov NS, Pavlov VV, Kuznetsova EV (2017) Strength and thermo-physical properties of concrete with polypropylene fiber under standard temperature regimes. Fire Explosion Safety 26(5):37–44. https://doi.org/10.18322/PVB.2017.26.05.37-44
12. Szeląg M (2019) Evaluation of cracking patterns of cement paste containing polypropylene fibers. Compos Struct 220:402–411
13. Flores Medina N, Barluenga G, Hernández-Olivares F (2014) Enhancement of durability of concrete composites containing natural pozzolans blended cement through the use of Polypropylene fibers. Compos B Eng 61:214–221
14. Varona FB, Baeza FJ, Bru D, Ivorra S (2018) Influence of high temperature on the mechanical properties of hybrid fibre reinforced normal and high strength concrete. Constr Build Mater 159:73–82
15. Haigh JN, Dargaville TR, Dalton PD (2017) Additive manufacturing with polypropylene microfibers. Mater Sci Eng, C 77:883–887. https://doi.org/10.1016/j.msec.2017.03.286
16. Nedviga E, Beresneva N, Gravit M, Blagodatskaya A (2018) Fire resistance of prefabricated monolithic reinforced concrete slabs of "Marko" technology. In: International scientific conference energy management of municipal transportation facilities and transport EMMFT 2017. Advances in intelligent systems and computing, vol 692, pp 739–749. https://doi.org/10.1007/978-3-319-70987-1_78
17. Bofang Z (2014) Temperature control of concrete dam in cold region. In: Thermal stresses and temperature control of mass concrete. Butterworth-Heinemann, pp 431–438. ISBN: 9780124077232
18. Korniyenko SV, Vatin NI, Gorshkov AS (2016) Thermophysical field testing of residential buildings made of autoclaved aerated concrete blocks. Mag Civil Eng 64(4):10–25. https://doi.org/10.5862/MCE.64.2

19. Vatin N, Korniyenko SV, Gorshkov AS, Pestryakov II, Olshevskiy V (2020) Actual thermophysical characteristics of autoclaved aerated concrete. Mag Civil Eng 96(4):129–137. https://doi.org/10.18720/MCE.96.11
20. Nassif AY, Petrou MF (2013) Influence of cold weather during casting and curing on the stiffness and strength of concrete. Constr Build Mater 44:161–167
21. Zach J, Sedlmajer M, Hroudova J, Nevařil A (2013) Technology of concrete with low generation of hydration heat. Proc Eng 65:296–301

Compensating Microaccelerations from a Temperature Shock

Andry Sedelnikov, Vadim Salmin, Yuri Lazarev, Denis Orlov, and Anastasia Taneeva

Abstract The problem of large elastic elements temperature shock is a serious problem of providing favorable conditions for the implementation of gravity-sensitive processes on board a small technological spacecraft. When a small spacecraft dives into and exits the Earth's shadow, its large elastic elements, primarily solar panels, experience a significant change in the temperature regime. The purpose is to analyze the possibility of implementing algorithms for controlling an electrothermal micro-engine, developed to eliminate the influence of temperature shock on the level of micro-acceleration of the internal environment of a small technological spacecraft. The research is carried out using methods of mathematical modeling, spacecraft parameters analysis, technical justification and analysis of advantages, operating conditions analysis, test results analysis, experimental dependencies construction. The study confirms that it is necessary to take into account the real law of changes in the electrothermal micro-engine thrust for a correct assessment of the decrease in the micro-accelerations magnitude from a temperature shock. Since the temperature shock is an extremely dynamic phenomenon, a detailed study of a turned on executive body thrust change law is necessary. The results of the work can be used in the design of a small technological spacecraft to meet the micro-accelerations requirements.

Keywords Small technological spacecraft · Electrothermal micro-engine · Temperature shock · Algorithms for controlling the movement of a small spacecraft

1 Introduction

The problem of temperature shock of large elastic elements is a serious problem of providing favorable conditions for the implementation of gravity-sensitive processes on board a small technological spacecraft [1]. When a small spacecraft dives

A. Sedelnikov · V. Salmin · Y. Lazarev · D. Orlov (✉) · A. Taneeva
Samara National Research University, Moskovskoye shosse, 34, Samara 443086, Russian Federation

A. Bieliatynskyi and V. Breskich (eds.), *Safety in Aviation and Space Technologies*, Lecture Notes in Mechanical Engineering,
https://doi.org/10.1007/978-3-030-85057-9_25

into and exits the Earth's shadow, its large elastic elements, primarily solar panels, experience a significant change in the temperature regime [2, 3]. At the same time, their temperature can change by several tens of degrees per second [3]. This leads to significant temperature deformations of the solar panel and loss of stability of the plane equilibrium shape [3, 4]. Depending on the flexibility of the elastic elements, their own vibrations can be excited. And they not only affect the microaccelerations of internal environment of the small spacecraft, but also its controllability [5, 6]. Gravity-sensitive processes require low-level accelerations for its successful implementation [7, 8]. It is shown in [2] that a temperature shock can disrupt favorable microacceleration conditions even for medium-class spacecraft (such as «Vozvrat-MKA», Fig. 1 [2]).

In [3], it is noted that for small technological spacecraft, temperature shock is a serious problem. In [3], theoretical algorithms for controlling the executive body of the orientation system and controlling the orbital motion of the spacecraft are proposed, that provide a significant (more than tenfold) reduction in the influence of temperature shock on the example of a promising technological medium-class spacecraft «OKA-T» (Fig. 2).

Fig. 1 The technological project «Vozvrat–MKA» small spacecraft (quoted in [2])

Fig. 2 General view of the «OKA-T» spacecraft with an opened molecular screen and solar panels

The purpose of this work is to assess the capabilities of a real executive body, taking into account the spread of its characteristics. As such an executive body, an electrothermal micro-engine is used in the work.

2 Unmanned Spacecraft Model

2.1 A Subsection Sample

To study the phenomenon of temperature shock, a spacecraft model with a single elastic element was chosen (Fig. 3) [3].

On the one hand, we can give examples of such spacecraft («Spot-4»9, «Electro-L»10, etc.). On the other hand, in the presence of more than one elastic element, the problem of studying the temperature shock becomes essentially stochastic. Other factors that are not related to the temperature shock become important in this analysis (the asymmetry of large elastic elements, the differences in the initial temperature fields of these elements at the time of the temperature shock, the characteristics of natural oscillations, etc.). Therefore, for the direct study of the temperature shock, the model shown in Fig. 3 was adopted. The main parameters of the simulated spacecraft are presented in Table 1.

The problem of estimating microaccelerations from a temperature shock in the framework of a one-dimensional model of thermal conductivity for the spacecraft, shown in Fig. 3, is solved in [3]. However, all conclusions about the possibility of reducing microaccelerations were made on the basis of theoretical analysis. In this

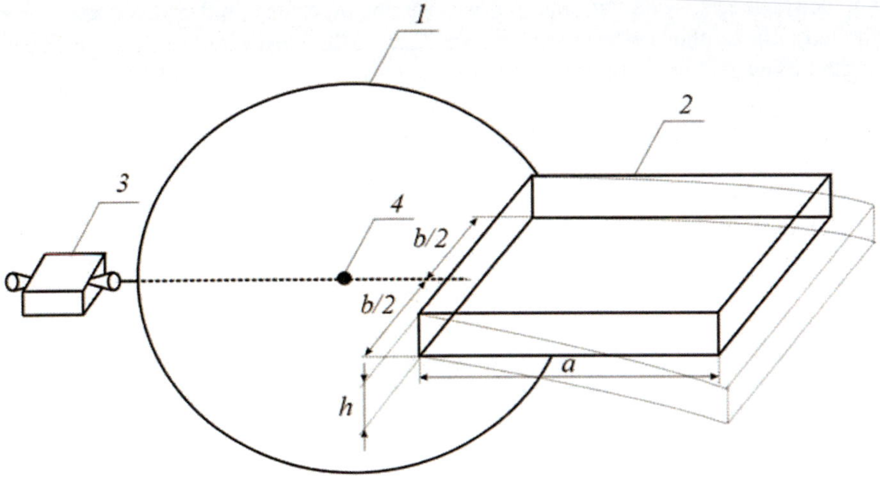

Fig. 3 Spacecraft model: 1—rigid body; 2—elastic element; 3—electrothermal micro-engine; 4—center of mass of the space-craft

Table 1 Basic parameters of the unmanned spacecraft

Parameter	Value	Dimension
Spacecraft mass	1000	kg
Material of the solar panel frame	MA2	–
Thermal conductivity	96.3	W/(m K)
External heat flow	1400	W/(m^2 K^4)
Initial temperature of the solar panel frame	200	K
Specific heat capacity	1130	J/(kg K)
Density	1780	kg/m^3
Thickness of the solar panel frame	6	mm
Length of the solar panel	2	m
Width of the solar panel	0.5	m

paper, the results of tests of a real executive body (an electrothermal micro-engine) are presented, that can be used to reduce microaccelerations caused by a temperature shock. The theoretical control law is compared with the characteristics of an electrothermal micro-engine obtained as a result of tests in a vacuum chamber.

3 Justification of the Choice of an Electrothermal Micro-engine as an Executive Body

When implementing long-term gravity-sensitive processes, it is advisable to use an orientation system, the main executive body of which is a complex of control engine-flywheels. Then the time interval of implementation in the ideal case can be limited only by the period of active existence of the smallest spacecraft, since the engine-flywheels do not consume the working body. However, this requires periodic reduction of the kinetic moment of the engine-flywheels with the help of other executive bodies. It is proposed to use electrothermal micro-engines as such bodies [11, 12]. According to research data [13], low-thrust liquid rocket engines have excessive thrust and will violate favorable conditions for gravity-sensitive processes for a long period of time. The appearance of an electrothermal micro-engine is shown in Fig. 4.

When implementing long-term gravity-sensitive processes, it is advisable to use an orientation system, the main executive body of which is a complex of control engine-flywheels. Then the time interval of implementation in the ideal case can be limited only by the period of active existence of the smallest spacecraft, since the engine-flywheels do not consume the working body. However, this requires periodic reduction of the kinetic moment of the engine-flywheels with the help of other executive bodies. It is proposed to use electro-thermal micro-engines as such bodies [11, 12]. According to research data [13], low-thrust liquid rocket engines have excessive thrust and will violate favorable conditions for gravity-sensitive processes

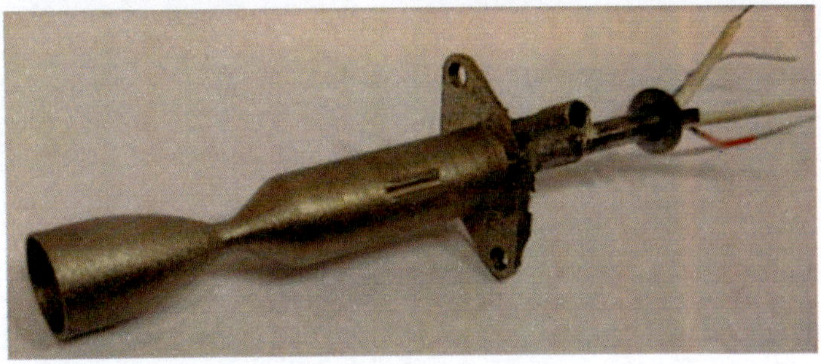

Fig. 4 ETME appearance

for a long period of time. The appearance of an electrothermal micro-engine is shown in Fig. 4.

The choice of using an electrothermal micro-engine can be justified by the fact that:

- unlike low-thrust liquid rocket engines, it has less thrust, which is enough to control a small spacecraft;
- unlike magnetic executive bodies, it creates a control moment in any direction, regardless of the induction vector of the Earth's magnetic field;
- it is able to effectively reduce the kinetic moment of engine-flywheels.

The main advantages of an electrothermal micro-engine include:

- low energy consumption for creating thrust (up to 4 W/mN);
- high efficiency in the class of electric rocket engines;
- a wide range of working bodies types used for the operation;
- simple engine design;
- high reliability of the engine;
- the possibility of periodic switching on during a long service life;
- high engine life (several hundred hours);
- high engine efficiency (up to 80%);
- the ability to change the amount of thrust in a wide range (10–200 mN);
- the ability to throttle the engine thrust.

4 The Main Characteristics of the Electrothermal Micro-engine According to the Test Data

The basis of an electrothermal micro-engine is a heating element. During the tests, a metal tubular heating element was used, which is shown in Fig. 5.

In the experimental scheme, a reliable insulation of the heater from the nozzle of an electrothermal micro-engine was used, which is able to withstand all real

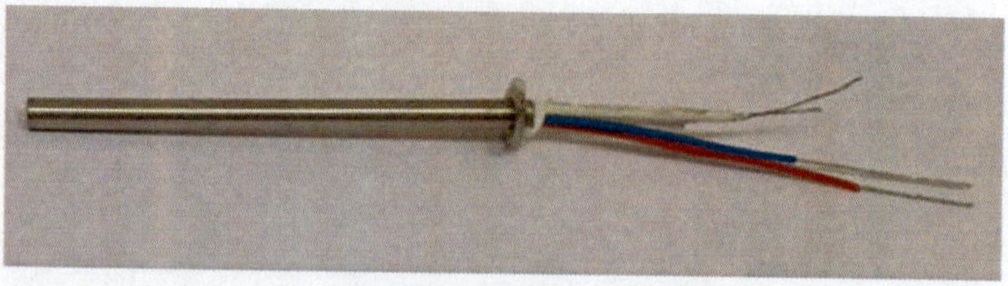

Fig. 5 Heating element of an electrothermal micro-engine

operational loads. The rated power of the heating element was 45 W at an operating voltage of 24 V. The electrothermal micro-engine was tested as part of a satellite platform.

The values of the voltage, applied to the heating element at different capacities, are shown in Table 2.

As a result of the tests, the main characteristics of an electrothermal micro-engine were selected at a power supply of 15 W (Table 3).

As a result of testing an electrothermal micro-engine in a vacuum chamber, an average dependence of the engine thrust on time was constructed, taking into account its spread (Fig. 6).

Table 2 Data for experimental studies of an Data for experimental studies of an electrothermal micro-engine

№	Parameters	Power (W)							
		3	4	5	10	15	20	25	30
1	Supply voltage (V)	6.84	7.90	8.83	12.49	15.30	17,66	19.75	21.63
2	Current (A)	0.44	0.51	0.57	0.80	0.98	1.13	1.27	1.39

Table 3 Data for experimental studies of an electrothermal micro-engine

№	Parameter	Value	Dimension
1	Applied voltage	15.30	V
2	Current strength	0.98	A
3	Thrust	255	mN
5	Specific impulse at the output of the mode	1025	m/s
6	Specific impulse in stationary operation	1351	m/s
7	Flow rate of the working body when entering the mode	246	g/s
8	Flow rate of the working body during stationary operation of the engine	190	g/s
9	Flow rate	346	W

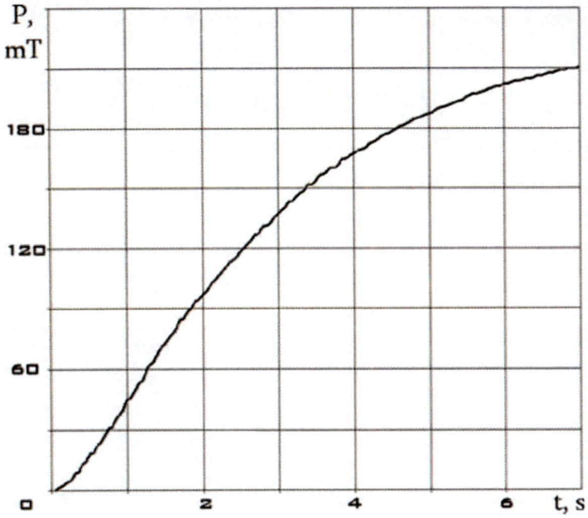

Fig. 6 Experimental dependence of the electrothermal micro-engine thrust during it's switching on

Let's use the experimental data to evaluate the feasibility of the theoretical algorithms developed in [3] using an electrothermal micro-engine, the main characteristics of which are presented in Table 3.

5 Simulation of the Electrothermal Micro-engine Operation in Order to Reduce the Influence of Temperature Shock

In [3], theoretical algorithms for controlling the executive body were developed to reduce the influence of the temperature shock of large elastic elements on the level of microaccelerations. The dependences of the executive body thrust on time are presented in Table 4 [3] and the magnitude of the decrease in microaccelerations

Table 4 There are basic control parameters and maximum micro-acceleration from temperature shock under various control modes (quoted from [3])

Mode	Number of plots	Thrust of the executive body (mN)				Maximum microacceleration (μm/s^2)
		[0; 0.24] *s*	[0.24; 1] *s*	[1; 7.03] *s*	[7.03; 15] *s*	
Without control	1	0				113.5
Constant thrust	4	0	255.0	255.0	0	42.0

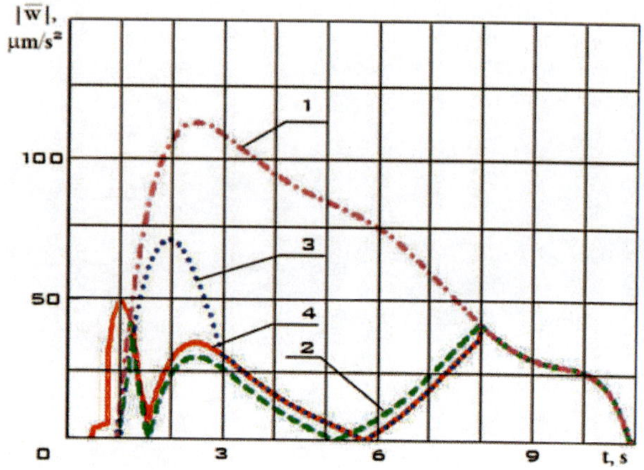

Fig. 7 The microaccelerations module from temperature shock: 1—without control; 2—with the management of the executive body of constant thrust; 3—real electrothermal micro-engine (switching on in the moment of thermal shock); 4—real electrothermal micro-engine (switching on 0.5 s before thermal shock)

caused by a temperature shock is estimated when using the theoretical control law with constant thrust (Fig. 7) [3].

Let's apply the electrothermal micro-engine considered in the previous sections for the implementation of the control law with constant thrust, taking into account the real law of change in its thrust, obtained as a result of its tests in a vacuum chamber (Fig. 7). Since the thermal shock is quite a fleeting phenomenon, for the correct estimation of microacceleration reduction values it is necessary to consider the transient output process of the engine to the operation mode.

Let's analyze the data presented in Fig. 7.

Curve 13 shows the dynamics of changes in microaccelerations from the longitudinal force that occurs during the all-round expansion of the elastic element (in the case of the spacecraft moving from the Earth's shadow to the illuminated part of the orbit). In this case, microaccelerations are not compensated and reach their maximum value of 113.5 μm/s^2 (Table 4) in about 1.5 s after the temperature shock. The maximum microaccelerations for technological vehicles should be about an order of magnitude lower (10 μm/s^2 for the «OKA-T» spacecraft (Fig. 2) [1, 6, 9] and «Tiangong-2» [14]). Therefore, it is necessary to actively combat microaccelerations from temperature shock, using the executive organs of the orientation and motion control system.

Curve 3 shows the dynamics of microaccelerations, taking into account the actual operation of an electrothermal micro-engine based on its tests (Fig. 6). In this case, the engine is switched on at the moment of temperature shock (t = 1 s). As can be seen from Fig. 8, the engine thrust increases significantly slower than the longitudinal force that occurs when the elastic element is comprehensively expanded during the temperature shock. As a result, the maximum value of microaccelerations reaches 69 μm/s^2. So, for a real executive body, the maximum value of microaccelerations increases by more than 60%. Therefore, it is necessary to look for a more optimal switch-on time of the electrothermal micro-engine.

Curve 4 shows the dynamics of microaccelerations for the same electrothermal micro-engine as curve 3. However, the engine itself turns on 0.5 s before the temperature shock. Up to t = 1 s, the engine creates microaccelerations, prolonging the unfavorable period for gravity-sensitive processes. However, at the time of the temperature shock, the engine thrust is significant, in contrast to curve 3. Therefore, the further dynamics of microaccelerations is close to ideal (curve 2). About 2 s after the temperature shock, curves 3 and 4 merge into one curve. Switching on the electrothermal micro-engine before the temperature shock allowed to reduce the maximum value of microaccelerations to 50 $\mu m/s^2$. However, this value exceeds by almost 20% the maximum microaccelerations under the ideal control law.

Accounting of the thrust spread of the electrothermal micro-engine, evaluated in the tests, shows that under ideal switch on 0.5 s before the temperature shock, the maximum accelerations will be 50 ± 5 $\mu m/s^2$.

6 Conclusions

Thus, based on the results of the conducted research, the following conclusions can be drawn.

1. For a correct assessment of the decrease in the magnitude of microaccelerations from a temperature shock with the help of a real executive body, it is necessary to take into account the real law of changes in its thrust. Since the temperature shock is an extremely dynamic phenomenon, a detailed study of the law of change in the thrust of the executive body when it is switched on is necessary.
2. The magnitude of the reduction of microaccelerations with the help of a real executive body in the vast majority of cases will be lower than with ideal control. This has been shown by the experimental studies and evaluations conducted. But the so-called exceptionally successful cases are not completely excluded, when the real executive body can provide a slightly larger reduction. However, it should be understood, that the probability of such a situation is extremely small. At the same time, the gain in the decline is also seen as insignificant.
3. Despite the more than twofold reduction in the maximum value of microaccelerations using an electrothermal micro-engine, this is not enough to ensure favorable conditions for microaccelerations. The control law with constant thrust cannot provide values close to 10 $\mu m/s^2$. In [3], a law with a piecewise linear change in thrust is constructed. However, this law has not yet been implemented on the basis of an electrothermal micro-engine, although there are prospects for throttling the engine.

Acknowledgements This work is supported by the Ministry of education and science of the Russian Federation in the framework of the State Assignments to higher education institutions and research organizations in the field of scientific activity (the project FSSS-2020-0017).

References

1. Sedelnikov AV (2016) Modeling of microaccelerations caused by running of attitude-control engines of spacecraft with elastic structural elements. Microgravity Sci Technol 28(5):491–498
2. Gorozhankina AS, Orlov DI, Belousova DA (2020) Problems of development motion control algorithms for a small spacecraft for technological purpose taking into account temperature deformations of solar panels. J Phys Conf Series 1546:012015
3. Sedelnikov AV, Orlov DI (2020) Development of control algorithms for the orbital motion of a small technologi-cal spacecraft with a shadow portion of the orbit. Microgravity Sci Technol 32(5):941–951
4. Belousova DA, Serdakova VV (2020) Modeling the temperature shock of elastic elements using a one-dimensional model of thermal conductivity. Int J Model Simul Sci Comput 11 (6):2050060
5. Chamberlain MK, Kiefer SH, Banik JA (2018) On-orbit structural dynamics performance of the roll-out solar array. American institute of aeronautics and astronautics spacecraft structures conference 2018: Proceedings of the AIAA spacecraft structures conference; 2018 Jan 8–12; Kissimmee. USA. American Institute of Aeronautics and Astronautics, Reston, pp 1–9
6. Sedelnikov AV, Potienko KI (2017) Analysis of reduction of controllability of spacecraft during conducting of active control over microaccelerations. Int Rev Aerosp Eng 10(3):160–166
7. Gaponenko Y, Gousselnikov V, Santos CIAV, Shevtsova V (2019) Near-critical behavior of fick diffusion co-efficient in taylor dispersion experiments. Microgravity Sci Technol 31(5): 475–486
8. Lyubimova T, Zubova N, Shevtsova V (2019) Effects of Non-Uniform temperature of the walls on the soret ex-periment. Microgravity Sci Technol 31(1):1–11
9. Sedelnikov AV, Molyavko DP, Potienko KI (2017) How Does Asymmetry of solar panels influence constructive component of microacceleration field of inner environment of space laboratory. Microgravity Sci Technol 29(4):305–311
10. Arakelov AS, Burov VA, Ochelkov YP (2013) Comparison of the simultaneous measurement results of SCR fluxes received by geostationary satellites «Electro-L» and «GOES». J Phys Conf Series 409:012164
11. Blinov VN, Shalay VV, Zubarev SI et al (2014) Studies of electrothermal micromotors of corrective propulsion systems of maneuvering small spacecraft. OmGTU, Omsk
12. Blinov VN, Fedyanin VV, Shalay VV, Yachmenev PS (2020) Experimental studies of arcjet thruster in pulsed mode of operation for corrective propulsion systems of small spacecraft. J Phys Conf Series 1441:012088
13. Anshakov GP, Belousov AI, Sedelnikov AV et al (2018) Efficiency estimation of electrothermal thrusters use in the control system of the technological spacecraft motion. Russ Aeronaut 61(3):347–354
14. Dong W, Duan W, Liu W, Zhang Y (2019) Microgravity disturbance analysis on Chinese space laboratory. npj Microgravity 5(18):1–6

Influence of Transport Process Participants on the Quality of Transport Services

Valerii Zubkov and Nina Sirina

Abstract This article examines one of the main issues, the organization, and management of integrated transport service in the field of freight traffic, as a key area for improving competitiveness in the transport services market and increasing the efficiency of production activities of all participants in the transportation process. An analysis of the responsibilities and areas of impact of participants of transport—supporting functions in the current model of the freight traffic market in railway transport is presented, the ability of influence and quality observance by the subjects of transport services, in the new Target model of integrated transport services are investigated. The basic criteria of consumers of a comprehensive transport service are formulated. The method is proposed for determining the degree of influence on the integrity of the quality of integrated transport service in different categories of transportation and various types of transport. The calculation of the coefficients of the quality of the transport service and the complex transport service was carried out, the use of which establishes the share of the contribution of each participant of the transport service in the implementation of the goods movement.

Keywords Transport services · Transportation process · Quality criteria · Responsibilities of subjects · Model of freight transportation · Economic performance · Quality factor

1 Introduction

The solution of the strategic task, an increase of competitiveness in the market of transport services in the field of cargo transportation [1], is provided through the effective impact on the main criteria of quality of cargo transportation [2].

V. Zubkov (✉)
JSC FGC, Masha Poryvaeva Street, 34, Moscow 107078, Russia

N. Sirina
Ural State University of Railway Transport, Kolmogorov street, 66, Ekaterinburg 620014, Russia

A. Bieliatynskyi and V. Breskich (eds.), *Safety in Aviation and Space Technologies*, Lecture Notes in Mechanical Engineering,
https://doi.org/10.1007/978-3-030-85057-9_26

The consumer of transport services is interested in the final result, i.e. the receipt of finished products to the consumer within the specified time. In the current model of freight transportation by rail the main criteria for consumers of the transport industry are:

1. Preservation of transported cargo [3] or goods, which affects the quality characteristics of the product (*Kpres*).
2. Safe transportation [3] of finished goods, which also affects the quality characteristics of the goods being transported (*Ksafe*).
3. Speed of delivery [3], goods (*Ksp*), influences the turnover of financial investments, the higher the speed of the cycle of financial investments.
4. Constancy and mass transport of products [4], which also affects the turnover of financial investments (*Kconst*).
5. Price availability [4] and rationality of transportation services (*Kprice*), provides a reduction in costs and, accordingly, an increase in revenues. The sum of all quality criteria [5] of transport service is calculated by the formula:

$$\sum K = K_{pres} + K_{safe} + K_{sp} + K_{const} + K_{price} \tag{1}$$

In the implementation of transport services, the considered criteria are mutually determined. This implies an increase in the competitiveness of both subjects of transport services and producers of products. At the same time, in the current conditions of freight transportation, the following for their solution:

1. What is the responsibility of the subjects of transport services what are their impact areas in the implementation of transport services?
2. What is the limit value of the quality degree that can be achieved by the subject while performing transport services?
3. What method determines the completeness of transport service quality performance by its concrete subject?
4. How can the participants of the transportation process influence the reduction of the added value of finished products?

Thus, it is advisable to consider the impact of the participants of the goods and cargo movement on the quality of the services provided and provide answers to the questions that have not yet been resolved.

2 Materials and Methods

Thus, the question of participants' influence on the transport market remains open. In the transport business there are often conflicts between the owner and subjects of transport services [5]. Contradictions arise in the provision of quality criteria., So it is necessary to determine the responsibility of subjects of transport services and the area of their impact according to the requirements of federal laws of Russia.

Table 1 presents the subjects of transport services, their responsibilities, and areas of their impact [6] in the current model of the market of freight transportation by rail.

It can see on Table 1 the rolling stock operator and forwarder have the least influence on the main quality criteria [6, 7]. The greatest impact on the main quality criteria for consumers of the transport industry is the owner of the infrastructure, the carrier, and the owner of the rolling stock [8], this confirms that the owner of the rolling stock [8]. The owner of the rolling stock can act as a guarantor of the implementation of a comprehensive transport service [8, 9], but this is ideal for the requirements of transport services. Based on the peculiarities of the quality of particular transport service, the area of influence of quality criteria may vary from greater to a lesser value.

To determine the quality of the transport service provided to the consumer by the concrete subject of the transport service. This is the quotient of the division of the sum of the quality criteria of transport service in the areas of influence of a particular subject ($\sum Ksub$) into the sum of all the quality criteria of the transport service ($\sum K$):

$$K_{qua}^{tr.s.} = \frac{\sum K_{sub}}{\sum K} \tag{2}$$

Table 1 shows that the owner of the infrastructure, the carrier, the owner of the rolling stock [9] guarantee the impact on all five indicators of the quality of transport services (if ideal). In this paper, the authors of any quality criterion is taken numerically as 1. Since only rail transport takes part in the described freight market model. So, the quality factor of transport service for these subjects is calculate as follows:

$$K_{qua}^{tr.s.} = \frac{5}{5} = 1$$

The consignor, consignee, owner of the railway track of no use [9] in the performance of transport services guarantee responsibility for the impact on the four main quality indicators, respectively, the calculation of the coefficient of quality of transport services for these participants looks like this:

$$K_{qua}^{tr.s.} = \frac{4}{5} = 0.8$$

For rolling stock operator and forwarder [9] quality factor is calculated in this way accordingly:

$$K_{qua}^{tr.s.} = \frac{2}{5} = 0.4$$

Table 1 Responsibilities and impact areas of subjects

Subjects of transport services	Reasons for liability	Impact area
Shipper	Statute of railway transport 1. Failure to use the submitted containers 2. Non-shipment of wagons, containers by the carrier for reasons dependent on the consignor 3. Shipper's refusal from the wagons and containers provided by the application Absence of own or rented wagons or containers, provided for in the application, for the reasons that depend on the consignor or the organization with which the consignor has an agreement regulating the provision of such wagons and containers	1. Security of transported cargo 2. Constancy and weight of cargo transportation 3. Price and transport accessibility 4. Safe transport
Consignee	Statute of railway transport 1. Calculations for transportation after the arrival of cargo, freight and equipment to the railway station of destination and notification of the consignee carrier about the arrival of cargo, freight and equipment to his address 2. Timely receipt, timely unloading of cargo at railway stations 3. After the unloading of cargo, freight wagons, containers should be cleaned inside and outside, the fixing devices should be removed from them	1. Security of transported cargo 2. Constancy and weight of cargo transportation 3. Price and transport accessibility 4. Safe transport
Owner of a railroad track of no public use	Statute of railway transport 1. Violation of the rights of owners of cars, containers 2. Delay of wagons, containers 3. Preservation of cargo on the ways of its non-use	1. Speed of delivery 2. Security of transported cargo 3. Price and transport accessibility
Owner of infrastructure	Statute of railway transport 1. Delay of transportation means under loading and unloading of cargo, violation of transportation means turnover terms 2. Non-performance or improper performance of obligations under the agreement on provision of infrastructure services	1. Speed of delivery 2. Security of transported cargo 3. Constancy and weight of cargo transportation 4. Price and transport accessibility 5. Safe transport
Carrier	Statute of railway transport 1. Failure to submit to the consignor for loading wagons, containers at the specified time or submission for loading of wagons, containers not suitable for cargo transportation	1. Speed of delivery 2. Security of transported cargo 3. Constancy and weight of cargo transportation

(continued)

Table 1 (continued)

Subjects of transport services	Reasons for liability	Impact area
	2. Safety of cargo, baggage, cargo-luggage after its acceptance for transportation and storage and before its delivery to the consignee, if it cannot prove that the loss, shortage or damage (deterioration) of cargo, cargo-luggage occurred due to circumstances that the carrier could not prevent or eliminate for reasons beyond its control	4. Price and transport accessibility 5. Safe transport
The owner of the rolling stock	Statute of railway transport 1. In case of unsecured carriage, in case of detection of a leak, damage or soaking up of the cargo due to the faulty condition of the wagon or container 2. Allowed losses from loss or damage to wagons 3. Provision of wagons and containers	1. Speed of delivery 2. Security of transported cargo 3. Constancy and weight of cargo transportation 4. Price and transport accessibility 5. Safe transport
Rolling Stock Operator	Resolution of the Government of the Russian Federation № 626 of 25.07.2013 1) To provide services in the provision of railway wagons and containers, for the carriage of goods by rail by any types of consignments	1. Constancy and weight of cargo transportation 2. Price transport accessibility
Forwarder	FZ-87 of 30.06.2003 (ed. 06.07.2016) 1. to provide services in accordance with the contract of transport expedition	1. Constancy and weight of cargo transportation 2. Price transport accessibility

As a result, guaranteed responsibility for the implementation of the main quality indicators in the current model is provided only on the railway transport, not providing coordination and the mutual influence on the quality criteria of transport services in other modes of transport in the general transport system, which does not suit consumers and creates prerequisites for the growth of the added value of finished products.

3 Results

Thus, in order to reduce the added value of finished products, a comprehensive reduction in the cost of transport services is require [9]. Therefore, it is necessary to integratively improve the economic performance of all resources of the transport industry. [9, 10], The resources of the transport industry include: infrastructure, rolling stock, labor and energy resources.

In the new Target Model of Integrated Transport Services in the sphere of freight transportation [10], the main criteria for consumers of the transport industry are considered not on the principle of their increase, but on the principle of improving the quality of integrated services. This requires changes in the current regulations of interaction, all participants of the transport service market [10]. Thus, its formation is based on the introduction of new subjects of transport services and identification of new areas of their impact. When using the new target model of complex transport service [11], the main criteria for consumers of complex transport service are:

1. Compliance with the complex transport cycle in production (K_{cycle}^{comp}).
2. Complexity satisfaction of transport service order (K_{ord}^{comp}).
3. Reliability of complex transport service (K_{serv}^{comp}).
4. Completeness of complex satisfaction of demand for transport service (K_{dem}^{comp}).
5. Price availability and transparency of complex transport service (K_{price}^{comp}).
6. Complex transport security of the territory ($K_{s.terr}^{comp}$).
7. Safety and environmental friendliness of the integrated transport service (K_{safe}^{comp}).
8. Complexity of transport service ($K_{tr.serv}^{comp}$).

The sum of all quality criteria of complex transport service is calculated by the formula:

$$\sum \mathrm{K} = K_{cycle}^{comp} + K_{ord}^{comp} + K_{serv}^{comp} + K_{dem}^{comp} + K_{price}^{comp} + K_{s.terr}^{comp} + K_{safe}^{comp} + K_{tr.serv}^{comp} \quad (3)$$

The quality coefficient of the complex transport service is defined as private from division of the sum of quality criteria of the complex transport service in the areas of influence of the concrete subject $\sum K_{sub}^{comp}$ into the sum of all quality criteria of the complex transport service ($\sum K^{comp}$):

$$\mathrm{K}_{tr}^{comp} = \frac{\sum K_{sub}^{comp}}{\sum K^{comp}}, \quad (4)$$

4 Discussion

On the basis of the basic platform for the formation of the target model [12], the authors took the subject of the transport service - the owner of the rolling stock [13], he is also an alternative owner of a complex transport service [13]. In order to exclude possible contradictions in the correctness of the decision made by the authors, it is advisable to analyze the area of influence of the participants of transport supporting functions [13] on the quality of a complex transport service provision.

The new subjects of services and areas of influence in the developed Task Model of Integrated Transport Service [13] are present in Table 2.

Table 2 Responsibility and impact areas of the subjects of the Target model of integrated transport service in the categories of transportation

Subjects of transport services	Shipping categories	Areas of influence
Integrated transport service division	Domestic transport	1. Compliance with the complex transport cycle in product manufacturing 2. Complexity satisfaction of transport service order 3. Reliability of complex transport service 4. Completeness of complex satisfaction of demand for transport service 5. Price availability and transparency of complex transport service 6. Integrated transport security of the territory 7. Safety and environmental friendliness of complex transport service
Integrated transport service division	Industrial transport	1. Complexity satisfaction of transport service order 2. Completeness of complex satisfaction of demand for transport service 3. Price availability and transparency of complex transport service 4. Reliability of complex transport service 5. Integrated transport security of the territory 6. Safety and environmental friendliness of complex transport service
Regional freight transport company	Regional shipping	1. Compliance with the complex transport cycle in product manufacturing 2. Complexity satisfaction of transport service order 3. Reliability of complex transport service 4. Completeness of complex satisfaction of demand for transport service 5. Price availability and transparency of complex transport service 6. Integrated transport security of the territory 7. Safety and environmental friendliness of complex transport service 8. Complexity of transport service
Large freight transport company	Interregional traffic	1. Compliance with the complex transport cycle in product manufacturing 2. Complexity satisfaction of transport service order 3. Reliability of complex transport service 4. Completeness of complex satisfaction of demand for transport service 5. Price availability and transparency of complex transport service 6. Integrated transport security of the territory 7. Safety and environmental friendliness of complex transport service 8. Complexity of transport service

(continued)

Table 2 (continued)

Subjects of transport services	Shipping categories	Areas of influence
International freight transport company	International traffic	1. Compliance with the complex transport cycle in product manufacturing 2. Complexity satisfaction of transport service order 3. Reliability of complex transport service 4. Completeness of complex satisfaction of demand for transport service 5. Price availability and transparency of complex transport service 6. Integrated transport security of the territory 7. Safety and environmental friendliness of complex transport service 8. Complexity of transport service
International freight transport company	Transit transport	1. Complexity satisfaction of transport service order 2. Reliability of complex transport service 3. Completeness of complex satisfaction of demand for transport service 4. Price availability and transparency of complex transport service 5. Integrated transport security of the territory 6. Safe and environmentally friendly transport service 7. Complexity of transport service

Table 2 shows that a regional freight transport company, a large-scale freight transport company, an international freight transport company guarantee the impact [13] on all eight indicators of quality of comprehensive transport service (if ideal), respectively, in regional, inter-regional, and international categories of transportation. The Integrated Transport Services Division [14] has an impact on seven and six quality indicators in domestic and industrial transport.

In the new target model, the level of complex transport service [14] depends on the participation of modes of transport in the scheme of transportation under the influence of one subject in different categories of transportation. In some cases, the service will be fully integrated with the interaction of the two modes of transport, and in some cases, more modes of transport. The main factors affecting the complexity are the possibility of providing transport services on different specialized infrastructure by different modes of transport and highly professional staff. In this paper, any of the main quality criteria [14] in the new target model is accepted in numerical terms for 2, as in the current model of complex transport services [14, 15], the authors consider the participation in the transport scheme of two modes of transport: railway and road. In case of participation in the integrated transport service [14, 15] more modes of transport, the numerical expression of the main quality criteria increases in proportion to their quantity. Therefore, the coefficient of quality of transport service for subjects of regional freight transport company, large scale freight transport company, an international freight transport company in the categories of regional, interregional and international transport, respectively, is calculated as follows:

$$K_{tr}^{comp} = \frac{16}{8} = 2$$

Integrated transport service division in the category of domestic transport and international freight companies in the category of transit transport guarantee the impact on seven main criteria, respectively, the calculation of the coefficient is as follows:

$$K_{tr}^{comp} = \frac{14}{8} = 1.75$$

For subdivision of complex transport service in the category of industrial transport the quality coefficient is calculated accordingly:

$$K_{tr}^{comp} = \frac{12}{8} = 1.5$$

As a result, guaranteed responsibility [15] for the implementation of the main quality indicators in the new target model is provided not in a single mode of transport, but in several types of transport involved in the complex transport service. Coordination and the mutual influence on quality criteria of transport services in other modes of transport in the common transport system are ensured, which satisfies consumers and creates prerequisites for reducing the added value of finished products.

5 Conclusion

The conducted analysis differentiates the responsibilities and areas of influence of subjects for the provision of quality services in the Task Model, it makes it possible to more effectively form a strategy for each participant of the complex transport service in terms of ensuring and increasing the quality of service. The presented methodology for calculating the quality factor makes it possible to assess the impact on the quality of service by each subject, but it does not allow to determine the amount of responsibility, this requires a detailed structure of the impact of the participants of the complex transport service on its quality level.

The target model of complex transport service reflects the needs of all participants of the market of transport services, it is the development of conditions for the provision and acquisition of quality services, the formation of norms and rules of business processes and their compliance, ensuring price availability and transparency of tariffs, ensuring the rationality of transport services. The target model defines the responsibility and areas of interaction of market participants with subsequent assessment of their impact on the quality of services.

References

1. Bakti IGMY, Rakhmawati T, Sumaedi S et al (2020) Public transport users' WOM: an integration model of the theory of planned behavior, customer satisfaction theory, and personal norm theory. Transp Res Proc 48:3365–3379
2. Vu TP, Grant DB, Menachof DA (2020) Exploring logistics service quality in Hai Phong, Vietnam. Asian J Shipping Logistics 36:54–64
3. Baklanov A, Cárdenas B, Masson V et al Integrated urban services: Experience from four cities on different continents. Urban Climate 32:100610
4. Jokinen J-P, Sihvola T, Mladenovic MN (2019) Policy lessons from the flexible transport service pilot Kutsuplus in the Helsinki Capital Region. Transp Policy 76:123–133
5. Popova O, Gorev A, Shavyraa C (2018) Principles of modern route systems planning for urban passenger transport. Transp Res Proc 36:603–609
6. Kavanagh P (2016) A case for negotiated performance-based contracting rather than competitive tendering in government public transport (bus) service procurementю Research in Transportation Economics 59:313–322
7. Canitez F, Alpkokin P, Black JA (2019) Agency costs in public transport systems: Net-cost contracting between the transport authority and private operators - impact on passengers. Cities 86:154–166
8. Shahin M, Baba MA, Chauhan MA (2020) Architectural design space for modelling and simulation as a service: a review. J Syst Softw 170:110752
9. Lovelace R, Parkin J, Cohen T (2020) Open access transport models: a leverage point in sustainable transport planning. Transp Policy 97:47–54
10. Walter J, Abendroth B (2020) On the role of informational privacy in connected vehicles: A privacy-aware acceptance modelling approach for connected vehicular services. Telematics Inform 49:101361
11. Weber KM, Heller-Schuh B, Godoe H, Roeste R (2014) ICT-enabled system innovations in public services: experiences from intelligent transport systems. Telecommun Policy 38:539–557
12. Narayanan S, Chaniotakis E, Antoniou C (2014) Shared autonomous vehicle services: a comprehensive review. Transp Res Part C Emerg Technol 111:255–293
13. Nowakowska-Grunt J, Strzelczyk M (2019) The current situation and the directions of changes in road freight transport in the European Union. Transp Res Proc 39:350–359
14. Brake J, Mulley C, Nelson JD, Wright S (2007) Key lessons learned from recent experience with flexible transport services. Transp Policy 14(6):458–466
15. Kiba-Janiak M (2019) EU cities' potentials for formulation and implementation of sustainable urban freight transport strategic plans. Transp Res Proc 39:150–159

Labor Conditions and Surface Quality of the Rolled for the Automotive Metal Products

Dmitriy Kozinov, Alexei Filippov, German Pachurin, Mariia Mukhina, and Zhanna Smirnova

Abstract Automotive fasteners, manufactured by cold upsetting from calibrated rolled metal, are among the most critical parts in mechanical engineering. The rolled products for cold forming are subject to fairly high requirements in terms of strength and plastic properties, upsetting tests, as well as special functional characteristics. These properties depend on the structure and mechanical characteristics that are formed during phase, structural transformations and drawing through the die. For a number of metalware products, for example, bolts of the car engine assembly, there are increased requirements for the presence of surface defects—the depth of surface defects, the presence of a decarbonized layer and the finish of the surface layer. The paper analyzes the defects of metallurgical blanks for the production of hot-rolled products and compliance with the requirements of regulatory and technical documentation in order to improve the quality of metalware, reduce metal consumption and increase the profitability of production.

Keywords Automotive fasteners · Hot-rolled stock · Cold heading · Metalware · Mechanical characteristics · Structure · Surface defects

1 Introduction

The most important task of the current stage of industrial production development [1, 2] is to increase the efficiency [3], reliability, durability [4], environmental friendliness [5, 6] of metal products, as well as to bring these indicators to world levels in order to ensure the competitiveness of Russian products on the world market [7–9].

D. Kozinov · A. Filippov · G. Pachurin
Nizhny Novgorod State Technical University named after R.A. Alekseev, Minin Street, 24, Nizhny Novgorod 603950, Russia

M. Mukhina (✉) · Z. Smirnova
Minin Nizhny Novgorod State Pedagogical University, Chelyuskintsev street, 9, Nizhny Novgorod 603138, Russia

A. Bieliatynskyi and V. Breskich (eds.), *Safety in Aviation and Space Technologies*, Lecture Notes in Mechanical Engineering,
https://doi.org/10.1007/978-3-030-85057-9_27

Among the most important parts in mechanical engineering are automotive fasteners [10, 11] which are made by cold upsetting from calibrated rolled metal [12–14]. One of the ways to increase the technology efficiency of the in the production of bolted products is to reduce the number of technological operations and the need to reduce labor costs. At the current stage of mechanical engineering development, products with a high level of physical and mechanical characteristics can be obtained either by using new materials, or through the development of new technological processes for their production. In metalware production, quite strict requirements are imposed on metals, especially in respect of the chemical composition, mechanical characteristics and operational durability. The requirements for cost reduction, material and energy resources saving are also relevant for any production enterprise.

Metal products or metalware include wire, wire products, fasteners and springs. A fairly wide range and a wide variety of mechanical characteristics of metalware are specific when these are used in various fields of mechanical engineering.

The main material for fasteners production [15, 16] is steel, which has the necessary set of properties with sufficient strength, ductility, fatigue resistance, wear and corrosion resistance.

Structural carbon steels with a carbon content of not more than 0.5% are used for fasteners manufacturing by the cold upsetting method. The grades, chemical composition and other properties of steels intended for cold forming are mainly regulated by GOST 10702-78. The optimal chemical composition of steels in this standard is established on the basis of generalization of existing experience in steel smelting and using them in the production of fasteners and studying thoroughly the influence of particular elements in it (carbon, silicon, phosphorus, manganese, aluminum, chromium) concerning the ability of the metal to cold and hot forming.

Carbon and low-alloy steels are produced by open-hearth and converter method, whereas medium-alloy steels by open-hearth and electrical melting. For cold heading of fasteners, rimming, semikilled and killed steels with a carbon content of up to 0.24% are used. Steels with a high content of carbon and alloying elements are smelted killed only. Killed steels deoxidized with aluminum are characterized by increased ductility, which is the most important factor in fasteners manufacturing using cold deformation. Such steels are practically not prone to cracking, while being ductile, highly resistant to bending stress, low crack-sensitive when heated and suddenly cooled, they have low ability to decarburization as well as increased deformability in the cold state.

In recent years, boron steels such as 20G2R and Z0G1R have been widely used for fasteners manufacturing. A special feature of these steels is the combination of high technological ductility and hardenability. In comparison with traditional chromium steel grades (35X, 38X, 40X), boron steels have a lower resistance to deformation during cold forming, which provides good formability and increases the resistance of cold-heading tools.

All physical, mechanical, chemical and metallographic characteristics of hot-rolled products for the production of fasteners are strictly controlled and regulated. Rolled metal products must be technologically efficient during processing, for example, having easily removable scale with a minimum mass loss, no surface defects, etc.

The main mechanical characteristics of rolled products are the temporary tear resistance and the narrowing of the cross-section during the tensile test, which reflect the possibility of hardening during calibration. The diameter, ovality, and sinking strain are strictly normalized. Rolled products must have a uniform structure [17, 18], so it is not allowed to have different grain sizes, the deposition of cementite and ferrite along the grain boundaries, the presence of martensitic structures, the proportion of granular perlite and the size of decarbonized layer on the surface of the rolled products are specified.

Metal structure has a great influence on the deformability of rolled products [19]. The microstructure should be uniform, the liquation zone should not exceed 25% of the metal cross-section and should not reach the surface layer. For cold upsetting, it is necessary to have the microstructure of the source metal after annealing 80–100% of the granular perlite.

The current technology, which is used for the processing of high-grade rolled metals, for example, 40X steel and the manufacture of car engine assembly bolts with M10 diameter (the initial size of hot-rolled rolled products is 12.0 mm, the finished size of calibrated rolled products is 9.65 mm), is presented in the form as follows:

- initial condition: hot rolled products with a diameter of 12.0 mm:
- annealing of hot-rolled products in chamber gas furnaces at a temperature of 770 °C (total annealing time is 28 h);
- etching of rolled metal products until complete removal of scale is ensured;
- coating of calibrated rolled products with a lubricant;
- drawing of rolled products from a diameter of 12.0 to a diameter of 10.25 mm;
- recrystallization annealing at 770 °C;
- etching of rolled metal products until complete removal of scale is ensured;
- coating of calibrated rolled products with a lubricant;
- drawing of rolled products from a diameter of 10.25 to a final diameter of 9.65 mm;
- quenching of bolt products;
- tempering of finished bolts.

The time-consuming technology for bolt products manufacturing is not only quite labor-and energy-consuming, but also provides for significant disadvantages that lead to a deterioration in working conditions and affect the service personnel due to the impact of a number of hazardous and harmful production factors.

1. During the heat treatment of hot-rolled products in chamber furnaces with a retractable hearth, the heat treater is exposed to infrared radiation (IR radiation). This is accompanied by the release of heat, and the heat is released both by chamber furnaces with a retractable hearth, and by coils of heated rolled metal. Staying near the heated metal, the heated surfaces of the equipment, the flame, the service personnel is exposed to IR radiation. The potential radiation hazard

is estimated by the energy flux density and exceeds 350 W/m^2, which does not meet the requirements of GOST 12.4.124-83 "Protection means against static electricity". The total level of irradiation at a distance of 0.7 m is more than 560 W/m^2. Therefore, the main measures aimed at reducing the negative impact of IR radiation on employees are as follows:

- move work stations from the source;
- make protective shielding of the source or the work station;
- make thermal insulation of hot surfaces or ensure cooling;
- use spot cooling systems;
- use personal protective equipment (workwear, glasses with light filters and shields);
- therapeutic and preventive measures (reasonable work-rest ratio).

2. The noise level resulting from gas combustion in the furnace nozzles and travel of lifting mechanisms in the areas of metal annealing in chamber furnaces exceeds the permissible standards according to GOST 12.1.003 SSBT "Noise. General safety requirements" resulting from combustion of gas in the furnace nozzles and travel of lifting mechanisms.
3. After the rolled metal coil etching, harmful waste solutions are present. The disposal of such solutions by chemical means is very time-consuming and involves high material and energy costs. Industrial waste that contains heavy metals is extremely dangerous for humans and the environment. Their neutralization is an important economic and environmental target for any metalware company. When the products of the interaction of acids with iron and other components that are part of the etched metal accumulate, the solution is entrapped and must be drained. The solutions might contain up to 100 g/l or more of acid and up to 300 g/l or more of ferrous (Fe_2O_3) and ferric (FeO) iron. The washing water contains the same contamination, but in much lower concentrations. Waste solutions are drained periodically (2–3 times a day). The washing water is drained relatively evenly. The most polluted and concentrated solutions are spent pickling solutions, the amount of which is from 5 to 7 m^3 per 1 ton of consumed acid. At the current level of acid consumption, large masses of spent pickling solutions are formed, amounting to hundreds of thousands of cubic meters per year.
4. Any metal coils that have been cleaned of scale from the surface must be washed with a jet of water. The waste water volume that is generated when washing the metal after the etching operation is not less than 3.0 m^3 per 1 ton of acid-treated metal. At the existing production facilities, the flow rate of washing water reaches a volume of 300–400 m^3 per hour. If an enterprise discharges polluted waste water into the water bodies, the concentration of harmful substances in these increases sharply, while a significant part of the substances is deposited near the place of release.
5. In the process of rolled metal etching, harmful discharges are formed, which are removed through the on-board vents during the entire technological process. After etching, the rolled products are washed in hot and cold water to remove

the etched sludge and acid. Washing in hot water is carried out at a temperature of 50–800 °C for 1–2 min. Cold washing is carried out at a pressure of 5–6 atm. for 1–2 min. To neutralize the sulfuric acid residues and reduce the coefficient of friction during drawing and cold forming, the rolled products are limed in a 3–5% lime solution. At the same time, there should be a solid film of lime on the surface of the rolled product. The acid can be neutralized in an aqueous solution with soap.

Harmful substances released during etching into the air of the working area are highly hazardous in terms of the degree of exposure and do not meet the requirements of GN 2.2.5.1313–03 "MPC of harmful substances in the air of the working area".

6. The chemical etching method has rather a low productivity. The duration of etching depends on the amount of scale on the rolled products surface and the composition of the acid solution concentration. When etching large coils of rolled metal, a long time and constant presence of service personnel at the workplace of a hazardous and harmful technological process is required.
7. In the conditions of the production process, etching is a physically difficult and hazardous operation for the working personnel. All processes take place in containers, rolled products in coils or bars are continuously transferred from one container to another. Heating of pickling and other baths is constantly carried out by hot steam. Since each operation takes place at temperatures of 40–1000 °C, there is an inevitable process of evaporation, which is accompanied by harmful odors at the permanent workplaces of metal picklers. The concentration of harmful vapors in the air of the working area often exceeds the values established by GOST 12.1.005-88 SSBT "General sanitary and hygienic requirements for the air in the working area". It is possible for the service personnel to come into direct contact with chemicals, hot materials and blanks that have a dangerous effect on them. In this case, emergency situations may occur. It is quite possible for harmful substances (acids and derived vapors) to pollute water and air basins.
8. The noise level resulting from process travel of lifting mechanisms, purging of rolled metal coils after washing with water and heating with supersaturated steam in the pickling areas exceeds the permissible standards that are specified by the requirements of GOST 12.1.003 SSBT "Noise. General safety requirements".

Heat treatment (quenching and tempering) of upset bolt products is often carried out in salt tank furnaces. The working medium for bolt products heating is molten salts. Heat treatment in salt baths is carried out in order to avoid warping of long-length bolt products. During heat treatment on this type of equipment, the heat treater is also exposed to IR radiation. Vapors of aerosol emissions of heated salts negatively affect the airways and cause irritation, which does not meet the requirements of GN 2.2.5.1313–03 "MPC of harmful substances in the air of the working area".

However, it is necessary to take into account the requirements ensuring the product technological properties and surface quality, as well as the issues of environmental friendliness of the process and labor safety when choosing a particular technological process for the preparation of hot-rolled products before the cold heading of metal products.

The technological properties of the metal subjected to upsetting include its ability to withstand the upsetting test under the influence of a workable tool having a flat surface without breaking (without the appearance of cracks and surface tears). In this case, the upset force and the deformation rate should be as constant as possible. Such tests determine both the deformability of the metal and the presence of surface and internal defects. According to GOST 10702-68, steels for cold upsetting are divided into groups according to the amount of deformation during cold upsetting, which are designated as follows: up to 1/2h, up to 1/3h and up to 1/4h, and samples that were subjected to spheroidizing annealing: up to 1/3h-bbI. The highest cold-upsetting capability of the steels is practically ensured if the sample withstands the upsetting test of up to 1/3h or more. The disadvantage of the sinking test is that the results obtained will not fully characterize the entire batch and even a single coil of the inspected metal. This happens because of the inability to control the entire batch of metal and the uneven distribution of defects along the entire length of the coil under the standards of sampling rates.

Surface defects of hot-rolled and calibrated rolled products include rolling gas blisters, hairline cracks, tears and laps [20, 21]. The main reasons for their occurrence are defects in metallurgical production. A permanent defect of the rolled metal surface is the formation of a decarburized layer due to the carbon partially burning off when the metal is heated, both at the rolling stage and during heat treatment before calibration. Decarburization and scale formation significantly reduce the mechanical properties in the surface layers of rolled metal. The surface becomes susceptible to the formation of notches, impact lines and scratches during rolling, calibration, and cold upsetting [4, 22, 23].

2 Study Methods

The paper analyzes the detection of surface defects of workpieces for further rolling of steel grades 10kp, 30, 35X and 38X affecting the surface quality of the finished hot-rolled products. The statistical analysis of the results of quality control of blanks for rolling and hot-rolled products obtained after rolling them has been carried out. The rolling stock is supplied according to TU 14-1-4492 with a surface category 2 (P) with the size of 125 mm × 25 mm × 420 mm. The quality control of the surface and macro-structure of the blanks of the listed steel grades was carried out on templates (transverse macro-sections) etched at a temperature of 60–70 °C in a 50% aqueous solution of hydrochloric acid.

The surface quality control of the hot-rolled products obtained from the rolling stock was carried out on samples taken from one end of the coil of each wire rod batch. After etching, washing and drying, the samples were visually examined, while micro-sections were cut out at the defect locations to determine and measure the depth of the defects using an optical microscope with a magnification power of x100.

The characteristics of surface defects are presented in accordance with GOST 21014-88 "Rolled ferrous metals. Terms and definitions of surface defects." A total of 1312 samples were tested.

3 Results and Discussions

The results of the hot-rolled product surface quality check are shown in Table 1. It is found that the main type of defects on the wire rod (92 samples, or 6.7% of the total number) are laps. Rolled gas blisters were found on one sample made of 38X steel, on 2 samples made of 10x steel and on 2 samples made of 30 steel, while scratch marks were detected on 10 samples (diameter 6.5; 10.0; 8.0 and 13.0 mm—10x steel and diameter 11.3 mm—38X steel). No defects were found in four batches of wire rod-steel of 10kp grade (diameter 11.0; 9.0; 13.0 mm) (289 samples).

It was discovered that rolled gas blisters were present on all templates with a depth of 0.5 to 2.0 mm (Fig. 1), the microstructure is dense, uniform, satisfactory.

It was discovered that the main defect type in the rolled wire (92 samples or 6.7% of the total number) are laps (Figs. 2 and 3).

On one steel 38XA sample (smelting 68,529), on 2 steel 10kp samples (smelting T522) and on 2 steel 30 samples (smelting T695) rolled gas blisters were found (Fig. 2), on 10 samples (diameter 6.5; 10.0; 8.0 and 13.0 mm—steel 10kp, smelting T522 and diameter 11.3 mm—steel 38XA, smelting 68,522) scratch marks were found (Fig. 4).

No defects were found in four batches of wire rod made of steel 10kp (smelting T522, diameter 11.0; 9.0; 13.0 mm) (289 samples).

The surface defects found on the samples of hot-rolled products—the laps, are obtained in the process of rolling the blanks to eliminate the rolled-in ridges, undercut, rough traces produced by cleaning the surface of the workpiece, as well as a result of rolling in gas blisters that were present on the surface of the workpiece.

The reason that the rolled gas blisters found on the surface of the templates make up an insignificant percentage of the number of surface defects found on the hot-rolled metal products (0.36%), explains their discontinuity along the length of the blank for rolling.

Thus, due to the increased requirements for calibrated rolled products intended for the further production of cold upset parts from them, the task of eliminating defects in the source hot-rolled metal products occurring during metallurgical and rolling production is of high relevancy.

Table 1 Hot-rolled stock

Steel grade/ sample diameter, mm	Smelting	Number of coils/number of samples, pcs	Number of coils/number of samples with defects	Defect nature and depth, mm
1	2	3	4	5
10kp/11.0	T-549	24/24	7/7	laps—0.10…2.0 mm
10kp/6.5	T-522	49/49	3/3	laps—0.5…1 mm—2 samples; scratch marks—0.07 mm—1 sample
10kp/13.0	-«-	78/78	3/3	laps—0.1…1.5 mm
10kp/11.0	-«-	63/63	None	None
10kp/10.0	-«-	43/43	2/2	laps—0.20 mm
10kp/9.0	-«-	136/136	6/6	laps - 0,0.2…2.0 mm
10kp/9.0	-«-	66/66	None	None
10kp/9.0	-«-	71/71	6/6	laps—0.15…2.0 mm—5 samples; rolled blisters—0.5 mm—1 sample
10kp/9.0	-«-	52/52	3/3	laps—0.5…2.0 mm - 2 samples; scratch marks—0.10 mm—1 sample
10kp/11.0	-«-	131/131	9/9	laps—0.15…2.0 mm - 8 samples; rolled blisters—0.6 mm—1 sample
10kp/8.0	-«-	44/44	12/12	laps—0.2…1.5 mm - 11 samples; scratch marks—0.05 mm—1 sample
10kp/13.0	-«-	50/50	2/2	Scratch marks—0.10 mm—2 samples
10kp/13.0	-«-	57/57	1/1	lap—0.20 mm
10kp/13.0	-«-	71/71	None	None
10kp/13.0	-«-	89/89	None	None
38XA/ 11.0	68,529	53/53	16/16	Rolled blisters—0.5 mm -1 sample; laps—0.10…2.0 mm —10 samples; scratch marks —0.10 mm—5 samples
30	T-695	98/98	16/16	Rolled blisters—2.2 mm—2 samples; laps—0.3…1.0 mm - 14 samples
35X/11,0	79,320	78/78	10/10	laps—0.15…2.0 mm
35X/11,0	-«-	37/37	10/10	laps—0.15…2.0 mm
35X/11,0	-«-	22/22	1/1	laps—1.0 mm
Total		**1312**	**107**	

Fig. 1 Rolled gas blister on the blank surface. A transverse micro-section. Etching with 4% HNO_3 in alcohol, x100

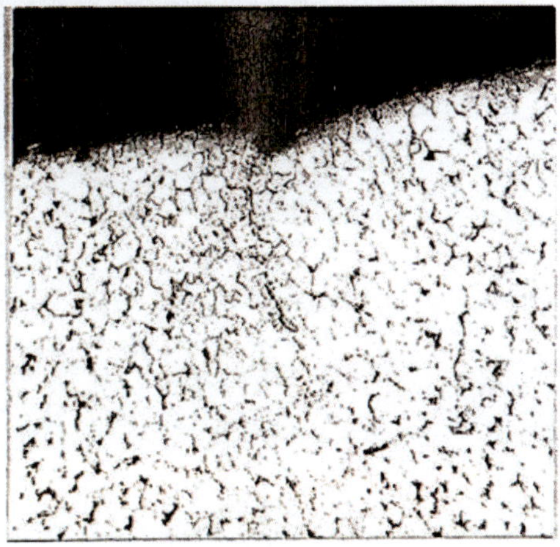

Fig. 2 Rolled gas blister on the hot-rolled metal surface. A transverse micro-section. Etching with 4% HNO_3 in alcohol, x100

4 Conclusions

The metallurgical blank for the production of hot-rolled products shall undergo a continuous chipping. The sinking strain group for hot-rolled products intended for further cold heading shall be at least 1/h-6b.

Upon receipt of hot-rolled products, a thorough input control of the macro-and microstructure, technical characteristics is required in order to identify defect signs and the control of compliance with the requirements of regulatory and technical documentation.

The elimination of the above-mentioned defects will help to improve the quality of hardware, significantly reduce the consumption of metal and increase the production profitability.

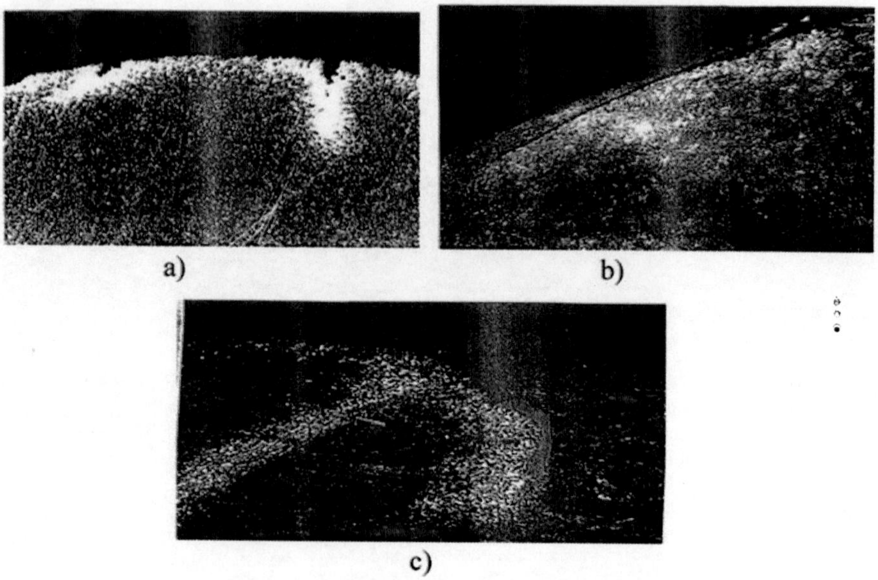

Fig. 3 Laps on the hot-rolled metal surface due to: **a** ridge rolled in; **b** undercutting; **c** rolled in scale. A transverse micro-section. Etching with 4% HNO_3 in alcohol, x100

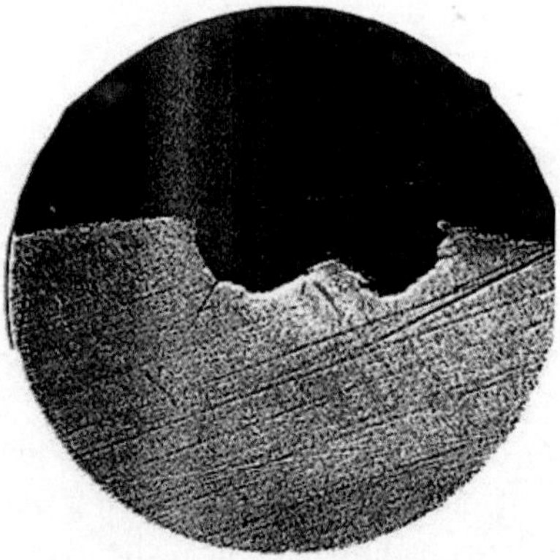

Fig. 4 Scratch marks on the hot-rolled non-etched metal surface, x100

References

1. Estrin Y, Vinogradov A (2013) Extreme grain refinement by severe plastic deformation: a wealth of challenging science. Acta Mater 61:782–817
2. Filippov A, Pachurin G, Naumov V, Kuzmin N (2016) Low-cost treatment of rolled products used to make long high-strength bolts. Metallurgist 59(9–10):810–815
3. Terentyev V, Petukhov A (2013) Fatigue of high strength metallic materials. In: TsIAM, p 515, IMET RAN, Moscow

4. Bobilev M, Grinberg V, Zakirov D, Lavrinenko Yu (1996) Preparation of structure during electrical thermal treatment of steels used for upsetting of high strength fasteners. Steel 11:54–58
5. Pachurin G, Goncharova D, Shevchenko S, Mukhina M, Kuzmin N, Pachurin V, Matveyev Yu, Kutepova L, Smirnova Z (2018) Development of fatigue test technology of sheet automobile materials. Eastern-Eur J Enterp Technol 5(12–95):31–37
6. Pachurin G, Filippov A (2008) Resource-saving and environmentally friendly surface treatment of rolled metal before cold heading. In: Ecology of industry in Russia, pp 13–15
7. Gurov V, Vladimirov A (2005) Improving of quality of fastening elements and reducing of metal during production. Steel 12:52–54
8. Romanovskaya E (2014) Creation of a new project on the basis of own research and development. Bull Minsk Univ 1(9)
9. Romanovskaya E (2014) Investigation of practical aspects of development of production technologies at machine-building enterprises. Bull Minsk Univ 4(9)
10. Ivchenko A, Ambrazhey M, Machuskaya N, Kokashinskaya G (2010) High strength fasteners of class 8.8. from thermomechanically hardened rolled wire. Metalware 01(22):58–63
11. Dziro T, Kinya V, Toshimichi M, Masaki M, Takafumi I (2008) Production of rolled wire with high capacity of scale elimination. Metalware 2(18):32–42
12. Amirov M, Lavrinenko Yu (1992) Basis of technology for automated cold upsetting. UAI, Russia, Ufa
13. Moiseichick E (2014) Operation of extended high strength bolts in the elements of steel structures and their delayed collapsibility. Bull MGSU 11:58–67
14. Pachurin G, Shevchenko S, Mukhina M, Kutepova L, Smirnova J (2016) The factor of structure and mechanical properties in the production of critical fixing hardware 38XA. Tribol Ind 38(3):385–391
15. Bunatyan G (2010) Fastening elements. Perspectives—in consolidation. Metalware 1(22): 12–15
16. Lavrinenko V (2007) Mathematical models of processes of cold forging of the fastening elements. Metalware 1:35–37
17. Galkin V (2014) Concerning the question of microstructural estimate of distribution of metal plastic deformations of cold upset fastening elements. Press Forging. Metal Form 8:11–14
18. Galkin V (2014) Structural and deformational estimation of metal hardening in multi-operational processes of cold deformation. Strengthening Techn Coat 8:8–12
19. Zinuti A, Saro J (2003) Drawing of wire at mills. Metalware 2(03):41–47
20. El Sherbini Gamal, Chatterjee U, Sircar S (1976) On the separation of initiation and propagation stages in transgranu1ar stress corrosion cracking of alphabrass. Indian Inst Metals 29(6):447–448
21. Gerasimova A, Radyuk A (2014) The improvement of the surface quality of workpieces by coating. CIS Iron Steel Rev 1:33–35
22. Pachurin G, Filippov A (1996) Rational reduction of hot-rolled 40X steel before cold upsetting. Steel Transl 38(7):522–524
23. Kulesha V (2000) Distinctive features of steel production for high quality metalware. The works of the third congress of rolling mill operator, Moscow

Mathematical Models for Assessment the Thermal Radiation of a Fireball During Bleve

Alexey Elizaryev, Elina Nasyrova, Carlo Cattani, Denis Tarakanov, Dmitrii Tarakanov, and Ilmir Khasanov

Abstract Intensive heating of oil and its products storage facilities can cause the BLEVE effect, with fireball formation. So that it is a fundamental problem to study the thermal effect of a fireball as accurately as possible and assess its consequences. This paper deals with the theoretical generalization and comparative analysis of three mathematical models for assessing fireball thermal radiation at BLEVE. The first model (MM1), proposed by M. W. Roberts, takes into account the time-dependent nature of the fireball thermal radiation. The second model (MM2) was given in the Appendix D of the project GOST 12.3.047-2012 of the National Standard of the Russian Federation. The third model (MM3) was given by Dhurandher, Kumar and Dhiman of the Indian Institute of Technology Roorkee. We will compare these three models by obtaining the following results: assumes that the bulk of the heat is emitted near the accident source. At the same time, at a long distance, thermal radiation intensity remains at a level similar to the levels MM2 and MM3. MM2 and MM3 do not have a fundamental difference in modeling of fireball parameters and calculating the thermal radiation intensity.

Keywords Thermal flux · Thermal radiation intensity · Fireball · BLEVE

1 Introduction

In case of accidents at oil and its products storage facilities, one of the main danger to be taken into consideration is the emergence of a combustion center. In this case, due to prolonged preparation for the localization and fire extinguishing, intense

A. Elizaryev · E. Nasyrova · D. Tarakanov (✉) · D. Tarakanov · I. Khasanov
Ufa State Aviation Technical University, Ufa 450008, Russia

A. Elizaryev
Water Problems Institute of the Russian Academy of Science, Moscow 119333, Russia

C. Cattani
Engineering School, Department of Economics, Engineering, Society and Enterprise, University of Tuscia, 01100 Viterbo, Italy

A. Bieliatynskyi and V. Breskich (eds.), *Safety in Aviation and Space Technologies*, Lecture Notes in Mechanical Engineering,
https://doi.org/10.1007/978-3-030-85057-9_28

heating of the storage objects occurs. As a consequence it can lead to the so-called boiling liquid expanding vapor explosions (shortly called BLEVE) and the formation of secondary hazards [1]. Many authors have devoted their research to this problem (see e.g. [1–3] and references therein). For example, in [2] experimental data on overpressure of BLEVE on small cylindrical tank are presented, and a mathematical model for forecasting the blast overpressures is proposed. In [3] a numerical study on the evaluation of blast waves at BLEVE is presented. In particular, in this paper the authors propose two approaches for determining the initial pressure. The results obtained on the basis of Computational Fluid Dynamics modeling approaches are consistent with experimental data and can be used for assessing the impact of explosion on structures.

When the BLEVE effect in tank occurs, then in 86% cases a fireball with fragments shattering is observed. A fireball is less likely to be observed in 19% cases, and the fragments shattering without a fireball is least likely 13% [4]. The probability of occurring of BLEVE in a tank is 87%.

So that it is necessary in advance to simulate the fireball thermal effect as accurately as possible and assess its consequences. Based on the obtained results of mathematical modeling, there follow some suitable measures for preventing and reducing the risk of fireball thermal effects. However, there is not yet available an efficient computational model for a reliable determination of fireball main parameters. Therefore, by using the known mathematical models for assessing the fireball thermal effect with the same initial data, a significant difference in the predictions are possible.

The objective of this work is a comparative analysis of mathematical models for assessing the fireball thermal radiation during BLEVE. The organization of this paper is as follows: in Sect. 2 materials and methods are given together with preliminary remarks on this problem and comparison of these three methods; results and discussion are given in Sect. 3.

2 Materials and Methods

In this work a comparative analysis of 3 mathematical models is carried out for assessing the fireball thermal radiation during BLEVE:

The mathematical model (MM1), proposed by M. W. Roberts to the United States Department of Energy [5], takes into account the time-dependent nature of the thermal radiation of a fireball. According to the results obtained by Martinsen and Marx [6], the maximum value of the surface heat flux is 300–450 kW/m^2. However, 400 kW/m^2 is proposed as the limit value. According to MM1, the time of fireball existence (t_d, s) is considered of two intervals (Fig. 1): phase I (growth phase) and phase II.

For the growth phase of the fireball, it is necessary to use the smaller value of the radiated surface heat flux of two: determined by Eq. (3) or the proposed 400 kW/m^2. During the 2/3 t_d, the radiated surface heat flux is determined by Eq. (4). The

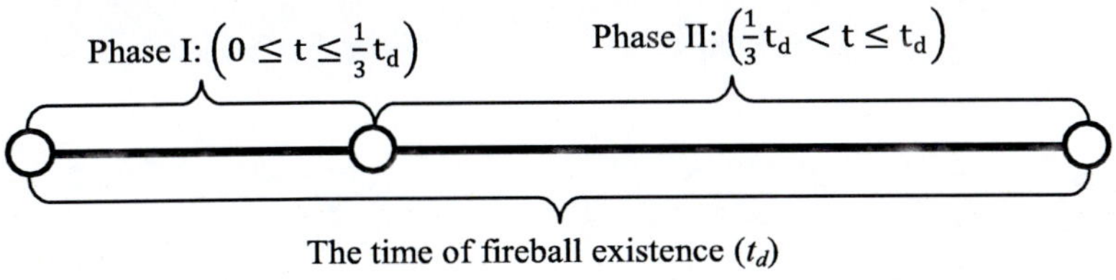

Fig. 1 The duration of the fireball according to the MM1 (t—the time at which is necessary to determine the intensity of thermal radiation, s; $t \in t_d$)

author in this work proposed to use the mathematical Eq. (5) of Table 1 for estimating the thermal radiation intensity.

The mathematical model (MM2) was presented in Appendix D of the National Standard of the Russian Federation GOST 12.3.047-2012 "Occupational safety standards system. Fire safety of technological processes. General requirements. "Control methods". On the basis of this model, the "Fireball" module in the "TOXI + Risk 5" program has been implemented, which is used for calculating the thermal radiation intensity. The MM2 suggests using the Eq. (6) as given in Table 2.

The mathematical model (MM3) was given by B. K. Dhurandher, R. Kumar and A. Dhiman of the Indian Institute of Technology Roorkee [7]. The mathematical model is partially based on the research of Roberts [8] and allows to estimate the thermal effect of a fireball according to the Eq. (9) of Table 3.

According to [7], the surface emissive power of thermal radiation is taken equal to the maximum surface heat radiation power of thermal radiation ($SEP_{act} = SEP_{max}$).

Analysis of mathematical models showed that MM1 and MM3 have the largest number of input parameters, which determines the accuracy of the results. MM1 also takes into account the time-dependent nature of the fireball's thermal radiation. This allows obtaining a dynamic model of a fireball and its corresponding characteristics (diameter and height of the center, the magnitude of the emitted surface heat flux and the thermal radiation intensity, etc.) at any time of the fireball existence. MM2 is a more simplified version of MM1 and MM3. It is obtained by using a constant value of the radiated surface heat flux (350 kW/m^2), which makes it possible to assess the thermal effect of a fireball.

Figure 2 shows the necessary initial parameters for the mathematical modeling. The general parameter for the considered mathematical models is the mass of the substance in the tank. The remaining parameters are common for both MM1 and MM3.

Using as a specific example of propane storage tank with a volume of 10,000 gal, filled to 80% (8000 gal), and a single initial data for calculation, a BLEVE with fireball was simulated. Also the comparative analysis of mathematical dependence of the thermal radiation intensity from the distance is carried out. Output parameters were calculated by mathematical models (MM1, MM2, MM3).

Table 1 MM1 Roberts model

Initial parameters	Interim parameters	Final equation
(R, P_V, x, t) (m, x, t) $(H_C,\ P_B,\ m,\ x,\ t)$	$\tau(x,t) = 2.02 \cdot \left\{R \cdot P_V \cdot \left[\sqrt{H(t)^2 + x^2} - \frac{D(t)}{2}\right]\right\}^{-0.09}$ (1) $F(x,t) = \frac{D(t)^2}{4 \cdot \left[H(t)^2 + x^2\right]}$ (2) $E_S(t) = E_{\max} = 0.0133 \cdot f \cdot H_c \cdot m^{\frac{1}{12}}$ for $0 \le t \le \frac{1}{3} \cdot t_d$ (3) $E_S(t) = E_{\max} \cdot \left[\frac{3}{2} \cdot \left(1 - \frac{t}{t_d}\right)\right]$ for $\frac{1}{3} t_d < t \le t_d$ (4)	$q(x,t) = \tau(x,t) \cdot F(x,t) \cdot E_S(t)$ (5)

Where:
q—the thermal radiation intensity, kW/m^2;
τ—the atmospheric transmissivity;
F—the geometric view factor;
E_S—the radiated surface heat flux, kW/m^2;
x—the distance from the object to the point on the earth's surface under the center of the fireball, m;
t—the time of at which it is necessary to determine the intensity of thermal radiation, s ($t \in t_d$, t_d—the time of fireball existence, s);
R—the fractional relative humidity, %;
P_V—the saturated vapor pressure of water at the ambient temperature, Pa;
D—the diameter of the fireball, m;
H—the height of the center of the fireball, m; m—the mass of the emitted flammable substance, kg;
H_C—the net heat of combustion of the flammable material, kJ/kg;
f—the radiant heat fraction;
P_B—the burst pressure of the vessel, Pa.

Table 2 MM2 model GOST (Russian Federation)

Initial parameters	Interim parameters	Final equation
(m, x) (m, x)	$\tau = \exp\left[-7.0 \cdot 10^{-4} \cdot \sqrt{x^2 + H^2} - \frac{D}{2}\right]$ (6) $F(x, t) = \frac{D^2}{4 \cdot [H^2 + x^2]}$ (7)	$q(x) = \tau(x) \cdot F(x) \cdot E_f$ (8)

Where:
E_f—the average surface heat radiation intensity, kW/m^2 (taken equal to 350 kW/m^2);
D_S—effective fireball diameter, m.

Table 3 MM3 model of Dhurandher Kumar and Dhiman

Initial parameters	Interim parameters	Final equation
(P_B, m, H_C) (m, x) (R, P_V, m, x)	$SEP_{\max} = f \cdot m' \cdot H_C$ (9) $F = \left(\frac{R_{FB}}{\sqrt{H^2 + x^2}}\right)^2$ (10) $\tau = 2.02 \cdot \left[P_w \cdot \left(\sqrt{H^2 + x^2} - R\right)\right]^{-0.09}$ (11)	$q(x) = \tau(x) \cdot F(x) \cdot SEP_{act}$ (12)

Where:
SEP_{act}—the surface emissive power, kW/m^2;
SEP_{max}—the maximum surface heat radiation power, kW/m^2;
m′—the mass burning rate, kg/m^2 • s;
P_w—the partial water vapor pressure in air, Pa;
R_{FB}—radius of the fireball, m.

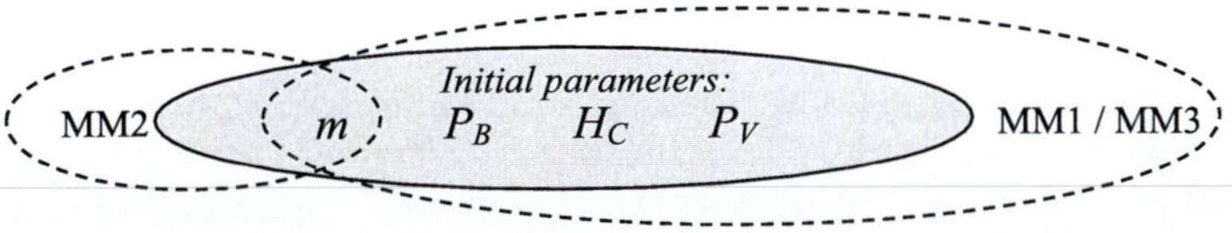

Fig. 2 Required initial parameters for the mathematical modeling

For the calculation of BLEVE with fireball, the following unified initial values of the parameters were used: m = 15,500 kg, PB = 2,210,000 Pa, HC = 46,390 kJ/kg, PV = 2490 Pa (at t = 21 °C), R = 70%.

3 Results and Discussion

Figure 3 shows the dynamics of changes in the surface radiated heat flux, the height of the fireball and its diameter with the duration t_d = 10 s for the BLEVE of the tank under consideration in accordance with MM1.

Figure 3 also shows how the surface radiated heat flux, the diameter and height of the fireball change during the time of its existence. At the time t = 0.5 s a fireball

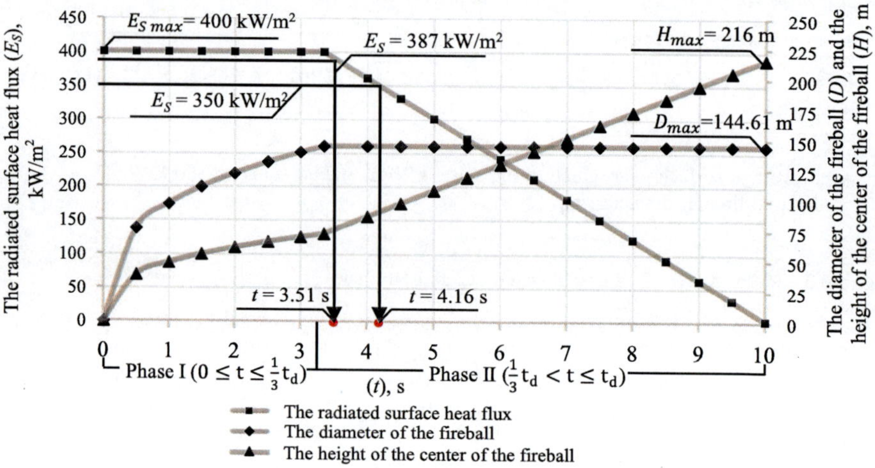

Fig. 3 Dynamics of changes in the surface radiated heat flux, diameter and height of the fireball in accordance with MM1

is formed: the diameter and height increase sharply, which is caused by a significant difference in atmospheric pressure and pressure in the tank. It is assumed that during 1/3 t_d, fireballs maximum development occurs, then the diameter of the fireball does not change significantly, and the height of its center increases linearly. The surface radiant heat flux during the first third of time is limited to 400 kW/m^2, then decreases linearly.

The surface radiant heat flux used by MM2 is 350 kW/m^2. According to the obtained dependence on Fig. 3 it is established that the value (Es = 350 kW/m^2) is reached at the time t = 4.16 s. Therefore, to compare the results of calculating by MM1 and MM2, we analyzed fireball model at 4.16 s. Table 4 shows the results of fireball modeling at BLEVE according MM1 (at t = 4.16 s) and MM2.

As can be seen from Table 4, a significant difference between the results of modeling MM1 and MM2 is the height of the fireball center—the value by MM1 is 59.22 m lower than the value by MM2. The differences in the diameter and time of fireball existence are insignificant—respectively 4.47 m and 0.43 s.

Figure 4 shows the dependence of the fireball thermal radiation intensity on the distance, calculated according to MM1 (at t = 4.16 s) and MM2. Mathematical

Table 4 The results of fireball modeling at BLEVE according MM1 (at t = 4.16 s) and MM2

The mathematical model	The diameter of the fireball D, m	The height of the center of the fireball H, m	The time of fireball existence t_d, s
MM1 (at t = 4.16 s)	144.61	89.86	10.04
MM2	149.08	149.08	10.47

dependences are plotted at a distance of 500 m, due to the similarity and insignificant level of the thermal radiation intensity. The distance is marked, corresponding to the degrees of radiation of a fireball for person (according to the values given in the National Standard of the Russian Federation GOST 12.3.047-2012). Distance values are determined graphically and differ depending on the degree of burn.

From Fig. 4 it can be seen that in a zone with a radius up to 150 m from a point on the earth's surface under the fireball center, there is a significant discrepancy in the values of the thermal radiation intensity. For III degree burns, the distances are identical—175 m, for II degree burns they have an insignificant difference—15 m, for I degree burns the distance values have a significant difference—45 m.

A similar comparison is made between the results of fireball modeling at BLEVE and the thermal radiation intensity according to MM1 and MM3. The surface radiant heat flux calculated according to MM3 (Eq. 9) is 387 kW/m^2. According to the obtained dependence in Fig. 3, this value is reached at time t = 3.51 s. For comparing the received results of calculations for MM1 and MM3, a fireball model was adopted, corresponding for t = 3.51 s. Table 5 shows the simulation results for MM1 (at t = 3.51 s) and MM3.

Comparative analysis of the data in Table 5 allows us to summarize that the simulation results by MM1 and MM3 correlate with results for MM1 and MM2 (Table 4). A significant difference between the results of modeling for MM1 and MM3 is the height of the center of the fireball—the value of MM1 is 68.33 m lower

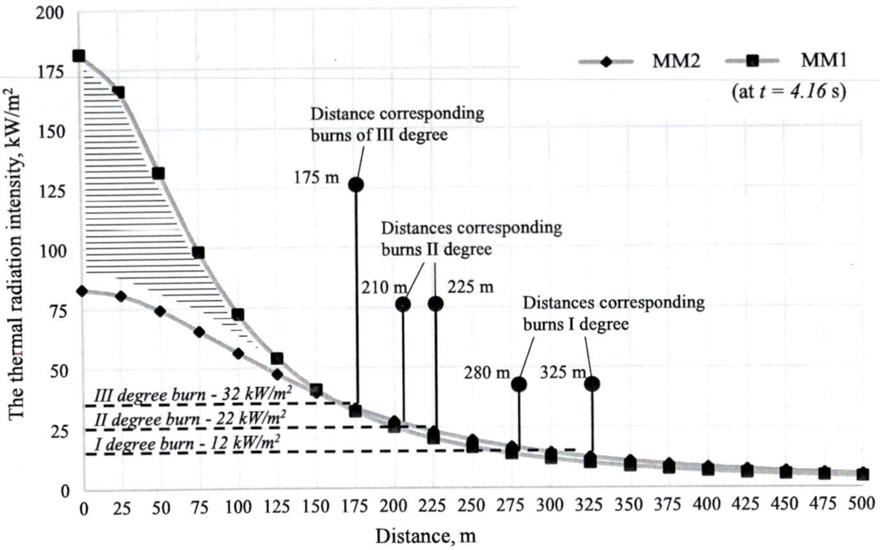

Fig. 4 The dependence of the fireball thermal radiation intensity on the distance, calculated according to MM1 (at t = 4.16 s) and MM2

Table 5 The results of fireball modeling at BLEVE according MM1 (at t = 3.51 s) and MM3

The mathematical model	The diameter of the fireball D, m	The height of the center of the fireball H, m	The time of fireball existence t_d, s
MM1 (at t = 3.51 s)	144.61	75.82	10.04
MM3	144.15	144.15	11.18

than the value of MM3. The differences in the values of diameter and time existence of the fireball are insignificant—respectively 0.46 m and 1.14 s.

Figure 5 shows the dependence of the fireball thermal radiation intensity on the distance, calculated according to MM1 (at t = 3.51 s) and MM3.

As can be seen from the Fig. 5 the dependence of the thermal radiation intensity calculated from MM1 and MM3 is similar to the dependence obtained from MM1 and MM2 interpreted in Fig. 4. Significant discrepancy in the values of the thermal radiation intensity in a zone with a radius of up to 200 m from a point on the earth's surface under the fireball center. Distances corresponding to different degrees of burn for a person (according to the values given in the National Standard of the Russian Federation GOST 12.3.047-2012), determined graphically. They have different discrepancies depending on the degree of burn: for III degree burns, the difference in distance is 55 m, for burns of I and II degrees—40 m.

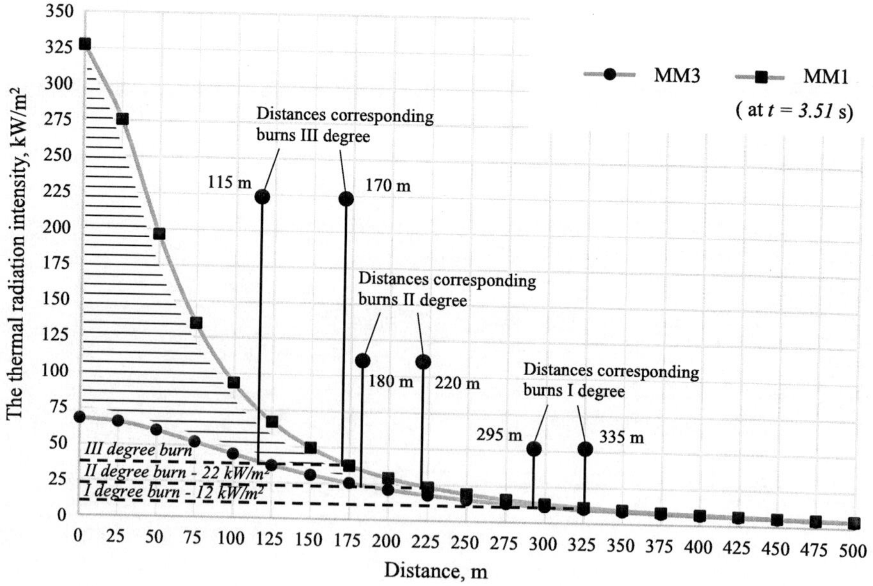

Fig. 5 The dependence of the fireball thermal radiation intensity on the distance, calculated according to MM1 and MM3

Table 6 The results of fireball modeling at BLEVE according MM2 and MM3

The mathematical model	The diameter of the fireball *D*, m	The height of the center of the fireball *H*, m	The time of fireball existence t_d, s
MM2	149.08	149.08	10.47
MM3	144.15	144.15	11.18

Based on the initial data, a fireball was simulated at BLEVE and the intensity of thermal radiation was calculated using MM2 and MM3. The results are presented in Table 6 and Fig. 6, respectively.

As can be seen from Table 6, the results of fireball modeling at BLEVE according to MM2 and MM3 have a minimum difference. The values of the center height and the diameter of the fireball according to MM2 are higher than MM3 by 4.93 m, the values of the fireball's time existence diverge by 0.71 s. This is due to the similarity of mathematical models for assessing the fireball parameters and calculating the thermal radiation intensity.

Figure 6 shows the dependence of the thermal radiation intensity of the fireball from the distance, calculated according to MM2 and MM3.

Figure 6 shows that the thermal radiation intensity calculated by the MM3 method is less than value calculated by the MM2. But at the same time, the

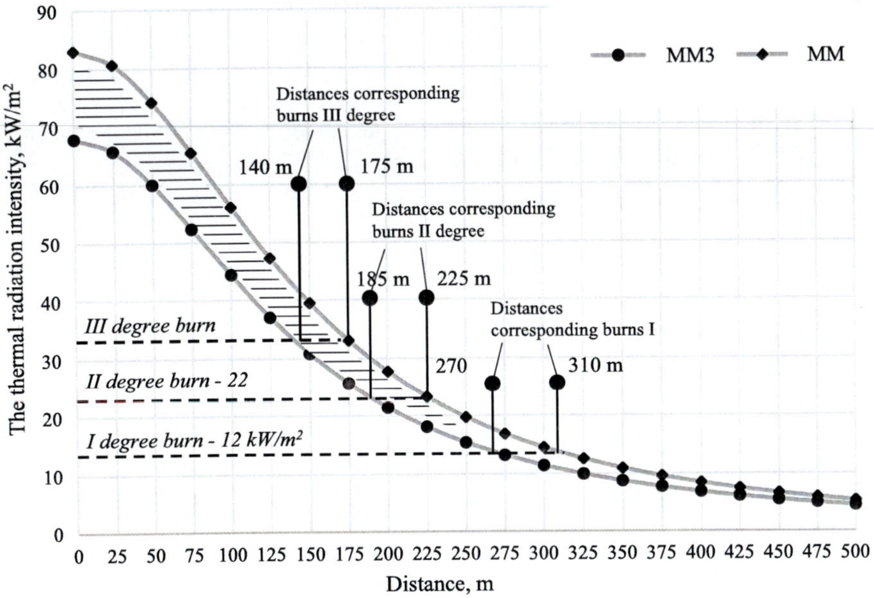

Fig. 6 The dependence of the fireball thermal radiation intensity on the distance, calculated according to MM2 and MM3

dependences obtained by both methods have a symbatic character of change. Distances corresponding to different degrees of burn for person have differences on average of 40 m.

The received results of mathematical models comparison for estimating the parameters and thermal radiation of a fireball during BLEVE allow us to summarize that the propagation of heat in MM1, as compared to MM2 and MM3, is of an impulsive nature. That is, according to MM1, it is assumed that the spread of heat to a greater extent falls on objects located near the source of the accident. At the same time, MM2 and MM3 do not compensate (equalize) this discrepancy at a more far distance from the accident source. We can say that the energy potential with the BLEVE effect on the tank, the estimated MM1 is incommensurable compared to MM2 and MM3. This is can negatively affect the choice of preventive measures and lead to harmful consequences [9–11]. MM2 and MM3 are in principle identical, but MM3 allows calculating the value of the radiated surface heat flux, in contrast to MM2, which implies this value to be constant.

4 Conclusion

A comparative analysis of mathematical models for assessing the fireball thermal radiation BLEVE has been carried out. According to the obtained results, MM1 assumes that the bulk of the heat is emitted near the source of the accident. At the same time, at a far distance, the value of the thermal radiation intensity remains at a level similar to the levels MM2 and MM3. MM2 and MM3 do not have a fundamental difference when simulating the fireball parameters and calculating the thermal radiation intensity. This is reflected in the graph of the value dependences of the thermal radiation intensity from the distance. However, the values obtained for MM3 are less than those for MM2.

References

1. Birk AM, Cunningham MH (1994) The boiling liquid expanding vapor explosion. J Loss Prev Process Ind 7:474–480
2. Birk AM, Eyssette R, Heymes F (2020) Analysis of BLEVE overpressure using spherical shock theory. Process Saf Environ Prot 134:108–120
3. Li JD, Hao H (2020) Numerical study of medium to large scale BLEVE for blast wave prediction. J Loss Prev Process Ind 65:10
4. Kumar A (1994) Gdelines for evaluating the characteristics of vapor cloud explosions, flash fires, and BLEVEs. Center for Chemical Process of the American Institute of Chemical Engineers, New York
5. Roberts MW (2000) Analysis of boiling liquid expanding vapor explosion (BLEVE) events at DOE sites, EQE International, Inc.
6. Martinsen WE, Marx JD (1999) An improved model for the prediction of radiant heat flux from fireballs. Center Chem Process Am Inst Chem Eng 605–621

7. Dhurandher BK, Kumar R, Dhiman A (2015) Impact assessment of thermal radiation hazard from LPG fireball. Procedia Earth Planet Sci 499–506
8. Roberts AF (1982) Thermal radiation hazards from releases of LPG from pressurized storage. J Fire Saf. 4:197–212
9. Elizaryev A, Tarakanov D, Tarakanov D, Mikhailova V, Elizareva E, Nasyrova E, Epimakhov N (2020) Fire and explosion measures for tanks protection selection algorithm based on quantitative assessment of big breathing. J Phys: Conf Ser 1614(1):012054
10. Heymes F, Aprin L, Birk AM et al (2013) An experimental study of an LPG tank at low filling level heated by a remote wall fire. J Loss Prev Process Ind 26:1484–1491
11. Wang K, He Y (2020) Three-Dimensional thermal field induced by gas pipeline explosion fireball based on predicted algorithm and experimental validation. High Temp 58:893–906

Road Accident Risk Modeling Based on the Machine Learning

Elena Pechatnova, Vasiliy Kuznetsov, and Sergei Pavlov

Abstract Road traffic accident is the serious socio-economic problem all over the world. The main solution is the road safety decision support system, which is based on road accident risk mathematical modeling. The study analyses the road accident risk due to the constant characteristics of the road, using the machine learning. The study was carried out on the road outside the human settlements A-322. It was analyzed 273 km outside large settlements. The research process included three stages. At the first stage, the correlation analysis was carried out. As the result, 12 main constant road characteristics for further research were identified. At the next stage, cluster analysis was realized using the k-means clustering algorithm. It allowed dividing road sections into 3 clusters and determining their accident rate. Analysis of variance confirmed the significance of the cluster difference. Discriminant analysis was implemented on the final stage. As the result, the classification functions were obtained. Therefore, it is possible to determine the accident risk rate depending on the values of constant road characteristics. The resulting model can be used in the development of road safety decision support system for the roads of the 2nd category or similar roads, as well as the roads design. In addition, the research results can be the basis in the development of traffic safety components for Intelligent Transportation Systems or software products.

Keywords Road accident risk modeling · Machine learning · Cluster analysis · Traffic safety

E. Pechatnova (✉)
Altai State University, 61, Lenina ave., Barnaul 656049, Russia

V. Kuznetsov
Altai State Agricultural University, 98, Krasnoarmeysky ave., Barnaul 656049, Russia

S. Pavlov
Polzunov Altai State Technical University, 46, Lenina ave., Barnaul 656038, Russia

A. Bieliatynskyi and V. Breskich (eds.), *Safety in Aviation and Space Technologies*, Lecture Notes in Mechanical Engineering,
https://doi.org/10.1007/978-3-030-85057-9_29

1 Introduction

Road traffic accidents are one of the important problems of health care, society and economy development [1–3]. More than 1 million people die in road accidents in the world every year. Systemic approach is required to solve this problem. For example, it is the development of a decision-making system in a short time interval. The basic requirement for its creation is the mathematical road accident risk modeling.

The promising basis for mathematical modeling is the use of machine learning methods. They allow analyzing data, studying their structure and creating mathematical models for predicting processes. Machine learning includes artificial intelligence techniques such as artificial neural networks, cluster analysis, and others.

Lots of researchers apply these methods in their work on accident rate modeling. The research work [4] is devoted to the comparison of different statistical and ma-chine learning methods for accident severity prediction; the study has shown high potential for use the Nearest Neighbor Classification in the similar problems. The combined method for determination the accident factors in the pedestrian crash is used in the research work [5]: the latent class clustering method is used to segment the analyzed accidents, and Multinomial Logit (MNL) models provide identification the main factors in the pedestrian crash severity. The research [6] is devoted to identifying scenarios for run-off-road situations based on hierarchical agglomerative cluster analysis. The research work [7] contains the results of development the structure of traffic crash severity prediction framework using deep learning at highway work zones. Russian authors applied cluster and discriminant analysis to develop the city streets classification based on traffic safety characteristics in the study [8].

The great attention of scientists to the use of machine learning methods in the accidents analysis confirms the high potential of their use in solving such problems. However, there is currently a lack of research in the field of assessment the road accidents risk, caused the road geometric design on various types of roads outside settlements using machine learning methods. At the same time, the roadway factor is recognized as important in many studies, in particular in [1, 9].

The purpose of our work is road accidents risk modeling as the result of cumulative effect of road parameters based on machine learning. The important condition is to obtain results that can be applied in practice and improve the decision support system.

2 Theoretical Research Hypothesis

In previous studies [10], we determined the general form of the mathematical model for short-term assessment the road accidents risk on roads outside settlements:

$$R^{i\times t} = R^{i}_{road} \cdot R^{t}_{e} \tag{1}$$

where

$R^{i\times t}$—is the risk of road traffic accident in the space-time cell i × t,
R^{i}_{road}—risk on the i-th section of the road due to the constant characteristics of the road,
R^{t}_{e}—risk in the t-th time, due to the influence of the external environment.

This study simulates the road accidents risk due to the constant road characteristics. Thus, we are talking about getting the multivariate function:

$$R^{i}_{road} = f\left(D^{i}\right) \tag{2}$$

where $D^{i} = \left[d^{i}_{p}\right]$—road performance vector $\left(d^{i}_{p}\right)$ i-th road section.

Lots of road sections are similar to each other, since they have similar or the same set of road characteristics values. This is due to the fact that the road characteristics is determined by its category. Most of the federal roads in Russia, which represent the backbone of the road network, belong to the II technical category. The characteristics of such roads are: intersections at the same level, lane width from 3.5 to 3.75 m, usually 2 lanes, traffic intensity from 6000 to 14,000 vehicles per day.

Similar or identical road sections should be characterized, respectively, by the single risk value R^{i}_{road}, since it is the risk caused by the set of permanent traffic characteristics. Therefore, the definition of function (2) can be implemented as the solution to the classification task.

3 Methods and Materials

The study is carried out on the example of the federal motorway A-322 Barnaul-Rubtsovsk—the state border with the Republic of Kazakhstan. Sections within the suburban zone of large cities, road sections of the III technical category, as well as road sections with long length of the settlement zone (over 2 km) are excluded. The total length of analyzed road sections is 273 km. The selected road is the typical representative of II roads category [11].

The road is divided into sections according to the mileage: the *i*-th road section is the segment between two adjacent kilometer signs on the road. This division allows assessing the accident rate, since the Russian accident location is usually indicated with an accuracy of a kilometer.

Initially, 18 road characteristics are selected. They have significant impact on the accidents appearance on the roads outside of settlements. Their choice is based on research [1, 12]. Their list is presented in the Table 1. The values of the selected variables and the value of the average accident rate are determined for each road section. The information about the evaluation index of the road can be obtained from the road design data and the traffic flow data. The average annual accident rate per kilometer of the road (accident/year) is calculated on the basis of information about fatal-and-injury crashes and property-damage-only accidents, recorded on the road in the period 2012–2019. The choice of the set of constant road characteristics for further modeling is carried out on the basis of correlation analysis.

The cluster analysis—unsupervised learning technique—is used for identification similar road sections. There is a variety of objects—road sections (kilometers)

Table 1 Initial set of variable

Variable	Constant characteristics of the road	Pearson/ Spearman correlation coefficients, p value
d_1	Generalized characteristic of the road section along the radius of curvature in the plan	0.27, $p < 0.05$
d_2	Generalized characteristics of the gradient and slope length	0.32, $p < 0.05$
d_3	Maximum value of the road gradient	0.34, $p < 0.05$
d_4	The length of the zone of inadequate sight distance (m)	0.40, $p < 0.05$
d_5	Lanes number	0.45, $p < 0.05$
d_6	The total power of the service facilities	0.57, $p < 0.05$
d_7	Summary characteristics of the intersection (points)	0.54, $p < 0.05$
d_8	Summary characteristics of the crosswalks and stops of public transport (points)	0.29, $p < 0.05$
d_9	The length of the fence (m)	0.25, $p < 0.05$
d_{10}	The length of the bridge (m)	0.38, $p < 0.05$
d_{11}	The length of the zone of possibility the animals' appearance (m)	0.31, $p < 0.05$
d_{12}	The average annual traffic volume	0.31, $p < 0.05$
d_{13}	Lane width (m)	−0.12, $p > 0.05$
d_{14}	The width of the hard shoulder (m)	−0.01, $p > 0.05$
d_{15}	Shoulder width (m)	0.05, $p > 0.05$
d_{16}	The length of a straight section (m)	−0.01, $p > 0.05$
d_{17}	The presence of an intersection in an area of 100 m from the site boundaries	−0.01, $p > 0.05$
d_{18}	Length of settlement area (m)	−0.05, $p > 0.05$

$Q = \{Q_1, Q_2, \ldots, Q_n\}$, $n = 273$. There is a variety of road characteristics $d = \{d_p\}_{p=1}^{12}$, which each Q_i has. The measurement result of p-th road characteristics on the i-th section Q_i is denoted as d_p^i. Multidimensional data vector $D_i = \left[d_p^i\right]$ is the road characteristics of each i-th road section. Thus, the variety of road characteristics for all road sections is formed $D = \{D_1, D_2, \ldots, D_n\}$.

The task is to split the set of objects Q into m clusters. Each road section Q_i must belong to the same cluster; and the clusters should be characterized by the same or similar objects in terms of road characteristics. The road accidents risk, caused by the set of road characteristics, is found in each cluster.

Discriminant analysis is chosen for determination the cluster finding rule for the i-th road section according to its road characteristics. Its task is to assign each of the n objects to one of the classes (clusters).

4 Results and Discussion

4.1 Definition the Group of Constant Road Characteristics

Pearson or Spearman correlation coefficients are calculated between the values of the road characteristics and the average accident rate on the road sections (Table 1). If the correlation coefficient is significant at the level $p < 0.05$, then this variable is selected for further modeling.

Thus, it is determined that 12 constant road characteristics have the significant influence: $d_1 \ldots d_{12}$.

4.2 Results of Cluster Analysis

The next step in solving this goal is the grouping of road sections based on the similarity of road characteristics. The cluster analysis is used for this.

K-Means clustering method is chosen for the cluster analysis realization. It is assumed that the number of clusters is $m = 3$. This choice allows dividing objects into the cluster with high, medium and low complexity of road conditions. An

Table 2 Clustering results

Cluster	Number of objects in the cluster	The level of risk (accidents/year), R^i_{road}
Cluster 1	199	0.87
Cluster 2	57	1.71
Cluster 3	17	3.73

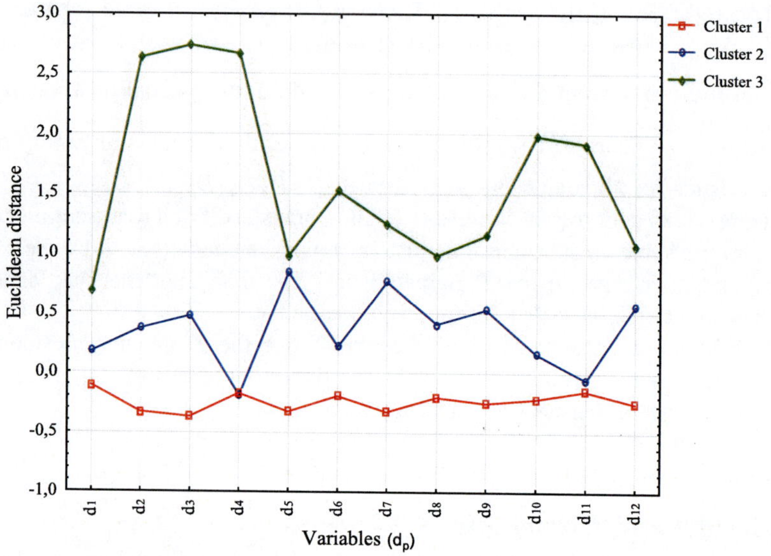

Fig. 1 Plot of means for each cluster

analysis of the graph of amalgamation schedule, performed during the tree clustering, also justifies this decision.

The data is standardized. Road sections Q_i are distributed across clusters. The number of objects in each cluster is presented in Table 2.

Figure 1 shows the graph of averages. The most complex combination of road characteristics is characterized by cluster 3, the least—by cluster 1.

Analysis of variance shows the difference between clusters at the acceptable level of significance ($p < 0.05$).

The distribution of accidents in each of the clusters is assumed normal based on the number of objects in the clusters and the assessment of the Kolmogorov-Smirnov test criterion. The level of risk is calculated based on the average annual accident rate in each cluster. The results of calculation are presented in Table 2. The analysis of variation proves the significance of the differences in the mean in the clusters.

Cluster analysis makes it possible to establish that sections with more difficult road characteristics have the higher level of road accidents risk.

4.3 Discriminant Analysis

The determination of the rule for choosing the cluster depending on the set of road characteristics values is implemented using discriminant analysis. It is carried out when there is information about the clusters. The standard procedure has been implemented: all variables (road characteristics) are included in the model.

The Wilks' lambda and the partial lambda values prove the significance of all road characteristics for modeling. The largest contribution to the model is made by the length of the zone of inadequate sight distance (d_4) and maximum value of the road gradient (d_3).

Canonical analysis is carried out to obtain information on the nature of discrimination. The maximum number of evaluated discriminant functions is equal to the number of variables or the number of classes of minus one, depending on which number is less. In our case, 2 functions are evaluated. It is found that both functions are significant. Based on the canonical analysis, the scatterplot of canonical scores is obtained (Fig. 2). It is determined that objects belonging to the same clusters are localized in the certain area.

Further, classification functions (3)–(5) are defined. Their values are calculated for each cluster during the classification. The road section will belong to the cluster for which the highest classification weight is calculated.

$$
\begin{aligned}
C_1 = 17.08 \cdot d_1 - 0.02 \cdot d_2 + 0.51 \cdot d_3 + 0.89 \cdot d_4 + 18.49 \cdot d_5 - 0.01 \cdot d_6 \\
- 0.31 \cdot d_7 - 1.60 \cdot d_8 + 5.46 \cdot d_9 + 97.20 \cdot d_{10} - 2.97 \cdot d_{11} + 0.06 \cdot d_{12} \\
- 25.73,
\end{aligned}
\tag{3}
$$

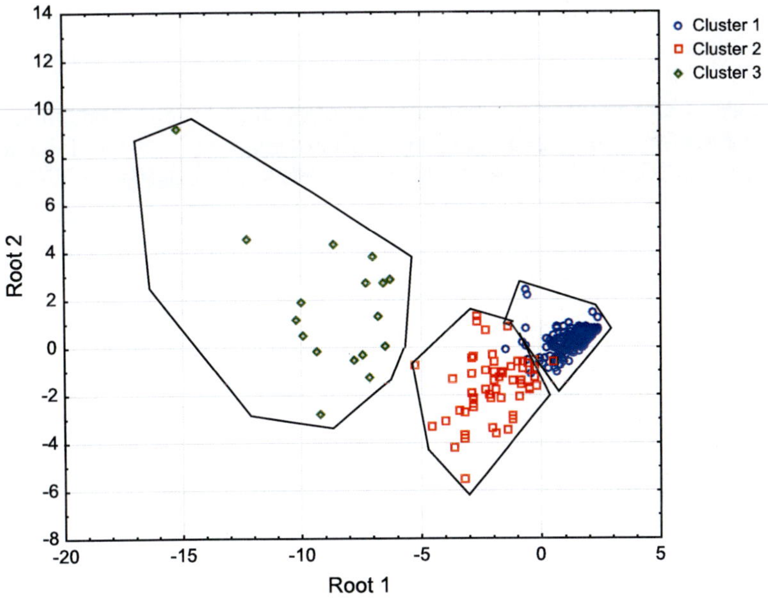

Fig. 2 Scatterplot of canonical scores

Table 3 Classification matrix

		The number of predicted objects by cluster		
	Percent correct	Cluster 1	Cluster 2	Cluster 3
Cluster 1	99.5	198	1	0
Cluster 2	89.5	6	51	0
Cluster 3	100.0	0	0	17

$$C_2 = 23.64 \cdot d_1 - 0.05 \cdot d_2 + 1.16 \cdot d_3 - 5.83 \cdot d_4 + 24.63 \cdot d_5 - 0.01 \cdot d_6 + 0.94 \cdot d_7 - 0.75 \cdot d_8 + 14.53 \cdot d_9 + 154.26 \cdot d_{10} - 1.24 \cdot d_{11} + 0.08 \cdot d_{12} - 52.95, \quad (4)$$

$$C_3 = 30.42 \cdot d_1 - 0.02 \cdot d_2 + 1.75 \cdot d_3 + 118.05 \cdot d_4 + 27.94 \cdot d_5 - 0.01 \cdot d_6 + 1.59 \cdot d_7 + 3.03 \cdot d_8 + 22.45 \cdot d_9 + 422.80 \cdot d_{10} + 25.94 \cdot d_{11} + 0.08 \cdot d_{12} - 110.6. \quad (5)$$

where C_1, C_2, C_3—classification functions.

Small values of the coefficients in the classification functions are associated with the large scale of measurements.

The evaluation of the discriminant analysis results is carried out on the basis of the classification matrix (Table 3). The total share of correctly predicted objects is 97.4%.

The set of classification functions (3)–(5) is the classification model that allows assigning the *i*-th road section to one of three clusters, and then determining the risk level R^i_{road} in accordance with Table 2. The resulting model based on machine learning methods can be used in the design, construction and reconstruction and roads repair. It supports policymakers in differentiating between risky and safe segments of the homogeneous motorway). In addition, the model can become the basis of the new software products development for road safety decision support system.

5 Conclusion

The combination of machine learning methods (unsupervised learning)—cluster analysis and supervised learning—discriminant analysis promote development the mathematical model that allows assessing the road traffic safety on the road section based on road characteristics. The resulting model differs from similar models by the classification approach to determination the risk of road accidents as the result of set of constant road characteristics. The model has acceptable accuracy. These results develop and complement the theory of road safety based on the theory of risk.

The results can form the basis of software applications in the field of decision-making. In addition, the model can become an element of the intellectual transport system.

The classification model can be used on roads of technical category II or other roads with similar characteristics with traffic volumes from 6000 to 14,000 vehicles/day. The prospect for the further research is obtaining the similar models for roads of other categories, roads with low or high traffic volume and mountain roads.

References

1. Islam MH, Teik Hua L, Hamid H, Azarkerdar A (2019) Relationship of accident rates and road geometric design. IOP Conf Ser: Earth Environ Sci 357:012040
2. Toval-Ruiz C, Rojas-Roque C, Hernández-Vásquez A (2021) Predictors and geographic analysis of road traffic accidents in Leon, Nicaragua. Revista Brasileira de Epidemiologia, 24
3. Tsubota T, Fernando C, Yoshii T, Shirayanagi H (2018) Effect of road pavement types and ages on traffic accident risks. Trans Res Procedia 34:211–218
4. Iranitalab A, Khattak A (2017) Comparison of four statistical and machine learning methods for crash severity prediction. Accid Anal Prev 108:27–36
5. Sun M, Sun X, Shan D (2019) Pedestrian crash analysis with latent class clustering method. Accid Anal Prev 124:50–57
6. Nilsson D, Lindman M, Victor T, Dozza M (2018) Definition of run-off-road crash clusters—for safety benefit estimation and driver assistance development. Accid Anal Prev 113:97–105
7. Md Adilur R, Hany MH (2021) A deep learning based traffic crash severity prediction framework. Accid Anal Prev 154:106090
8. Korchagin VA, Klyavin VE, Suvorov VA (2017) Classification of urban strees and roads based on road safety characteristics. Mod Technol Syst Anal Model 2(54):142–149
9. Shah SAR, Ahmad N, Shen Y, Kamal MA, Basheer MA, Brijs T (2019) Relationship between road traffic features and accidents: an application of two-stage decision making approach for transportation engineers. J Safety Res 69(1):201–215
10. Pechatnova EV, Kuznetsov VN (2021) Assessment of the conditions for allocating independent road safety ITS subsystem. In: International scientific conference energy management of municipal facilities and sustainable energy technologies EMMFT 2019, advances in intelligent systems and computing, vol 1258, pp 136–145. Springer, Cham
11. Pechatnova EV, Kuznetsov VN (2020) Impact investigation of road characteristics on the accident rate based on the neural network modelling. In: International conference on information technology in business and industry (ITBI 2020), Journal of Physics Conference Series 1661:012069
12. Papadimitriou E, Filtness A, Theofilatos A, Ziakopoulos A, Quigley C, Yannis G (2019) Review and ranking of crash risk factors related to the road infrastructure. Accid Anal Prev 125:85–97

Technical Diagnostics of Equipment Using Data Mining Technologies

Evgeniya Tsarkova, Alexander Belyaev, Yaroslav Lagutin, Elena Andreeva, and Yuri Matveev

Abstract The paper considers the problem of implementing a practical approach to predictive maintenance (maintenance based on the actual technical condition). With this type of service, the state of the system is analyzed continuously or periodically. Based on the data obtained, a forecast of the technical condition of the equipment for a certain period of time is carried out, programs and maintenance plans are formed and, if necessary, adjusted. The aim of the work is to develop a generalized approach to building a predictive service system based on data of some complex technical object, collected by the SCADA system, with their further processing using computer modeling and machine learning methods. Implementation of this approach minimizes the likelihood of an unplanned system shutdown.

Keywords Predictive diagnostics and repair · Maintenance · SCADA systems · Failure prediction · Machine learning methods · Computer modeling

1 Introduction

1.1 A Subsection Sample

The most widely used approach to the maintenance of technical systems during their operation is to carry out planned activities, the schedule of which is prepared by specially trained personnel based on the performance characteristics of the equipment, data from manufacturers and guidelines [1]. In many cases, the system of scheduled preventive maintenance can be taken as the basis for servicing simple

E. Tsarkova (✉) · A. Belyaev · Y. Lagutin · Y. Matveev
Research Institute of Information Technologies of the Federal Penitentiary Service of Russia, Tver, Russian Federation

E. Tsarkova · E. Andreeva
Tver State University, Tver, Russian Federation

Y. Matveev
Tver State Technical University, Tver, Russian Federation

A. Bieliatynskyi and V. Breskich (eds.), *Safety in Aviation and Space Technologies*, Lecture Notes in Mechanical Engineering,
https://doi.org/10.1007/978-3-030-85057-9_30

machines and mechanisms, but for the main non-standby equipment its use is impractical. Therefore, further development of the repair system should include:

1. the establishment of differentiated criteria for assessing the resource of parts, assembly units and items of mining equipment, taking into account the specific conditions of their operation;
2. the establishment of specific terms and volumes of work during the repair of mining equipment, depending on the actual technical condition of its parts, assembly units and items.

One of the most effective ways to maintain the health of the system is technical diagnostics of the state of critical units and components, carried out in real time. However, the implementation of this approach is hampered by the complexity of the software and hardware and their cost.

Predictive diagnostics of production equipment allows predicting the onset of an emergency based on the analysis and monitoring of its current state and predicting failures. As a result, the enterprise can take early action to correct the problem or mitigate the adverse effect. An additional effect of predictive diagnostics can be a shift from preventive maintenance and repairs based on events to maintenance based on the actual condition of the equipment, which reduces the number of production downtime and operating costs.

The creation of predictive analytics and diagnostics systems covers a number of stages:

- primary collection of operational data of the equipment, including the history of its work;
- data analysis and construction of predictive models (based on mathematical algorithms, training neural networks, pattern recognition methods);
- verification and check of the accuracy of the models.

It is possible to minimize negative factors by using modern technologies, for example, the Arduino platform and a set of sensors compatible with it, which allow recording vibration load, temperature, pressure, humidity and other important parameters. With the help of additional extension modules, it is possible to transmit the obtained values both via wired (Ethernet module) and wireless data transmission channels (Bluetooth and WiFi modules) for centralized storage of logs and analysis of values using large computational capacities relative to the microcontroller.

Work [2] describes an example of the implementation of a compact system for online monitoring of the state of an electric motor based on a bundle of an Arduino microcontroller, temperature and vibration sensors, as well as a Bluetooth module for communication with smartphones, which is necessary for processing the received data. The authors concluded that the use of such devices can significantly reduce the cost of diagnostics, and also noted the ease of connection.

In [3], a description of a system based on an Arduino microcontroller and a compatible accelerometer for vibration diagnostics of the state of a vehicle gearbox is given. The authors of the work also noted the advantages of the Arduino-based solution in comparison with industrial devices, which were lower cost, a variety of sensors and expansion modules available for connection, as well as good prospects for modernizing the program code for different tasks.

2 Materials and Methods

In practice, there are cases of operation of equipment beyond the originally designed service life, service life up to the 1st repair, interrepair life and overhaul life in hours.

Failures of functional systems of equipment during its long-term operation are largely caused by the processes of degradation of their properties, leading to the loss of their performance, depletion (remaining life), and their prevention largely depends not only on the efficiency of the technical operation system, but also on the completeness and quality of scientific and technical support, including:

- periodic monitoring of the technical condition, the studied equipment park (for example, industrial in the form of machine tools, furnaces, conveyors) and equipment (transport);
- identification of patterns, trends and dynamics of changes in the main quantitative and qualitative indicators of equipment reliability;
- operational and mid-term planning of organizational and technical measures to improve reliability and maintain the equipment fleet at a given level of operability.

In view of this, studies on a comprehensive assessment of reliability are becoming increasingly important both in terms of maintaining the operability of the equipment fleet according to resource reserves and service life, and in terms of the need for periodic analysis of the dynamics of changes in values.

SCADA (Supervisory Control And Data Acquisition).

The structure of technical solutions in industry and transport has significantly evolved over the past decades. And the principles of building mobile transport and the organization of production lines in industry implies the division of the entire facility into structural elements depending on the tasks performed, which allows ensuring the performance of individual functions at a higher level. When considering the processes of industrial production, it can be seen that many subsystems are involved in ensuring the functioning of a separate unit. This includes production automation tools, management information systems and decision support systems, operator workstations, etc. All of these units generate a significant amount of useful data that needs to be processed. Integration of these subsystems into a single, basic one allows their interaction in order to achieve a higher level of automation of production processes, which is the main goal of the CIM approach.

Based on the history of formation and development for SCADA systems, the following classification can be presented:

Monolithic. These are independent systems that operate separately without a common network for communication using special network protocols for exchanging data between microcontrollers;

Distributed. They are distinguished by the addition of a LAN interface (organization into a local network), which unites a set of systems for joint processing of control commands. Solutions of this class are designed to work in the local network of an enterprise. Most often, non-standardized communication protocols with an insufficient level of security were used;

Network. They are a further development of the class of distributed systems, in which complex systems were divided into separate executive modules with control over the network (LAN–PCN). These systems have already incorporated the functionality for the management and coordination of geographically distributed objects. This class also implies the ability to use HMI (Human-Machinery Interface) in a browser on various devices, as well as making changes to a «live» project without the need to stop and create intermediate copies;

Internet Of Things (IOT). The emergence of systems of this class is caused by the spread of wireless communication methods and cloud technologies, which in practice means further steps towards decentralization and unification (controllers, communication methods, data collection and processing).

Examples of ready-made industrial SCADA systems of the 4th generation are: AggreGate SCADA/HMI [4], Mango Automation, PROMOTIC SCADA Visualization software, WINLOG Lite, IGSS, FREESCADA and others [5, 6]. Most of them are cross-platform, use NoSQL solutions as a DBMS for managing and storing data on controlled systems (for the possibility of storing large logs), support various communication protocols with microcontrollers and industrial standard sensors, and also have the ability to provide information in different interfaces—in a browser window, in an application on a mobile device, in a desktop application.

The figures below show the graphical interface displaying the parameters of controlled systems, which is provided by SCADA systems. Often, the appearance of these software solutions is identical, not replete with artistic design, because when it was created, requirements were laid to preserve visibility even with a large number of objects, ensure performance and reduce the minimum requirements for the machine on which the software will be installed (Figs. 1, 2, 3 and 4).

The graphical interface of SCADA systems (both a developer's tool and a working instance) allows operating with mnemonic diagrams, which are used to visualize controlled processes, simplify the definition and finding of the necessary information, and facilitate decision-making. It displays a diagram of the system as a whole and the relationship between the main objects, and also provides data on the state of individual components.

To determine the place occupied by SCADA systems in automated process control systems (APCS), one should refer to the APCS architectures, which can be divided into three levels [4]:

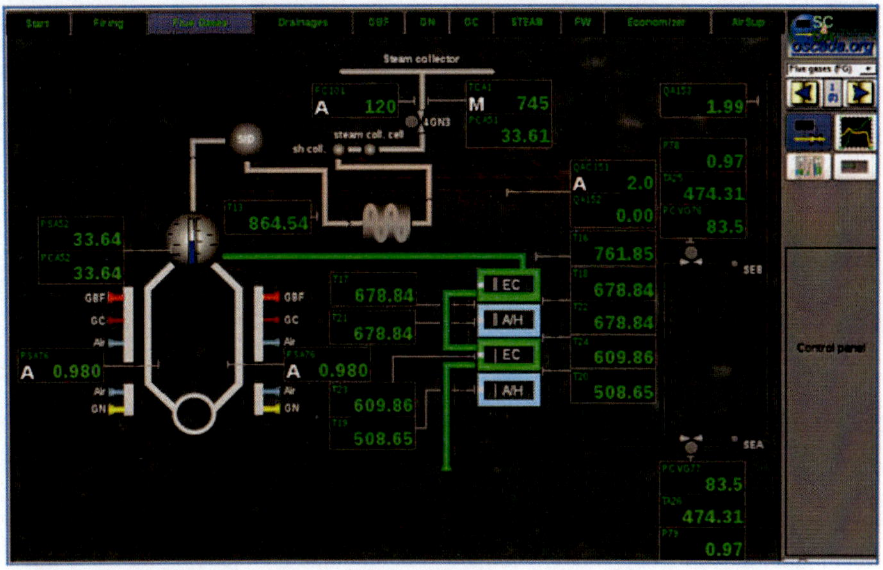

Fig. 1 Open SCADA, demo project, graphical interface

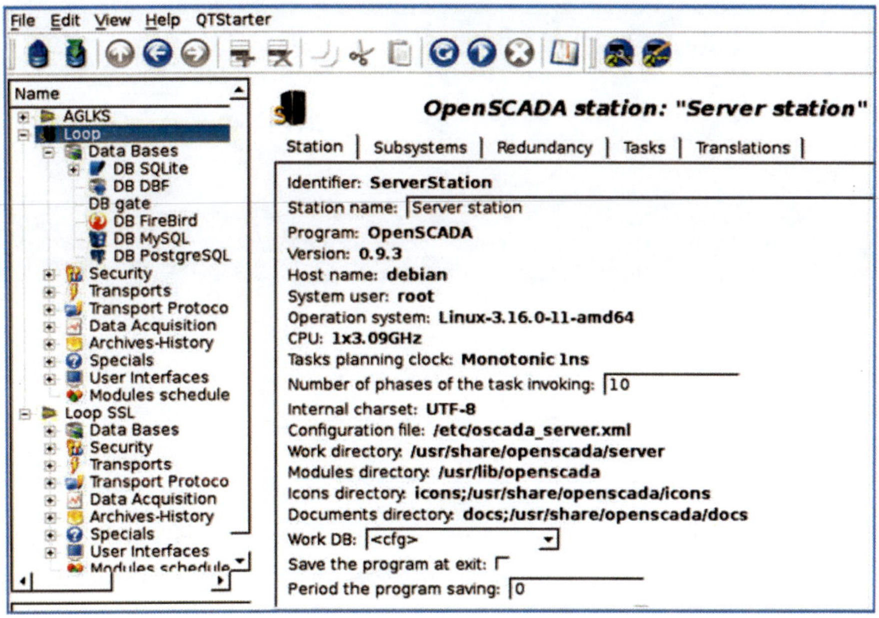

Fig. 2 Open SCADA, graphical interface of the main menu

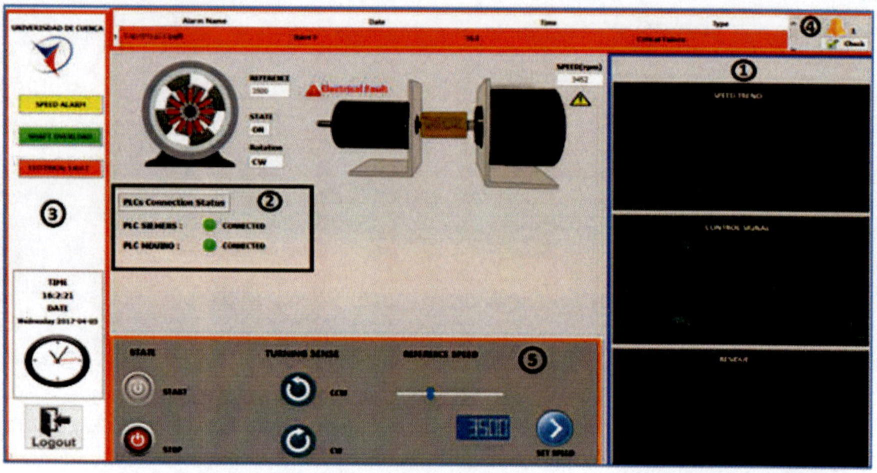

Fig. 3 Open source SCADA System [7], demo project, graphical interface

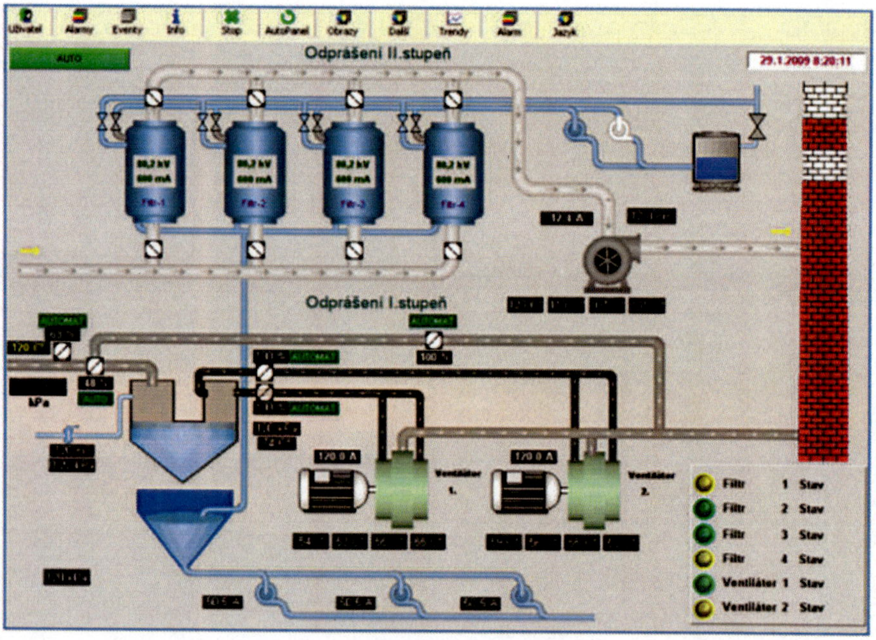

Fig. 4 PROMOTIC SCADA Visualization software, demo project, graphical interface

The upper level is the Human-Machinery Interface (HMI), SCADA, operator panels, communication with the PLC.

The middle level—programmable logic controllers (PLC), regulators, programmable relays, counters, etc. This level is responsible for the operational control and management of the technological object. It includes equipment designed to perform the following tasks: receiving (various status signals from lower-level devices), processing (controllers generate control signals according to the programmed program), transmission (control signals to lower-level devices) and aggregation (collection and transmission to devices of the middle and upper levels) of data. This includes PLCs, communication controllers, computer-process interface (CPI), data collection and transmission unit (DCTU), regulators, programmable relays, etc.

Lower level—sensors, actuators.

Figure 5 shows a typical diagram of the components involved in the description of an automated process control system using SCADA systems.

SCADA systems are used to solve such problems as:

Data exchange in real time between the control object, the managing server (or controller), and the data collection server (for analytics);

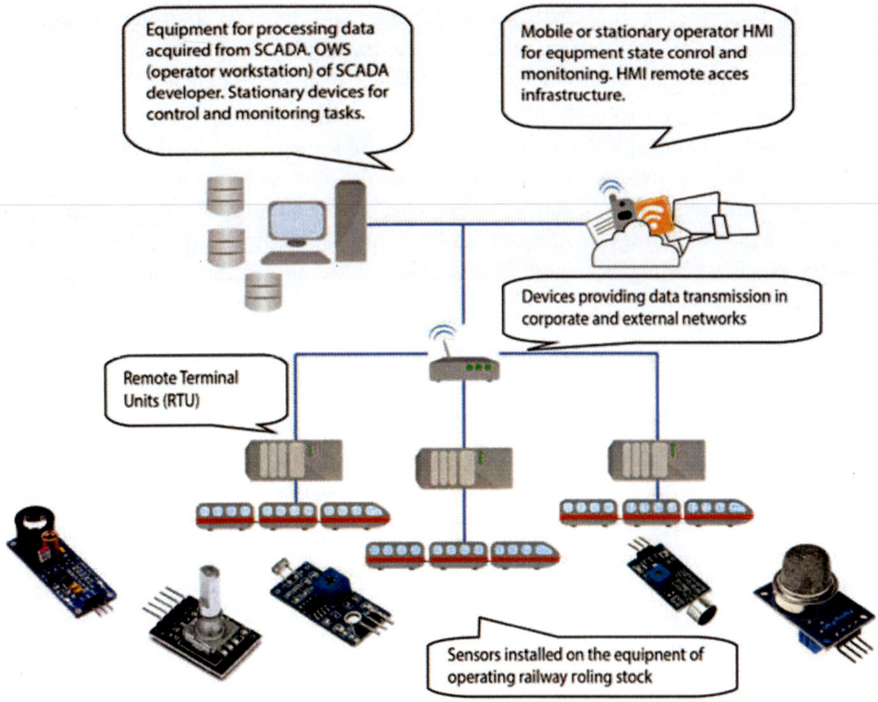

Fig. 5 The structure of the monitoring and control system using SCADA systems

Information processing and the formation of control actions (in the case of automatic control) or options for the operator's actions (in the case of DSS and automated control) in real time;

Display of information in a form suitable for human perception;

Handling emergency situations (signaling, automatic switching off/switching on by the mechanism, protection systems);

Formation of reporting files on the progress of the controlled processes;

Communication with external software.

The functioning of the automation of most systems in the industrial and transport industry involves the participation of a person (dispatcher, operator). The module that ensures the interaction of a person and a system is called a Human-Machinery Interface (HMI).

Timely detection of anomalies in the technical processes of individual elements is a paramount task to ensure the efficient and safe functioning of the entire system. As a rule, during the operation of a complex technical object, a certain amount of historical data on the state of important hosts, obtained using SCADA systems, is accumulated. These data are most often used in the analysis of the main causes of emergency situations that have already occurred and in the analysis of the consequences.

Presumably, a higher level of understanding of the true causes of equipment failures can be achieved by analyzing the same data, but using more advanced analytical methods, such as artificial intelligence and machine learning. In the long term, this should allow the transition from the paradigm «diagnostics—mitigation» (analysis of root causes) to a more desirable, from a practical point of view, approach «detection—forecasting—analysis of the forecast—prevention». Machine learning methods can be used to analyze data obtained from SCADA systems in order to identify situations that do not necessarily manifest themselves as an alarm from a security system, but, nevertheless, can lead to emergency system behavior.

When it comes to the use of machine learning and artificial intelligence methods to solve predictive diagnostics problems, it should be understood that, in fact, there are several tasks:

1. Detection of malfunctions (anomalies in work).
2. Classification of malfunctions (failures).
3. Estimation of the residual life, the very forecasting of future failure.

Fault detection is the main step in preventive maintenance, in which data on the condition of the operating object is used to determine if the system is operating in a normal or abnormal state. This process is often referred to as anomaly detection or outlier detection in machine learning. To detect anomalies, special frameworks and program code libraries are used, such as ELKI [8] or Scikit-Learn [9], containing implementations of special algorithms. The most used mathematical methods for detecting anomalies are the principal component analysis (PCA) and its modifications—the PCA-T2 basic component estimation graph and graph that takes into account the squared prediction error (PCA-SPE), the K-Means method, autoencoders based on a special ANN topology (Auto Encoders) and Gaussian mixture models (GMM).

3 Results

In [10], using a program developed in the R language, an analysis of the operation of a metal valve that closes the bypass channel between the compartments in a certain technical device is carried out. Four different states are recorded (system at rest, normal operation, and two unbalanced starts), the results of measurements taken during startup are split into four different csv files. First, the data is read using the read.csv() method. The figure below shows a piece of data where time is the time between samples, ax is the x-axis acceleration, ay is the y-axis acceleration, az is the z-axis acceleration, and aT is the control signal indicating the generated force. Data was collected at a sampling rate of 100 Hz (Table 1).

It is possible to immediately get statistical information about the read data (Table 2).

Then it is necessary to apply methods to detect anomalies in the sample (in this case—PCA-T2) (Fig. 6).

Analyzing this graph, we can see that the values recognized as atypical are below the line. A similar result can be achieved using the K-means method. Its purpose is to divide the dataset into predefined clusters depending on the metric (a measure of the distance between the objects in question). In the example, Euclidean distance was used (Fig. 7).

After determining the very fact of the presence of an anomaly in the work, one should proceed to determining its type, to the classification task, i.e. assigning an object to one of the species on the basis of its distinctive features [10]. Linear classification models include logistic regression and support vector machines. In a nonlinear, more complex classification, the most commonly used decision tree model is the k nearest neighbors (kNN) method.

For the further use of the results obtained in the previous steps from trained models, the capabilities of the Scikit-Learn module for Python can be used as data for failure prediction methods, as it shown in [7, 10–12].

Table 1 Data snippet used in examples

Time	ax	ay	az	aT
0.002	−0.3246	0.2748	0.1502	0.451
0.009	0.6020	−0.1900	−0.3227	0.709

Table 2 Statistical information

Time	ax	ay	az	aT
Min: 0.002	Min: −2.11880	Min: −2.143600	Min: −4.1744	Min: 0.032
lst Qu: 16.507	lst Qu: −0.41478	lst Qu: −0.625250	lst Qu: −0.7359	lst Qu: 0.848
Median: 33.044	Median: 0.02960	Median: −0.-22050	Median: −0.1468	Median: 1.169
Mean: 33.037	Mean: 0.01233	Mean: 0.008697	Mean: −0.1021	Mean: 1.277
3rd Qu: 49.535	3rd Qu: 0.46003	3rd Qu: 0.641700	3rd Qu: 0.4298	3rd Qu: 1.579
Max: 66.033	Max: 2.096203	Max: 2.003000	Max: 4.9466	Max: 5.013

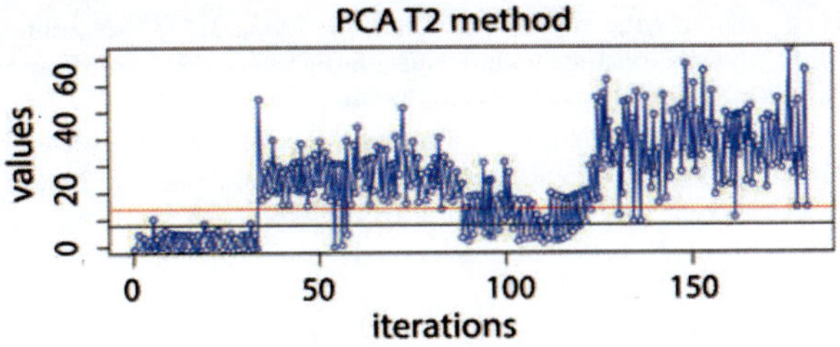

Fig. 6 The result of the PCA-T2 algorithm

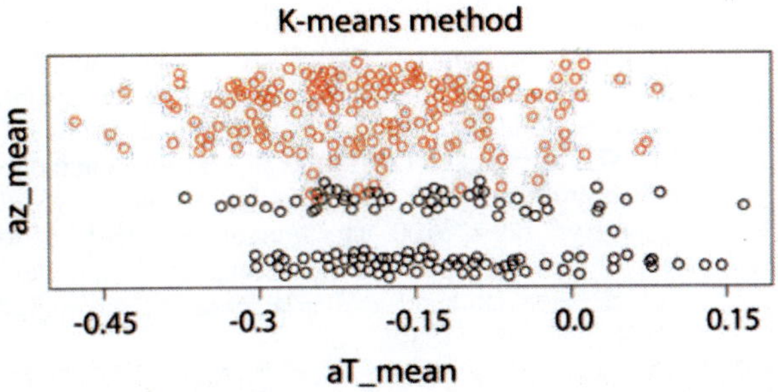

Fig. 7 Result of the K-means method

4 Discussion

As a result, it should be noted that to predict the remaining useful lifetime of the unit (RUL), both linear and nonlinear models are studied, including parametric and nonparametric types. Various transformations for the output data are tested in order to select the best prediction form. The best form is chosen based on the predictive characteristics of the models. Performance is measured by root mean square error (RMSE) for predictions using test data. For this data, the inverse $\frac{1}{\mathrm{RUL}}$ is the best form of output. The input data has been standardized in order to exclude the influence of predictor data blocks on prediction models.

Once the RUL estimates are acceptable, they can then be integrated into dashboards used by operators or into intrusion alarm systems under the supervision of maintenance personnel. This will allow responding to changes in equipment condition faster, with less impact on the workflow.

References

1. Vorobiev AA, Spitsyn IN, Anisimov AV, Bryukhanov IS, Chmyrev IA (2015) Technical diagnostics of technological equipment based on the Arduino controller. Reshetnev's Readings, 19
2. Gordienko VV (2019) Analysis of malfunctions of gearboxes of MTZ family tractors and methods of their diagnostics based on the ARDUINO microcontroller board//Scientific review. Tech Sci 6:44–49

3. Almas MS, Vanfretti L, Gjerde J et al (2014) Open source SCADA implementation and PMU integration for power system monitoring and control applications. IEEE, 1–5. https://doi.org/10.1109/PESGM.2014.6938840
4. Shravan IV (2017) Top opensource data mining software. OSFY. OSFY Magazine February 2017. Top OpenSource Data Mining software
5. Merchan D, Peralta J, Vázquez-Rodas A, Minchala I, Astudillo D (2017) Open source SCADA system for advanced monitoring of industrial processes. https://doi.org/10.1109/inciscos.2017.9
6. Lukov DK (2019) Automated process control systems (ACS TP). European Sci 2(44):19–21
7. Churakov D, Tsarkova E, Grechishnikov E, Marchenko N (2018) Analysis of methods of processing of expert information by optimization of administrative decisions. J Phys: Conf Series 012030. https://doi.org/10.1088/1742-6596/973/1/012030
8. Amruthnath N, Gupta T (2018) A research study on unsupervised machine learning algorithms for early fault detection in predictive maintenance. In: 2018 5th international conference on industrial engineering and applications (ICIEA), pp 355–361. IEEE
9. Kraus M, Feuerriegel S (2019) Simple classification models in Python. Forecasting remaining useful life: interpretable deep learning approach via variational Bayesian inferences. Decis Support Syst 125:113100. https://doi.org/10.1016/j.dss.2019.113100
10. Kraus M, Feuerriegel S (2019) Forecasting remaining useful life: interpretable deep learning approach via variational Bayesian inferences. Decis Support Syst 125:113100. https://doi.org/10.1016/j.dss.2019.113100
11. Tsarkova E, Belyaev A, Churakov D, Andreeva E (2020) Reliability forecasting for optimal planning of measures for maintenance of security systems of transport infrastructure facilities. IOP Conf Series: Mater Sci Eng 8:012090. https://doi.org/10.1088/1757-899X/918/1/012090
12. Taha H, Sakr A, Yacout S (2019) Aircraft engine remaining useful life prediction framework for industry 4.0. In: Proceedings of the 4th North America conference on Industrial Engineering and Operations Management, Toronto, ON, Canada, pp 23–25

Methodology for Assessing the Effectiveness of Work in the Road Sector

Maria Mironova, Ulia Mironova, Tatiana Konovalova, Sofia Nadiryan, and Artem Litvinov

Abstract Current and forecast economic conditions imply the use of new theoretical provisions in the field of organization of personnel policy and personnel management system. This trend can be traced not only in large manufacturing enterprises, but also in the road sector. In this regard, personnel management should be based on methods that are objectively inherent in effective production management. Evaluation of the effectiveness of work in the road sector is necessary to compare the tasks assigned to the personnel and the results of their practical implementation. The article proposes a method for constructing a system for assessing the effectiveness of the work carried out by road branches. The most important component of the system for assessing the effectiveness of the work performed is the assessment rules that are uniform for all performers. The set of procedures for determining and assessing compliance indicators of the estimated volume of work performed must objectively reflect the results obtained, which determine the effectiveness of the road branch. The proposed method for assessing the performance of road branches allows management personnel to build an objective system of reward for the fulfilled value of the established standard for the range of works.

Keywords Road facilities · Labor efficiency assessment · Personnel management system · Transport infrastructure · Road branch

1 Introduction

In modern conditions, the concept of assessing the effectiveness of work in the field of road facilities is positioned as a choice of means and methods for assessing the practical implementation of tasks assigned to personnel in order to increase the efficiency of work on traffic management.

M. Mironova (✉) · U. Mironova · T. Konovalova · S. Nadiryan · A. Litvinov (✉)
Kuban State Technological University, Krasnodar, 2a, Moskovskaya st.,
Krasnodar 350020, Russia

A. Bieliatynskyi and V. Breskich (eds.), *Safety in Aviation and Space Technologies*,
Lecture Notes in Mechanical Engineering,
https://doi.org/10.1007/978-3-030-85057-9_31

Assessment of labor productivity is complex in nature, since the estimated decisions depend on a large number of factors (variables). Personnel performance appraisal is a labor-intensive process undertaken by an organization to determine how satisfactorily the personnel are performing their assigned tasks [1–3].

The effectiveness of the performance of tasks by personnel is largely determined by motivation. The motivation of labor is a condition for ensuring an increase in its effectiveness. Employees of the enterprise can be motivated for high performance and quality of work. Among the motives for the professional development of personnel are usually distinguished:

- the desire to keep the job;
- desire to get a promotion or take a new position;
- interest in the process itself;
- the desire to receive bonuses for labor.

To date, the following methods for assessing labor productivity are distinguished, presented in Fig. 1.

Evaluation of the effectiveness of work in the road sector is carried out in order to:

- determination and assessment of the level of "tension" of the plan of labor productivity;

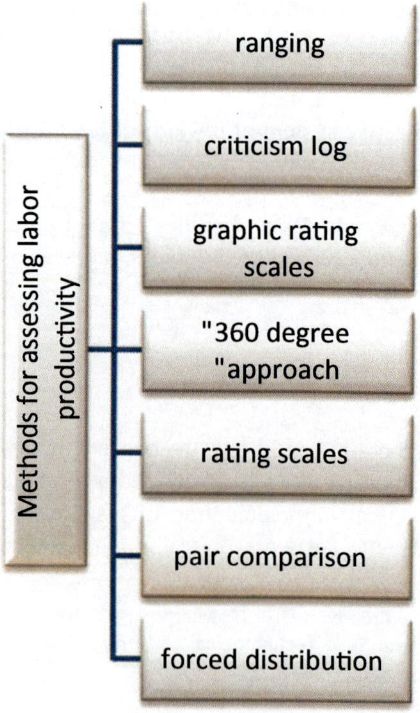

Fig. 1 Methods for assessing labor productivity

- determination of the dynamics of the level of labor productivity and its comparison with previous years;
- determination of factors contributing to changes in labor productivity in certain periods;
- determination of directions for increasing the efficiency of the use of labor resources of the enterprise and the growth of labor productivity.

The assessment of the work done by the road branches in the Krasnodar Territory is carried out on the basis of an integrated approach. When performing the assigned tasks, the quality of both individual elements (indicators) and the entire system of indicators as a whole is assessed. Timeliness and compliance of the volume and composition of the work performed with the requirements of regulatory documents, along with the completeness and objectivity of the reflection of the results obtained, determine its effectiveness. The implementation of the unified concept of compliance of the indicators of the point-rating system is carried out on the basis of an updated base of the current regulatory and technical documentation, design and working documentation, technological regulations containing requirements for the performance and assessment of the work performed. The use of modern updated methods, taking into account the provisions of documents regulating the implementation of state control (supervision), attracting qualified personnel allow ensuring the objectivity of the assessment [3–5].

Let us consider in more detail the structure of management of the operation of highways, Fig. 2.

The powers of the Federal Highway Agency include:

- taking measures for the arrangement of public highways of federal significance UDS in accordance with the documentation for the planning of the territory, the organization of their work;
- taking into account indicators of the state of road traffic safety in terms of length, technical condition of highways and the presence of SLM on them;
- informing road users about the presence of SLM and safe traffic conditions on the relevant road sections.

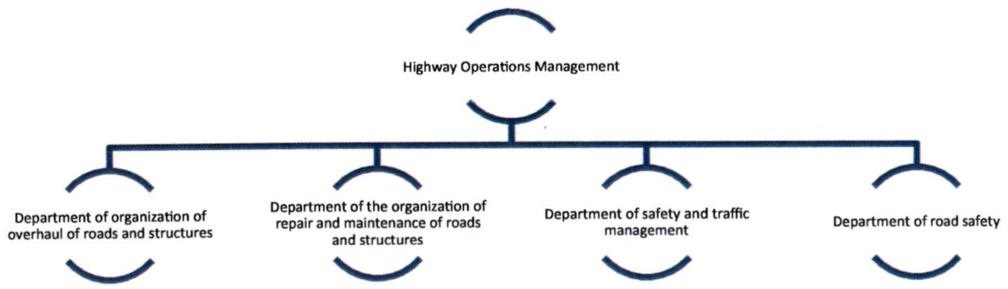

Fig. 2 The structure of the management of the operation of highways

The activities of the Avtodor Group of Companies include:

- taking measures to provide the State Company highways with road service facilities, organizing their use;
- monitoring the state of road service facilities within the boundaries of right-of-way and roadside strips of the State Company's highways, as well as determining the needs for the construction and reconstruction of highways of the State Company and the specified road service facilities;
- information support for users of the State Company's motor roads, including informing about the availability of road service facilities within the boundaries of right-of-way and roadside lanes of the State Company's motorways and about the safe conditions for vehicles to travel on the State Company's roads;
- performing the functions of a customer in the design, construction, reconstruction, overhaul, repair and maintenance of the SLM (and abutments) within the boundaries of the right-of-way and roadside strips of the State Company highways;
- The quality management of road works on federal public highways is expedient to be considered as a three-level system, including strategic, tactical and operational levels.

The subjects of quality management are:

- at the I level—the Federal Road Agency (Rosavtodor);
- at the II level—PKU, performing the functions of road management bodies (functions of the Customer);
- at the III level—contractors performing road works.

Quality assurance of road works is possible if the following conditions are met:

- availability of a quality system, including regulatory management procedures;
- implementation by the participants of the quality system of the relevant regulatory procedures;
- monitoring the implementation by the participants of the quality system of the regulatory management procedures.

The road works quality management scheme is shown in Fig. 3.

The basic principles of the personnel management system can be formulated as follows:

- leaders ensure the unity of purpose and direction of the organization;
- involvement of employees in solving problems;
- the effectiveness of achieving the required result is significantly increased when managing activities and related resources as a process subject to the same rules and aimed at achieving a single goal;
- identification and management of interrelated processes as a system contributes to improving the quality, efficiency and effectiveness of road activities;
- continuous improvement of the processes of organization, production and management;

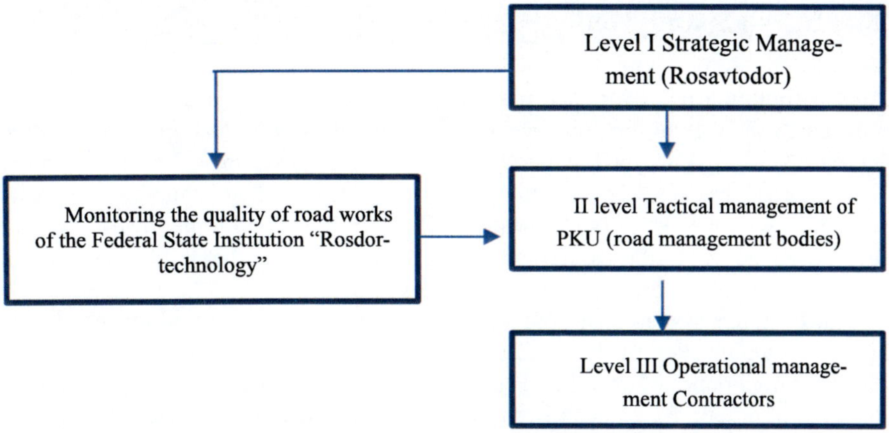

Fig. 3 Scheme of the road works quality management system

- making management and corrective decisions based on the results of quality assessment and monitoring of road works performed;
- competent management of relationships with stakeholders (suppliers);

The assessment system of road branches in the Krasnodar Territory is based on the construction of a point-rating system with the assignment of penalty points [6]. The assessment criteria for the main indicators are presented in Table 1:

Table 1 Criteria for evaluating road branches

Criteria for evaluation	
Disciplinary action (violation of labor discipline, etc.)	Points are awarded for each disciplinary sanction in the current month: remark—1 point, reprimand—2 points
Control over the maintenance of facilities under government contracts. Observations identified during on-site inspections by departments of the institution (s) or ministry	3 points are awarded for each established case for each government contract
Late delivery of acceptance of road works	3 points are awarded for each identified case for each district
Errors in as-built documentation during acceptance, Failure to fulfill orders, untimely and inaccurate provision of information, Lack of prescriptions, absence or inaccurate information on the implementation of prescriptions	1 point is given for each established case for each government contract
Inconsistency of the assessment of the level of real	3 points are awarded for each established case for each government contract

2 Materials and Methods

After assigning a point for each indicator, the overall rating of the road branch is reduced by the sum of the points of the assessed indicators. The highest total of the scored penalty points corresponds to the mark not satisfactory. Further, a rating scale of road branches is constructed in accordance with the number of points scored from the lowest to the highest [7]. The road branch with the lowest value of the sum of points in the assessment rating corresponds to the given value of the work performance efficiency (Table 2).

However, this system does not take into account the degree of importance of each criterion. Optimization of the calculation of the total penalty point according to the degree of importance of the criterion can be realized by using a weighting factor, then the table of criteria for evaluating road branches will take the following form:

Figure 4 clearly shows the methodology for assessing the efficiency of road branches in the form of an algorithm.

Table 2 Criteria for evaluating road branches

Assessment indicators	Criteria for evaluation		Weight coefficient
Disciplinary action (violation of labor discipline, etc.)	Points are given for each disciplinary sanction in the current month: remark—1 point, disciplinary sanction—2 points, reprimand—3 points	0—chorus, 1–2—satisfactory, 3—failed	0.2
Control over the maintenance of facilities under government contracts. Observations identified during on-site inspections by departments of the institution (s) or ministry	3 points are awarded for each established case for each government contract	0—chorus, 1–2—satisfactory, 3—failed	0.15
Late delivery of acceptance of road works	3 points are awarded for each identified case for each district	0—chorus, 1–2—satisfactory, 3—failed	0.2
- Errors in as-built documentation upon acceptance, - Failure to fulfill orders, untimely and inaccurate provision of information, - Lack of prescriptions, lack or inaccurate information on the implementation of prescriptions	1 point is given for each established case for each government contract	0—chorus, 1–2—satisfactory, 3—failed	0.12 0.10 0.08
Inconsistency of the assessment of the level of real	3 points are awarded for each established case for each government contract	0—chorus, 1–2—satisfactory, 3—failed	0.15

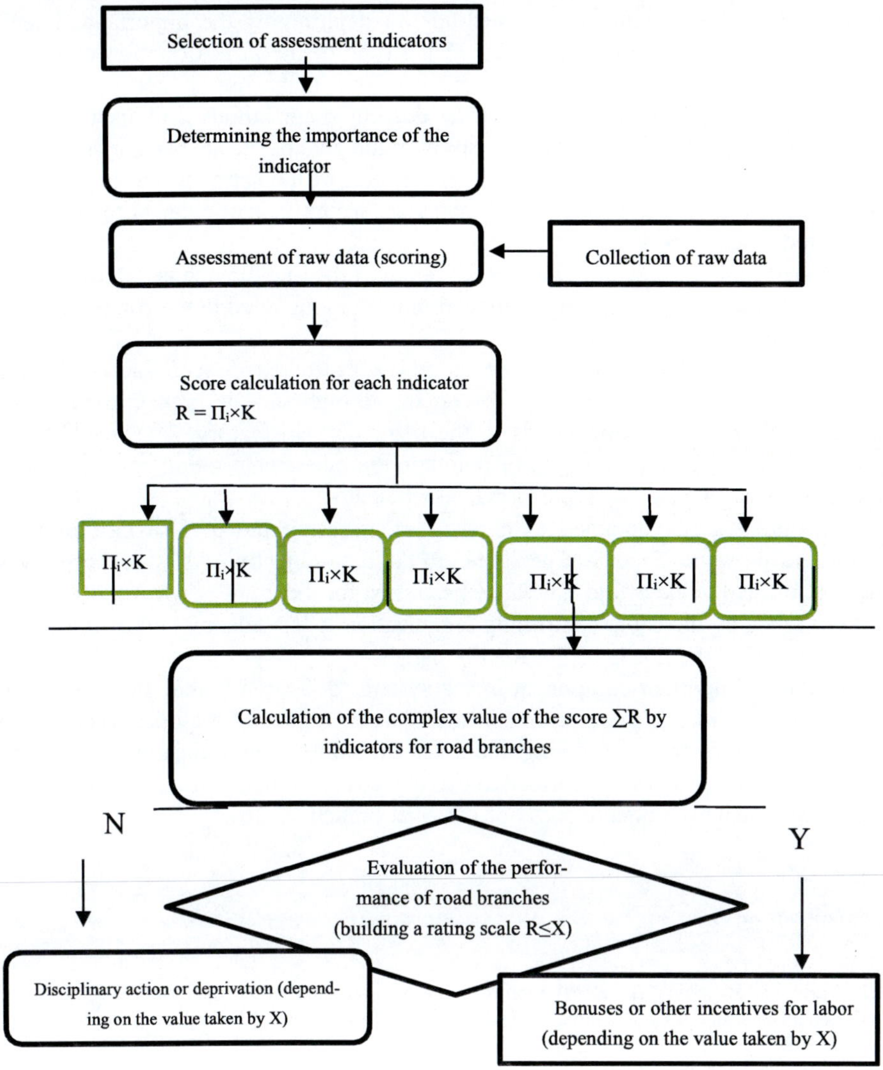

Fig. 4 Algorithm of the methodology for assessing the effectiveness of the work of road branches

3 Conclusion

The content of the optimized methodology: each district included in the road branch is assigned a point for the assessed indicators, after assigning a point for each indicator, it is calculated using a weight coefficient that reflects the degree of importance of a particular indicator when performing a range of works.

Currently, the classification of methods for determining the importance coefficients exceeds 20 methods that differ in the type of initial information about the requirements and methods of processing it to obtain the importance coefficients. The content of particular criteria for evaluating the methods and means under consideration is determined by the meaning of the goal of the analysis, contained in the formulation of the goal. Based on the essence of the set goal, the importance, weight or value of the values of estimates according to particular criteria is also measured when they are generalized.

Further, for each indicator, the overall rating of the road branch is reduced by the sum of the average score of the estimated indicators included in the road branch of the contracting areas.

Further, a rating scale of road branches is constructed in accordance with the number of points scored from the lowest to the highest. The highest total of the scored penalty points corresponds to the mark not satisfactory. The road branch with the lowest value of the sum of points in the assessment rating corresponds to the given value of the work performance efficiency.

The rating scale, obtained as a result of assessing the performance of road branches, allows management personnel to build an objective system of incentives for the fulfilled value of the established standard for the range of works. The value of X is taken by the equal controlling organization independently and is brought to the attention of the heads of the road branches.

The most important component of the system for assessing the effectiveness of the work performed is the assessment rules that are uniform for all performers. The set of procedures for determining and assessing compliance indicators of the estimated volume of work performed must objectively reflect the results obtained, which determine the effectiveness of the road branch [8–10].

References

1. Indira R, Bharatish A (2020) Optimization of ventilated brake disc rotor geometry for enhanced structural characteristics. J Meas Eng 8:98–106. https://doi.org/10.21595/jme.2020.21399
2. Balaev EYu, Konovalova TV, Litvinov AE, Nadiryan SL, Senin IS (2020) Optimization of freight traffic in cities using the forecast transport model of freight traffic. In the collection: IOP conference series: materials science and engineering. VIII international scientific conference transport of Siberia, 012045
3. Konovalova TV, Nadiryan SL, Litvinov AE (2018) Transport systems of cities in the processes of new industrialization. Adv Soc Sci Educ Humanit Res 590–593
4. Konovalova TV, Kotenkova IN (2013) Transport infrastructure. Kuban State Technological University, Krasnodar
5. Sukharev OS (2015) Industrial policy and the development of transport infrastructure in Russia. Nat Interests: Priorities Secur 11(1)(286):2–20
6. Pugachev I, Kulikov Y, Markelov G, Sheshera N (2016) Factor analysis of traffic organization and safety systems. In: 12th International conference organization and traffic safety management in large cities, SPbOTSIC-2016, 28–30 September 2016, pp 529–535. St. Petersburg, Russia

7. Kravchenko P, Oleshchenko E (2016) Mechanisms of functional properties formation of traffic safety systems. In: 12th International conference "organization and traffic safety management in large cities", SPbOTSIC-2016, 28–30 September 2016, pp 367–372. St. Petersburg, Russia
8. Konovalova T, Zarovnaya L (2016) The assessment model for economic efficiency of traffic safety improvements. In: 12th International conference organization and traffic safety management in large cities, SPbOTSIC-2016, 28–30 September 2016, pp 311–315. St. Petersburg, Russia
9. Kurakina E, Evtyukov S (2018) Results of studying road construction parameters condition. Archit Eng 3(1):29–37
10. George R, Jana I, Joseph K (2013) International student's traffic engineering project seminar meps. World Trans Technol Mach 4(43):106–111

Chamber Air Flow Control Before the Radiator

Alexander Petrov

Abstract The passage of cooling air through the engine compartment is accompanied by a significant increase in aerodynamic drag of the car. On the other hand, forming of air flow due to the operation of the fan is associated with additional energy costs. To reduce fuel consumption and emission of harmful substances, it is necessary to reduce the power consumption for supplying cooling air. At the same time, the possibility of rational use of the oncoming air flow and fan power should be taken into account. The easiest way to solve the problem is a fan installation with the case, which area is less than the front area of the radiator. The main disadvantage of this design is the air circulation from the engine compartment of the engine. In this work there is a conducted research on reducing air circulation. The simplest method has been studied, a partition has been installed in the pre-radiator chamber separating the free zone of the radiator and the zone of the fan case. A good effect has been recorded, but only if the resistance of the engine compartment is less than the resistance of the radiator.

Keywords Radiator · Fan · Air circulation · Pre-Radiator chamber · Air flow control · Computational fluid dynamics

1 Introduction

The main source of environmental pollution is exhaust gases from motor vehicles. The amount of emissions of harmful substances per unit of time depends on several factors, on the perfection of the engine, the design features of the car and on operating conditions. The more economical the vehicle is, the less environmental damage it causes. Let us consider some of such factors when the design features of the vehicle affect its efficiency in different operating conditions.

Aerodynamic perfection of the shape of the vehicle body allows reducing its aerodynamic drag, thereby reducing fuel consumption. Given that at a vehicle speed

A. Petrov (✉)
Polytechnical Institute, Kurgan State University, Tomina str., 40, Kurgan, Russia

A. Bieliatynskyi and V. Breskich (eds.), *Safety in Aviation and Space Technologies*, Lecture Notes in Mechanical Engineering,
https://doi.org/10.1007/978-3-030-85057-9_32

of 100 km/h, about 70% of the energy is spent on overcoming air resistance, it becomes obvious that special attention should be paid to this problem. But the other part has to do with the fact it has long been known that the aerodynamic drag of a vehicle largely depends on the flow of air that passes through its interior. This air is necessary for blowing the radiator of the engine-cooling system and the condenser of the air-conditioning system of the passenger compartment. The amount of this air is quite large. According to some data, the percentage of the inner air flow in the aerodynamic drag of a passenger car can be 8–10% [1–6]. It is stated that there is a correlation between the cooling resistance and the mass flow rate of cooling air [7, 8].

The decrease of the opening area of the front grille can limit air flow through the engine compartment and, therefore, can reduce aerodynamic drag [3, 4, 6]. However, reducing the flow of cooling air leads to the problems connected with the requirements of engine cooling, heating, ventilation and air conditioning, as well as thermal requirements, especially at low speeds or at idling speed. In order to compensate for the loss of cooling capacity, a larger power of the fan of engine cooling is often required.

The more air passes through the vehicle interior, the more influence it will have on the coefficient cx. Therefore, it is necessary to try to reduce the amount of air passing through the engine compartment of the car, while maintaining the effectiveness of the cooling and air conditioning systems.

The air intake openings, located in the front of the vehicle bonnet, play an important role in the aerodynamics of a passenger car. Since the passage of air through the engine compartment of the vehicle leads to an increase in its aerodynamic drag, it can be concluded from this that it is necessary to limit the flow of excess air and, on the other hand, it is necessary to use cooling air more efficiently.

The front of the car is crucial, both in terms of the overall aerodynamics of the vehicle and the inner aerodynamics, connected with the cooling system. As the studies show, the in-coming part has a large effect on the efficiency of the cooling system [9]. From a theoretical point of view, the inlet diffuser (pre-radiator chamber) at certain expansion angles and a large in-coming part would allow slowing down the flow gradually and preventing the probable separation of the flow from the channel borders. On the other hand, a large in-coming part increases the level of airflow, which obviously leads to the improvement of the cooling process. As a result, maximization of the in-coming part for a vehicle with a duct cooling system is beneficial from the point of view of its technical functionality.

In [10] there is a research on the air distribution and unevenness through full-size Ford Falcon radiator. The objective has been to find the optimal radiator grille design. The results of these tests have shown that the best design option is an air intake opening that elongated horizontally. This design allows guaranteeing uniform air flow over the surface of the radiator. In addition, this has made it possible to obtain a higher average air flow rate through the radiator, which leads to better cooling. The unevenness affects engine cooling system, radiator heat transfer efficiency is becoming worse.

The cooling system must be able to provide engine operation in two rare and extreme conditions: firstly, moving at maximum speed and, secondly, climbing ability in a full throttle, towing a trailer with a maximum authorized mass. The rest of the time, the system operates below maximum capabilities, while at the same time influencing the resistance.

The work studies various layout designs of the air intake openings and cooling units. Sometimes in vehicles, there is a layout design with the air conditioning condenser, being placed next to the radiator. In a cold season, when the condenser is not used, it is possible to block the condenser from the side of cooling air flow completely. A symmetric version of the grilles and the lower air intake opening has been considered, one of the halves has been cut off.

The flow of cooling air in front of the radiator is very heterogeneous. Air flows coming through the upper and lower grilles are mixed with each other [11]. Recirculation of the cooling air flow can be observed directly in front of the upper and lower parts of the radiator. On the radiator in the corners there is a zone where the air speed is very low. This is because of the cut-off of the fan case.

Williams and others [12, 13] conduct extensive experimental studies of the flow under the vehicle bonnet of type Ford Taurus. They show that the current lines on the front of the vehicle are very different between the stationary state and the one in motion. When the car is standing and the fan is running, air intake through the air intake openings occurs from the entire area around these openings and the current lines form a semicircle. In road conditions, at speed, the current lines run in a denser stream, at first straight towards the car and then they are slightly curved in front of the in-coming part.

The elements that make up the engine compartment and that are in the way of the cooling flow may create recirculation zones or uneven flow distribution, this must be avoided. Uneven flow distribution on the front of the radiator can lead to inefficient cooling [14].

The radiator grille is installed on the front of the vehicle and is an important element affecting the aerodynamic characteristics and thermal efficiency of cooling [15]. The studies have shown that the larger the size of the grille openings is, the greater the flow of the cooling air is. It matters where the grille is located, the higher the grill is located, the lower the air flow is and vice versa. It has been stated that an increase or decrease in air flow through heat exchangers leads to an increase or decrease in the vehicle aerodynamic drag proportionally.

An important element for cooling and aerodynamic efficiency is the directing channel of cooling air between the radiator grille and the radiator. With properly designed channels, all the air that goes inside the air intake openings passes through coolers [16].

One of the dilemmas in the development of vehicles is a compromise between the dimensions of the front grille and the power of the engine cooling fan. A smaller front grille opening provides better vehicle aerodynamics, but requires more power of the engine cooling fan. In this configuration, additional fuel is consumed by the engine cooling fan to cool at low speed or at idling speed. On the other hand, a larger grille opening results in lower power consumption of the engine cooling fan, but increases aerodynamic drag. In this case, additional fuel is consumed to overcome aerodynamic drag at high speed. The optimal configuration of the front grill opening and the power level of the engine cooling fan cannot be achieved without understanding the aerodynamic and thermal characteristics of the vehicle under various driving conditions [17].

Many factors influence the rate profile of cooling air through heat exchangers, including the position and size of the intake and exit openings, respectively. The areas with the highest intensity are immediately behind the grilles. Since the heat exchangers are in close proximity to the grilles, the air flow cannot be evenly distributed over their surface. As a result, there are areas with low mass flow through heat exchangers. As air approaches the heat exchanger interface, it also changes direction. Since the porous medium makes the passage of air difficult, most of the air goes to the side (up and down). Due to the presence of two openings, air is more evenly distributed over the interface of the condenser or radiator, since air can come from both the upper and lower intake opening at the same time [18].

In work [19] it is noted that the while designing air inlets, compromises between the aerodynamic and thermal requirements must take place so that components are soundly functioning. For example, an efficient air inlet favors the underhood thermal environment (improves the convective heat transfers) but at the same time it increases the aerodynamic drag due to cooling systems. Furthermore, it is worthy to be cognizant that Automobiles whose engines placed in front have two types of intakes, namely: high grille air intake, and low grille air intake. The design of air intakes is based on the aforementioned types and can also be classified as:—Two air intakes design: Air intake of high grille type (Superior); and Air intake of low grille type (Inferior)—One air intake design: Air intake of high grille or low grille type From the aerodynamic point of view, the air inlet has a heavy impact on the aerodynamic tensor of the vehicle, particularly on the cooling air drag in frontal or lateral wind conditions. Indeed, the cooling air drag is essentially a function of the underhood geometry, the design and layout of the air inlet and outlet, pressures at the inlet and outlet, and the direction and speed of the penetrating air. The study is determined to examine explicitly the components architectural arrangements' of underhood compartments and present simplified models of the compartments so that they can serve as aerothermal baseline models in the early stages of design. Hence, the different components in the underhood compartment are classified with respect to aerothermal orders of impact. Followed, two geometrical models, simple and advanced, are developed based on the impact orders. These models serve for future aerothermal studies (analytical, numerical and experimental). Finally, a simplified vehicle model is designed and implemented as underhood compartment model. The structurally simplified model is already utilized in experiments and can be used afterward in both numerical and experimental analyses.

In work [20] it is noted that the cooling air drag can be around 5–15% of the total drag, and a lot of research has been done over the years on how to decrease it. Cooling drag is defined as the force acting in x-direction from the cooling air flowing through the engine bay. The air is let in through the grilles to cool down the engine and escapes through different outtakes usually below the vehicle. Been done on which design parameters that affect the cooling air drag and how air ducts could be designed to better transport the flow to and from the radiators without loosing to much in energy. The first study showed that air through the wheelhouses increases the drag as well as having air entering only in the upper grille and travelling down through floor and wheelhouse. In the second study, the area between the grilles and cooling package was sealed and inlet ducts were created to control the flow from the grilles to the cooling package. When just adding inlet ducts, the mass flow through the grilles was decreased, but the mass flow through the cooling package was increased due to less separation of the air, which lead to a drag reduction of 0.2%. An air duct was also designed to lead the flow after the cooling package fan to the outlets. The ducts purpose was to prevent the air from loosing in energy when rising to the roof of the engine bay, the duct compresses the air and leads it efficiently over the engine with a drag force decrease on 0.3%. In conclusion there is much that can be done to improve the cooling air drag. It is most favorable to have a sealed volume with inlet air ducts before the cooling package, have outtakes aligned with the free stream flow, minimize cooling air to escape through the wheelhouse outlets and to minimize the height of the engine bay as much as possible.

The author concludes that in order to design a vehicle to have improved cooling air flow would recommended.

- Design a sealed air guide volume in between grilles and cooling package
- Design inlet air ducts that controls the flow to the cooling package and avoids lost energy through separation
- Decrease the engine bay height as much as possible or put in an after fan duct that leads the air above the engine and down to the floor outlet.

It has been proven in this study that designing an inlet air duct to control the flow to the cooling package is most useful. The air duct allows an increase of air through cooling package which allows a decrease of the size of the upper grille. For future models, smaller upper grilles could be designed that are lower in height with similar width as today. Designing an air duct could possibly be done even better if the grilles are better aligned with cooling package erad and brad so that the ducts could expand more exponentially. This would most probably allow some further decrease of the drag.

For hybrid vehicles there is still a lot of parts needed in an engine bay and also a higher cooling need for the engine compared to a Battery Electric Vehicle (BEV). For a BEV there would be a lesser need of cooling air and it would therefore probably be possible to decrease the size of the inlet grilles. For a battery electrical vehicle with lesser cooling air need the best steps to decrease drag. Lower the height of the cooling package, possibly put it in serial placement, which allows the vehicle to have less grille height and total area.

2 Numerical Experiment

As the analysis of the presented works has shown, the air flow necessary for the operation of the cooling system leads to a significant increase in aerodynamic drag. To reduce these losses, it is necessary to try to reduce air flow through the engine compartment, but the amount of air should be sufficient to maintain optimal engine temperature in all operating driving conditions and in severe climatic conditions. To solve this problem, it is necessary to make a compromise between the use of the incoming air flow during movement and the fan power. In both cases, a certain energy is spent. Usually both methods interfere into each other.

The easiest way to combine the ability of using the oncoming flow and the fan is a fan case, the sizes of which are smaller than the front surface of the radiator. Thus, the free zone of the radiator remains, which is used for the smooth oncoming air flow. In this case, if necessary, the fan is turned on and the air will pass through the radiator zone, closed with the case. This method has a major drawback, the circulation of air flow through the free zone of the radiator significantly decreases the efficiency of the cooling system at low speed, and this is the very motion mode during which the main source of creating air supply through the radiator is a fan. The compromise between the efficiency of using the oncoming flow and the fan is achieved by choosing the necessary proportions of the free zone and the sizes of the case and the power of the fan.

Air recirculation is due to the fact that air tends to take the path of least resistance. In case of incomplete coverage of the radiator with a case, air can go through the radiator in two ways. When the fan is operating, part of the air passes through the radiator area covered with the case, and the second part goes through the free zone of the radiator in the opposite direction. Depending on the relationship between the resistance of the engine compartment and the resistance of the radiator, the degree of circulation will be different.

If you somehow increase the resistance of the part of the flow, caused by circulation, the circulation will decrease and due to this, the flow of "cold" air through the air intake openings will increase. This fact will increase the efficiency of the cooling system. Radically, forced closing of the free zone of the radiator can be the way out, but this way is quite complicated, because this requires a special device and control mechanism. It is possible to increase the resistance for the passage of circulating air with the help of a partition installed in a pre-radiator chamber, located along the boundary line of the free zone of the radiator and the case.

To test this possibility, the numerical (CFD) experiment has been conducted. In the pre-radiator chamber a partition is installed, dividing the chamber into two parts. The partition divides the chamber throughout its height. On the front panel there is a horizontal air intake opening. The experiment has been carried out with partitions of 100, 150 and 200 mm deep. The partition of 200 mm deep blocked up the pre-radiator chamber completely, i.e. reached the front panel of the chamber. In the studies, the fan case occupied 72% of the surface of the radiator; this option is often used in practice.

The simulation was carried out with the use of a simplified 3D model of the engine compartment of a passenger car (Fig. 1). The air intake opening (0.8 × 0.07 m) was located on the front panel, the way of the air out of the engine compartment was through the opening in the lower panel behind the power unit (0.8 × 0.105 m). The geometric parameters of the model corresponded to the average characteristics of the car. The geometric parameters and aerodynamic characteristics of the radiator and electric fan (Fig. 2) corresponded to the medium car. The radiator size was 0.48 × 0.34 m. The electric fan model was represented by the PQ curve, without swirling the air flow, diameter 0.28 m. The fan casing in transverse projection was of 0.345 × 0.34 size. The static pressure of the environment was set at the inlet and outlet. The heat characteristics of the radiator were not simulated.

The model with the above boundary conditions is simulated using the following settings:

- turbulence is simulated, using the k-ε model;
- turbulence intensity at the inlet is 2%;
- oncoming air has the temperature of the environment;
- mesh is of 1.2 million cells.

The studies with the turned-on fan have shown that the partition reduces air circulation. Figure 3 shows a pressure distribution pattern in the pre-radiator chamber and in the engine compartment without a partition in front of the radiator (original option). As one can see, in the pre-radiator chamber, rarefaction of air is formed, and is distributed relatively evenly throughout the entire chamber. Under the influence of this rarefaction, the air from the under-bonnet space flows back into the pre-radiator chamber, causing circulation.

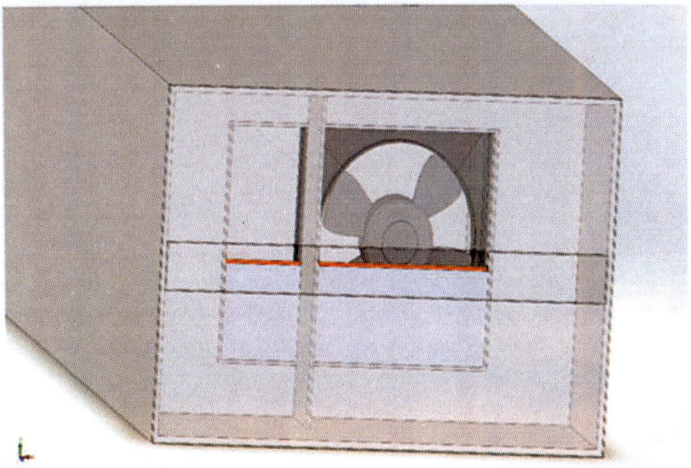

Fig. 1 Engine compartment layout, with a partition of 100 mm

Fig. 2 Pressure loss as a function of air speed for radiator (**a**), fan characteristic curve (**b**)

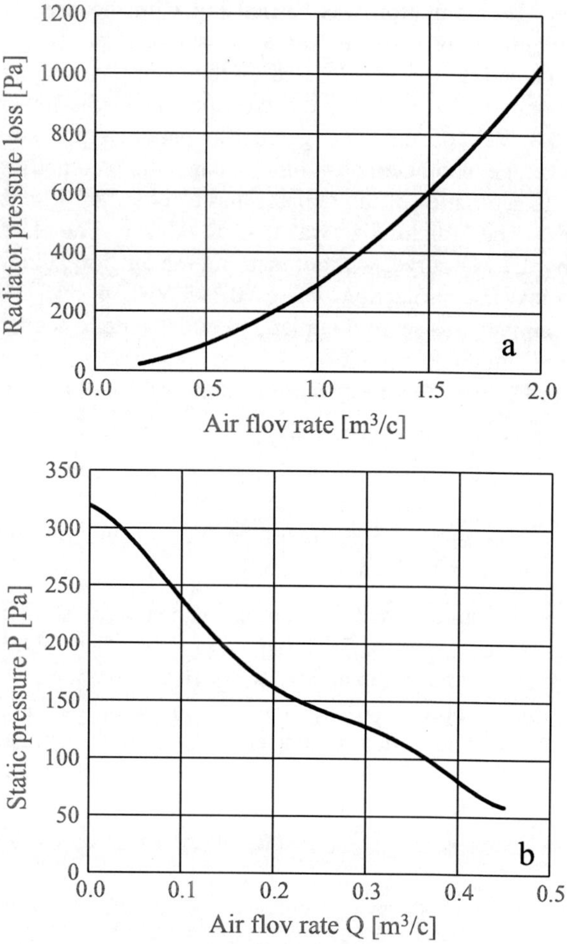

When installing the partition, the pressure distribution pattern changes significantly. Figure 4 shows the distribution of pressure when the partition of 100 mm deep has been installed in the pre-radiator chamber. As you can see, the rarefaction in the right side of the radiator has increased, but it has not spread to the left side, it has become smaller there. Because of it air circulation decreases by 7%, it decreases by 16% with the partition of 150 mm deep and by 24% with the partition of 200 mm deep. This is because the flow is forced to overcome the additional resistance, going around the partition. The degree of circulation is calculated by the amount of air, passing through the free zone from the engine compartment back to the pre-radiator chamber, to the amount of the flow, passing through the radiator from the pre-radiator chamber to the engine compartment.

Since the incomplete coverage of the radiator with the case is a compromise between using a fan and using oncoming flow to provide air through the radiator of

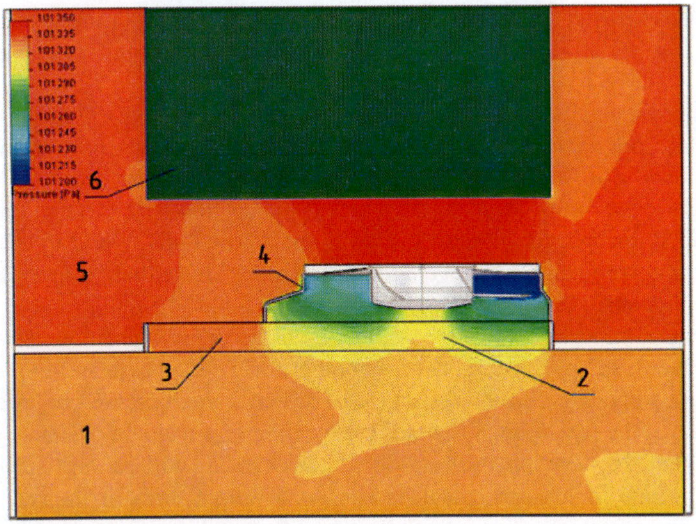

Fig. 3 Pressure distribution pattern without partition. 1—pre-radiator chamber, 2—radiator, 3—free zone, 4—fan case, 5—engine compartment, 6—engine

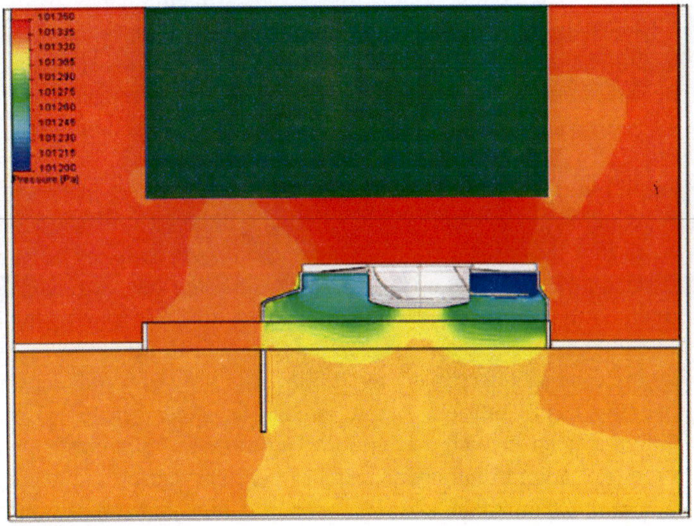

Fig. 4 Pressure distribution pattern with partition in pre-radiator chamber

the cooling system, it has been necessary to check the influence of the partition in the pre-radiator chamber on the efficiency of using incoming flow.

The next experiment has been different in the following way: instead of simulating the fan work, the air flow was created under the influence of the air the vehicle pushes aside. As these studies have shown, the partition in the pre-radiator chamber had practically no effect on the efficiency of using oncoming flow. The thing is that the resistance of the radiator zone covered with the case and the fan unit itself is approximately equal to the resistance of the free zone of the radiator. The air flow has not been redistributed between these zones and has been going in two parallel flows. For this reason, the presence of the partition practically has not changed the direction of air flow. As a result, the air flow rate through the radiator has remained unchanged, moreover, both in the whole radiator and in the zones of the radiator covered and non-covered with the case. This situation persisted at any depth of the partition.

In practice, another option is possible when the degree of the coverage of the radiator case is different, when there is another form of the case and fan, etc. Then the air flow, depending on the resistance of different sections, will be forced to go either towards the free zone or towards the location of the fan unit. That is why, to a certain extent the partition in the pre-radiator chamber will prevent such a redistribution, which in its turn can decrease air flow rate through the radiator, and therefore, the efficiency of using oncoming air flow will decrease. In this situation, it will be necessary to find the optimal width of the partition, which, on the one hand, would reduce air recirculation, and, on the other hand, would have a reasonable influence on the oncoming flow.

The resistance of the engine compartment affects the efficiency of the partition use. When the resistance of the compartment is greater than the resistance of the radiator, the partition reduces air circulation noticeably; otherwise, the effect will be weak. At the same time it doesn't matter which part of the compartment has a lot of resistance—the air intake openings or the rest of the engine compartment.

In order to verify this circumstance, a second series of the numerical experiments has been carried out. In this case, the resistance of the engine compartment has been increased by reducing the size of the exit opening by 50% (0.8×0.055 m). Thus, it has been found that the resistance of the engine compartment has become greater than the resistance of the radiator. At the same time, the degree of air circulation during the work of the fan has increased from 8.6% to 22%. Due to the increased resistance of the engine compartment, the air flow rate through the radiator zone, covered with the case, has decreased; in addition, the flow of "cold" air through the air intake opening has decreased. Active circulation is seen in Fig. 5, this is significantly different from the original option. Under these conditions, the partition, installed in the pre-radiator chamber, does not give any noticeable decrease in the degree of the circulation of air flow through the free zone of the radiator. No noticeable effect occurs even with a complete separation of the pre-radiator chamber, i.e. when the depth of the partition is equal to the depth of the chamber. The circulation remains almost the same as without any partition. In this case, the circulation flow goes around the partition through the air intake opening on the outside part of the pre-radiator chamber. Figure 6 shows the movement of air flow with the presence of the partition of 100 mm wide in front of the radiator.

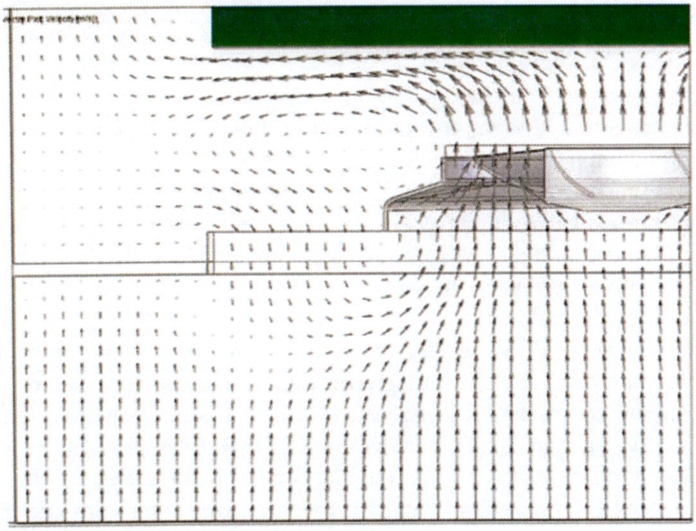

Fig. 5 Vectors of air movement without partition, with increase of resistance of engine compartment

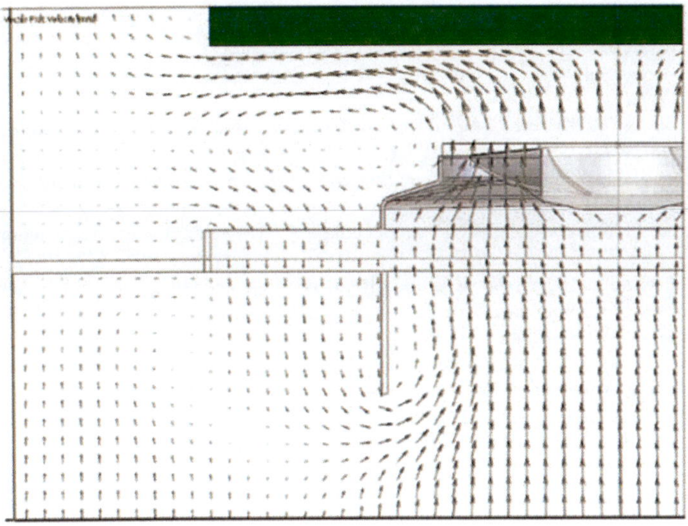

Fig. 6 Vectors of air movement with partition in pre-radiator chamber and great resistance of engine compartment

3 Conclusions

One of the easiest ways of rational combination of the possibility of air supply in the cooling system, due to the use of oncoming flow and the fan, is the use of the fan with an incomplete coverage of the radiator with the case. The disadvantage of this design is the low efficiency of the fan because of the air circulation. To a certain extent, this problem can be solved in a rather simple way, through the use of a partition in the pre-radiator chamber along the boundary of the case. But this method is effective only when the resistance of the engine compartment is less than the resistance of the radiator.

The air flow, formed by the fan, goes in two directions. Most of the air comes out of the engine compartment through the environment outlets, the rest of the air from the engine compartment goes back to the pre-radiator chamber through the free zone of the radiator. As it is known, the flow follows the path of least resistance. When the resistance of the engine compartment increases, the air flow through the outlets decreases, whereas the circulation through the free zone increases. This can happen because of the tight arrangement of the engine compartment or due to the insufficient size of the outlets. In this case, the efficiency of the partition in the pre-radiator chamber decreases.

References

1. Hucho W H (1998) Aerodynamics of road vehicles, 4th Edition SAE International. ISBN 0-7680–0029-7
2. Carr GW (1995) The influence of engine cooling airflow on car performance and stability. C496/079/95 491–498
3. Baeder D, Indinger T, Adams N, Unterlechner P (2012) aerodynamic investigation of vehicle cooling-drag. SAE Technical Paper 2012-01-0170. https://doi.org/10.4271/2012-01-0170
4. Jama H, Watkins S, Dixon C (2006) Reduced drag and adequate cooling for passenger vehicles using variable area front air intakes. SAE Technical Paper 2006-01-0342. https://doi.org/10.4271/2006-01-0342
5. Williams J (1985) An automotive front-end design approach for improved aerodynamics and cooling. SAE Technical Paper Series 850281, International Congress & Exposition, Detroit, Michigan. https://doi.org/10.4271/850281
6. Williams J (2003) Aerodynamic drag of engine-cooling airflow with external interference. SAE Technical Paper 2003-01-0996. https://doi.org/10.4271/2003-01-0996
7. D'Hondt M (2010) Theoretical, experimental and numerical study of the cooling airflow and its effects on the aerodynamics of road vehicles. Engineering Sciences. Universities Orleans
8. Baeder D, Indinger T, Adams NA, Unterlechner P, Wickern G (2011) Interference effects of cooling air-flows with external aerodynamics. Int J Autom Eng 2(4):115–121. https://doi.org/10.20485/jsaeijae.2.4_115
9. Barnard RH (2000) Theoretical and experimental investigation of the aerodynamic drag due to automotive cooling systems. In: Proceedings of the institute of mechanical engineers (IMechE). https://doi.org/10.1177/095440700021400810

10. Jama H, Watkins S, Dixon C, Ng E (2004) Airflow distribution through the radiator of a typical australian passenger car. School of Aerospace, Mechanical & Manufacturing Engineering. RMIT University, VIC, 3081 AUSTRALIA. In: 15 th Australasian fluid mechanics conference, The University of Sydney, Sydney, Australia 13–17 December
11. Wang Z, Tuszynski K, Tummescheit H, Alajbegovic A (2009) Integrated thermal management simulation: evaluating the effect of underhood recirculation flows on AC-system performance. In: Proceedings 7th Modelica conference, Como, Italy, 20–22 Sep. https://doi.org/10.3384/ecp09430098
12. Williams J, Oler J, Hackett J, Hammar L (1991) Water flow simulation of automotive underhood airflow phenomena. SAE 910307. https://doi.org/10.4271/910307
13. Williams J, Karanth D, Oler W (2002) Cooling inlet aerodynamic performance and system resistance. SAE 2002-01-0256. https://doi.org/10.4271/2002-01-0256
14. Alajbegovic A, Xu B, Konstantinov A, Amodeo J, Jansen W (2007) Simulation of cooling airflow under different driving conditions. SAE Technical Paper Series 2007-01-0766, World Congress, Detroit, Michigan. https://doi.org/10.4271/2007-01-0766
15. Kim J, Cho H, Kim T, Kim M, Kim Y, Kim YA (2015) Numerical study of the effect of small passenger car's grille shape on the aerodynamic performance. Trans Korean Soc Autom Eng 23:74–87. https://doi.org/10.7467/KSAE.2015.23.1.074
16. Simmonds N, Pitman J, Tsoutsanis P et al (2017) Complete body aerodynamic study of three vehicles. In: Proceedings of WCX17: SAE world congress experience, 4–6 April 2017, Detroit, MI, Paper number 2017-01-1529. https://doi.org/10.4271/2017-01-1529
17. Xu B, Leffert M, Belanger B (2013) Fuel Economy impact of grille opening and engine cooling fan power on a mid-size Sedan. SAE Technical paper 2013-01-0857. 2 https://doi.org/10.4271/2013-01-0857
18. Johan N (2017) CFD study of optimal under-hood flow for thermal management of electric vehicles. Master's thesis in Applied Mechanics. Göteborg, Sweden. https://hdl.handle.net/20.500.12380/251549
19. Khaled M, Ramadan M, El-Hage H et al (2014) Review of under-hood aerothermal management: towards vehicle simplified models. ApplTherm Eng 73(2014):842–858. https://doi.org/10.1016/j.applthermaleng.2014.08.037
20. Haglöf J (2019) Cooling air management for hybrid electric vehicles using combined 3D aerodynamic & thermodynamic CFD. Master Thesis. Lulea University of Technology

Research of Influence of Air Intake Openings and Pre-radiator Chamber Design on Airflow Distribution

Alexander Petrov

Abstract A large amount of cooling air is required to keep the cooling system of a passenger car engine running. The air intake openings and the pre-radiator chamber play a very important role in forming the flow of cooling air through the heat exchangers. In order to design these elements purposefully, it is necessary to know the mechanism of interaction of the air intake openings with the oncoming flow and the parameters affecting the organization of the required air flow through the air channel of the engine compartment of the cooling system. For this, design options are considered with a symmetrical and asymmetric position of the air intakes and pre-radiator chamber with different depths.

Keywords Cooling system · Air intake openings · Pre-radiator chamber · Cooling airflow · System efficiency · Airflow distribution · Mechanism of airflow formation · Dynamic pressure · Fan

1 Introduction

To keep the cooling system of the engine, air conditioner and other devices of a passenger car running, a sufficiently large amount of cooling air is required, but at the same time, excessive air intake must not be allowed, since this will lead to an increase in the aerodynamic drag of the car.

In passenger cars, the air inlet is carried out through the front panels of the body and the outlet is into the underbody space. This is the most rational way of organizing the air flow, because most of the air flow is due to the oncoming air flow. In this case, the air pressure is used, while slowdowning on the front surface of the hood and uncharging in the underbody space.

The air intake openings (radiator grille) have several functions: they are an important design element, often as an element of corporate identity, as a protective

A. Petrov (✉)
Kurgan State University, st. Sovetskaya, 63 bld. 4, Kurgan city, Russia

A. Bieliatynskyi and V. Breskich (eds.), *Safety in Aviation and Space Technologies*, Lecture Notes in Mechanical Engineering,
https://doi.org/10.1007/978-3-030-85057-9_33

element of the radiator, as an element that regulates the even distribution of cooling air over the radiator surface, and finally, as a "diaphragm" that restricts air intake.

There are many different air intake openings: openings in the form of a slot, grille openings with horizontal or vertical ribs, mesh openings, etc.

In the scientific literature, there is uncertainty about the role of air intakes and their influence on the organization of air flow through the heat exchangers of the engine cooling system and the air conditioning system of a passenger car. Various combinations and designs of air intake openings are considered, but the role and mechanism of air flow through them is not fully studied [1–5].

There have been many studies on the role of air intake openings in the formation of air flow passing through the heat exchangers of a cooling and air conditioning systems.

Olson [6] and Williams [7] note that the profile of the air flow velocity at the border of the radiator is greatly distorted and, as a result, the volume of oncoming air, passing through the radiators, decreases, as a result, the total area of the radiators is used ineffectively. To achieve optimal radiator efficiency, it has been stated that the distribution of air velocities through the radiators should be as equal as possible.

In the study [8], a numerical analysis was carried out for three-dimensional heat transfer and air flow of the vehicle cooling system. The flow field was analyzed in the space between the radiator grille and the radiator. The influence of the radiator grille angles was analyzed in the upper and lower air intakes, located in the bumper. The results show that, depending on the change in the angle in the grille of the upper and lower openings, the air flow rate varied over wide range in the direction of the decrease and increase from the initial condition. The optimal combination of the angles of the grille of the upper and lower openings was found, which made it possible to increase the air flow rate by 9.5%.

Most of the air comes in through the bumper inlet opening, which goes directly through the radiator, but some of the air covers longer before passing through the radiator. This part causes break-off and turbulence so that the air cannot flow smoothly. Only a part of the radiator surface is located opposite the intake openings, there are radiator sections located above or below, therefore, smooth or unsmooth flow onto these areas will depend on the angle of the grille. This leads either to an increase or to a decrease in air flow rate.

It was found that at certain combinations of the angles of the upper and lower grilles, the air flows will begin to interact with each other, which will lead to some "throttling"of flows. As a result, air consumption will be significantly reduced.

The radiator grill, installed at the front of the vehicle, is designed to allow air flow to come into the engine compartment [9]. It is an important device that affects the aerodynamic characteristics and cooling performance of a vehicle engine. Consequently, grilles and grille meshes are one of the design goals for improving vehicle performance. The purpose of this work was to analyze the resistance change of the grille and the study the air flow. Three variants of the shape of the intake openings were considered as well as three variants of the shape of the grille mesh.

Studies were performed with the use of numerical CFD method on a simplified model of a compact passenger vehicle with a simplified front to ensure simplicity of modeling.

Parts inside the engine compartment were removed, and the engine compartment was a rectangular box. The air, flowing inside, went out through a bulkhead near the front wheels of the vehicle underbody. All the air passed through the grille and the module of condenser, radiator and fan. The grille of the vehicle is divided into three areas: a grille, installed at the bottom under the bumper, and two grilles, installed above the bumper.

The sizes of the upper and middle openings, their width and height, varied, which led to changes in the area within the range from −44% to +44% from the initial condition. The location of the grille was also changed, its position varied according to height within the range from −20 to +20 mm from the center.

With the increase in the size of the middle opening, the air flow through it increases, but this causes stagnation of flows through the upper and lower grilles. At the same time, the total air flow through the module of condenser, radiator and fan increases.

Depending on the change in the position of the middle opening in height, the air flow through the upper and lower openings varies. When moving upwards, the flow rate through it decreases, while moving downwards, the flow rate increases. While moving the opening downwards, the total flow through the openings increases. These results demonstrate that there is interaction between flows.

As one can see from the results of work [9], the processes occurring in the front of the car are very complex and the efficiency of the cooling and air conditioning systems depends on a great many parameters, related to the front of the car and the design of the air intake openings.

Work [10] presents the analysis of the cooling system of the engine compartment of a car using CFD method. As a result, the flow field and temperature fields in the engine compartment were calculated. Some improvement schemes have been proposed, basing on the analysis results. Baffle plates were installed to change flow condition in the engine compartment, preventing turbulence and ensuring smooth flow.

During the simulation vehicle geometry was simplified, details irrelevant to the study were removed, and the research was focused on airflow in the engine compartment.

The research has shown that the flow in the compartment is complicated. Turbulence occurs between the intake grille and the condenser. In the standard design, turbulence is formed between the upper and lower grilles, and because of that the air flow is reduced. All these disadvantages affect the cooling capacity of the radiator group. To streamline the air flow in the pre-radiator chamber, baffle plates were installed opposite the upper and lower radiator grilles.

The task was to choose the best angle for the baffle plates. As a result, the optimal model was chosen. It has been found that the baffle plates can prevent air flow turbulence. These measures allowed the air speed in the radiator to be increased by 3.61% compared to the original model. To remove the turbulence between the upper and lower grille, a middle baffle plate was installed. Due to this, an additional effect was produced.

The results of the work show that it is possible to control the flow in the pre-radiator chamber by various means.

In work [11] it is noted that the distribution of air flow over the surface of heat exchangers is usually very uneven, which leads to an increase in the resistance of the heat exchanger. The uneven speed profile caused by the bumper, grille and front structure increases the pressure drop in the radiator, which reduces intake efficiency. Intake efficiency is influenced by the size and location of the intakes in relation to the heat exchangers.

The dynamic pressure recovery on the radiator depends on how much the radiator is exposed to air through the intake. The oncoming air flow affects the open part of the radiator, but the part of the heat exchanger opposite the bumper is shielded and there is no shock pressure there. Since the heat exchangers in the vehicle are located very close behind the grille and bumper, the air of the oncoming flow is not able to dissipate before it comes into contact with the radiator.

Bumper and grille marks on the radiator surface are very sensitive to geometry. The degree of intake flow distortion in a vehicle is related to the clearance between the bumper and the radiator and the flow balance between the intakes at the top and bottom of the bumper.

In real vehicles, the air flow distribution at the intake is never equal. When flow distortion occurs, performance at the intake begins to get worse. Unequal input flow is one of the external interferences, affecting the vehicle cooling system.

The value of the frontal projection of the grille and bumper onto the radiator surface, as well as the unequal flow, plays a major role in assessing the pressure recovery at the intake. The influence of unequal flow on pressure loss in the heat exchanger can be significant.

Work [12] researches the reasons for the loss of air energy when passing through the air intake openings for cooling in the front of the car. This is necessary to design more efficient vehicle cooling systems. The study was carried out in a wind tunnel with simplified car models in order to study the factors, affecting the energy loss in the cooling hole. An algorithm has been developed. It can predict the behavior of a combination of cooling holes accurately, with respect to the behavior of each of the holes working individually.

The energy supply to the cooling system usually comes from two independent sources: the fan and the oncoming air flow due to the forward movement of the vehicle. Accordingly, both external and internal flow conditions must be considered to optimize the cooling system. Moreover, the fields of the external and internal flows interact with each other to some extent. The interfaces between the two flow fields are the air intakes and the lower part of the engine compartment. The change in energy due to the loss of total pressure in the openings of the vehicle front is the first line of energy change in the cooling air circuit. Other energy changes are due to the effects of the radiator, condenser, fan, fan case and blocking of motor.

The main purpose of this study was to determine the relationship between pressure loss at the intake openings and cooling air flow rate, vehicle speed, as well as the size and location of openings. To identify these relationships more clearly, the experimental model included only the front of the vehicle. This approach was chosen to exclude the influence of any irrelevant unique aerodynamic elements, associated with other geometric features of the real car. Experiments were carried out in an open test section of a wind tunnel.

As a result of the research, correlation dependences were obtained. They determine the relationship between the geometric parameters of the air intake openings with their aerodynamic efficiency. The effectiveness of a combination of cooling openings can be accurately predicted by correlating the separate openings of that combination. Several factors, influencing efficiency, have been identified. For example, the recovery of the ram-air flow for the bottom opening can be improved by increasing the spoiler height, decreasing the ground clearance and decreasing the distance of the spoiler from the front edge of the vehicle.

The analysis of the presented materials shows that the air intake openings and the pre-radiator chamber play a very important role in the formation of the cooling air flow through the heat exchangers. In order to design these elements purposefully, it is necessary to know the mechanism of interaction of the air intake openings with the oncoming flow and the parameters, influencing the organization of the required air flow through the air channel of the engine compartment of the cooling system. The main difficulty in optimizing vehicle cooling systems is the inability to predict the cooling performance of a new vehicle accurately before a production prototype is built and tested. That is why at this stage of the development changes in the cooling system can only be made at great cost.

Work [13] researches investigates optimal flow under the hood for temperature control of electric vehicles. To improve fuel economy of conventional vehicles and the driving range of electric vehicles, reducing the aerodynamic drag is of particular interest. As electric vehicles typically have lower cooling power requirement, reducing the aerodynamic drag induced by cooling air is of utter importance. The results show that a similarly low cooling drag can be obtained from positioning heat exchangers both in series and in parallel. When heat exchangers are positioned in parallel, the results indicate high mass flow rate and low drag potential for side positioned inlets including air outlets located at the front arc of the wheel houses. Furthermore, results indicate that low positioned air inlets are more drag efficient than those positioned higher on the front bumper. The results of this study should serve as an suggestion towards under-hood cooling flow solutions for battery electric vehicles.

In work [14] it is noted that the while designing air inlets, compromises between the aerodynamic and thermal requirements must take place so that components are soundly functioning. For example, an efficient air inlet favors the underhood thermal environment (improves the convective heat transfers) but at the same time it increases the aerodynamic drag due to cooling systems. Furthermore, it is worthy to be cognizant that Automobiles whose engines placed in front have two types of intakes, namely: high grille air intake, and low grille air intake. The design of air intakes is based on the aforementioned types and can also be classified as:—Two air intakes design: Air intake of high grille type (Superior); and Air intake of low grille type (Inferior)—One air intake design: Air intake of high grille or low grille type From the aerodynamic point of view, the air inlet has a heavy impact on the aerodynamic tensor of the vehicle, particularly on the cooling air drag in frontal or lateral wind conditions. Indeed, the cooling air drag is essentially a function of the underhood geometry, the design and layout of the air inlet and outlet, pressures at the inlet and outlet, and the direction and speed of the penetrating air. The study is determined to examine explicitly the components architectural arrangements' of underhood compartments and present simplified models of the compartments so that they can serve as aerothermal baseline models in the early stages of design. Hence, the different components in the underhood compartment are classified with respect to aerothermal orders of impact. Followed, two geometrical models, simple and advanced, are developed based on the impact orders. These models serve for future aerothermal studies (analytical, numerical and experimental). Finally, a simplified vehicle model is designed and implemented as underhood compartment model. The structurally simplified model is already utilized in experiments and can be used afterward in both numerical and experimental analyses.

In work [15] it is noted that the cooling air drag can be around 5–15% of the total drag, and a lot of research has been done over the years on how to decrease it. Cooling drag is defined as the force acting in x-direction from the cooling air flowing through the engine bay. The air is let in through the grilles to cool down the engine and escapes through different outtakes usually below the vehicle. Been done on which design parameters that affect the cooling air drag and how air ducts could be designed to better transport the flow to and from the radiators without loosing to much in energy. The first study showed that air through the wheelhouses increases the drag as well as having air entering only in the upper grille and travelling down through floor and wheelhouse. In the second study, the area between the grilles and cooling package was sealed and inlet ducts were created to control the flow from the grilles to the cooling package. When just adding inlet ducts, the mass flow through the grilles was decreased, but the mass flow through the cooling package was increased due to less separation of the air, which lead to a drag reduction of 0.2%. An air duct was also designed to lead the flow after the cooling package fan to the outlets. The ducts purpose was to prevent the air from loosing in energy when rising to the roof of the engine bay, the duct compresses the air and leads it efficiently over the engine with a drag force decrease on 0.3%. In conclusion there is much that can be done to improve the cooling air drag. It is most favorable to have a sealed volume with inlet air ducts before the cooling package, have outtakes aligned with the free stream flow, minimize cooling air to escape through the wheelhouse outlets and to minimize the height of the engine bay as much as possible.

The author concludes that in order to design a vehicle to have improved cooling air flow would recommended.

- Design a sealed air guide volume in between grilles and cooling package
- Design inlet air ducts that controls the flow to the cooling package and avoids lost energy through separation

- Decrease the engine bay height as much as possible or put in an after fan duct that leads the air above the engine and down to the floor outlet

It has been proven in this study that designing an inlet air duct to control the flow to the cooling package is most useful. The air duct allows an increase of air through cooling package which allows a decrease of the size of the upper grille. For future models, smaller upper grilles could be designed that are lower in height with similar width as today. Designing an air duct could possibly be done even better if the grilles are better aligned with cooling package erad and brad so that the ducts could expand more exponentially. This would most probably allow some further decrease of the drag.

For hybrid vehicles there is still a lot of parts needed in an engine bay and also a higher cooling need for the engine compared to a Battery Electric Vehicle (BEV). For a BEV there would be a lesser need of cooling air and it would therefore probably be possible to decrease the size of the inlet grilles. For a battery electrical vehicle with lesser cooling air need the best steps to decrease drag. Lower the height of the cooling package, possibly put it in serial placement, which allows the vehicle to have less grille height and total area.

2 Air Flow Distribution Mechanism

The distribution of air flow along the front surface of the radiator depends on the size, design and location of the air intake openings. The geometrical sizes and shape of the pre-radiator chamber also play an important role in the formation of the air flow. The influential parameters of the pre-radiator chamber in this case are: its depth, the location of the enclosing walls along the length and height in relation to the sizes of the radiator, and its shape. The combination of various parameters of the air intake openings and the parameters of the pre-radiator chamber in each specific case gives an individual result that affects the air flow rate and its distribution along the front of the radiator. The air intake openings and the pre-radiator chamber affect the air flow through the radiator, in both cases: when using the oncoming flow and when the fan is running.

It is difficult to study the physical nature of the formation and distribution of the air flow in the pre-radiator chamber and along the front of the radiator using a natural experiment. The problem is that it is difficult to visualize and measure the local air speeds on the radiator. Measuring the speed in front of the radiator is difficult due to the fact that the vectors of the air flow in front of the radiator are directed at different angles with the front surface of the radiator. Even the use of microsensors is difficult under these conditions. Numerical experiment is free from these shortcomings.

The studies were carried out, using CFD software module. Two models of the air channel of the engine compartment were used in the calculations. The first model had a horizontal air intake of 70 mm high on the front panel, located symmetrically in relation to the radiator, and corresponding to the sizes of the chamber in width. The second model differed only in the fact that a similar opening was shifted downward, and its lower edge was located at the level of the lower edge of the radiator. In addition, models with different depths of the pre-radiator chamber were used. The passage of air through the air channel, causing by the action of a fan and the action of an oncoming flow, was studied. Local air speeds along the front of the radiator were determined at the outlet of the radiator. The experiment was carried out using a fan system with 72% coverage by the radiator case, that is, the fan case was smaller than the width of the radiator. Thus, the part of the radiator had a free zone.

In order to determine the mechanism of the air flow formation through the heat exchangers, it was necessary to exclude the mutual influence of the flows. Therefore, in the experiments, one air intake opening was carried into effect. In the first case, we considered an opening, which was symmetrical in relation to the radiator. It provides the most favorable conditions for the air flow. In the second case, on the contrary, the opening was located in the lower zone with the most unfavorable conditions. There was no influence of the roadbed. The geometric sizes of the car engine were symmetrical in relation to the engine compartment. The boundary conditions in all the calculations were consistent.

2.1 Use of Oncoming Air Flow

Figure 1 shows the movement of the air through the chamber of the engine compartment under the effect of the oncoming flow. Since the air intake is located symmetrically, the air flow in the vertical plane is distributed symmetrically over the radiator surface. Since the cross-section of the pre-radiator chamber is larger than the cross-section of the radiator, the air flow experiences turbulence when it suddenly expands from the air intake. Expansion and depression of the flow in the pre-radiator chamber inevitably leads to the loss of pressure in this place.

The air flow pattern with an asymmetric arrangement of the air intake opening changes significantly (Fig. 2). As in the first case, a turbulent air flow is observed in the pre-radiator chamber, in addition, the flow undergoes a significant distortion, as a result of which the resistance of the pre-radiator chamber increases noticeably in comparison with the symmetric flow. As a result, the air flow through the radiator is reduced by 6%. The inequality of the velocity field along the front of the radiator increases by 13%.

The figure shows clearly that the flow at the inlet to the radiator is directed at an acute angle, and in some areas it goes almost along the surface of the radiator, that is why it is almost impossible to measure the real air speed in front of the radiator accurately during a natural experiment.

Figure 3 shows the profiles of air speeds in the vertical plane for the two variants of the location of the air intake openings. In Fig. 3a there is a speed profile in the zone of the radiator covered by the case (graph 1) with the use of a symmetrical

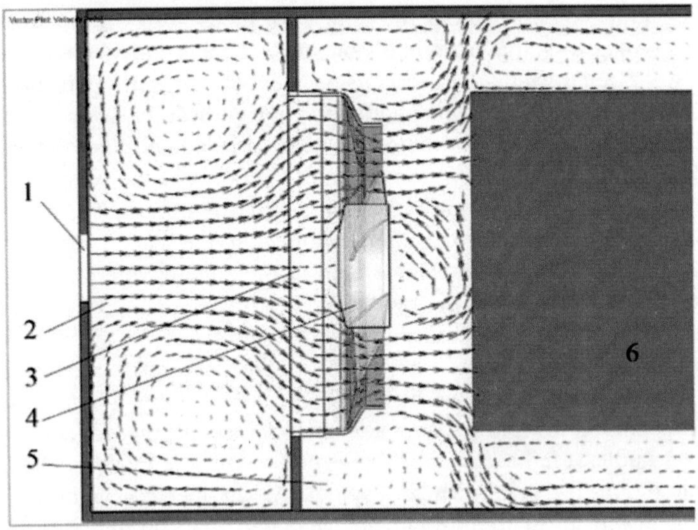

Fig. 1 Vectors of air movement under the effect of the oncoming flow, with a symmetrical arrangement of the air intake opening (view in the vertical longitudinal plane): 1—air intake opening; 2—pre-radiator chamber; 3—radiator; 4—fan system; 5—engine compartment; 6—engine

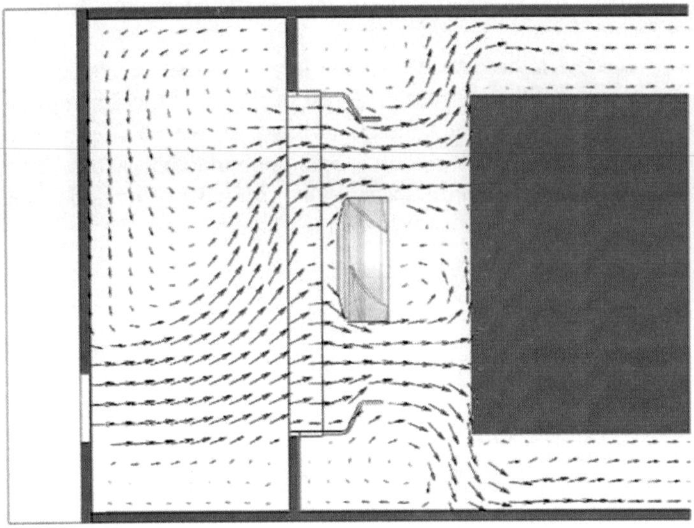

Fig. 2 Vectors of air movement under the effect of the oncoming flow, with an asymmetric arrangement of the air intake opening (view in the vertical plane, chamber depth 200 mm)

opening. Contrary to expectations, the impact of the air flow in the zone opposite the opening (zones 8–9), has a "failure", but this is understandable, since its effect is levelled by the fan system, also located on the flow way. It "deforms" the flow greatly, the fan center is just in the middle part, it significantly slows down the flow. Around the center, the flow, on the contrary, is concentrated, and the speed is the highest there. In the free zone of the radiator, the effect of the air intake is clearly visible, in the middle part of the radiator the speed is noticeably higher (Fig. 3b, graph 2). In the case of using an asymmetric opening (the second variant), the effect of the air intake opening is noticeable both in the zone covered by the case, and even more in the free zone of the radiator.

This experiment shows that even against the background of the influence of the fan system on the distribution of air speed along the front of the radiator, the role of air intake openings remains quite high.

As the depth of the pre-radiator chamber decreases, the situation gets even worse. The studies of the effect of the depth of the pre-radiator chamber were carried out for an air path with an asymmetric air intake opening. In the first experiment, the depth of the pre-radiator chamber was 200 mm (Fig. 2). With a decrease in the depth of the chamber to 150 mm, the air flow remained practically the same, only the inequality of the speed field increased by 8%. When a chamber depth was 100 mm, the air flow decreased by 3%, the inequality of the speed field increased by 7% compared to a chamber 200 mm deep (Fig. 4).

Under the influence of dynamic pressure, most of the air flow goes directly through the radiator area, located opposite the air intake opening, this can be seen in Figs. 3 and 5 (sections 1–5 in the lower part of the radiator). The highest air speed is observed precisely in these sections, both located in the free zone of the radiator and in the area covered by the case. The rest of the air, experiencing a sharp turn, passes along the surface of the radiator, in this case the air flow is drawn into a complex turbulence that occurs in the vertical and horizontal planes. Air turbulence is observed throughout the entire space of the pre-radiator chamber, between the

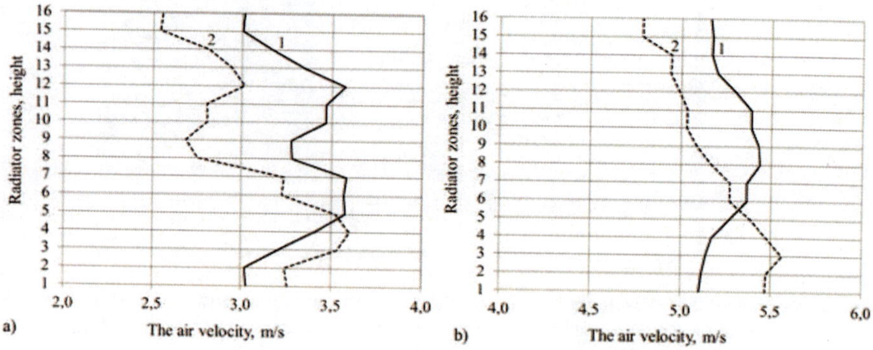

Fig. 3 Average value of the speed of air, passing through the radiator in the vertical plane: **a** in the area of the radiator, covered by the case; **b** in the free zone of the radiator; 1—with a symmetrical position of the air intake; 2—with an asymmetrical opening pattern

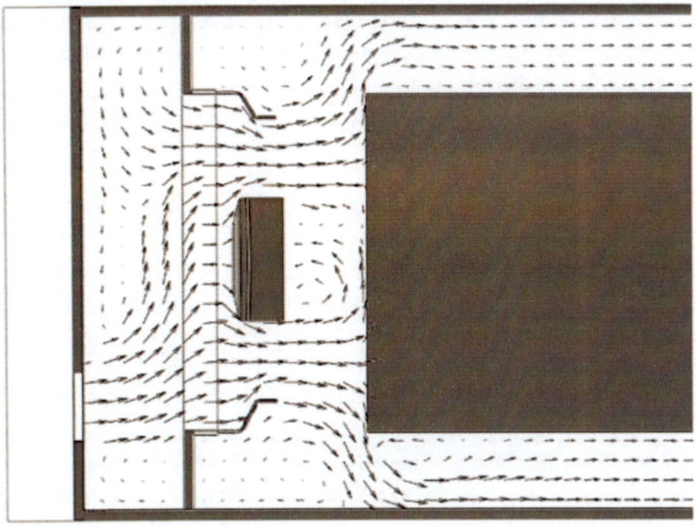

Fig. 4 Vectors of air movement under the effect of the oncoming flow, with an asymmetric arrangement of the air intake opening (chamber depth 100 mm)

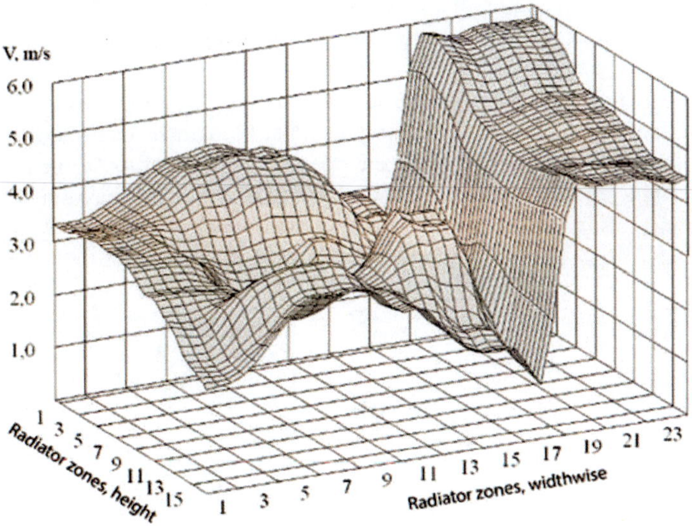

Fig. 5 Epure of the field of air speeds along the front of the radiator under the effect of the oncoming flow, with an asymmetric arrangement of the air intake opening (chamber depth 100 mm)

front panel and the radiator and the side walls of the chamber. The flow gets into the rest of the radiator sections along a complex trajectory, redistributing both along the height and along the width of the chamber.

2.2 Radiator Use

The role of the openings and the pre-radiator chamber was also considered while using a fan. It should be noted that a fan system with an incomplete coverage by the radiator case has such a negative feature when the fan is running. This negative phenomenon is observed in the circulation of "hot" air through the free zone of the radiator from the engine compartment to the pre-radiator chamber. Thus, "cold" air comes to the radiator through the intake openings, and part of the air gets in from the engine compartment. This reduces the efficiency of the cooling system.

When generating an air flow with a fan, the influence of the location of the air intake and the depth of the pre-radiator chamber is in less evidence. The flow of "cold air" through the air intake opening with its asymmetrical arrangement compared to a symmetrical arrangement is reduced by only 1%. Air circulation with an asymmetric opening increases by 3%, the inequality of the speed field increases by 5%. Figure 6 shows the pattern of the air flow through the pre-radiator chamber and the distribution of the air flow over the radiator surface with a symmetrical and asymmetric arrangement of the air intake openings. In both cases, the air stream, formed by the air intake openings, is visible. Air flow also hits the surface of the radiator and then spreads out. In this case, there are turbulent in the pre-radiator chamber.

With a decrease in the depth of the pre-radiator chamber from 200 to 100 mm, the flow of cold air through the air intake opening decreases by 6%, air recirculation from the engine room increases by 7%, the inequality of the speed field increases by 14%.

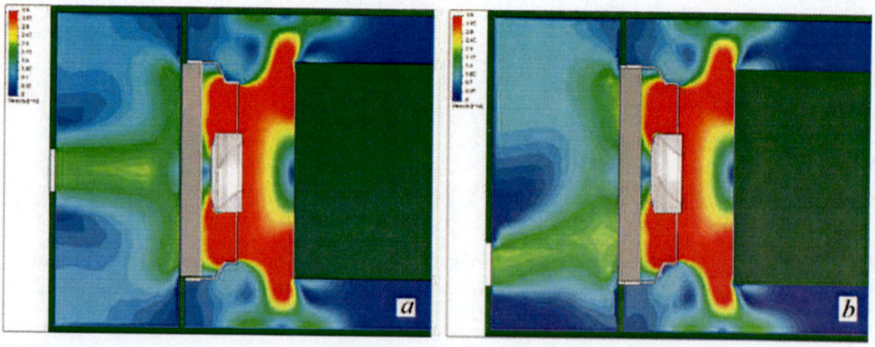

Fig. 6 Pattern of the distribution of the speeds of the air flow, generated by the fan, with a symmetrical arrangement of the air intake opening: **a** symmetrical arrangement of the opening; **b** asymmetrical opening arrangement

3 Conclusions

The carried-out studies show the patterns of air flow and its distribution over the radiator surface. This article presents only a part of the materials, but even they can lead to several conclusions. As it has already been mentioned, the air intake openings and the pre-radiator chamber are of great importance for the supply of cooling air.

If to consider the oncoming air flow, the main parameter that must be controlled is the inequality of the air flow distribution over the radiator surface. In this situation, an increase in air flow is not a goal in itself. Additional air supply is regulated by additional air intake openings and their area of passage. Cooling efficiency must be achieved by more even air distribution, with minimum air flow. Otherwise, efficiency will be provided by increasing the aerodynamic drag of the car when additional air passes through the interior space.

If to consider the use of the fan, the situation with the air intake openings is somewhat different. In this case, not only the distribution of the air flow is important, but also the resistance of the air intake openings and the pre-radiator chamber. Cooling efficiency will be related to the amount of air drawn in and the energy required to drive the fan.

In these studies, the number of factors influencing the formation of the air flow is minimized. If the resistance of the radiator is changed even with the same geometric parameters of the system, the figure of the formation of the air flow will change significantly. This suggests that studies of the passage of the air flow through the air channel of the cooling system should be carried out with a limited number of variable parameters.

References

1. Williams J (2003) Aerodynamic drag of engine-cooling airflow with external interference. SAE Tech Paper 01:0996. https://doi.org/10.4271/2003-01-0996
2. Jama H, Watkins S, Dixon C, Ng E (2004) Airflow distribution through the radiator of a typical australian passenger car. In: 15th Australasian fluid mechanics conference, The University of Sydney, Sydney, Australia, 13–17 December 2004
3. D'Hondt M, Gilliéron P (2010) Aerodynamic drag and flow rate through engine compartments of motor vehicles. In: 28th AIAA applied aerodynamics conference 28 June–1 July 2010, Chicago, Illinois
4. Şener Y, Koray E, Umut C (2014) Önden motorlu bir araçta aktif gril kapatma sistemi geliştirilmesi. Otekon'14. 7. Otomotiv Teknolojileri Kongresi 26–27 Mayıs 2014, BURSA
5. D'Hondt M (2010) Etude theorique, experimentale et numerique de l'e coulement de refroidissement et de ses effets sur l'aerodynamique automobile. Engineering Sciences. Universite d'Orleans, 2010. French
6. Olson ME (1976) Aerodynamic effects of front end design on automobile engine cooling systems, SAE Technical Paper 760188
7. Williams J (1985) An automotive front-end design approach for improved aerodynamics and cooling. SAE Technical Paper 850281

8. Hsieh C-T, Jang J-Y (2007) 3-D thermal-hydraulic analysis for airflow over a radiator and engine room. Int J Automot Technol 8:659–666
9. Kim J, Cho H, Kim T, Kim M, Kim Y, Kim Y (2015) A numerical study of the effect of small passenger car's grille shape on the aerodynamic performance. Trans Korean Soc Automot Eng 23:74–87. https://doi.org/10.7467/KSAE.2015.23.1.074
10. Wei Y, Wu Sh-N, Wang W, Yang Y, Lu L (2016) Automobile Engine compartment parameters optimization based on CFD simulation. In: Proceedings of the 3rd international conference on material engineering and application (ICMEA 2016). https://doi.org/10.2991/icmea-16.2016.15
11. Williams J, Karanth D, Oler W (2002) Cooling inlet aerodynamic performance and system resistance. SAE Tech Paper 01:0256. https://doi.org/10.4271/2002-01-0256
12. Roseberry C (2011) Automotive cooling airflow correlations. http://hdl.handle.net/2346/8741
13. Johan N (2017) CFD study of optimal under-hood flow for thermal management of electric vehicles. In: Master's thesis in applied mechanics. Chalmers University of Technology
14. Khaled M, Ramadan M, El-Hage H et al (2014) Review of under-hood aerothermal management: Towards vehicle simplified models. ApplTherm Eng 73:842–858. https://doi.org/10.1016/j.applthermaleng.2014.08.037
15. Haglöf J (2019) Cooling air management for hybrid electric vehicles using combined 3D aerodynamic & thermodynamic CFD. Master Thesis. Lulea University of Technology

Metal Removal and Surface Roughness Formation in Robotic Finishing

Aleksei Aleksandrov, Michail Vartanov, and Dmitry Mironov

Abstract This article deals with the mechanisms of metal removal and surface roughness formation during robotic finishing. To analyze the process mechanism, a single abrasive body with a single abrasive grain protruding during its interaction with the machined surface was considered. The destruction of a solid body surface by a single abrasive grain depends on the impact nature of the latter on the machined surface, i.e., on the resulting pattern of destructive stresses. Embodiment zones of various mechanisms of metal removal are considered. In experimental research influence of such factors as the spindle speed (n), the feed rate (S), the cutting depth (t) and the abrasive grit size (z) on the axial components of the cutting force and the torque along the X axis was estimated in a full-factor experiment. As a result, when the spindle speed increases, elastic movements and oscillations arise in the technological system, which conditionally increase the different height arrangement of abrasive grains on the working surface of the tool and lead to an increase in the roughness of the machined surface.

Keywords Robotic finishing · Metal removal · Micro-cutting · Cutting force

1 Introduction

Finishing is one of the most promising areas of robotized production [1, 2]. Using industrial robots with six degrees of freedom and articulated design for finishing makes it possible to reach any surface of the part. This can be relevant, for example, in the machining of aluminum parts for aircraft components, which have both complex shapes and surfaces that are difficult to access and have high quality requirements.

A. Aleksandrov (✉) · M. Vartanov · D. Mironov
Moscow Polytechnic University, Moscow, Russia

A. Bieliatynskyi and V. Breskich (eds.), *Safety in Aviation and Space Technologies*, Lecture Notes in Mechanical Engineering,
https://doi.org/10.1007/978-3-030-85057-9_34

The main mechanisms of metal removal and surface formation during finishing are micro-cutting and abrasion [3]. The shearing process occurs when abrasive grains collide directly with the surface of the machined workpiece. This is possible during machining with free abrasives, but it is practically impossible when machining with abrasive tools.

2 State of the Matter

The micro-cutting process takes place in sufficiently favorable conditions for chip formation, i.e., at certain cutting force ratios. In this case it is possible to remove chips as a result of element shear. The analysis of grinding waste carried out by E. N. Maslov confirms the presence of chips with clearly visible shear lines in them. Such chips resemble (with sufficient magnification) the chips removed during very fine high-speed milling.

The process of abrasion occurs at the mutual movement of parts and abrasive grains at a small angle, i.e. at parallel movement. The difference between abrasion mechanism and micro-cutting was studied in detail by Maslov [4], who conducted experiments on preheated polished plates of steel 40 to obtain a blue annealing color. The experiments were carried out on the upgraded Martens-type device. A diamond cone with an apex angle of 90° and rounding radius apex of 6 microns was used as a scratching tool. This device most accurately simulated the action of an abrasive grain also having an average angle of sharpening close to 90°. The cone was put under a certain load, and the modernization of the device allowed the horizontal force of cutting—scratching to be measured. When tested by the scratching method, the actual negative front angles have maximum values in the area of especially thin skimmed layers (several micrometers or less), thereby cutting is impossible and metal bearing occurs. In this case, the result is not a scratch, but a flattened groove, and the destruction characteristic of scratching is absent. The metal is compressed without the chips being removed, so the metal undergoes large deformation in the vertical direction, and the bottom of the groove retains its blue color. Repeated passes of the diamond cone along the groove bottom lead to cold-hardening and destruction of the metal surface layers. The number of passes depends on the physical and mechanical properties of the material machined. In that case the wear products had the form of the thinnest flaky particles. Thus, destruction from abrasion is not a cutting process.

Despite the fact that the intensity of metal removal from abrasion is very small and is not decisive in machining, this process allows to achieve small values of surface roughness, as it leads to smoothing of irregularities and metal flowing from protrusions into hollows. This changes the shape and direction of the irregularities and allows to reduce their size to limiting values [5]. Traces of strong plastic flow of metal on the surface of the sample can be seen, for example, in Fig. 1.

The largest volume of metal removal occurs in micro-cutting, so the main mechanism of metal removal in abrasive finishing is the process of cutting with the

Fig. 1 Traces of strong plastic flow of metal on the surface of the sample (magnification × 1300) [6]

formation of the finest chips. Both mechanisms of metal removal act simultaneously, but depending on the conditions of external forces one particular mechanism of metal removal may prevail or completely dominate.

3 Theoretical Research

To analyze the process mechanism, consider a single abrasive body with a single abrasive grain protruding during its interaction with the machined surface. The destruction of a solid body surface by a single abrasive grain depends on the impact nature of the latter on the machined surface, i.e., on the resulting pattern of destructive stresses.

To create large tangential stresses, ensuring metal fracture from shearing, normal stresses of a certain value are necessary, i.e. the presence of normal forces, providing pressing of the cutting tool to the processed surface. Thus, the magnitude and ratio of normal and tangential stresses arising during the action of a single abrasive grain on the machined surface, entirely determines the nature of fracture and, consequently, the technological productivity of the process and surface microrelief.

According to the studies of cutting conditions during grinding [4], the impact of the elementary contacting protrusion of the abrasive grain on a sufficiently plastic material when changing the normal component of the force acting on the machined surface, sharply changes the nature of deformation of the material. Sliding of the abrasive grain on the surface of the material occurs at a small value of the normal component of the force.

If the normal component of the force P_y increases with increasing thickness of the skimmed layer a_z, the conventional stress changes suddenly:

$$p_y = \frac{P_y}{f_c},$$

where f_c—cross-sectional area of the deformed layer.

This change is characterized by the ABC curve (Fig. 2), on which two essentially different sections: AB and BC can be defined. These sections indicate the following regularities of the contacting element impact on the deformed metal.

At the AB section (in zone I), the plastic deformation of the material by the contacting protrusion takes place. In this case, the increase in the degree of material plastic deformation with an increase in the thickness of the deformed layer points out a corresponding increase in the conventional stress. In this zone, the abrasion mechanism dominates.

At the BC section (in zone II) micro-cutting occurs with shearing of metal elements. This section is characterized by the decrease of engineering stress when the load (force P_y) increases to a certain value as a result of the deformed element cutting.

The process of micro-cutting will be carried out at a certain ratio of forces acting on a single abrasive grain, while the process of abrasion occurs in the presence of any forces acting on a single grain. The ratio of these forces will determine only the intensity of metal removal.

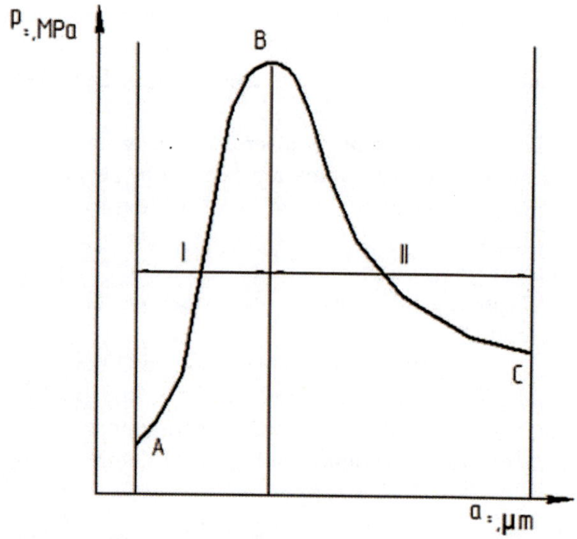

Fig. 2 Change of conventional stress when the load on the contacting element increases [4]

To construct a system of forces acting on a single contacting element, it is advisable to introduce some constraints:

- exclusion of the temperature-velocity factor influence, as the thickness of the removed layer does not exceed a few micrometers;
- the hardness of the abrasive grain is much greater than the hardness of the processed materials;
- the machined material is sufficiently plastic and fractures from the maximum tangential stresses;
- the frictional forces are proportional to the normal load.

The apex of a single protrusion of an abrasive grain is characterized by the radius ρ, although in real conditions the abrasive grain may have a rough surface with numerous micro-cuts and even with worn places.

The external force P with the longitudinal component P_z, which cuts the chip and the normal component P_y, which presses the abrasive grain to the machined surface, acts at an arbitrary angle of encounter λ on the abrasive pellet, interacting with the surface through a single grain (Fig. 3). As a result, the forward motion is performed.

Elementary reactive normal forces and elementary reactive frictional forces act on the front surface of the abrasive grain, the effect of which can be completely replaced by the total reactive normal and frictional forces, since the thickness of the removed layer does not exceed several micrometers and the forces are applied on a very small contact area.

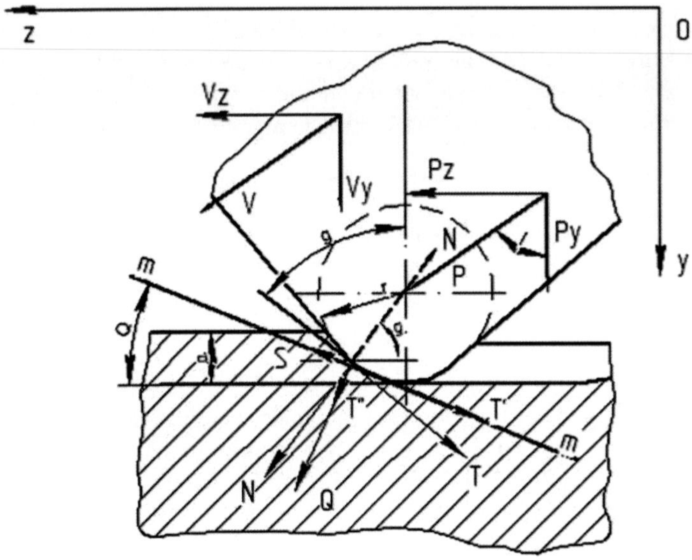

Fig. 3 System of forces acting on a single contacting element

The ratio of forces P_y and P_z at the adopted scheme of interaction of the abrasive grain with the machined surface is characterized by the ratio

$$\frac{P_y}{P_z} = \frac{\sin\gamma + \mu \cdot \cos\gamma}{\cos\gamma + \mu \cdot \sin\gamma}; \tag{1}$$

The value of the dynamic friction coefficient μ for steel, studied in a wide range by Maslov [4], remains constant on average and can be equal to 0.5, which is significantly greater than the value of the coefficient in friction-sliding.

The formula (1) shows that the ratio P_y/P_z increases rapidly with increasing the value of the actual negative rake angle γ and at a value of 60° tends to infinity. This practically means that there is no micro-cutting at large values of the negative rake angle. Since the value of the actual front angle at the apex radius of the protruding grain ρ depends on the thickness of the removed layer, micro-cutting, all other conditions being equal, can occur when the thickness of the removed layer is not less than a certain thickness a_{min}.

When the P_y/P_z ratio does not provide the micro-cutting process, the metal is removed by abrasion. And, finally, when the force P_y prevails significantly over the force P_z, the hardening of the processed material surface due to plastic deformation of surface layers with no metal removal or shear, when processing brittle materials.

Thus, the minimum value of the actual cutting angle at which abrasion is the prevailing mechanism of metal removal can be deduced from formula (1)

$$\begin{aligned} &\cos\gamma - \mu \cdot \sin\gamma = 0; \\ &\mu = \cot\gamma; \end{aligned} \tag{2}$$

If μ = 0.5 for steel, $\gamma_{min} = 63°25'$ is determined by formula (2).

Taking into account that the thickness of the skimmed layer "a" is only a few micrometers, the normal and tangential components of the force are assumed to be applied to half the thickness of the layer. Then from the scheme (Fig. 3) we can write:

$$\gamma \arcsin\frac{\rho - \frac{a_{max}}{2}}{\rho}_{min} \tag{3}$$

where

$$a_{max} = 0,2 \cdot \rho$$

The possible values of the actual cutting angle γ are determined by formula (3) and depend on the depth of cutting and the rounding radius of the abrasive grain tips ρ.

The cross-sectional area of a scratch can be defined as the area of a circular segment of radius ρ, height a and central angle α:

$$\alpha = 90 - \gamma.$$

Then:

$$f_c = \frac{\rho^2}{2} \cdot \left(\frac{\pi}{90^\circ} \cdot \arccos\frac{\rho - a}{\rho} - \sin 2\arccos\frac{\rho - a}{\rho}\right); \quad (4)$$

Taking the scratch profile as an isosceles triangle with the angle at the apex equal to 2γ, formula (4) can be simplified:

$$f_c = a^2 \cdot \tan\gamma = a \cdot \rho \cdot \cos\gamma, \quad (5)$$

by substituting the expression (3), expression (5) will be the following:

$$f_c = a \cdot \rho \cdot \cos\gamma = a \cdot \rho \cdot \cos\arcsin\frac{\rho - \frac{a}{2}}{\rho}. \quad (6)$$

The formula (6) shows that with equal values of the cross-sectional area of a single scratch, the depth of cut will be the greater, the sharper the grain is, that is, the smaller the value of ρ. In this case, the depth of cut is limited by the value of the protruding grain and, if using coarse-grained fillers, the surface roughness will deteriorate in proportion to the value of the grain.

The value of the actual cutting angle γ is entirely determined by the properties of the abrasive, and abrasion becomes the dominant mechanism of metal removal with significant blunting of the grain tips.

The effective zone of force application P, which ensures the predominance of the abrasion mechanism, is defined by a spatial cone with the angle between the formative and the normal in the interval:

$$\pi/4 \leq \lambda < \pi/2;$$

Thus, the intensity of the abrasion process depends on the magnitude of the external force P and the angle of attack λ. Figure 4 schematically shows the embodiment zones of different mechanisms of metal removal depending on the angle of attack λ, that is, the angle of encounter of the abrasive particle with the machined surface.

To study the mechanism of abrasion, the value λ = 67°30′ is taken for P_3.

Metal removal from the machined surface will occur at the contact points of the single abrasive grains with this surface under the action of the radial component of the force P_y, which deepens the top of the protruding grain into the material to a depth of h.

Under the action of the tangential component of force Pz, balanced by friction force T (see Fig. 3), the particle will move, leaving scratches of area F'_c on the surface.

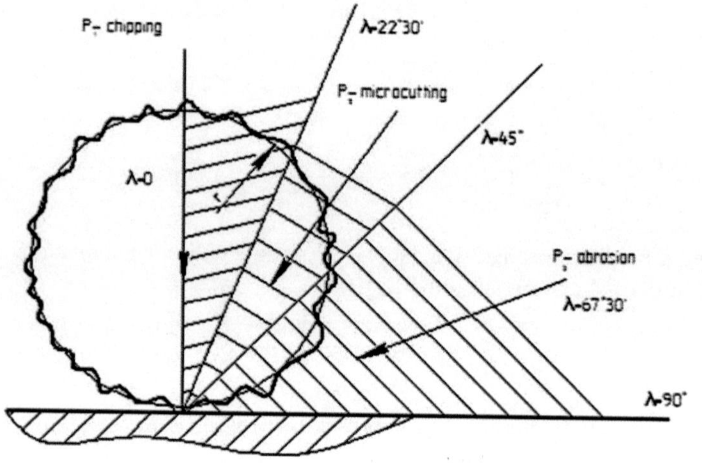

Fig. 4 Embodiment zones of different metal removal mechanisms [3]

The shear area from the impact of a single abrasive grain protrusion per pulse will be determined by the expression:

$$F'_c = \frac{P_y}{\sigma_m} = b \cdot l,$$

where P_y—radial component of the force normal to the machined surface; σ_m—yield strength of the part material; b and l—width and length of the scratch, respectively.

In turn, the cutting area F'_c is determined by assuming that a single scratch has the shape of an isosceles triangle with a depth h and width b, with the angle at the apex equal to the actual cutting angle γ.

Then the width of a single scratch will be:

$$b = 2h \cdot \tan \gamma;$$

and:

$$h = \frac{b}{2 \cdot \tan \gamma} = \frac{F'_c}{2 \cdot l \cdot \tan \gamma} = \frac{P_y}{2 \cdot \sigma_m \cdot l \cdot \tan \gamma}; \quad (7)$$

It follows from formula (7) that as the sharpness of the abrasive grains increases (the angle γ decreases) and the vertical component of the force P_y decreases, the surface roughness deteriorates. The depth of scratches also increases when the strength characteristics of the machined material decrease.

In [5] it was found that roughness reduction is possible with the increase of such characteristics as surface grain concentration on the working surface of the tool, grinding width and rotation speed of the abrasive tool. On this basis, using fine abrasive grinding heads in sizes comparable to the machined surfaces and high-speed spindles is promising for robotic finishing.

4 Experimental Findings

An ABB IRB-140 robot was chosen for the experimental studies. Its positioning accuracy is ± 0.03 mm, lifting capacity is up to 6 kg, and its active radius is 810 mm (Fig. 5).

The robot was equipped with a 0.8 kW spindle with a maximum speed of 24,000 rpm, a frequency of 400 Hz and a weight of 3 kg. A Schunk axia 80 zc22 force-torque sensor was also installed on the robot.

The influence of such factors as the spindle speed (n), the feed rate (S), the cutting depth (t) and the abrasive grit size (z) on the axial components of the cutting force and the torque along the X axis was estimated in a full-factor experiment.

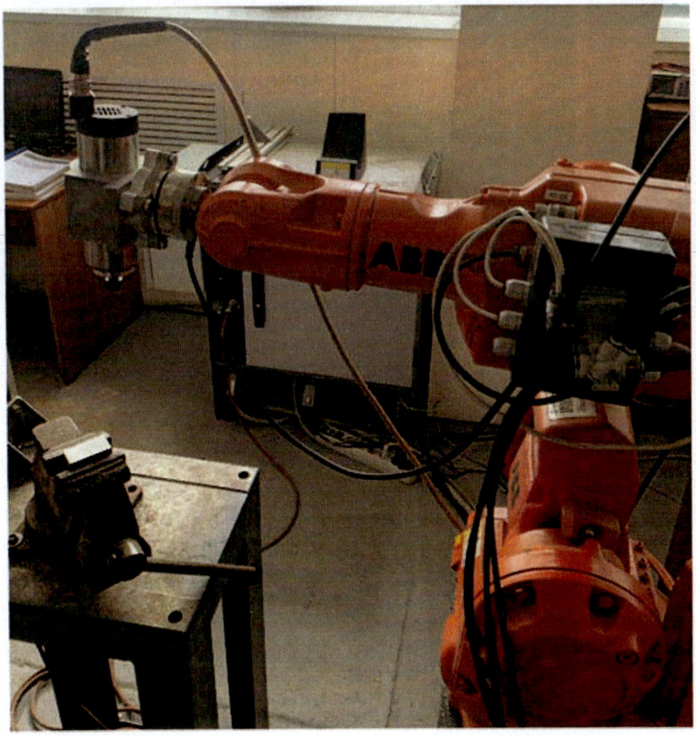

Fig. 5 Experimental installation based on the ABB IRB-140 robot and Schunk axia 80 zc22 force-torque sensor

Table 1 Estimation of factors influence on the parameters (%)

		Factors									
		n	S	t	z	n_S	nt	nz	St	Sz	tz
Parameter	P_x			81.76			7.21		10		
	P_y	32.21		36.45	23.3	8.04					
	P_z	46.61							51.4	1.98	
	T_x		25						75		

This influence in percentages is presented in the Table 1:

As the Table shows, an increase in the spindle speed provides an increase of the cutting force component along the Y and Z axes. As a result, when the spindle speed increases, elastic movements and oscillations arise in the technological system, which conditionally increase the different height arrangement of abrasive grains on the working surface of the tool and lead to an increase in the roughness of the machined surface.

5 Conclusion

The cause of oscillations in robotic finishing is the low rigidity of modern industrial robot manipulators compared to conventional machines [7]. It is possible to compensate the negative impact of the low rigidity of the robot and to avoid the occurrence of oscillations by using the control of the interaction force with the workpiece and the appropriate correction of the trajectory of motion [8]. Further research will focus on the search for optimal values of machining modes in order to achieve minimum roughness, taking into account the vibration resistance support.

References

1. Gienke O, Pan Z, Yuan L, Lepper T, Van Duin S (2019) Mode coupling chatter prediction and avoidance in robotic machining process. Int J Adv Manuf Technol 104:2103–2116
2. Schneider U, Drust M, Ansaloni M, Lehmann C, Pellicciari M, Leali F, Gunnink JW, Verl A (2016) Improving robotic machining accuracy through experimental error investigation and modular compensation. Int J Adv Manuf Technol 85:3–15
3. Aleksandrov AV (2004) Determination of parameters of spatial pendulum vibromachines, providing the reduction of surface roughness in high-performance vibroabrasive machining: dissertation of candidate of technical sciences, 05.03.01, Stary Oskol, Russia
4. Maslov EN (1962) Basic regularities of high-performance grinding. High-performance grinding, pp 4–7. Academy of Sciences, Moscow
5. Ryabenkov IA, Novikov VV, Andilakhai AA (2016) Laws of surface roughness formation during abrasive machining and conditions of its reduction. In: Bulletin of Priazov State Technical University. Series: Technical Sciences, pp 144–150

6. Samodumsky YuM, Trunin VB (1973) Electron microscopic study of surfaces machined by vibroabrasive grinding. Finishing, finishing-strengthening and shaping treatment of metals. pp 22–26. Rostov-on-Don, Russia
7. Vartanov MV, Vlasov AS (2020) Robotization of part finishing in mechanical engineering. Stankoinstrument 3:56–61
8. Chen F, Zhao H, Li D, Chen L, Tan C, Ding H (2019) Contact force control and vibration suppression in robotic polishing with a smart end effector. Robot Comput Integr Manuf 57:391–403

Designing a Single-Way Metal Crossing for Railway Transport

Alexey Loktev, Vadim Korolev, Irina Shishkina, Jalil Shukurov, Vladimir Tkachenko, and Alexander Mikhal'chenkov

Abstract One of the most characteristic aspects of our time is the continuous development of a network of permanent communication routes. Transport links provide for the needs of industry, agriculture and trade, contribute to the international division of labor and rapprochement between countries and peoples. In this regard, it is necessary to build a large number of the road, railway, as well as the city bridges. Along with reinforced concrete bridges, which have recently become especially widespread, metal bridges are being built, which make it possible to block large spans most quickly and effectively. The development of metallurgy, the continuous improvement of the quality of construction steels and the improvement of methods for the manufacture of metal structures provide all the conditions for the successful and economically justified use of metal in the construction of bridges. Metal bridges have great production advantages due to their industrial manufacture, as well as the wide possibilities of quick mechanized assembly and installation of the structure in place. The need to save metal and reduce labor intensity in the manufacture and installation of span structures are closely related to the creation of advanced structures that meet all the requirements of operation, as well as today's technology for mass production at the bridge structure factories. The paper presents the design results of a single-track railway bridge, analyzes the location of the designed bridge. Three versions of the bridge crossing over the river were considered, the main structural elements of the bridge crossing were developed. The authors designed the main structural solutions of joints and abutments of elements.

Keywords Railway transport · Bridge crossing · Design · Railway bridge · Span structure · Bonding

A. Loktev · V. Korolev · I. Shishkina (✉) · J. Shukurov · V. Tkachenko
Russian University of Transport (MIIT), Chasovaya Str., 22/2, Moscow 125190, Russia
e-mail: shishkinaira@inbox.ru

A. Mikhal'chenkov
Bryansk State Agrarian University, Sovetskaya Str., 2a, Bryansk Region, Kokino Village, Vygonichsky District 43365, Russia

A. Bieliatynskyi and V. Breskich (eds.), *Safety in Aviation and Space Technologies*, Lecture Notes in Mechanical Engineering,
https://doi.org/10.1007/978-3-030-85057-9_35

1 Introduction a Subsection Sample

In tectonic terms, the region is the southeastern part of the folded Donbass, where high-thickness Carboniferous rocks form large linear layers elongated in the southeast direction.

Rocks of carbonate, Cretaceous, Tertiary and Quaternary ages take part in the geological structure to the depth of existing knowledge. Carboniferous rocks are represented by a thickness of bulking shales, coals, limestones, marls. Their deposits are represented by sands, sandstones, marls and clays [1–4]. The thickness of these deposits is variable from several meters to the range of 200–250 m. From the surface, tertiary deviations are covered by quaternary loess-like loams, with a thickness of 12–15 m, sandy, clay and silty-clay formations with a thickness of 13–30 m. The working settlement of Ust-Donetsky is located at the mouth of the Seversky Donets River and its branches—the Sukhoi Donets River. Both rivers wash the territory of the village on the south and southeast sides. The Seversky Donets River is navigable. The territory of the village is of a different nature:

- southwestern region (industrial transport hub) is a floodplain flooded by the Sukhoi Donets River;
- southern and southeastern regions are high terraces of the right bank of the Seversky Donets River, cut by numerous gulches;
- central, northern and northeastern regions are a weakly wavy, fairly flat surface with a common slope in the direction of the Seversky Donets River and a significant rise to the northwest. The surface of the site decreases from northeast to southwest with a slope of about 5–15% towards the Sukhoi Donets River, which ensures a good runoff of surface waters. There are green spaces along the terraced bank of the Seversky Donets River, along the slopes of the Dubovaya gully and in the form of forest belts in the northern part of the territory. The soils are mainly of the chernozem type [5–8].

Soils in the territories of watershed spaces are cover loams with groundwater bedding at a depth of 2–20 m; on the slopes of watersheds are cover loams with groundwater bedding at a depth of 1.5–15 m; in floodplain areas and low terraces are sands, sandy loam and loam. The entire floodplain part of the Sukhoi Donets River is flooded with spring floods. In the area where the village is located, the maximum flood horizon is at an absolute elevation of 10 m. The general type of relief can be characterized as a plateau slightly sloping to the south, the surface of which is a treeless steppe crossed by valleys of gullies steeply breaking to the Seversky Donets River [9–12].

The design of modern bridge structures requires high professional training of bridge engineers, who are well familiar with computing technology and the latest achievements of construction science, able to successfully solve the problems of accelerating technological progress using mathematical theory for applied purposes [13–16].

2 Study Methods and Results

In accordance with the geological structure of the region, water content is established in deposits of Carboniferous, Cretaceous, Tertiary and Quaternary ages.

According to the conditions for the formation of groundwater, the Ust-Donetsk upland plain is part of the Donetsk hydrogeological folded region. Here can be distinguished the following units of hydrogeological stratification:

- aquifer of Upper-Quaternary and modern alluvial-deluvial sediments of small rivers and gulches (in the floodplain of the Sukhoi Donets);
- aquifer of the Middle-Upper-Quaternary and modern alluvial sediments of the first to third above-floodplain and floodplain terraces of the Seversky Donets River;
- aquifer below the Upper Quaternary aeolian-deluvial sediments;
- water-resistant layer of Scythian clays of the Upper Pliocene;
- aquifer complex of Sarmatian deposits (pontic deposits are partially eroded on the territory of the city);
- aquifer complex of Cretaceous sediments (upper section);
- aquifer complex of middle-upper Carboniferous sediments.

Of greatest interest for economic use is the aquifer of the Middle-Upper-Quaternary and modern alluvial deposits of the above-floodplain and floodplain terraces of the Sev. Donets River. The horizon is penetrated by numerous wells and bores and comes to the surface in the form of springs. Water-bearing rocks are represented by mixed-grained sands with inclusions of pebbles and gravel in the middle part of the section. The source of power for the horizon is atmospheric precipitation, flood waters; unloading occurs during drainage into gulches and rivers, during summer evaporation, and water intake for economic purposes [1–4].

The aquifer of lower–Upper Quaternary aeolian-deluvial deposits developed on watersheds and slopes at a depth of 0.5–20 m. The waters of this horizon are confined to light and medium loams, sometimes with sand interlayers up to 18 m thick. The aquifer is fed mainly by the infiltration of atmospheric precipitation, and is unloaded with a flow of underground water into the underlying horizons, drainage by the river and gulch network, and with evaporation from the soil surface. Due to low water content and water quality, the described horizon is not recommended for operation with large water intakes.

Sarmatian deposits are represented by interlayering of different-grained quartz sands and clays, the depth of water in the watershed areas reaches 90–94 m. The horizon is fed due to the infiltration of precipitation on the watersheds and their slopes, where water-permeable Sarmatian rocks come to the surface or under the low-power thickness of Quaternary cover loams. The horizon is unloaded into erosional incisions in the form of numerous descending springs, or during economic use by the population. The aquifer of Middle-Upper Carboniferous deposits is ubiquitous in the studied area. Waters lie at various depths from 7 to 48 m. Aquifers are either weakly or free-flowing. The thickness of the aquifer ranges from 1 to

28 m [17–19]. Having studied the conditions of the area of the bridge crossing, and taking into account such factors as: a river of class III, under-bridge dimensions of 120 m and 80 m, bridge opening of 462 m, the number of railway tracks of 1, it was decided to design the following three design options for a single-track metal railway bridge crossing [20–23].

3 First Option

According to the design condition, it was required to provide for two navigable dimensions of 120 m and 80 m. Since the main and adjacent scaffold dimensions are large, we overlap them with two trusses with an estimated span of l_p = 88 m and l_p = 132 m. The floodplains of the river are also covered with metal trusses with a calculated span l_p = 88 m, as well as two metal solid-wall spans. Based on this, we outline the following bridge scheme: 33.6 + 88.0 + 88.0 + 132.0 + 88.0 + 88.0 + 27.0.

Pile-type loose abutments are used. The wing length of the abutment is 6.7 m. Monolithic hollow reinforced concrete supports. The height of the edge of the embankment and the rail sole depend on the navigable dimension, the building height of the truss and the estimated navigable level [24–26].

Theoretical length of the bridge is

$$L_T = l_0 + n \cdot b + 3 \cdot H_{emb} + 2 \cdot a = 1.05 \cdot 462 + 2 \cdot (53 - 46 + 2) \\ + 3 \cdot (65 - 53 - 1) + 10 = 546.1 \text{ m}$$

Actual length of the bridge is

$$L_A = 33.6 + 88.0 + 88.0 + 132.0 + 88.0 + 88.0 + 27.0 = 544.6 \text{ m}$$

Deviation from theoretical length of the bridge is

$$\frac{546.1 - 544.6}{544.6} \times 100 = 2.75\%$$

Advantage of this option is the use of standard spans. Disadvantages of this option is inconvenience of installation. Characteristics of the pile foundation of the supports: diameter d = 1.5 m, length l = 22 m, bearing capacity on the ground P_g = 5200 kN. Channel and floodplain supports are assumed in the form of monolithic structures. The volume of reinforced concrete:

- for floodplain supports No. 1, 5 is 361.4 m^3
- for channel supports No. 2, 3, 4 is 649.9 m^3

The volume of a 3.5 m high grillage made of monolithic reinforced concrete, with dimensions in plan for supports 1.5 to be 8.5 × 11.5 with bevels (to improve flow-around conditions) 1.0 m each, is 342.2 m^3.

The volume of a 3.5 m high grillage made of monolithic reinforced concrete, with dimensions in plan for supports 2, 3, 4 to be 8.5 × 16 with bevels (to improve flow-around conditions) of 1.0 m each, is 476 m^3. Collecting loads at the foundation level:

$$N = N_{pr} + N_{m.p.} + N_{tr} + N_{vr} + N_{vr,tr} + N_{op}; \tag{1}$$

$$N_{pr} = 1.1 \cdot 1.1 \cdot 0.5 \cdot (754 + 907) \cdot 10 = 10049.1 \text{ kN};$$

$$N_{m.p.} = 2 \cdot 16.7 \cdot 1.2 \cdot 0.5 \cdot (88 + 132) = 4849.8 \text{ kN};$$

$$N_{tr} = 6 \cdot 1.1 \cdot 0.5 \cdot (88 + 132) = 798.6 \text{ kN};$$

$$N_{vr,v} = 2 \cdot 1 \cdot 1 \cdot 9.807 \cdot 11 \cdot 1.7 \cdot 121 = 44380.6 \text{ kN};$$

$$N_{vr.tr} = 1.96 \cdot 1.2 \cdot 2 \cdot 0.7 \cdot 0.5 \cdot (88 + 132) = 398.5 \text{ kN};$$

- determining the support type for the span structure of 132 m:

$$\text{N} = 10049.1 + 4849.8 + 798.6 + 4438.6 + 398.5 = 60476.6/2 \text{ kN};$$

Thus, Type 14 (h_s = 1210/850 mm) is assumed;

- determining the support type for the span structure of 88 m:

$$N = N_{pr} + N_{m.p.} + N_{tr} + N_{vr} + N_{vr,tr} = 2997.5 + 1763.5 + 290.4 + 16673.3 + 144.9 = 21869.6/2 \text{ kN};$$

Thus, Type 8 (h_s = 965/750 mm) is assumed;

$$N_{op} = 24 \cdot 1.2 \cdot (365.6 + 300 + 317.1 + 365.6) = 26389.5 \text{ kN}$$

$$N = 10049.1 + 4849.8 + 798.6 + 44380.6 + 398.5 + 26389.5 = 63854.1 \text{ kN};$$

Required number of piles in pile foundation—n:
n = 1.8 N/P_g;
n = 1.8·63,854.1/5200 = 14.95,

thus, we take n = 15.

We assume the grillage size: 16 by 8.5 m. Calculation of abutments.

$$\sum N = \gamma_{f1} G_{ab} + \gamma_{f2} G_{pr.str.} + \gamma_{f3} G_{m.p.} + N_{vr.} \quad (2)$$

$$G_{ab} = V_{ab} \times \gamma_{\text{r.c.}} = 65.4 \times 2.5 = 163.5 \text{ t}$$

$$G_{\text{pr.str.}} = P \times l_1/2\text{x}B' = 0.3 \times 17.1 \times 4.9 = 25.13 \text{ t}$$

$$N_{\text{vr}} = \gamma_f \times (1 + \mu) \times P_{\text{ref}} \times \lambda/2 = 1.3 \times 1.15 \times 17.66 \times 17.1 = 451.4 \text{ t}$$

$$G_{\text{m.p.}} = h_{bm} \times B_{bm} \times \gamma_{bm} \times l/2 = 0.5 \times 4 \times 2 \times 17.1 = 68.4 \text{ t}$$

$$\begin{aligned}\sum N &= \gamma_{f1} G_{ab} + \gamma_{f2} G_{pr.str.} + \gamma_{f3} G_{m.p.} + N_{vr.} \\ &= 1.1 \cdot 163.5 + 1.1 \cdot 25.13 + 1.5 \cdot 68.4 + 451.4 = 761.5\text{t}.\end{aligned}$$

The bearing capacity of the piles is determined by the length of the selected piles and the class of underlying soils. Applicable to our option, sandy-gravelly soils belong to the class of solid soils.

Foundations of abutments are piles with d = 0.6 m.

We will determine the required number of abutment piles on which a single-track span of 34.2 m is based.

$$n = \frac{\sum N}{F_d} = 761.5/120 = 5.3 \text{ pcs}$$

We place 6 shell piles with a diameter of 0.6 m and a length of 15 m in the grillage.

The total cost of the bridge is determined in Table 1.

Table 1 Total cost of bridge, calculated according to option 1

Name of structural elements	Number of same type elements	Cost of one element, thousand rubles	Total cost, thousand rubles
Span structure 88.0 m	4	355.86	1423.4
Span structure 132.0 m	1	503.4	503.4
Span structure 34.2 m	1	49.5	49.5
Span structure 27 m	1	34.75	34.75
Intermediate support 1, 5	2	114.71	229.42
Intermediate support 2, 3, 4	3	173.8	521.4
Abutment	2	21.2	42.4
Total cost of bridge according to option 1			2804.27

4 Second Option

In the second option, we get rid of the trusses and accept metal solid-wall span structures with a length of 45, 33.6, 55 m. The navigable dimension is also covered by through trusses with estimated spans of l_p = 88 m and l_p = 132 m. Based on this, we plan the following bridge scheme: 33.6 + 45.0 + 45.0 + 88.0 + 132.0 + 55.0 + 55.0 + 55.0 + 33.6.

The abutments are used to the loose pile abutments. The length of the abutment wing is 6.7 m. The supports are monolithic hollow reinforced concrete.

Theoretical length of the bridge is

$$L_T = l_0 + n \cdot b + 3 \cdot H_{emb} + 2 \cdot a = 1.05 \cdot 462 + 2 \cdot (53 - 46 + 2) + 3 \cdot (65 - 53 - 1) + 10 = 546.1 \text{ m}$$

Actual length of the bridge is

$$L_A = 33.6 \cdot 2 + 45.0 \cdot 2 + 88.0 + 132.0 + 55.0 \cdot 3 = 548.2 \text{ m}.$$

Deviation from theoretical length of the bridge is

$$\frac{548.2 - 546.1}{548.2} \times 100 = 1.99\%$$

Characteristics of the pile foundation of the supports:

- diameter d = 1.5 m,
- length l = 22 m,
- bearing capacity on the ground P_g = 5200 kN.

Channel and floodplain supports are assumed in the form of monolithic structures. The volume of reinforced concrete:

- for floodplain supports No. 1, 5 is 361.4 m^3
- for channel supports No. 2, 3, 4 is 649.9 m^3

The volume of a 3.5 m high grillage made of monolithic reinforced concrete, with dimensions in plan for supports 1, 5 to be 8.5 × 11.5 with bevels (to improve flow-around conditions) 1.0 m each, is 342.2 m^3.

The volume of a 3.5 m high grillage made of monolithic reinforced concrete, with dimensions in plan for supports 2, 3, 4 to be 8.5 × 16 with bevels (to improve flow-around conditions) of 1.0 m each, is 476 m^3.

Collecting loads at the foundation level of the 1st, 2nd, 7th, 8th supports:

$$N = N_{pr} + N_{m.p.} + N_{tr} + N_{vr} + N_{vr,tr} + N_{op}; \quad (3)$$

$$N_{\text{pr}} = 1.1 \cdot (57.0 + 241.6) \cdot 10/2 = 1642.3\,\text{kN};$$

$$N_{\text{m.p.}} = 2 \cdot 16.7 \cdot 1.2 \cdot 0.5 \cdot (23.6 + 45.7) = 1388.8\,\text{kN};$$

$$N_{\text{tr}} = 6 \cdot 1.1 \cdot 0.5 \cdot (23.6 + 45.7) = 228.7\,\text{kN};$$

$$N_{\text{vr,v}} = 2 \cdot 1 \cdot 1 \cdot 10.5 \cdot 11 \cdot 1.7 \cdot 34.65 = 13607.1\,\text{kN};$$

$$N_{\text{vr.tr}} = 1.96 \cdot 1.2 \cdot 2 \cdot 0.7 \cdot 0.5 \cdot (23.6 + 45.7) = 114.1\,\text{kN};$$

- determining the support type for the span structure of 45.7 m:

$$\begin{aligned} N &= N_{\text{pr}} + N_{\text{m.p.}} + N_{\text{tr}} + N_{\text{vr}} + N_{\text{vr,tr}} \\ &= 313.5 + 473 + 77.9 + 6112.3 + 38.9 = 7015,6/4 \end{aligned}$$

Thus, Type 3 (h_s = 570 mm) is assumed;

- determining the support type for the span structure of 55.6 m:

$$\begin{aligned} N &= N_{\text{pr}} + N_{\text{m.p.}} + N_{\text{tr}} + N_{\text{vr}} + N_{\text{vr,tr}} \\ &= 1328.8 + 915.9 + 150.9 + 9921.8 + 75.3 = 12392.7/4\,\text{kN}; \end{aligned}$$

Thus, Type 3 (h_s = 570 mm) is assumed;

$$N_{\text{op}} = 24 \cdot 1.2 \cdot (28.7 + 172.7 + 146.8 + 166.4) = 14820.5\,\text{kN};$$

$$N = 1642.3 + 1388.8 + 228.7 + 13607.1 + 228.7 + 14820.5 = 31916.1\,\text{kN};$$

Required number of piles in pile foundation—n:
n = 1.8·N/P_g;
n = 1.8·31,916.1/4800 = 11.97;
thus, we take n = 12.
Supports 3, 4, 5 are the same as in the previous option.
We will take the required number of piles in the pile foundation of 15 pieces. We assume a grillage with dimensions: 16 by 8.5 m.
Calculation of abutments.

$$\sum N = \gamma_{f1} G_{ab} + \gamma_{f2} G_{pr.str.} + \gamma_{f3} G_{m.p.} + N_{vr.} \tag{4}$$

$$G_{ab} = V_{ab} x \gamma_{\text{r.c.}} = 65.4 \times 2.5 = 163.5\,\text{t}.$$

$$G_{\text{pr.str.}} = P \times l_1/2xB' = 0.3 \times 17.1 \times 4.9 = 25.13\,\text{t.}$$

$$\text{Nvr.} = \gamma\text{f} \times (1+\mu) \times \text{Pref.} \times \lambda/2 = 1.3 \times 1.15 \times 17.66 \times 17.1 = 451.4\,\text{t.}$$

$$G_{\text{m.p.}} = h_{bm} \times B_{bm} \times \gamma_{bm} \times l/2 = 0.5 \times 4 \times 2 \times 17.1 = 68.4\,\text{t.}$$

$$\sum N = \gamma_{f1} G_{ab} + \gamma_{f2} G_{pr.str.} + \gamma_{f3} G_{m.p.} + N_{vr.} = 1.1 \cdot 163.5 + 1.1 \cdot 25.13 + 1.5 + \cdot 68.4 + 451.4 = 761.5\,\text{t.}$$

The bearing capacity of the piles is determined by the length of the selected piles and the class of underlying soils. Applicable to our option, sandy-gravelly soils belong to the class of solid soils. Foundations of abutments are piles with $d = 0.6$ m.

We will determine the required number of abutment piles on which a single-track span of 34.2 m is based.

$$n = \frac{\sum N}{F_d} = 761.5/120 = 5.3\,\text{pcs}$$

We place 6 shell piles with a diameter of 0.6 m and a length of 15 m in the grillage.

The total cost of the bridge is determined in Table 2.

Table 2 Total cost of bridge, calculated according to option 2

Name of structural elements	Number of same type elements	Cost of one element, thousand rubles	Total cost, thousand rubles
Span structure 88.0 m	1	355.86	355.8
Span structure 132.0 m	1	503.4	503.4
Span structure 55.8 m	3	118.026	354.06
Span structure 45.8 m	2	84.85	169.7
Span structure 34.2 m	2	49.5	99
Intermediate support 3, 4, 5	3	169.36	508.08
Intermediate support 1, 2, 6, 7, 8	5	114.71	573.55
Abutment	2	21.2	42.4
Total cost of bridge according to option 2			2606.09

5 Third Option

In the third option, we will cover the main and adjacent dimensions with a continuous truss, with an estimated span of $l_p = 2 \times 132.0 = 264$ m.

In floodplain areas, we accept metal solid-wall spans with a length of 45 m. Based on this, we outline the following bridge scheme: 45.0 + 45.0 + 45.0 + 264.0 + 45.0 + 45.0 + 45.0.

Pile-type loose abutments are used. The wing length of the abutment is 7.2 m. Supports are assumed to be monolithic hollow reinforced concrete. Theoretical length of the bridge is: $L_T = l_0 + n{\cdot}b + 3{\cdot}H_{emb} + 2{\cdot}a = 1.05{\cdot}462 + 2{\cdot}(53 - 46 + 2) + 3{\cdot}(65 - 53 - 1) + 10 = 546.1$ M.

Actual length of the bridge is.

LA = 264.0 + 6·45.0 = 534 m. Deviation from theoretical length of the bridge is:

$$\frac{546.1 - 534}{546.1} \times 100 = 2.21\%$$

Characteristics of the pile foundation of the supports:

- diameter $d = 1.5$ m,
- length $l = 22$ m,
- bearing capacity on the ground $P_g = 5200$ kN.

Channel and floodplain supports are assumed in the form of monolithic structures. The volume of reinforced concrete:

- for floodplain supports No. 1, 2, 6, 7 is 361.4 m^3
- for channel supports No. 3, 4, 5 is 649.9 m^3

The volume of a 3.5 m high grillage made of monolithic reinforced concrete, with dimensions in plan for supports 1, 5 to be 8.5 × 11.5 with bevels (to improve flow-around conditions) 1.0 m each, is 342.2 m^3.

The volume of a 3.5 m high grillage made of monolithic reinforced concrete, with dimensions in plan for supports 2, 3, 4 to be 8.5 × 16 with bevels (to improve flow-around conditions) of 1.0 m each, is 476 m^3.

Collecting loads at the foundation level of the 1st, 2nd, 6th, 7th supports:

$$\mathrm{N = Npr + Nm.p. + Ntr + Nvr + Nvr, tr + Nop} \tag{5}$$

$$N_{pr} = 1.1 \cdot (57.0 + 241.6) \cdot 10/2 = 1642.3\,\mathrm{kN};$$

$$N_{m.p.} = 2 \cdot 16.7 \cdot 1.2 \cdot 0.5 \cdot (23.6 + 45.7) = 1388.8\,\mathrm{kN};$$

$$N_{\rm tr} = 6 \cdot 1.1 \cdot 0.5 \cdot (23.6 + 45.7) = 228.7\,{\rm kN};$$

$$N_{\rm vr,v} = 2 \cdot 1 \cdot 1 \cdot 10.5 \cdot 11 \cdot 1.7 \cdot 34.65 = 13607.1\,{\rm kN};$$

$$N_{\rm vr.tr} = 1.96 \cdot 1.2 \cdot 2 \cdot 0.7 \cdot 0.5 \cdot (23.6 + 45.7) = 114.1\,{\rm kN};$$

– determining the support type for the span structure of 45.7 m:

$$\begin{aligned} N &= N_{\rm pr} + N_{\rm m.p.} + N_{\rm tr} + N_{\rm vr} + N_{\rm vr,tr} \\ &= 313.5 + 473 + 77.9 + 6112.3 + 38.9 = 7015,6/4 \end{aligned}$$

Thus, Type 3 (h_s = 570 mm) is assumed;

– determining the support type for the span structure of 55.6 m:

$$\begin{aligned} N &= N_{\rm pr} + N_{\rm m.p.} + N_{\rm tr} + N_{\rm vr} + N_{\rm vr,tr} \\ &= 1328.8 + 915.9 + 150.9 + 9921.8 + 75.3 = 12392.7/4\,{\rm kN}; \end{aligned}$$

Thus, Type 3 (h_s = 570 mm) is assumed;

$$N_{\rm op} = 24 \cdot 1.2 \cdot (28.7 + 172.7 + 146.8 + 166.4) = 14820.5\,{\rm kN};$$

$$N = 1642.3 + 1388.8 + 228.7 + 13607.1 + 228.7 + 14820.5 = 31916.1\,{\rm kN};$$

Required number of piles in pile foundation - n:$n = 1.8 \cdot N/P_g$;

$$n = 1.8 \cdot 31916.1/4800 = 11.97;$$

Thus, we take n = 12.

Supports 3, 4, 5 are the same as in the previous option. We will take the required number of piles in the pile foundation of 15 pieces. We assume a grillage with dimensions: 16 by 8.5 m.

Calculation of abutments

$$\sum N = \gamma_{f1} G_{ab} + \gamma_{f2} G_{pr.str.} + \gamma_{f3} G_{m.p.} + N_{vr.}$$

$$G_{ab} = V_{ab} \times \gamma_{\rm rc} = 65.4 \times 2.5 = 163.5\ {\rm t}$$

$$G_{\text{pr.str.}} = P \times l_1/2\text{x}B' = 0.3 \times 17.1 \times 4.9 = 25.13\ \text{t}$$

$$N_{\text{vr}} = \gamma_f \times (1+\mu) \times P_{\text{ref}} \times \lambda/2 = 1.3 \times 1.15 \times 17.66 \times 17.1 = 451.4\ \text{t}$$

$$G_{\text{m.p.}} = h_{bm} \times B_{bm} \times 2_{bm} \times l/2 = 0.5 \times 4 \times 2 \times 17.1 = 68.4\ \text{t}$$

$$\sum N = \gamma_{f1} G_{ab} + \gamma_{f2} G_{pr.str.} + \gamma_{f3} G_{m.p.} + N_{vr.} = 1.1 \cdot 163.5 + 1.1 \cdot 25.13 + 1.5 \cdot 68.4 + 451.4 = 761.5\ \text{t}.$$

The bearing capacity of the piles is determined by the length of the selected piles and the class of underlying soils. Applicable to our option, sandy-gravelly soils belong to the class of solid soils. Foundations of abutments are piles with $d = 0.6$ m.

We will determine the required number of abutment piles on which a single-track span of 34.2 m is based.

$$n = \frac{\sum N}{F_d} = 761.5/120 = 5.3\ \text{pcs}$$

We place 6 shell piles with a diameter of 0.6 m and a length of 15 m in the grillage. The total cost of the bridge is determined in Table 3.

Comparison of all three options considered is given in Table 4.

Grid elements of the main trusses of modern bolted span structures are attached by vertical sheets to the nodal overlapping plates using high-strength bolts with one friction plane. The calculation of the attachment of the element shall be made according to the maximum force in it. In all cases, the load-bearing capacity of the attachment must be at least 75% of the load-bearing capacity of the attached element, determined depending on the nature of its work by the calculation of stability, endurance or strength [27, 28]. Accordingly, the number of bolts in attaching the element to the gussets is determined as follows:

Table 3 Total cost of bridge, calculated according to option 3

Name of structural elements	Number of same type elements	Cost of one element, thousand rubles	Total cost, thousand rubles
Span structure 264.0 m	1	1282.05	1282.05
Span structure 45.8 m	6	84.85	509.1
Intermediate support 3, 4, 5	3	169.36	508.08
Intermediate support 1, 2, 6, 7	4	114.71	458.8
Abutment	2	21.2	42.4
Total cost of bridge according to option 3			2800.43

Table 4 Comparison of options

Option no	Bridge scheme	Total cost, thousand rubles
I	33.6 + 88.0 + 88.0 + 132.0 + 88.0 + 88.0 + 27.0	2804.27
II	33.6 + 45.0 + 45.0 + 88.0 + 132.0 + 55.0 + 55.0 + 55.0 + 33.6	2606.09
III	45.0 + 45.0 + 45.0 + 264.0 + 45.0 + 45.0 + 45.0	2800.43

$$n = \frac{N}{Q_{bh}}, \text{but not less } \frac{0.75 A_p R_d}{Q_{bh}} \tag{6}$$

The results of the calculations are summarized in Table 5.

When calculating the attachment of the grid elements, it is necessary to check the strength of the knot gussets "for gouging," that is, for the absence of dangerous over-tension in them. The strength of the gusset is checked by the probable section of its destruction. This section typically extends along the outermost row of bolts at the end of the attachment element and the shortest distances from the centers of the outermost bolts in this row to the edge of the gusset (Fig. 1).

The design resistance of the gusset material in individual sections of the dangerous section is taken to be $R_a = 0.675 \cdot m \cdot R_y \cdot (0.212 \cdot \alpha + 1)$, where α is the angle in radians between the considered area and the axis of the attached element.

The gusset thickness t is assigned from the condition

$$N \leq (R_L \cdot l_L + R_2 \cdot l_2 + R_3 \cdot l_3) \cdot t \tag{7}$$

Check is performed for the most loaded node. The results of the calculations are summarized in Table 6.

Overlapping of joints is carried out by means of overlays. Within the height of the belts, the role of overlays is also performed by nodal gussets. It is necessary to strive to ensure that each part of the belt section is attached according to the force coming into it (Table 7).

Table 5 Results of bridging element attachment calculations

Elements	Force N, tf	Q_{bh}, tf	n, pcs
N0-V1	726.42	10.85	38
V1-N2	552.832	10.85	30
N2-V3	371.4	10.85	20
N0-N1 and N1-N2	452.64	10.85	24
N2-N3	956	10.85	50
V1-V2 and V2-V3	730	10.85	38
V1-N1	205.94	10.85	12
B2-H2	180.03	10.85	10

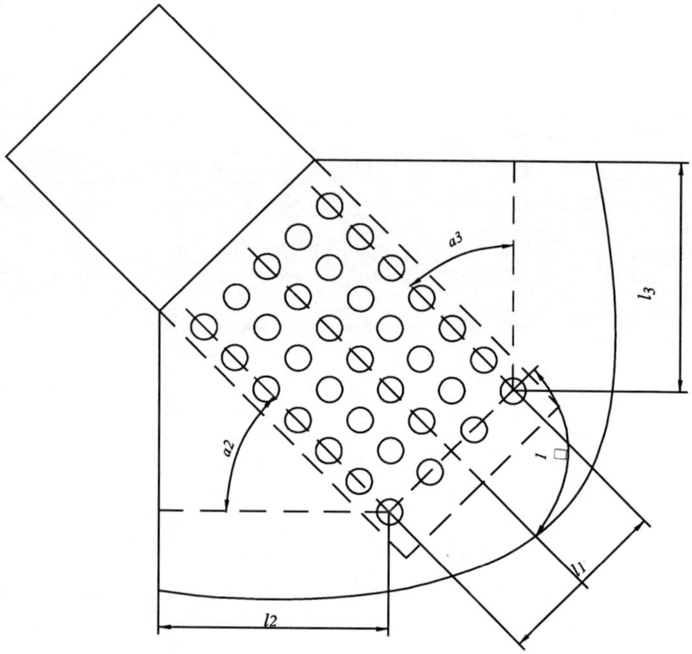

Fig. 1 Diagram of gusset gouging

The selection of the section is carried out based on the maximum flexibility of the element $\lambda_{max} = 130$. Cross-section of lower links is taken in the form of welded bevel, upper links in the form of composite I-beam. The results of the calculations are summarized in Table 8.

6 Conclusions

The conducted research made it possible not only to consider and analyze several options for a bridge crossing for railway transport, but also to develop the main structural elements of the bridge, carry out calculations of structures for the main effects on strength, stability, select sections of the main elements and check them in accordance with the current regulatory documents. Also, the main design solutions of joints and abutments of elements were considered, including the nodes of support of the span structure beams on the girder beams of the supports, which are modeled by two types of viscoelastic dampers.

Table 6 Results of gusset check calculations

Node	Force N, tf	Angle, rad			Section length, m			Design resistance, tf/m^2				Thickness
		α1	α2	α3	l_1	l_2	l_3	R_y	R_1	R_2	R_3	t, mm
1	2	3	4	5	6	7	8	9	10	11	12	13
N0-V1	727	1.57	0.63	0.97	0.8	0.27	0.369	29,500	23,889.18	20,314.81	21,606.5	12

Table 7 Belt section part selection

Node	N_{max}	Overlay no	Section	A, cm	ΔA	A_b	$\Sigma A_b m R_y$	N_i	Semi-overlay	Q_{bh}	n
									No		
N0	452	1	2F 1075*12	129	16.8	112.2	530	194.2	1	10.85	20
		2	2N 540*12	129.6	21.6	108		186	1 + 2	10.85	40
		3	1N 490*10	49	7.5	41.5		71.8	3	9.45	9
			Total	308.6	45.9	261.7					
N2	956	1	2F 1600*12	348	40.8	307.2	1212,53	643	1	10.85	95
		2	2N 540*12	129.6	21.6	108		226	1 + 2	10.85	24
		3	1N 490*10	49	7.5	41.5		86.9	3	9.45	9
			Total	526.6	69.9	456.7					
V1	730	4	2F 880*12	211.2	42	169.2	841.6	404	1	10.85	74
		5	2N 700*12	168	31,5	136.5		325.9	1 + 2	10.85	34
		3	1N 490*10	49	7.5	41.5		86.9	3	9.45	8
			Total	428.2	78.5	346.2					

Table 8 Results of section calculations based on element's ultimate flexibility

Section	Area F, m^2	Moment of inertia I, m^4	Gyration radius r, m	Free length l, m	Flexibility λ	Maximum flexibility λ_{max}
v.l. 200 × 20	0.008	0.0000377	0.0686	8.0	117	130
g.l. 200 × 20		0.0000135	0.041	4.0	98	
2L 220 × 220 × 20	0.0312	0.000163	0.0723	8.0	111	
g.l. 600 × 30		0.000149	0.219	4.0	18	

References

1. Chechevichkin AV, Vatin NI, Samonin VV, Grekov MA (2017) Purification of hot water by zeolite modified with manganese dioxide. Mag Civ Eng 76(8):201–213. https://doi.org/10.18720/MCE.76.18
2. Chechevichkin V, Vatin N (2014) Megacities land drainage and land runoff features and treatment. Appl Mech Mater 641–642:409–415. https://doi.org/10.4028/www.scientific.net/AMM.641-642.409
3. Chechevichkin VN, Vatin NI (2014) Specifics of surface runoff contents and treatment in large cities. Mag Civ Eng 50(6):67–74. https://doi.org/10.5862/MCE.50.7
4. Dražić J, Vatin N (2016) The influence of configuration on to the seismic resistance of a building. Proc Eng 165:883–890. https://doi.org/10.1016/j.proeng.2016.11.788
5. Loktev AA, Korolev VV, Poddaeva OI, Chernikov IYU (2018) Mathematical modeling of antenna-mast structures with aerodynamic effects. In: IOP conference series: materials science and engineering, vol 463. https://doi.org/10.1088/1757-899X/463/3/032018
6. Loktev AA, Korolev VV, Shishkina IV (2018) High frequency vibrations in the elements of the rolling stock on the railway bridges. In: IOP conference series: materials science and engineering, vol 463. https://doi.org/10.1088/1757-899X/463/3/032019
7. Glusberg B, Korolev V, Shishkina I, Loktev A, Shukurov J, Geluh P, Loktev D (2018) Calculation of track component failure caused by the most dangerous defects on change of their design and operational conditions. In: MATEC web of conferences, vol 239. https://doi.org/10.1051/matecconf/201823901054
8. Korolev V, Loktev A, Shishkina I, Zapólnova E, Kuskov V, Basovsky D, Aktisova O (2019). Technology of crushed stone ballast cleaning. In: IOP conference series: earth and environmental science, vol 403. https://doi.org/10.1088/1755-1315/403/1/012194
9. Gluzberg B, Korolev V, Loktev A, Shishkina I, Berezovsky M (2019) Switch operation safety. In: E3S web of conferences, vol 138. https://doi.org/10.1051/e3sconf/201913801017
10. Loktev A A, Korolev VV, Gridasova EA (2019) Influence of high-frequency cyclic loading on mechanical and structural characteristics of rail steel under extreme conditions. In: IOP conference series: materials science and engineering, vol 687. https://doi.org/10.1088/1757-899X/687/2/022036
11. Lyudagovsky A, Loktev A, Korolev V, Shishkina I, Alexandrova D, Geluh P, Loktev D (2019) Energy efficiency of temperature distribution in electromagnetic welding of rolling stock parts. In: E3S web of conferences, vol 110. https://doi.org/10.1051/e3sconf/201911001017
12. Savin AV, Korolev VV, Shishkina IV (2019) Determining service life of non-ballast track based on calculation and test. In: IOP conference series: materials science and engineering, vol 687. https://doi.org/10.1088/1757-899X/687/2/022035

13. Bonić Z, Ćurčć GT, Trivunić M, Davidović N, Vatin N (2015) Some methods of protection of concrete and reinforcment of reinforced-concrete foundations exposed to environmental impacts. Proc Eng 117(1):419–430. https://doi.org/10.1016/j.proeng.2015.08.189
14. Buka-Vaivade K, Serdjuks D, Goremikins V, Pakrastins L, Vatin NI (2018) Suspension structure with cross-laminated timber deck panels. Mag Civ Eng 83(7):126–135. https://doi.org/10.18720/MCE.83.12
15. Buka-Vaivade K, Sliseris J, Serdjuks D, Pakrastins L, Vatin NI (2018) Rational use of HPSFRC in multi-storey building. Mag Civ Eng 84(8):3–14. https://doi.org/10.18720/MCE.84.1
16. Campian C, Chira N, Iuhos V, Pop M, Vatin N (2016) Structural upgrading of steel columns for overhead power lines. Proc Eng 165:876–882. https://doi.org/10.1016/j.proeng.2016.11.787
17. Dudin MO, Vatin NI, Barabanshchikov YG (2015) Modeling a set of concrete strength in the program ELCUT at warming of monolithic structures by wire. Mag Civ Eng 54(2):33–45. https://doi.org/10.5862/MCE.54.4
18. Dudin MO, Vatin NI, Barabanshchikov YG (2015) Modeling a set of concrete strength in the program ELCUT at warming of monolithic structures by wire. Mag Civ Eng 54(2):33–45. https://doi.org/10.5862/MCE.54.4
19. Garifullin M, Bronzova M, Sinelnikov A, Vatin N (2016) Buckling analysis of cold-formed c-shaped columns with new type of perforation. Advances and trends in engineering sciences and technologies. In: Proceedings of the international conference on engineering sciences and technologies, ESaT 2015, pp 63–68
20. Garifullin M, Vatin N, Jokinen T, Heinisuo M (2017) Numerical solution for rotational stiffness of RHS tubular joints. Advances and trends in engineering sciences and technologies II. In: Proceedings of the 2nd international conference on engineering sciences and technologies, ESaT 2016, pp 81–86
21. Goremikins V, Rocens K, Serdjuks D, Pakrastins L, Vatin N (2015) Cable truss topology optimization for prestressed long-span structure. Advances in civil engineering and building materials iv—selected and peer reviewed papers from the 2014 4th international conference on civil engineering and building materials, CEBM 2014, pp 363–368
22. Goremikins V, Serdjuks D, Buka-Vaivade K, Pakrastins L, Vatin N (2017) Prediction of behaviour of prestressed suspension bridge with timber deck panels. Baltic J Road Bridge Eng 12(4):234–240. https://doi.org/10.3846/bjrbe.2017.29
23. Gorshkov A, Vatin N, Nemova D, Shabaldin A, Melnikova L, Kirill P (2015) Using life-cycle analysis to assess energy savings delivered by building insulation. Proc Eng 117(1):1080–1089. https://doi.org/10.1016/j.proeng.2015.08.240
24. Gorshkov A, Vatin N, Nemova D, Tarasova D (2014) Definition of the overturning and holding moments for floor-by-floor leaning walls made from aerated concrete blocks. Appl Mech Mater 633–634:897–903. https://doi.org/10.4028/www.scientific.net/AMM.633-634.897
25. Gorshkov AS, Rymkevich PP, Vatin NI (2014) Simulation of non-stationary heat transfer processes in autoclaved aerated concrete-walls. Mag Civ Eng 52(8). https://doi.org/10.5862/MCE.52.5
26. Gorshkov AS, Rymkevich PP, Vatin NI (2014) Simulation of non-stationary heat transfer processes in autoclaved aerated concrete-walls. Mag Civ Eng 52(8):38–66. https://doi.org/10.5862/MCE.52.5
27. Harmati N, Jakšić Z, Vatin N (2015) Energy consumption modelling via heat balance method for energy performance of a building. Proc Eng 117(1):786–794. https://doi.org/10.1016/j.proeng.2015.08.238
28. Hirkovskis A, Serdjuks D, Goremikins V, Pakrastins L, Vatin NI (2015) Behaviour analysis of load-bearing aluminium members. Mag Civ Eng 57(5). https://doi.org/10.5862/MCE.57.8

Parameters Ensuring Smooth Passage of the Rolling Stock Along Turnouts of Railway Transport

Boris Glusberg, Alexey Loktev, Vadim Korolev, and Irina Shishkina

Abstract The paper considers the formation of approaches to standardizing parameters that ensure a given level of impact on passengers and cargo when passing rolling stock through turnout switches, considers the dynamics of interaction between rolling stock and switches in the horizontal plane. There is now a need to accelerate the design and implementation of turnouts on the operated railway network, which will significantly increase the permissible train speeds, especially on the side track. One of the most important tasks in designing such turnouts is to provide comfortable conditions for passengers when the train moves to the side track by limiting the values of the so-called centrifugal acceleration and increment (changing) of centrifugal acceleration per unit of time (second). If it is necessary to realize the speed of movement on the side track above 50 km/h, the main factor in determining the radius of the switch curve according to comfort conditions is to limit the magnitude of the increment (change) of centrifugal acceleration per unit time (second). Based on this, when designing switches for high speeds, it is advisable to use curves of variable radius of rolling stock action and switches in the horizontal plane as a switch curve.

Keywords Turnout switch · Transition smoothness · Dynamic and kinetic characteristics · Widening · Counter-rail · Technique

1 Introduction

The unification of requirements for turnout switches and other types of turnout products used by the OSShD member countries is designed to ensure the best conditions for rolling stock circulation along the way and should facilitate inte-

B. Glusberg
JSC «Research Institute of Railway Transport», 3-ya Mytischinskaya Str., 10, 129626 Moscow, Russia

A. Loktev · V. Korolev · I. Shishkina (✉)
Russian University of Transport (MIIT), Chasovaya Str. 22/2, 125190 Moscow, Russia
e-mail: shishkinaira@inbox.ru

A. Bieliatynskyi and V. Breskich (eds.), *Safety in Aviation and Space Technologies*, Lecture Notes in Mechanical Engineering,
https://doi.org/10.1007/978-3-030-85057-9_36

gration processes within the organization, as well as the implementation of its immediate and promising tasks.

The purpose of the design of turnout switches (or other types of turnout products) is to develop structures that ensure compliance with safety, comfort and specified indicators of operational reliability.

The requirements for the transfer smoothness of rolling stock along turnout switches belong to the group of requirements for comfort and traffic safety. Their implementation is ensured by compliance of dynamic and kinematic characteristics of a motion with the specified or standardized values.

When the railway carriages pass the turnout switch, there should be no sharp shifts of wheelsets, jolts and impacts that violate the trajectory of movement. The force factors associated with the peculiarity of movement along the turnout switch should not exceed the values that ensure the comfortable state of passengers and the permissible effect on the cargo.

2 Research Methods and Results

Due to the fact that the direction of movement along the turnout switch can be different, it is necessary to distinguish between movement in the forward direction and movement along the turnout curve. The dynamic and kinematic requirements for each of these types of motion are different [1–4].

When moving in the straight direction of the turnouts, the changes in the horizontal trajectories of the wheelsets of the rolling stock occur in the area of the turnout, due to the widening of the track necessary to provide shelter for the point rail and the point of crossing with fixed elements, due to the presence of a counter rail.

To minimize the effect of track widening at the turnout point rail on the smoothness of a motion along the turnout switch, it is necessary to standardize the value of track widening at the point rail—ΔS and the retraction intensity of track widening—i_1 and i_2.

$$\Delta S = S_o - S,$$

$$i_1 = \Delta S/l_1, \quad i_2 = \Delta S/l_2, \tag{1}$$

wherein:

S_o—track width at the point rail;

S—track width at the stage tracks;

l_1—length of length of the track widening retraction section towards the front joint of the stock rails;

l_2—length of the track widening retraction section towards the switch frog.

Intensity of track widening tap i1 and i2, is usually standardized through the section lengths l1 and l2, whereon track widening taps are made.

Ensuring the smooth transit of the rolling stock along the connecting part of the turnouts when moving on a straight track does not require any restrictions.

In order to ensure the safety of rolling stock movement along the frogs with such cross-pieces as parts of their design, counter rails are used that guide the wheelsets of rolling stock when passing the dead space of the frog. Studies have shown that, for example, at 1520 mm track of turnout switches, up to 40% of wheelsets interact with a counter rail when moving along a crossing point [5–8]. At the same time, transverse shifts of wheelsets arise, which violate the smoothness of the rolling stock passing through the turnout switch.

The safety and comfort of movement through the area of the counter rail should be ensured by limiting the loss of kinetic energy when the wheels run over the counter rail and the speed of the transverse displacement of the wheelsets. In the practice of designing turnouts for these purposes, a parameter associated with the loss of kinetic energy is used—W_{κ}, measured in [m/s], or in [km^2/h^2]. The limiting value W_{κ} is determined on the basis of operating experience and dynamic-and-strength tests.

The speed of lateral displacement of the wheelsets depends on the speed of movement along the turnout switch [9–12]. On the counter rail, the displacement speed of the wheelsets is limited by the angle of retraction of the bent part of the counter rail—γ_{κ}, which is determined from the formula

$$\sin \gamma_{\kappa} = W_{\kappa}/V, \tag{2}$$

wherein: V—passing speed of the rolling stock along the turnout switch.

Similarly, smooth movement along the switch frog itself is ensured. The angles of the frog components are standardized. The angles of the wing rail retraction between the throat and the area of the wheels rolling over the wing rail to the point of crossing—γ_I—are standardized.

$$\sin \gamma_1 = W_1/V \tag{3}$$

The angles of the wing rail retraction from the side of crosspiece tail-end

$$\sin \gamma_y = W_2/V \tag{4}$$

In this case, the limiting values W_1 and W_2 are determined in the same way as for the counter rail on the basis of operating experience and the dynamic and strength tests. The angles of the deflecting wing rail retraction and counter rails are determined by the same method. The conditions of the rolling stock transition along the frogs with a continuous rolling surface, from the point of view of the movement smoothness, are close to the conditions of the turnout switch transition, therefore, for this case, the requirements stated above can be used.

Due to the fact that the angles of the wheels running on the crossing point are small, in the practice of designing turnouts for marks up to 1/34, widening of the track on the frog may not be provided. The conditions of movement along the turnout curves differ significantly from the conditions of movement along the track curves.

On the turnout switches, in contrast to the curves on the tracks, there are neither elevation of the outer rail no transition curves. To hide the switch point, a protecting housing is used at the switch-point tip, which violates the smoothness of the conjugation of the switch curve with the straight component adjacent to it [13–16]. On turnouts, in the design of which a switch with secant type points is used, there is no smooth conjugation of a straight section with a curvilinear one. The conjugation of the switch curve with the adjacent straight part is carried out at an angle (initial switch angle).

At the same time, carriages along the turnout switch curve pass within a few seconds, therefore, due to inertial qualities, the time of action of the forces affecting passengers or cargo is relatively short. Under these conditions, the results of the action are largely determined by the changing rate of the acting forces, which is characterized by the values of the derivatives of the displacement of the second and higher order than the second, that is, the magnitudes of accelerations and the rate of their change.

At speeds up to 60 km/h, the factors that determine the smoothness of movement along the turnout switch are the values of horizontal lateral accelerations. The necessary parameters of smoothness of movement along the turnout are provided by standardizing the values of the suddenly appearing (when the wheel runs over the point) acceleration—j_0 and constantly acting acceleration when moving along the turnout curve—γ.

Based on these considerations and the design speed of movement along the turnout switch—V, the curvature parameters of the turnout curve are determined as:

$$R_1 = V^2/j_0, \quad R_2 = V^2/\gamma \tag{5}$$

wherein: R_1 is the radius of the turnout curve in the area of possible running-on of the wheels on the point; R_2 is the radius of the turnout curve outside the area of possible running-on of the wheels on the point.

The initial turnout angle for turnout switches with points of secant type is determined taking into account the standardized value of the kinetic energy loss factor—W_o and the biggest possible gap between the wheel flange and the stock rail—δ

$$sin\beta_n \leq \frac{1}{V}\sqrt{W_0^2 - 2\delta j_o} \tag{6}$$

At the motion speeds higher than 100 km/h along turnout curves for ensuring carriage transition smoothness on the turnout switches the compound turnout curves containing as a rule varying curvature sections are used [17–20]. When designing them, they proceed from a detailed analysis of the carriage movement conditions along the curved section. Based on the general provisions of mechanics, the radius vector characterizing the position of a point at each moment in time is equal to

$$\vec{r}(t) = x(t)\,\vec{i} + y(t)\,\vec{j} + z(t)\,\vec{k}, \tag{7}$$

wherein: *x(t), y(t), z(t)*—functions characterizing the change in the position of a moving point in time in the considered coordinate system. Acceleration vector

$$\vec{a} = \frac{dv}{dt} = \frac{d}{dt}(v\vec{\tau}) = \frac{dv}{dt}\vec{\tau} + v\frac{d\vec{\tau}}{dt}, \tag{8}$$

wherein: *v*—carriage motion speed along the turnout switch. Acceleration rate ($\vec{\mathrm{h}}$-vector)

$$\vec{h} = \frac{d\vec{a}}{dt} = \frac{d}{dt}(a_\tau)\vec{\tau} + a_\tau\frac{d}{dt}(\vec{\tau}) + \frac{d}{dt}(v^2)q\vec{n} + v^2\frac{d}{dt}(q)\vec{n} + v^2q\frac{d}{dt}(\vec{n}). \tag{9}$$

$\vec{\mathrm{h}}$-vector components

$$\vec{h} = \left(\frac{da_\tau}{dt} + v^3q^2\right)\vec{\tau} + \left(3va_\tau q + v^3\frac{dq}{dl}\right)\vec{n} + (v^3c'q)\,\vec{b} \tag{10}$$

longitudinal component | transverse component | vertical component

wherein: *q*—curve curvature (*q* = *1/R*); *c*—curve curvature in a vertical plain.

Of the three impact components, there is of our interest the component acting in the horizontal transverse direction, which at a constant motion speed along the turnout switch is equal to

$$h = v^3\frac{dq}{dl} \tag{11}$$

If we consider the motion not of a point, but of a carriage with a base *d*, then when entering the curve from a straight line, we get:

$$h = \frac{v^3}{Rd} \tag{12}$$

Values of effective and standardized values of accelerations $\vec{a}$ and $\vec{h}$—vectors determine selection and radii of turnout curves ensuring smooth transition along the turnout switch.

Two basic concepts are used in the design of turnout switches [20–23].

When moving along curves, centrifugal acceleration occurs, depending on the speed and radius of the curve. To compensate for the forces caused by it, an elevation of the outer rail is arranged in the curves. In the turnout curve, the elevation is not arranged, so the uncompensated part of the acceleration can be expressed in the form of elevation deficiency.

The value of the radius, based on the permissible value of the elevation deficiency, can be determined from the expression:

$$R_0 = \frac{12,5v_{\max}^2}{H} \tag{13}$$

wherein: $v_{\max}$—maximum motion speed; H—permissible value of elevation deficiency.

The design of such turnout curves is as follows. The entry to the turnout curve, within the area of running the wheels on the point, is performed with a constant radius. The initial turnout angle and the angle of impact of the wheels in the deflecting tongue are determined based on the standardized values of the loss of kinetic energy and sudden acceleration by conventional methods.

The R_1 radius is minimal. Outside the area of running wheels on the deflecting tongue, the radius of the turnout curve gradually increases to the value $R_2 = \infty$ in the middle of the crossover or at the tail-end of the frog. Straight portion of crossing may not be arranged. Curvature retraction, as a rule, is performed with a change in curvature along a cubic parabola or clotoid. The retraction rate is equivalent to the retraction rate of the outer rail elevation. Based on the permissible value of this indicator, the length of retraction is determined by the formula:

$$L = Hv_{\max}/(\Delta i/\Delta t) \tag{14}$$

At high speeds, the dynamic parameters of the interaction between the track and the carriage are proportional to the second and third-order derivatives of the displacement vector (horizontal transversal acceleration and the rate of its change).

When studying the kinematics of motion with speeds over 60 km/h, the $\vec{h}$-vector is decisive (for turnouts). If the speed of motion along the turnout switch is considered constant, then the formulas (11 and 12) should be used to calculate the motion. In principle, these relations allow projecting the curvature of the turnout curve according to the standardized value of the modulus of the $\vec{h}$--vector. Formula (11) corresponds to motion along the turnout curve of a point carriage. Therefore, in the practical determination of the parameters, it is necessary to proceed from the fact that the element that perceives the rounding of the track is the carriage chassis [24–27].

Formula (12) makes it possible to determine the initial radius of the switch points and, together with (11), to select the turnout curve. However, the analysis of practical implementations of this technique shows that in the practical analysis of the entry to the turnout switch, it is necessary to introduce simplifications in the calculation of the maximum allowable angle of impact.

$$k \approx \frac{\tau}{d}, \quad \frac{dk}{dl} \approx \frac{\Delta k}{\Delta l} \approx \frac{\tau}{d^2} \tag{15}$$

wherein τ—angle of impact at switch point.

As a result, we obtain the relation

$$h \approx v^3 \frac{\tau}{d^2} \tag{16}$$

From this ratio, the maximum allowable angle of impact into the switch point is found.

In the currently used turnouts, different options of the switch point geometry are used. The most common switch points are made according to one of the following schemes:

- switch points with constant or abrupt changing curvature, single-radius, double-radius;
- clotoid with an increase in curvature, with a decrease in curvature;
- combined;
- switch points in the form of a combination of a clotoid and a circular curve with an increase in curvature or with a decrease in curvature;
- switch points with a cosine change in curvature with zero initial curvature or non-zero initial curvature.

Each of these types of switch points can be made along a tangent or secant pattern. In addition, for each type, various subrange types are possible, associated with the design of the actual switch toes.

Implementation of dynamic and kinematic requirements ensuring smooth transition along the turnout switch can be ensured by using almost all the abovesaid schemes, therefore, during the design of the turnout switch and especially the exit ramp, design solutions of all these types should be considered and compared [28–31]. The best option is one that, when all conditions are met, ensures the minimum length and cost of a turnout switch or exit ramp.

The design of a turnout curve at the switch point may be done separately. The general mathematical formulation of the optimization problem, suitable for all options, has the form

$$\min\{L_0(k,n),\ L_c(k,n)\},$$

$$\left|\frac{dv}{dt}\right|_b \le a_b,$$

$$\left|\frac{dv}{dt}\right|_r = a_y,$$

$$\left|\frac{d^2v}{dt^2}\right|_k \le h,$$

$$\left|\frac{d^2v}{dt^2}\right|_d \leq \psi, \tag{17}$$

wherein:

L_0—switch point length;

L_c—adzing length;

n—tick (for secant type switch points);

a_b is the limit value of a sudden acceleration;

a_n—limiting value of the steady-state acceleration for options with sections of constant curvature or the greatest acceleration for a curve of variable curvature;

ψ—standardized acceleration rate at the entry to the turnout curve.

Selection of limiting values determines the quality of the technical solution. Sudden acceleration and ultimate acceleration values should be taken on the basis of existing design experience.

The turnout curve behind the switch heel continues the curve within the turnout switch. Its parameters are determined by the value of the standardized acceleration rate and the limiting value of steady-state acceleration. After reaching the limit value of acceleration (steady state), the curve can have a section of constant curvature. The areas adjacent to it have a variable curvature, which, as already indicated above, depends on the acceleration rate. Then follows a section of the curve that ensures symmetric, or not worse, than on the turnout switch, conditions for entering the curve [32–35]. After that, within the limits of the turnout, there is a straight portion of crossing (it may not be therein), which connects the adjacent turnout switches of the crossover.

The length of the straight portion of crossing is taken as a function of the motion speed.

$$\min L_{bc} = m\,v \tag{18}$$

The coefficient m depends on the speed and constriction of driving conditions. The value L_{bc} may be limited from below by the conditions of minimum length of the rail element or by other conditions. There are recommendations to arrange straight portions of crossing with a length of at least the carriage chassis.

Since different combinations of sections constituting the crossover are possible, the task of designing the turnout curve outside the switch point is also an optimization problem. Its mathematical formulation is as follows:

$$\min \sum_j L,$$

$$\alpha_i\left(v, \frac{dv}{dt}, \frac{d^2v}{dt^2}\right) \angle [\alpha]_i, \quad i = 1, \ldots 4$$

$$L_{bc}^{3}\min L_{bc}$$

$$L_k \geq \frac{a_y - a_b}{n}, \tag{19}$$

wherein: $\sum L$—turnout switch (crossover) length;
L_{bc}—length of straight portion of crossing;
L_k—length of clotoid section;
α_i—dynamic and kinematic criteria;
$(a_y - a_b)$—increment of accelerations in the section of the curve of variable curvature.

Already now, there is a need to speed up the design and implementation of turnouts on the operated railway network, allowing to significantly increase the permissible train speeds, especially on the side track. One of the initial points in the design of turnouts for high speeds is to provide passengers with a comfortable ride when the train moves on the side track by limiting the values of the so-called centrifugal acceleration a and the increment (change) of centrifugal acceleration per unit of time (second) ψ.

Based on the conditions for providing passengers with a comfortable ride when the carriages move along the turnout switch to the side track, the turnout curve must meet the requirements so that the emerging centrifugal accelerations, as well as their increment (change) per unit of time (second), do not exceed certain established values:

$$a \leq a_{inc}$$

$$\psi \leq \psi_{inc}$$

wherein a—value of uncompensated centrifugal acceleration, m/s^2; ψ—value of the change in the uncompensated centrifugal acceleration per unit of time (second), m/s^3.

On the basis of research, to check the ride comfort conditions when a passenger train moves along a turnout to a side track, it can be considered acceptable in the car body: centrifugal acceleration a = 1.0 m/s^2, and its increment (change) per unit time ψ = 0, 8 m/s^3, since the impact of these values on passengers within the turnout curves will be short-term and repeated at relatively long intervals.

On our railway network, turnouts with constant radius turnout curves are widely used. In the turnout switch of the last design made of P50 rails with a 1/18 cross-piece grade, the radius of the switch points up to a section of 40 mm is 1698.0 m, and then along the switch point and the turnout curve—960.0 m. The permissible speed of trains on the side track of this turnout, based on the conditions for providing passengers with a comfortable ride, according to experimental data carried out by the VNIIZhT Turnout Switch Economy laboratory, can be set within 80 to 85 km/h.

Professor G. M. Shakhunyants, when considering the issue of the speeds of movement along the turnouts to the side track, pointed out that the permissible speeds of movement should be assigned in such a way that, as a result of the interaction of the track and the rolling stock, unacceptable effects on both the rolling stock and passengers are not caused. As set permissible values of the uncompensated centrifugal acceleration, changes in the centrifugal acceleration per unit of time and the loss of kinetic energy per impact at the entry to the turnout, the speed of the train and the radius of turnout curve are interdependent and determine the main parameters of the turnout switch.

In the calculations below, a given train speed is assumed, and the radius of the turnout curve is determined as a function of the speed. As you know, the value of the uncompensated centrifugal acceleration when the carriage moves in the curve of radius R, which has an elevation of the outer rail h, can be determined by the formula:

$$a_{ou} = \frac{V^2}{R} - \frac{gh}{S_1} \tag{20}$$

In absence of elevation of an outer rail, which usually occurs in turnout curves, the value of centrifugal acceleration will be determined by the formula:

$$a_{ou} = \frac{V^2}{R} \tag{21}$$

Taking $a = 1.0$ m/s^2 and a conversion factor of 3.6^2, we obtain an expression for determining $R_{\min}$ from the conditions for limiting the magnitude of centrifugal acceleration

$$R_{\min} = \frac{V^2}{3,6^2 \cdot 1,0} = 0,0772V^2 m \tag{22}$$

Increase in centrifugal acceleration at the entry to the turnout switch (at R = const) prof. P. G. Koziychuk [8] proposed to determine by the formula:

$$\psi = \frac{V^3}{Rb} \tag{23}$$

Taking $\psi = 0.8$ m/s^3, $b = 17.0$ m (distance between the axes of rotation of rotary bogies in all-metal cars) and also introducing a conversion factor of 3.6^3 for ease of use, we obtain an expression for determining $R_{\min}$ from the conditions for limiting the increment value (changes) of centrifugal acceleration per unit of time (second).

$$R_{\min} = \frac{V^3}{3,6^3 \cdot 0,8 \cdot 17} = 0,00158V^3 \text{ m} \tag{24}$$

Thus, the minimum value of radius of a turnout curve (in case of R = const) at a given speed, according to the conditions of ride comfort, will be determined by:

1. Under conditions of limiting the magnitude of centrifugal acceleration—from the expression $R_{\min} = 0.0772\ V^2$ m;
2. Under conditions of limiting the magnitude of the increment (change) of centrifugal acceleration per unit of time (second)—from the expression $R_{\min} = 0.00158\ V^3$ m.

Table 1 shows the results of calculating the values of minimum radii depending on the train speed on the side track.

According to the data of Table 1, a graph (Fig. 1) of $R_{\min}$ values is drawn depending on the train speed along the turnout curve at the given a = 1.0 m/s^2 and ψ = 0.8 m/s^3.

It can be established from the graph that at a motion speed along the side track of up to 50 km/h, the possible minimum radius of the turnout curve is determined by limiting the magnitude of centrifugal acceleration, and at a speed of more than 50 km/h, by limiting the magnitude of the increment (change) of centrifugal acceleration per unit time. From this, we can conclude that on turnout switches for high speeds, the turnout curve parameters will be primarily determined by limiting the magnitude of the centrifugal acceleration change per unit time. In this case,

Table 1 Results of calculating values of minimum radii depending on train speed onto side track

Train side track speed, km/h	Values	
	From conditions of centrifugal acceleration limited magnitude (a = 1.0 m/s^2)	From conditions of limiting magnitude of increment (change) of centrifugal acceleration per unit of time (ψ = 0.8 m/s^3)
40	123.52	101.12
50	193.00	197.12
60	277.92	341.28
70	378.28	541.94
80	494.08	808.96
90	625.32	1151.82
100	772.00	1580.00
110	934.12	2102.98
120	1111. 68	2730.24
130	1304.68	3471.26
140	1513.12	4335.52
150	1737.00	5332.50
160	1976.32	6471.68

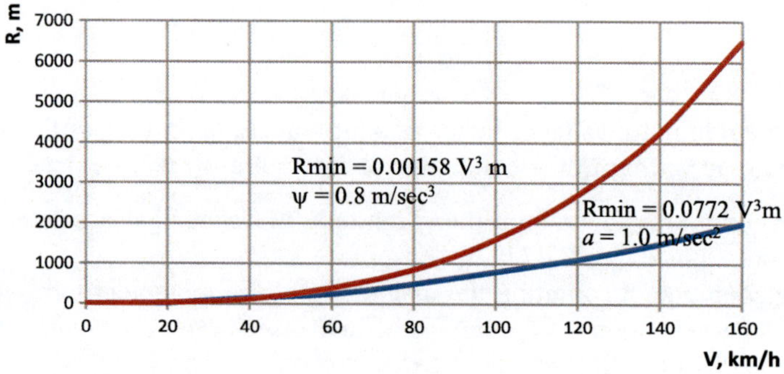

Fig. 1 Graph of calculating results of minimum radii values depending on train speed onto side track

variable curvature curves whose variation in curvature meets the requirements of monotone increment (change) of centrifugal acceleration magnitude are particularly useful as turnout curves.

In a curve of variable curvature that meets these requirements, the effect of factors (changes in accelerations over time) that especially unpleasantly affect the human body, which is very significant when implementing high speeds of passenger trains, is provided less sharply than in the curve of constant radius.

The feasibility of using variable radius curves in turnout switches for high speeds of motion is indicated in their studies by prof. P. G. Koziychuk [8], prof. S. V. Amelin, Candidate of Engineering, Sciences G. I. Ivashchenko and other authors. It is known that a variable curvature turnout curve in the flat model turnout switches may consist of one or two branch turnout curves or may be composed of two turnout curves and a constant radius curve inserted therebetween. In the turnout curve of variable radius, varying from $\rho = 0$ to $\rho = \text{R}$, the minimum value of the radius from the condition of limiting the magnitude of centrifugal acceleration will be determined, as for the curve of constant radius.

Therefore, the values of $R_{\min}$ for curves of variable radius, determined from the conditions for limiting the magnitude of centrifugal acceleration, will correspond at different speeds to the data given in Table 1 (column 2). Change in centrifugal acceleration per unit of time ψ in curves of variable radius, within which the centrifugal acceleration a increases from $a = 0$ to some finite value $a = V^2/R$, prof. BN Vedenisov [13] proposed to define v^2 as follows:

$$\text{if } a = V^2/\rho, \quad \text{and} \quad \psi = da/dt$$

Then

$$\psi = \frac{d\left(\frac{V^2}{\rho}\right)}{dt} \tag{25}$$

At $\rho = R$, the magnitude of the change in centrifugal acceleration per unit of time will be determined from the expression:

$$\psi = \frac{V^2}{Rt} \tag{26}$$

wherein t = $\frac{l}{V}$, herein l—theoretical length of branch turnout curve.

Then $\psi = \frac{V^3}{Rt}$, wherefrom $l = \frac{V^3}{\psi R}$

Taking ψ = 0.8 m/s^3 and introducing a conversion factor of 3.6^3 for the convenience of using the formula, we get:

$$l = \frac{V^3}{3,6^3 \cdot 0,8 \cdot R} = 0,0268\frac{V^3}{R}\,\text{m} \tag{27}$$

Wherein V—train speed, km/h, R—minimum radius value of turnout curve, m.

For different train speeds and the corresponding $R_{\min}$ from the conditions for limiting the magnitude of centrifugal acceleration, assigned according to Table 1 (column 2), the length of the branch turnout curve was calculated from the conditions for limiting the magnitude of the increment (change) of centrifugal acceleration per unit time. The calculation results are presented in Table 2.

Table 2 Results of calculating length of branch turnout curve

Train motion speed onto side track, km/h	$R_{\min}$ from conditions of centrifugal acceleration limited magnitude (a = 1.0 m/s^2), m	Theoretical length of branch turnout curve from conditions for limiting magnitude of increment (change) of centrifugal acceleration per time unit (ψ = 0.8 m/s^3), m
1	2	3
40	124	13.83
50	193	17.36
60	278	20.82
70	378	24.32
80	494	27.78
90	625	31.26
100	772	34.71
110	934	38.19
120	1112	41.65
130	1305	45.12
140	1513	48.60
150	1737	52.07
160	1976	55.55

From Table 2 it follows that with the assigned values of R_{min} from the conditions for limiting the magnitude of centrifugal acceleration, the theoretical length of the branch turnout curve, determined from the conditions for limiting the magnitude of the increment (change) of centrifugal acceleration per unit of time, with an increase in the speed of movement, increases significantly less than the radius value of circular curve (Table 1, column 3). This indicates expediency of using curves of variable curvature in turnouts for high speeds of motion as a turnout curve.

It is recommended to take angle of switch in ordinary and symmetrical turnouts for high speeds in stages: 1/18, 1/22 and 1/36 [14, 15]. The following conclusions can be drawn from the above.

3 Conclusions

The proposed method allows further forming an approach to unification of requirements for turnout switches and other types of turnout products. This will make it possible to increase the safety, comfort and specified operational reliability indicators for new turnout designs.

If it is necessary to realize the motion speed along the turnout switch to the side track above 50 km/h, the main factor in determining the radius of the turnout curve according to the conditions of ride comfort is to limit the magnitude of the increment (change) of centrifugal acceleration per unit of time (second).

Based on this, when designing turnouts for high speeds, it is advisable to use curves of variable radius as a turnout curve.

References

1. Hirkovskis A, Serdjuks D, Goremikins V, Pakrastins L, Vatin NI (2015) Behaviour analysis of load-bearing aluminium members. Mag Civ Eng 57(5):86–117. https://doi.org/10.5862/MCE.57.8
2. Isaev SA, Baranov PA, Vatin NI, Zhukova YV, Sudakov AG (2014) Suppression of the Karman vortex street and reduction in the frontal drag of a circular cylinder with two vortex cells. Tech Phys Lett 40(8):653–656. https://doi.org/10.1134/S1063785014080057
3. Isaev SA, Vatin NI, Baranov PA, Sudakov AG, Usachov AY, Yegorov VV (2013) Development and verification of multiblock computational technologies for solution of unsteady problems of high building aerodynamics in the framework of URANS approach. Mag Civ Eng 36(1):103–109. https://doi.org/10.5862/MCE.36.13
4. Isaev SA, Vatin NI, Guvernyuk SV, Gagarin VG, Basok BI, Zhukova YV (2015) Drag reduction of energy-efficient buildings and wind energy extraction due to bleeding effect. High Temp 53(6):873–876. https://doi.org/10.1134/S0018151X15040136
5. Isaev SA, Vatin NI, Lebiga VA, Zinoviev VN, Chang K-C, Miau J-J (2013) Problems and methods of numerical and experimental investigation of high rise constructions' aerodynamics in the coastal region "sea-land." Mag Civ Eng 37(2). https://doi.org/10.5862/MCE.37.8

6. Jakšic Z, Ladjinović D, Trivunić M, Harmati N, Vatin N (2015) Masonry construction remedial measures in case of a multi-story housing facility caused by floor extension process. Proc Eng 117(1):502–515. https://doi.org/10.1016/j.proeng.2015.08.252
7. Kanyukova S, Vatin N, Leybman D, Sazonova T (2016) Dynamic control method of design terms in underground construction. Proc Eng 165:1918–1924. https://doi.org/10.1016/j.proeng.2016.11.942
8. Khodzhaev D, Abdikarimov R, Vatin N (2018) Nonlinear oscillations of a viscoelastic cylindrical panel with concentrated masses. In: MATEC web of conferences, vol 245. https://doi.org/10.1051/matecconf/201824501001
9. Savin A, Korolev V, Loktev A, Shishkina I (2020) Vertical sediment of a ballastless track. Adv Intell Syst Comput 1115:797–808. https://doi.org/10.1007/978-3-030-37916-2_78
10. Korolev V (2020) Switching shunters on a slab base. Adv Intell Syst Comput 1116:175–187. https://doi.org/10.1007/978-3-030-37919-3_17
11. Loktev A, Korolev V, Shishkina I, Illarionova L, Loktev D, Gridasova E (2020) Perspective constructions of bridge crossings on transport lines. Adv Intell Syst Comput 1116:209–218. https://doi.org/10.1007/978-3-030-37919-3_20
12. Glusberg B, Savin A, Loktev A, Korolev V, Shishkina I, Alexandrova D, Loktev D (2020) New lining with cushion for energy efficient railway turnouts. Adv Intell Syst Comput 982:556–570. https://doi.org/10.1007/978-3-030-19756-8_53
13. Loktev A, Korolev V, Shishkina I, Chernova L, Geluh P, Savin A, Loktev D (2020) Modeling of railway track sections on approaches to constructive works and selection of track parameters for its normal functioning. Adv Intell Syst Comput 982:325–336. https://doi.org/10.1007/978-3-030-19756-8_30
14. Korolev V (2020) Guard rail operation of lateral path of railroad switch. Adv Intell Syst Comput 1115:621–638. https://doi.org/10.1007/978-3-030-37916-2_60
15. Glusberg B, Savin A, Loktev A, Korolev V, Shishkina I, Chernova L, Loktev D (2020) Counter-rail special profile for new generation railroad switch. Adv Intell Syst Comput 982:571–587. https://doi.org/10.1007/978-3-030-19756-8_54
16. Gridasova E, Nikiforov P, Loktev A, Korolev V, Shishkina I (2020) Changes in the structure of rail steel under high-frequency loading. Adv Intell Syst Comput 1115:559–569. https://doi.org/10.1007/978-3-030-37916-2_54
17. Klyuev S, Klyuev A, Vatin N (2018) Fine-grained concrete with combined reinforcement by different types of fibers. In: MATEC web of conferences, vol 245. https://doi.org/10.1051/matecconf/201824503006
18. Klyuev SV, Klyuev AV, Vatin NI (2018) Fiber concrete for the construction industry. Mag Civ Eng 84(8):41–47. https://doi.org/10.18720/MCE.84.4
19. Korneeva EA, Vatin NI, Dontsova AE (2019) Mechanical properties of the Crimean limestone, treated with material based on silicic acids. Mag Civ Eng 85(1):59–70. https://doi.org/10.18720/MCE.85.6
20. Korniyenko SV, Vatin NI, Gorshkov AS (2016) Thermophysical field testing of residential buildings made of autoclaved aerated concrete blocks. Mag Civ Eng 64(4):10–25. https://doi.org/10.5862/MCE.64.2
21. Korsun V, Kalmykov Y, Niedoriezov A, Korsun A (2015) The influence of the initial concrete strength on its deformation under triaxial compression. Proc Eng 117(1):959–969. https://doi.org/10.1016/j.proeng.2015.08.190
22. Korsun V, Korsun A, Volkov A (2013) Characteristics of mechanical and rheological properties of concrete under heating conditions up to 200°C. In: MATEC web of conferences, vol 6. DOI:https://doi.org/10.1051/matecconf/20130607002
23. Korsun V, Vatin N, Franchi A, Korsun A, Crespi P, Mashtaler S (2015) The strength and strain of high-strength concrete elements with confinement and steel fiber reinforcement including the conditions of the effect of elevated temperatures. Proc Eng 117(1):970–979. https://doi.org/10.1016/j.proeng.2015.08.192

24. Korsun V, Vatin N, Korsun A, Nemova D (2014) Physical-mechanical properties of the modified fine-grained concrete subjected to thermal effects up to 200 °C. Appl Mech Mater 633–634:1013–1017. https://doi.org/10.4028/www.scientific.net/AMM.633-634.1013
25. Korsun VI, Kalmykov YY, Makarenko SY (2017) A version of the failure criterion modification for plane concrete. Key Eng Mater 755:300–321. https://doi.org/10.4028/www.scientific.net/KEM.755.300
26. Kovačič B, Kamnik R, Pustovgar A, Vatin N (2016) Landscape interventions and appreciations with program software RO. Proc Eng 165:918–925. https://doi.org/10.1016/j.proeng.2016.11.801
27. Kovačič B, Kamnik R, Solovev V, Vatin N (2016) The comparison of measuring and computing results of construction vertical displacement. Proc Eng 165:926–935. https://doi.org/10.1016/j.proeng.2016.11.802
28. Kovačič B, Kamnik R, Štrukelj A, Vatin N (2015) Processing of signals produced by strain gauges in testing measurements of the bridges. Proc Eng 117(1):795–801. https://doi.org/10.1016/j.proeng.2015.08.249
29. Kovačič B, Kamnik R, Vatin N, Ishkov A (2016) The vertical displacement measurement of concrete plates with special emphasis on rough error. Proc Eng 165:936–946. https://doi.org/10.1016/j.proeng.2016.11.803
30. Lanko A, Vatin N, Kaklauskas A (2018) Application of RFID combined with blockchain technology in logistics of construction materials. In: MATEC web of conferences, vol 170. https://doi.org/10.1051/matecconf/201817003032
31. Lazarevska M, Cvetkovska M, Knezevic M, Trombeva Gavriloska A, Milanovic M, Murgul V, Vatin N (2014) Neural network prognostic model for predicting the fire resistance of eccentrically loaded RC columns. Appl Mech Mater 627:276–282. https://doi.org/10.4028/www.scientific.net/AMM.627.276
32. Melović B, Mitrović S, Djokaj A, Vatin N (2015) Logistics in the function of customer service-relevance for the engineering management. Proc Eng 117(1):802–807. https://doi.org/10.1016/j.proeng.2015.08.255
33. Milajić A, Beljaković D, Davidović N, Vatin N, Murgul V (2015) Using the big bang—big crunch algorithm for rational design of an energy-plus building. Proc Eng 117(1):911–918. https://doi.org/10.1016/j.proeng.2015.08.178
34. Mirsaidov MM, Abdikarimov RA, Vatin NI, Zhgutov VM, Khodzhaev DA, Normuminov BA (2018) Nonlinear parametric oscillations of viscoelastic plate of variable thickness. Mag Civ Eng 82(6):112–126. https://doi.org/10.18720/MCE.82.11
35. Mrdak I, Rakočević M, Žugić L, Usmanov R, Murgul V, Vatin N (2014) Analysis of the influence of dynamic properties of structures on seismic response according to montenegrin and european regulations. Appl Mech Mater 633–634:1069–1076. https://doi.org/10.4028/www.scientific.net/AMM.633-634.1069

Technology for Recovery of Defective Rail Bars

Alexander Savin, Alexey Loktev, Vadim Korolev, and Irina Shishkina

Abstract The continuous welded rail track is the main design of the track used by the railways and can be operated in accordance with the technical requirements established by the railways on the tracks of all classes. The paper presents the calculation of the possibility of laying a continuous welded rail track. Analysis of works on temporary and final restoration of welded rail strings is carried out. At the final restoring, rail strings can be welded by electric contact or aluminum thermite welding. Depending on the location of cropped rail or the place of temporary restoration relative to the ends of the restored rail string, the final restoration by electric contact welding can be carried out with tightening of the welded rail string or with its preliminary bending. Final restoration of rail strings by aluminum-thermite welding can be carried out immediately after identification of a dangerous defect requiring cutting or after temporary restoration of the string. Laying of continuous welded track with or without replacement of the assembled rails and sleepers should be carried out in accordance with the projects and technologies approved by the railways. When planning work, it is necessary to take into account daily and long-term forecasts of the rail temperature. During the work, continuous monitoring of the rail temperature, carried out with the help of portable rail thermometers, should be organized.

Keywords Continuous welded rail track · Restoration of rail strings · Electric contact welding · Aluminum-thermite welding

1 Introduction

On the railways of the Russian Federation, a temperature-stressed structure of a continuous welded track is used. The main difference between the operation of a continuous welded track and a conventional link track is that significant longitu-

A. Savin · A. Loktev · V. Korolev · I. Shishkina (✉)
Russian University of Transport (MIIT), Chasovaya Str., 22/2, Moscow 125190, Russia
e-mail: shishkinaira@inbox.ru

A. Bieliatynskyi and V. Breskich (eds.), *Safety in Aviation and Space Technologies*, Lecture Notes in Mechanical Engineering,
https://doi.org/10.1007/978-3-030-85057-9_37

dinal forces caused by temperature changes act in the rail strings. When the temperature of the rail strings rises in comparison with the fixing temperature, longitudinal compressive forces arise in them, which can create a danger of the track ejection. With a decrease in temperature, tensile forces appear, which can cause the rail string to break and form a large gap, dangerous for the train transition, or rupture of the rail joint due to the shear of the bolts. An additional effect on the continuous welded track is exerted by the forces arisen during track adjusting, track lining, cleaning of rubble and other repair track works.

2 Research Methods and Results

The possibility of laying a continuous welded rail track in specific conditions is established by comparing the permissible temperature amplitude [T] for these conditions with the amplitude of temperature fluctuations T_A actually observed in a given area.

If $T_A < [T]$, then the continuous welded rail track can be laid. The T_A value is defined as the algebraic difference between the highest t_{maxmax} and the lowest t_{minmin} rail temperatures observed in a given area (taking into account that the highest rail temperature in open areas exceeds the highest air temperature by 20 °C):

$$T_A = t_{\mathrm{maxmax}} - t_{\mathrm{minmin}} \tag{1}$$

Amplitude of allowable rail temperature changes:

$$[T] = [\Delta t\_y] + [\Delta t\_s] - [\Delta t\_f], \tag{2}$$

wherein

$[\Delta t_y]$ is the permissible increase in the temperature of the rails in comparison with the temperature of their fastening, determined by the stability of the track against ejection under the action of compressive longitudinal forces;

$[\Delta t_s]$ permissible decrease in the temperature of rail strings in comparison with the temperature of fastening, determined by their strength under the action of tensile longitudinal forces;

$[\Delta t_f]$ the minimum temperature range in which the rail strings are finally fixed; according to the conditions of work for calculations, it is usually taken equal to 10 °C, but if necessary, it can be reduced to 5 °C if we provide for the fastening of the rail strings in autumn, in cloudy weather, in the early morning or evening hours, when the temperature of the rails changes slowly during the fastening process, or when the rail strings are planned to be taken in operation into the calculated temperature range using compulsory means (stretching devices, heating installations).

The permissible temperature rise of rail strings [Δt_y] is established on the basis of theoretical and experimental studies of track stability.

The permissible decrease in the temperature of the rail strings is determined by calculating the strength of the rails, based on the condition that the sum of the tensile stresses arising from the action of the rolling stock and from temperature changes should not exceed the permissible stress of the rail material:

$$k_p \sigma_k + \sigma_t \leq [\sigma] \quad (3)$$

wherein k_p is the safety factor (k_p = 1.3 for rails of the first service life and old-year rail strings that have undergone diagnostics and repairs in stationary conditions or profile grinding and diagnostics on the way; k_p = 1.4 for rails that have carried passing by the standard tonnage or were re-laid without grinding);

σ_k stresses in the edges of the rail base from bending and torsion under load from the wheels of the rolling stock, MPa;

σ_t stresses in the cross-section of the rail from the action of tensile temperature forces arising when the temperature of the rail decreases in comparison with its temperature during fastening, MPa; [σ]-admissible stress (for heat-strengthened rails [σ]—400 MPa).

Stresses in the rail base sк are determined according to the rules for calculating the upper structure of the track for strength. In this case, the moduli of elasticity of the under-rail base in winter with wooden sleepers (U_w^w) are taken equal to 40 and 50 MPa; with reinforced concrete sleepers (U_{rc}^w) with rubber and rubber-cord gaskets—120 and 130 MPa (respectively, at 1840 and 2000 sleepers per 1 km).

Temperature stress arising in the rail due to a failed change in its length when the temperature changes:

$$\sigma_t = \alpha \cdot E\Delta t \approx 2.5\Delta t, \quad (4)$$

wherein

α coefficient of linear expansion of rail steel (α = 0.0000118 1/deg);

E modulus of elasticity of rail steel (E = 210 GPa = 2.1 × 105 MPa);

Δt is the difference between the temperature at which the stress is determined and the temperature at which the rail string is fixed on the sleepers, °C.

The maximum permissible decrease in the temperature of the rail string in terms of rail strength condition compared to its temperature when fixed:

$$[\Delta t_p] = \frac{[\sigma] - k_p \sigma_k}{\alpha \cdot E} = \frac{[\sigma] - \kappa_p \cdot \sigma_k}{2.5} \quad (5)$$

In accordance with the specified calculation procedure, the temperature of the rail strings admissible by the strength condition [Δt_p] was determined in comparison with the temperature of their fastening for a continuous-welded track with rails of the first service life type on reinforced concrete sleepers and crushed stone or asbestos ballast, depending on the type of locomotives in circulation, the realized motion speed and radii of curves.

Work on the temporary restoration of the rail strings should be carried out in special “time windows”.

Before temporarily restoring a rail string, it is necessary to determine its temperature and the actual fixing temperature and prepare a temporary rail with three bolt holes at the ends of at least 8 m in length.

After the first cut, the replaceable end of the cropped rail string is released from the intermediate rail fasteners before the planned second cut and to be tapped with hammers. The position of the second cut is marked after the ends of the rail string have been completely moved in the place of its first cut.

When the temperature of the rail string is higher than the temperature of its fixation, compressive forces act in it, the values of which are determined by the difference between the indicated temperatures. In such cases, it is necessary to first cut a piece of rail 10–20 cm long at the point of the defect with a gas cutter, and then cut the rail ends of the rail strings at the required distances with a rail cutting machine. Before cutting out a piece of rail, it is necessary to check the fastening of the strings with intermediate rail fasteners in the section of cutting out the defective rail. If the tightening of the bolts and screws does not correspond to the standard meanings, then they must be tightened. To ensure that the cut piece of rail is not pinched during the cutting process, to ensure the safety of workers, it should be cut sequentially. First, the whole head and neck are cut to half height along the ends of the cut piece, then the part of the rail located between two vertical slots is removed by horizontal cutting, after which two parts of the sole on both sides of the neck are cut out on the same length of 10–20 cm. The remaining part of the neck is gradually burned out by means of strictly vertical movements of the cutter until a free gap [1–4] is formed.

The final restoration of rail strings consists in cutting out a part of the string with a defect, including taken into the lining, or in removing the rail laid during temporary restoration, and welding in their place a rail insert, which is the same in terms of wear and quality of metal with the rail string, and does not exceed it on the operating time of more than 100 million gross tons.

At the final restoration, the rail strings can be welded by electric contact or aluminum thermite welding. The restoration of the rail strings by welding is carried out at the temperature of their fixing. It is allowed to restore the rail strings when the rail temperature is higher than its fastening temperature, but not more than 10 °C. The permissible temperature drops of the rails relative to the fastening temperature during the restoration of the strings are determined in each specific case, depending on the production technology, but should not be lower than 0 °C for electrical contact and −5 °C for aluminum thermite welding of rails.

At the final restoration of cropped rail string integrity, if the temperature of the rail string exceeds its fastening temperature, or with a temporary rail in the absence of butt gaps at its ends, it is necessary to first remove the stresses, in the first case—in the rail string at the work site, and in the second—in temporary rail. Relief of stresses in the rail string and in the temporary rail is performed by cutting a piece of rail [5–8].

After completion of welding in the rail string at the work site, its fixing temperature must be restored. Works on final restoration of rail strings integrity by welding and restoration of their fixation temperature in the work area shall be performed in accordance with the Technological Processes.

Final restoration of rail strings by electric contact welding is carried out by mobile rail welding machines (RPSM) with welding heads of type K-922, MSR-120.01, etc., ensuring tightening forces up to 120 tons, or of type K-900, MSR-8001, K-355, etc., ensuring tightening forces from 65 to 80 tons. Taking into account the fact that when the weld joint is compressed at the final moment of welding, forces of about 50 tons are required, in the first group of RPSM directly for tightening (stretching) of rail strings it can be used not more than 70 ton, in the second group with tightening forces of 80 ton—not more than 30 ton, and 65 ton—not more than 15 ton.

During final restoration of the cropped rail string with defect taken into the fish pads, a piece of rail with the defect is cut out from it and a prepared rail insert (without bolt holes) with a length of at least 8 m is welded in. In final restoration by welding the rail string with a temporary rail, the string ends with bolted holes are cut off with temporary rail and the rail insert is welded instead. Prior to welding of rail insert and the ends of the RPSM rail string, control welding of two joints of similar rails shall be performed and required metal reserve for welding of two joints shall be determined [9–12].

Depending on the location of the cropped rail or the place of temporary restoration relative to the ends of the restored rail string, the final restoration by electric contact welding can be carried out with tightening of the welded rail network or with its preliminary bending. Welding with tightening of welded rail string is performed at location of recovery point at the distance not more than 150 m from the end of the rail string. When welding with tightening, a short part of the string (up to 150 m long) is unlocked and hung on each 15th sleeper on paired plates or rollers with a diameter of 20–22 mm.

After cutting a piece of a rail with a defect or trimming the ends of the strings with bolt holes in the places of temporary restoration of the string, removing the cut rail with a defect or a temporary rail with cut ends of the strings with bolt holes and placing a previously prepared rail insert instead of them, it is welded the string with the ends of adjacent strings.

Welding of a rail insert with a trackable string, and then a hung short string (up to 150 m long) with the other end of the rail insert is carried out by pulling. If the length of the welded string is more than 150 m, the strings are welded with preliminary bending of the welded string [13–16].

When welding with preliminary bending, only part of the string is loosened. On the section BC of 5 m long for facilitation of movement of a string the nuts of the KB-65 fastening bolts are unscrewed on several turns, fastening clips of ZhBR-65 type are overturned on 180° and to be tightened with a force of 100–150 N·m, and the ARS-4 fastening mono-regulator is set on the first position. On the CD section of 40 m length, the KB-65 and ARS-4 fastening clips are removed, and the ZhBR type fastening clips are installed in the mounting position.

After preparing the rail insert, the length of which should correspond to the distance between points A, B ($l_{A\text{-}B}$), plus the margin for welding of 2 joints on the spine of two cuts (2 mm) and plus the margin for the residual bend boom (5–6 mm), i.e.

$$L_{ins} = l_{A-B} + 2\delta + 2 + 5,\ \text{mm}$$

The rail insert is welded to the ends of the string in section A. After welding in section A, the stage of rail string relative to the rail insert (Δl_{st}) must correspond to:

$$\Delta l_{st} = \delta + 1 + 5 = \delta + 6,\ \text{mm}$$

The unfastened part of the string is hung out and bent in a horizontal plane on straight sections towards the axis of the track or outwards, and on curves only to the outside of the curve. The bend of the string should be completed when the end of the string coincides with the end of the previously welded rail insert.

During the welding process, the curved part of the string is straightened under the action of the longitudinal force generated by the welding machine. At the end of welding, the string should not occupy its original position—the residual bend boom should remain within 15–25 cm. If the residual boom (the greatest distance from the inner edge of the sole relative to the bend of the string to the lining flange farthest from the rail (with lining fasteners) or support bracket (with fasteners of the ZhBR type)) goes beyond the specified limits, the welded joint must be rejected and cut out of the string. The actual residual bend boom should be recorded by the welder in a shift log book.

After cooling of the closing joint (2–3 min after the end of welding), the remaining curved part of the rail string is straightened by applying a transverse force. Fastening of the string at the bend section must be performed in the direction from the closing joint [17–20].

The displacement of the string in the CD section during its bending before welding, in the welding process and during straightening after welding should occur along three metal side beams, evenly distributed over the bending section, while the free movement of the string along them should be ensured without high friction.

If the restoration of a rail string by welding is carried out at a temperature of the rails higher than its fastening temperature, but not more than 10 °C, then in the section including the site for the restoration of the string plus 100 m on both sides, stress adjustment must be performed, after which the strings should be fastened. The logbook of the service and the temperature regime of the rail strings must contain the date of the final restoration of the string, the temperature of the string during the work performance and the limits of stress adjustment.

If the restoration of the string by welding is carried out at a rail temperature below its fastening temperature, then before welding it is necessary to calculate the

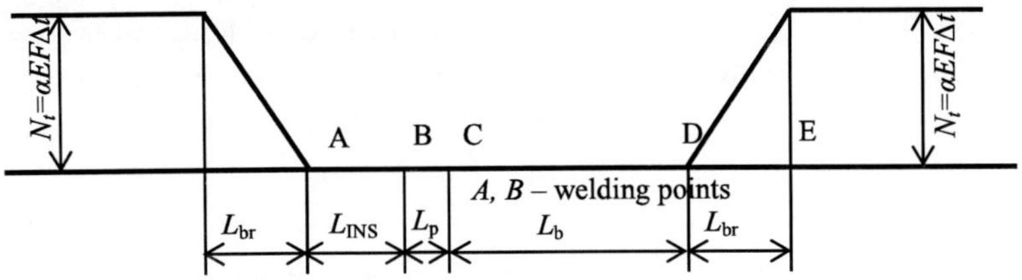

Fig. 1 Diagram of temperature forces in rail string in area of work site before welding and stretching one of string ends for restoring temperature of string fastening

elongation of the string at the work site for restoration within its temperature forces (string fastening temperature).

Figure 1 shows a diagram of the temperature forces in the string at the area of work site before its restoration by electric contact welding.

Figure 1 shows the following designations:

L_{ins} is length of rail insert to be welded (from 8 to 12.5 m);
L_p section with free rail string slipping (without pressing rail) 5 m long;
L_b string section, released from fastening to form loop when second joint welding (section of string bend);
$L_b + l_p = l_{CB}$ total length of free change in string length;
L_{br} is breathing string section, determined from condition:

$$l_{br} = N_t/r = \alpha EF\Delta t/r, \tag{6}$$

wherein

N_t temperature force occurring in rail string at temperature difference relative to temperature of its fixation;
Δt differential temperature of rail string during work with respect to its fixing temperature;
r linear resistance to longitudinal shift of track along one rail string, taken for stabilized ballast equal to 12 kN/m, for unstabilized to 7 kN/m.

The length of the welding work area where it is necessary to restore the temperature force (N_t), will eventually be equal to

$$l = l_{INS} + l_{CB} + l_{br}. \tag{7}$$

Value of required elongation of this section is determined from the condition:

$$\Delta l = \alpha(l_{INS} + l_{CB} + l_{br})\Delta t. \tag{8}$$

Lengthening of the string at the work site is carried out by unfastening the end of the string at the section l_p, adjacent to the first welded joint (point A), and stretching it with a hydraulic tensioner (GNU). In order to restore the temperature of the rail string fixing in the welding area, i.e. a temperature force equal to N_t, it is necessary to stretch the end of the unfastened part of the string with a force

$$N_t + \Delta N, \text{wherein } \Delta N = N_t(l_{INS} + l_{CB} + I_{br})/(l_p - l_{br}). \tag{9}$$

It follows from this that to enter string at work site into fixing temperature, GNU tensile forces must be at least:

$$N_{GNU} \geq N_{tot} = N_t + \Delta N. \tag{10}$$

The length of the section to be unfastened l_p must be at least:

$$l_p \geq \frac{(l_{INS} + l_{CB} + l_{br})N_t}{N_{GNU} - N_t} + l_{br}. \tag{11}$$

The value of the initial elongation of the string Δl_{tot} is determined taking into account its elongation to restore N_t in the section l_p, the section of welding work l and the "rollback" of the stretched and fixed end of the string Δl_o at point A.

$$\Delta l_{tot} = \Delta l_m + \Delta l + \Delta l_o + 5, \text{mm} \tag{12}$$

wherein

5 mm—elongation of the stretched end of string, necessary to compensate for the compressive stresses that appear during the elimination of the residual bending boom when welding the 2nd joint.

With close values of N_{GNU} and N_{tot}, the "rollback" value is determined from the condition:

$$\Delta l_o = N_{GNU}^2 / 2EFr. \tag{13}$$

It is possible not to take into account in the calculations the magnitude of the lengthening of the straightened end of the string (Δl_p), but for this, before it is unfastened, it is necessary to fix the string end position [21–24]. In the process of stretching, it is necessary to stretch the string until its end coincides with the fixed point, and then by (Δl + Δl_o + 5) mm.

After fixing the stretched part of the string l_p, removing the GNU and performing welding, including welding the first joint at point B by the pull-up method, and the second at point C by the preliminary bending method, the diagram of temperature forces in the section (l + l_p) will have the form presented in Fig. 2.

After adjusting the temperature forces in the AE section, their values at the working site for string restoration and on the approaches to it must correspond to N_t.

As an example, let us consider restoring the string integrity by welding with RPSM and restoring its fixing temperature at the work site, provided that the fixing

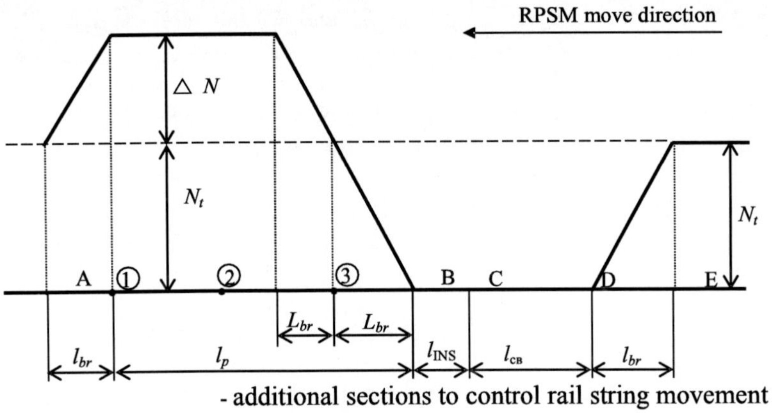

Fig. 2 Diagram of temperature forces before stress adjustment (solid line) and after adjustment (dashed line)

temperature of the string is +35 °C, and the work is performed at a rail temperature of +15 °C, i.e. the difference between them is $\Delta t = t_f - t_s = 20$ °C.

Before work commencement, proceeding from the condition $N_{GNU} \geq N_t + \Delta N$, substituting the corresponding values into it, we determine N_{GNU}.

$$N_{GNU} = N_t + \Delta N = \alpha EF\Delta t + N_t(l_{LNS} + l_{CB} + l_{br})/(l_p - l_{br}).$$

To determine N_t, we substitute the values of α, E, F (for P65 rails) and Δt, and to determine ΔN, we substitute the values l_{INS}, l_{CB}, l_{br}, and N_t.

The length of the rail insert is equal to the distance between the cut ends of the string plus metal reserve for welding two joints and the residual bending boom (5 mm). We assume that the length of the rail insert is 10 m.

$$L_{CB} = l_b + l_p = 40 + 5 = 45 \text{ m};$$

$$l_{br} = N_t/r = \alpha EF\Delta t/r = 40916/7 = 58.5 \text{ m};$$

$$l_p = \frac{(l_{INS} + l_{CB} + l_{br})N_t}{N_{GNU} - N_t} + l_{br} \tag{14}$$

Taking that $N_{GNU} = 70$ t and substituting the known values of l_{INS}, l_{CB}, l_{br}, N_t, we determine l_p:

$$l_p = \frac{(10.0 + 45.0 + 58.5)40.9}{70.0 - 40.9} + 58.5 = 218.02 \text{ m}.$$

$$\Delta l_p = \alpha l_p \cdot \Delta t = 51.4 \text{ mm}$$

For known values of N_t, l_{INS}, l_{CB}, l_{br} and l_p, the additional tensile forces ΔN required to restore the temperature of fixing the string at the work site will be equal: $\Delta N = N_t \quad (l_{INS} + l_{CB} + l_{br})/(l_p - l_{br}) = 40.9 \quad (10.0 + 45.0 + 58.5)/(218.0 - 58.5) = 29.1$ t, i.e. in the amount of $N_t + \Delta N = 40.9 + 29.1 = 70$ t, which corresponds to the tensile forces (N_{GNU}) of the assigned GNU.

Taking into account that the string part unfastened in the section l_p, when stretched, should return to its original position, fixed before it was unfastened, the necessary additional elongation is determined to restore the temperature of the string fastening at the work site

$$\Delta l = [\alpha(l_{INS} + l_{CB} + l_{br})\Delta t] + 5 = 31.8\ \text{mm}$$

Taking into account the partial reverse movement of the end of the stretched string after its fastening and removal of the GNU ("rollback"), determined from the condition:

$$\Delta l_o = N_{GNU}^2 / 2EFr = 20.2\ \text{mm},$$

we determine l_{tot} in section B.

Before "rollback" $\Delta l_{tot} = \Delta l_p + \Delta l + \Delta l_o = 51.4 + 31.8 + 20.2 = 103.4$ mm.

After the "rollback" $\Delta l_{tot} = \Delta l = 83.2$ mm.

To control the string shifting during stretching in section A–B and adjust the stresses in section A–E, control sections are assigned every 50 m in section l_p (1, 2, 3) and in sections close to points B, C, D and E (±0.5 m). When stretching the string in the section $l_p + l'_{br}$, the displacements of the string end in the section B should be equal to 50 ÷ 51 m, and in sections 3, 2 and 1, respectively, 37 ÷ 38; 25 ÷ 27 and 12 ÷ 14 mm. Taking into account the "rollback" margin, the displacements of the string end at point B increase by 20 mm, and at points 3, 2 and 1, respectively, by 15, 10 and 5 mm. Taking into account the stretching of the string to restore the temperature of its fixing at the work site B–E, the displacements of the string end at point B will increase by 31 ÷ 32 mm, and at points 3, 2 and 1, respectively, by 23 ÷ 24, 15 ÷ 16 and 7 ÷ 8 mm.

After fixing the rail string in the section A–B, removing the GNU as a result of partial return ("rollback") of the stretched part of the string, the displacements at point B will be 82 ÷ 83 mm, and at points 3, 2 and 1, respectively, 60 ÷ 61, 40 ÷ 41 and 20 ÷ 21 mm. After welding the string in sections B and C, hanging it in the section A–E on paired plates and additional impact on the string with an impact mechanism in section B in the direction of point E, the string in section A–E should shift in the direction of the action line of the impact mechanism [25–28].

The adjustment is considered complete if the displacements of the string at points 1, 2, 3 and B are close to their initial values obtained when stretching the string to recover its fixing temperature in the section $l_p + l'_{br}$, i.e. the values of the string displacements in sections 1, 2, 3 and B should be close, i.e. correspond to 12 ÷ 14, 25 ÷ 27, 37 ÷ 38 and 50 ÷ 52 mm with an accuracy of ±3 mm.

The displacements at point D are determined from the condition $\Delta l_{br} = \Delta l \cdot \frac{l_{br}}{l_o}$, and at point C from the condition $\Delta l_c = \Delta l \cdot \frac{l_{br} + l_{CB}}{l_o}$.

Restoration of the string by electric contact welding at rail temperatures below the fixing temperature can be performed without the GNU, but using PRSM with pulling forces up to 120 tf, of which no more than 70 tf is used to stretch the string to recover its fixing temperature at the work site.

To perform work, including the restoration of the rail string by welding and recovery of the temperature of its fixing in the work area, it is necessary to determine the length of the welding work area 1, where it is necessary to recover the temperature of the string fixation, (Fig. 3)

$$1 = l = l_{\mathrm{INS}} + l_{\mathrm{br}} \tag{15}$$

The length of the rail insert is determined from the condition

$$L_{\mathrm{INS}} = l' + \delta,$$

wherein

l' is the distance between the cut string ends;
δ rail metal reserve for welding one joint.

When the temperature difference between the fixing of the string relative to its temperature during welding by the value Δt = 10; 15; 20 and 25 °C l_{br} are, respectively, equal to 29.2; 43.8; 58.4 and 70.0. For further calculations, we take Δt = 20 °C and, accordingly, l_{br} = 58.4 m. Hence, the length of the welding area $l = l_{\mathrm{INS}} + l_{\mathrm{br}}$ = 10 + 58.4 = 68.4 m.

In order to enter the string at the welding site into the temperature of its fixing, the site must be lengthened by a value $\Delta l = \alpha l \Delta t$. At Δt = 20 °C Δl = 16 mm.

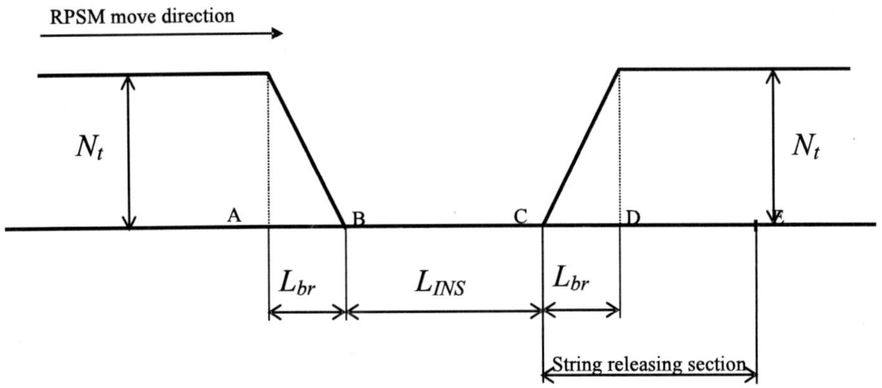

Fig. 3 Diagram of temperature forces at the ends of strings adjoining on both sides to work site before welding

Then we unfasten the string end, adjacent to the welding point of the second joint at point C at a length l_p equal to $l_p = \frac{l \cdot N_t}{N_{RPSM} - N_t} + l_{br}$, wherein $N_t = \alpha EF\Delta t =$ 40.9 tf, N_{PRSM} = 70 tf and we get $l_p = \frac{68.4 \cdot 40.9}{70 - 40.9} + 58.4 = 81.4 + 58.4 = 149.8$ m.

After unfastening the string in the section l_p, we lay the rail insert and weld it to the end of the string at point B.

We determine the total elongation of the unfastened string end, which is necessary to recover the fastening temperature in the welding area and in the area where the string is unfastened.

$$\Delta l_{tot} = \Delta l + \Delta l_p \tag{16}$$

$$\Delta l_p = \alpha l_p \Delta t = 11.8 \cdot 10^{-6} \cdot 149.8 \cdot 20 = 35.5 \text{ mm}$$

$$\Delta l_{tot} = 16.0 + 35.5 = 49.5 \text{ MM}.$$

Based on the results of control welding, it was obtained that when welding one joint, the costs of welding and metal melting are about 40 mm. The difference between the required string lengthening and the cost of metal for welding the first joint is 9.5 mm, so we stretch the end of the PRSM string by 9.5 mm, i.e. to the end of the rail insert and weld. After completion of welding, we maintain the joint for 6 min in a compressed state. Then the welding head is removed, the welded joint is processed. The rail string along the entire length of AE is unfastened, tapped, and then fixed again with the standard tightening of the bolts (screws). Upon completion of the specified work, the welded joints are checked with a flaw detector and marked in accordance with the requirements, at a distance of at least 25.0 cm from the welding point of each joint, their serial numbers are recorded.

The date of welding, the temperature of the rail during welding, the length of the unfastening section and the elongation of the string to recover the temperature of its fastening at the work site are recorded in the Register of the service and the temperature regime of the rail strings.

Aluminum thermite welding of rails during the final restoration of the strings must be carried out in accordance with the Specifications.

The final restoration of the strings by aluminum thermite welding is carried out at the temperature of the rail string, corresponding to its fixing temperature ±5 °C. It is allowed to restore the strings by aluminum thermite welding at a string temperature above or below their fixing temperature with subsequent redistribution or stress relief by methods similar to those for PRSM welding.

The final restoration of strings by aluminothermic welding can be carried out immediately after revealing a dangerous defect requiring cutting or after temporary restoration of the lash [29–32].

In cases where the string is restored immediately after the detection of a defect, then at the temperature of the rail string above its fixing temperature, a piece of rail is cut from the string, in the area of the detected defect, with an autogenous or gas cutter, then the string ends are cut with rail saws with the creation of a distance

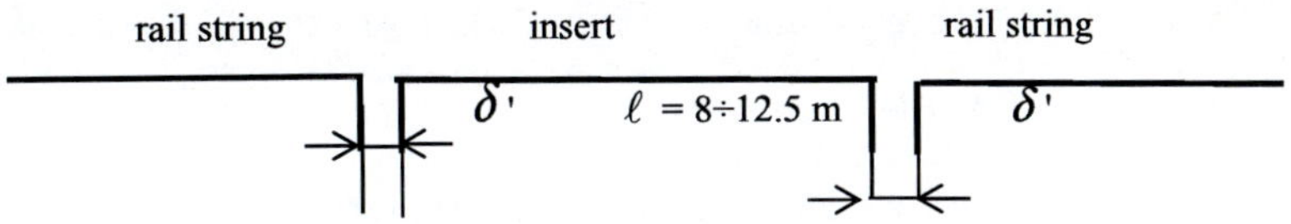

Fig. 4 Scheme of welding rail insert by aluminum thermite welding

between them equal to the length of the insert to be welded (8 ÷ 12.5) and two gaps ($\delta' = 25 \pm 1$ mm) for aluminothermic welding (Fig. 4).

Each welded joint is assigned a serial number, which is applied at a distance of at least 500 mm from the welded joint. The date of welding, the temperature of the rails during welding, the calculated and actual elongation of the stretched end of the string are recorded in the Register of service and temperature conditions of rail strings, and the number of joint, the date of welding and the code of the enterprise are recorded in the Register of aluminum thermite welding [33–36].

Each joint welded with aluminum thermite is marked with white indelible paint in the form of two vertical stripes 10 mm wide, applied at a distance of 100 mm on both sides of the weld. In addition, the stripes must be duplicated on the overlay. When changing the linings, the stripes are restored.

3 Conclusions

The considered methods and calculations make it possible to compare and choose the use of one of the welds, namely: electric contact or aluminum-thermite for the restoration of rail strings. The use of aluminum thermite welding will save not only time, but also the means of enterprises serving the railway track, while eliminating defective spots in the rail string.

References

1. Muravyeva L, Vatin N (2014) Application of the risk theory to management reliability of the pipeline. Appl Mech Mater 635–637:434–438. https://doi.org/10.4028/www.scientific.net/AMM.635-637.434
2. Muravyeva L, Vatin N (2016) Elaboration of the method for safety assessment of subsea pipeline with longitudinal buckling. Adv Civ Eng. https://doi.org/10.1155/2016/7581360
3. Muravyeva L, Vatin N (2014) Pipelines stability under extreme hydrodynamic conditions. Appl Mech Mater 635–637:451–456. https://doi.org/10.4028/www.scientific.net/AMM.635-637.451
4. Muravyeva L, Vatin N (2014) Risk assessment for a main pipeline under severe soil conditions on exposure to seismic forces. Appl Mech Mater 635–637:468–471. https://doi.org/10.4028/www.scientific.net/AMM.635-637.468

5. Muravyeva L, Vatin N (2014) Study of integrity and interaction of a non-buried marine subsea pipeline with soil. Appl Mech Mater 633–634:1042–1046. https://doi.org/10.4028/www.scientific.net/AMM.633-634.1042
6. Muravyeva L, Vatin N (2014) The safety estimation of the marine pipeline. Appl Mech Mater 633–634:958–964. https://doi.org/10.4028/www.scientific.net/AMM.633-634.958
7. Murgul V, Vuksanovic D, Vatin N, Pukhkal V (2014) Decentralized ventilation systems with exhaust air heat recovery in the case of residential buildings. Appl Mech Mater 680:524–528. https://doi.org/10.4028/www.scientific.net/AMM.680.524
8. Murgul V, Vuksanovic D, Vatin N, Pukhkal V (2014) The use of decentralized ventilation systems with heat recovery in the historical buildings of St. Petersburg. Appl Mech Mater 635–637:370–376. https://doi.org/10.4028/www.scientific.net/AMM.635-637.370
9. Nazmeeva TV, Vatin NI (2016) Numerical investigations of notched C-profile compressed members with initial imperfections. Mag Civ Eng 62(2):92–101. https://doi.org/10.5862/MCE.62.9
10. Nemova D, Murgul V, Pukhkal V, Golik A, Chizhov E, Vatin N (2014) Reconstruction of administrative buildings of the 70's: the possibility of energy modernization. J Appl Eng Sci 12(1):37–44. https://doi.org/10.5937/jaes12-5610
11. Nikolskiy S, Vatin N, Pertseva O (2015) Possible rapid method for the determination of concretes frost resistance. J Appl Eng Sci 13(1):11–18. https://doi.org/10.5937/jaes13-6553
12. Ognjenovic S, Donceva R, Vatin N (2015) Dimensioning of the speed-transition lanes at the entering ramps on the motorway and urban road intersections. Proc Eng 117(1):544–550. https://doi.org/10.1016/j.proeng.2015.08.210
13. Ognjenovic S, Donceva R, Vatin N (2015) Dynamic homogeneity and functional dependence on the number of traffic accidents, the role in urban planning. Proc Eng 117(1):551–558. https://doi.org/10.1016/j.proeng.2015.08.212
14. Ognjenovic S, Ristov R, Vatin N (2015) Designing of rehabil 1:568–573. https://doi.org/10.1016/j.proeng.2015.08.215
15. Orlovich RB, Nowak R, Vatin NI, Bespalov VV(2018) Strength evaluation of the Prussian vaults made from brick aggregate concrete. Mag Civ Eng 82(6):95–102. https://doi.org/10.18720/MCE.82.9
16. Glusberg B, Loktev A, Korolev V, Shishkina I, Alexandrova D, Koloskov D (2020) Calculation of heat distribution of electric heating systems for turnouts. Adv Intell Syst Comput 982:337–345. https://doi.org/10.1007/978-3-030-19756-8_31
17. Korolev, V. (2020). The study of rolling stock wheels impact on rail switch frogs. In: E3S web of conferences, vol 164, p 03033. https://doi.org/10.1051/e3sconf/202016403033
18. Glusberg B, Korolev V, Lokteva O, Shishkina I (2020) Contact stresses in point frogs. In: E3S web of conferences, vol 164, p 03034. https://doi.org/10.1051/e3sconf/202016403034
19. Loktev A, Korolev V, Shishkina I, Savin, A, Illarionova L, Lokteva O, Kuskov V (2020) Dynamic behavior model of the ballastless railroad track segment considering wave processes. In: E3S web of conferences, vol 164, p 03035. https://doi.org/10.1051/e3sconf/202016403035
20. Loktev A, Korolev V, Shishkin, I, Lokteva O, Gridasova E (2020) Span operational aspects under offsetting the axis of the track panel. In: E3S web of conferences, vol 164, p 03037. https://doi.org/10.1051/e3sconf/202016403037
21. Korolev V, Shishkina I, Lokteva V (2020) Basic stages of creating a BIM model for transport infrastructure objects. In: IOP conference series: materials science and engineering, vol 918, p 012014. https://doi.org/10.1088/1757-899x/918/1/012014
22. Loktev D, Korolev V, Tkachenko V, Klekovkina N (2020) Calculating model allowing to keep the authenticity of transport infrastructure objects. In: IOP conference series: materials science and engineering, vol 918, p 012013. https://doi.org/10.1088/1757-899x/918/1/012013
23. Gluzberg B, Korolev V, Shishkina I, Berezovsky M, Tregubchak P, Zverkova N (2020) Reliability indicators of railway joints and crossings. In: IOP conference series: materials science and engineering, vol 918, p 012147. https://doi.org/10.1088/1757-899x/918/1/012147

24. Petrichenko M, Vatin N, Nemova D, Kharkov N, Korsun A (2014) Numerical modeling of thermogravitational convection in air gap of system of rear ventilated facades. Appl Mech Mater 672–674:1903–1908. https://doi.org/10.4028/www.scientific.net/AMM.672-674.1903
25. Petrichenko M, Vatin N, Nemova D, Kharkov N, Staritcyna (2014) A EOR (oil recovery enhancement) technology using shock wave in the fluid. Appl Mech Mater 627:297–303. https://doi.org/10.4028/www.scientific.net/AMM.627.297
26. Ponomarev A, Knezević M, Vatin N, Kiski S, Ageev I (2014) Nanosize scale additives mix influence on the properties of the high performance concretes. J Appl Eng Sci 12(3):227–231. https://doi.org/10.5937/jaes12-6161
27. Popović SG, Vukanić S, Komatina DF, Alihodzic-Jasarevic E, Vatin N (2015) Models of landscape shaping in exploited quarries of urban area. Proc Eng 117(1):609–615. https://doi.org/10.1016/j.proeng.2015.08.221
28. Priadko IN, Mushchanov VP, Bartolo H, Vatin NI, Rudnieva IN (2016) Improved numerical methods in reliability analysis of suspension roof joints. Mag Civ Eng 65(5):27–41. https://doi.org/10.5862/MCE.65.3
29. Pukhkal V, Tanić M, Vatin N, Murgul V (2015) Studying humidity conditions in the design of building envelopes of "passive house" (in the case of Serbia). Proc Eng 117(1):859–864. https://doi.org/10.1016/j.proeng.2015.08.152
30. Pukhkal V, Vatin N, Murgul V(2014) Central ventilation system with heat recovery as one of the measures to upgrade energy efficiency of historic buildings. Appl Mech Mater 633–634:1077–1081. https://doi.org/10.4028/www.scientific.net/AMM.633-634.1077
31. Radovanović Z, Sindić R, Dimovska S, Serdar N, Vatin N, Murgul V (2015) The mechanical properties of masonry walls—analysis of the test results. Proc Eng 117(1):865–873. https://doi.org/10.1016/j.proeng.2015.08.155
32. Radović G, Murgul V, Cvetkovska M, Aronova E, Vatin N (2014) Solar power supply for remote objects of historical heritage in montenegro. J Appl Eng Sci 12(4):277–284. https://doi.org/10.5937/jaes12-6162
33. Rybakov VA, Kozinetc KG, Vatin NI, Velichkin VZ, Korsun VI (2018) Lightweight steel concrete structures technology with foam fiber-cement sheets. Mag Civ Eng 82(6):103–111. https://doi.org/10.18720/MCE.82.10
34. Shishkin A, Shishkina A, Vatin N (2014) Low-shrinkage alcohol cement concrete. Appl Mech and Mater 633–634:917–921. https://doi.org/10.4028/www.scientific.net/AMM.633-634.917
35. Stamatovic B, Upadhyay R, Vatin N (2015) Cellular automata in modeling of porous building materials and solis porous media. Proc Eng 117(1):655–662. https://doi.org/10.1016/j.proeng.2015.08.227

Multi-rotor (24-Rotor) Tiltrotor Discplane, as Version of a Circular Wing (Flying Saucer)

Alexander Nakhaba

Abstract The object of this study is a multi-engine multi-rotor tiltrotor discplane (multi-rotor (24-rotor) tiltrotor of disc-shaped form), as version of a circular wing (flying saucer) with a decentralized system of separate parallel control of its engines. The construction of this aircraft provides the possibility of a smoother vertical takeoff and landing directly from sidewalks, roads and conventional car parking lots, and does not require special helicopter pads. The multi-rotor scheme gives the possibility of absolutely stable hovering and jewelery precise maneuvering on small heights, provides a change in the inclination of the aircraft (circular wings) during its transition from the phase of vertical flight to the phase of horizontal flight and vice versa without the need for extremely dangerous turns of engines of the vertical traction, such a transition occurs only by changing the speed of part of engines of vertical traction. The technology of completely closed and hidden under the case and protective lattices of air screws allows to use this discplane safely in the conditions of densely populated city quarters. The disc-shaped form of the aircraft allows you to fly at higher speeds in the dense layers of the atmosphere (due to the thinning of air above (in front of) the aircraft) and reduces energy consumption during flight by using the Coanda-effect—creating increased pressure (airbag) behind and under the discplane, and a simultaneous reduction in air pressure and aerodynamic drag at the front and top of the discplane.

Keywords Multi-rotor (24-rotor) tiltrotor discplane · Circular wing · Aircraft of disc-shaped form · The Coanda-effect · Functional antigravity

1 Introduction

Over the past 10 years, is increased the interest of civilians and military in aircraft with vertical takeoff and landing and high-precision maneuvering in difficult flight conditions, to solve a variety of tasks (from aerial photography and automatic

A. Nakhaba (✉)
National Aviation University, Liubomyra Huzara ave., 1, Kyiv 03058, Ukraine
e-mail: nakhaba91@ukr.net

A. Bieliatynskyi and V. Breskich (eds.), *Safety in Aviation and Space Technologies*, Lecture Notes in Mechanical Engineering,
https://doi.org/10.1007/978-3-030-85057-9_38

delivery of goods for various purposes—to transport passengers in unfavorable for landing conventional aircraft site and conducting complex rescue operations for evacuation of victims from the most dangerous places of road accidents, catastrophes, emergencies or zones of armed conflicts). Among the known types of aircraft, multi-rotor aircraft with a polycopter flight control algorithm are the newest and least studied and improved aircraft. Unlike other types of aircraft, multi-rotor aircraft with a polycopter control algorithm have the following differences and advantages: the possibility of the smoothest vertical takeoff and landing; the possibility of almost motionless hanging in the air at different heights; possibility of "jewelry" precise maneuvering at different heights; the ability to instantly change the direction of movement without additional deployment maneuvers; the possibility of flying at higher speeds in the dense layers of the atmosphere (due to the rarefaction of air over (before) the aircraft); the possibility of more efficient and safer flight of almost any aircraft (classic aircraft, helicopter, or discplane) in convertible mode, i.e. with a non-perpendicular location of the engines relative to the horizontal plane of the aircraft used in convertible flight mode. Most of these advantages are due to the following technical features: more number of resistance points in the air (i.e., a more number of vertical take-off engines (engine of the vertical thrust); reduced air pressure and aerodynamic drag over the discplane and increase air pressure and lift under the discplane); more advanced polycopter mathematical algorithms for controlling the flight of the aircraft, which in contrast to the classic aircraft and helicopter control algorithms do not depend of the mechanical components of the aircraft (ailerons, rudders, rudders or skew helicopter, helicopter propeller), and are determined only by the speed of engines of vertical traction and thus such algorithms are safer, more fault-tolerant, less dependent on the mechanical factor, more precise in control, more flexible, more versatile and suitable for control of almost any aircraft of traditional and non-traditional scheme.

2 Literature Review and Problem Statement

1. Transport with one degree of freedom of movement in space (one-dimensional movement along a broken line)—road and rail transport [1–4].
2. Transport with two degrees of freedom of movement in space (two-dimensional movement in space within one plane)—annual, sea, ocean transport, snowmobiles, transport on hovercraft, ekranoplanes [5–7].
3. Transport with three degrees of freedom of movement in space (three-dimensional movement in space)—submarines, air transport (planes [8], helicopters [9, 10], policopters, [11–13]), space transport [14, 15] (rockets, space ships, artificial satellites of the Earth, space stations).

Of all the above types of transport, the fastest is air transport, because first it has the maximum possible degree of freedom of movement in three dimensions, on the other hand it has the ability to move from point A to point B in a straight line at a speed of 500–1100 km/h, which is almost 20 times faster than the average speed of urban vehicles and could allow the most efficient solution of transport problems of the above emergency municipal services with minimal loss of time. Unfortunately, the conditions of urban neighborhoods do not allow the use of most modern modes of air transport, because planes and helicopters require runways and helicopter pads, which most urban neighborhoods are not equipped with. In this regard, there is a real need to develop a fundamentally new generation of air transport that can fly outside the city at jet speeds (500–1100 km/h), within the city to fly with helicopter capabilities at speeds of 0–500 km/h, but which, unlike its predecessors, will have ten times more precise maneuvering, the ability to absolutely stable and stationary hovering at low altitudes, the ability to maximize vertical takeoff and landing directly from roads, sidewalks, and conventional car parks, which are equipped with modern cities in sufficient quantities, i.e. transport that will have all the advantages of helicopter and plane, and will be easy to park as usual car.

As the carrying capacity of mankind increased, the next problem arose: to increase the carrying capacity of the aircraft only by increasing the power of its engines became unprofitable and very limited. Unfortunately, any best aircraft engine and rotor cannot be increased indefinitely, when a certain size of rotors is reached, or their speed of rotation, supersonic flows appear at the edges of the rotor, which further reduces the efficiency of this engine. Therefore, there is a need to increase the load capacity not by increasing the power of one engine, but by increasing the number of identical standard engines with optimal parameters and the highest efficiency for these conditions of use. This is how multi-engine air transport systems and even multi-engine space rocket systems appeared. All the world's heaviest aircraft ("Ruslan", "Dream", "Hercules" of Howard Hughes) and space rockets ["Energy", "Angara", "R7" (which delivered Yuri Gagarin into space and still continues to deliver the astronauts of NASA to the International Space Station)] —are multi-engine. But most heavy multi-engine aircraft have significant disadvantages and limitations in further increasing their size and load capacity, as they require very large, very strong and costly runways. Therefore, there is a need to improve and develop multi-engine multi-rotor aircraft systems with the possibility of smooth vertical takeoff and landing on any surface, including water.

3 The Aim and Objectives of Research (Research Tasks)

The aim of the study is to provide and further optimize a more compact and more advanced model of a multi-rotor unmanned aerial system with greater specific traction, improved flight characteristics, with the possibility of perfectly smooth vertical takeoff and landing, with the possibility of stationary hovering in the air, with the possibility of "jewelry" precise maneuvering in automatic mode at different altitudes in any weather and seasonal conditions, with the possibility of safer and trouble-free flight with multiple redundancy of functionally important elements.

To achieve the goal, the following tasks were set:

- development of a more compact, more reliable and more fault-tolerant aircraft of polycopter scheme with the possibility of smooth vertical take-off and landing from a normal sidewalk, high-precision maneuvering and horizontal flight within normal city quarters, and most importantly a vehicle that can continue safe flight in the event of an emergency failure of one of more vertical takeoff engines or propellers;
- to maximize the aerodynamic properties of the body of this aircraft using a disco-shaped form, i.e. development of the discplane (flying saucer) with a circular wing to be able to move at high speeds and at low altitudes;
- to develop such a structure of the body of aircraft, which provides a closure of dangerous propellers with a special casing and protective grilles for flight (traffic) safety within the city quarters;
- to increase fault tolerance—to develop a decentralized system of separate parallel control of engines of the aircraft and to develop a possibility of using spare duplicate synchronously operating autopilots with a system of fast switching from one autopilot to another.

4 The Solution of the Problem

To solve this problem developed this aircraft (Fig. 1a, b). Proposed device consists of a discplane of parabolic (disc-shaped) circular wing of the form of biconvex lens (1), consisting of 8 functional sectors, each of which is an independent arm of the lever for tilting the discplane in a certain direction (forward, backward, left, right, left-forward, left-back, right-forward and right-back), 24 vertical take-off motors of vertical thrust (2), distributed in 8 functional sectors of the discplane wing (3 identical in function engines in each functional sector), and telescopic supports (4) (chassis) of the discplane (Fig. 1a, b). To the disc-shaped (parabolic) wing of the discplane of the form of a biconvex lens (Fig. 1a, b) along its entire plane, attached the engines of the vertical take-off (of the vertical thrust) (2) (in this case, 24 (twenty-four), but their number may be different, but always a multiple of 2. These engines are directed up and forward (i.e. the longitudinal axis of the vertical takeoff engines (5) is located at an angle of 60° to the longitudinal axis (4) of the discplane (can also be from 10° to 80°—depending on the maximum of the acceleration of the discplane). Half of the vertical take-off engines rotate clockwise, the other half rotate counterclockwise (to compensate for rotation along the yaw axis). Instead of the usual heavy wheeled chassis, telescopic supports (9) are installed, and the front supports are longer than the rear supports, so that when standing on the Earth, the longitudinal axis of the discplane is at a certain (from 10° to 80°) angle to the Earth's plane (6) (often at an angle of 30°), and the front of the discplane (nose) is above its rear (tail). If necessary, the number of telescopic supports may be different depending on the size and weight of the discplane. If the discplane is not very heavily loaded, and if large accelerations are required (for example for a high-speed and highly maneuverable discplane), the vertical take-off engines are directed as far forward as possible (at an angle of 10°–45°s to the longitudinal axis of the discplane) and the angle between the longitudinal axis (4) Earth (6) at parking and takeoff is 80°–45°. If the discplane is heavily loaded, and if large accelerations are not required (for example for a freight transport discplane)—vertical take-off engines are directed upwards (at an angle of 45°–80° to the longitudinal axis of the aircraft), and the angle between the longitudinal axis of the discplane and the Earth at parking and takeoff is 45°–10°. If the discplane is of a different layout and of a different load, the number of vertical takeoff engines may be different, but always a multiple of 2, so that the same number of engines of the same power is located on symmetrical sections of the parabolic wing of the discplane, and half of these engines rotate clockwise and the other half of them rotated counterclockwise. The types of engines used can also be different (piston, turbofan turbojet plasmic, ionic, photonic, inertial, gravitational and others). This aircraft is equipped by 24 (twenty-four) electric engines of vertical takeoff (vertical thrust) increased by 2–4 times the power, i.e. their power is 2–4 times greater than the power required to rotate conventional double-bladed propellers (this is necessary to rotate additional blades). Six-bladed screws are installed on the shafts of these motors with increased power instead of two-bladed screws (if it is necessary to reduce the maximum specific rotor thrust four-bladed screws are installed, eight-bladed screws are installed to increase the specific thrust of rotors), instead of one eight-bladed—four ordinary two-bladed screws, instead of one six-bladed—three ordinary two-bladed screws, and instead of one four-bladed screw—two ordinary two-bladed screws. Installation of additional 2, 4, or 6 blades of the same diameter on one shaft of the more powerful electric motor allows without increasing dimensions of the aircraft in 2–4 times to increase thrust of the given rotor and other similar rotors and thus to increase specific thrust of all aircraft that allows to increase compactness of the aircraft several times. But this method of increasing the specific thrust in contrast to increasing it by increasing the diameter of the propeller has one drawback—with increasing the number of blades of the same diameter, the maximum speed to which can accelerate the aircraft such multi-bladed rotors will not be increased, but will be the same as operation of a conventional two-bladed rotor of the same diameter (for example, for a rotor with a diameter of 11 inches at a speed of 8000 rpm, the maximum speed is approximately 60 km/h, for a rotor 18 inches at a speed of 5000 rpm, the maximum speed is 72 km/h, and etc.). For densely populated areas of the megalopolis speed of 60 or 72 km/h is more than enough, if necessary, you can increase the maximum speed of such multi-bladed rotors to 700–900 km/h, to increase the diameter of the blades, speed, power and size of the aircraft about the size of a small helicopter, which is not a problem. The angle of attack of the screws is greatest in the central parts of the blades and gradually decreases and the smallest on their periphery. These propellers are hidden under protective grilles to avoid collisions of propellers with other objects when flying in densely populated neighborhoods of the metropolis. To supply current to the motor windings, speed regulators with a power 3–4 times greater than the maximum power of the electric motors used are installed (for example, for electric motors with a power of 400 watts, speed regulators with a maximum power of 1200–1600 watts are used). This increase in the maximum power of the transistor switches of the stroke regulators does not significantly affect the weight of the aircraft, but is essential to increase the speed of climb and most importantly to increase the descent speed of a fully loaded aircraft and rapid braking and hanging at the end of such a rapid protracted descent by increase the current and total electric power on the windings of electric motors in 3–4 times the nominal. If such fast descents and fast braking in the Earth to do on regulators of a course (ESC) of usual (nominal), not increased in 3–4 times power at attempt to brake sharply fast descent (controlled falling) of a discplane due to an overload transistor keys do not sustain and burn out, which leads to an instantaneous stop of the rotors and to the fall and damage of the aircraft. If you do not use high acceleration and braking and do not use high speeds of ascent and descent, you need to spend 3–4 times more time on such maneuvers and, accordingly, more electricity of power batteries. At the bottom of the aircraft are powerful white LEDs directed downwards to illuminate what is under the aircraft at night, and powerful white light LEDs directed forward to illuminate what is in front of the aircraft at night. Red LED position lights are installed on the rear of the aircraft, and blue LED position lights are installed on the front of the aircraft to indicate the rear and front of the aircraft at night. All power components (power cables, power batteries, regulators and motors) are interconnected by soldering, to provide a more reliable power supply at high currents. All wiring is placed inside the perforated duralumin or composite tubes of the frame of this aircraft. Autopilot and electronics for radio communication are located in a sealed plastic case, transmitting and receiving antennas are brought out and located on the rear and lower parts of the discplane frame, barometric altimeter of the autopilot must be covered with a piece of foam 5 × 5 × 10 mm (to protect it from airflows more correct registration of air pressure and barometric altitude without erroneous (false, false) pressure fluctuations associated with the operation of propellers).

In Fig. 1a schematically shows the proposed device (discplane) at the time of its parking and during the execution of phases 1 and 4 of the flight (side view), where: 1—parabolic (dome-shaped) wing of the discplane in the form of a biconvex lens; 2—vertical takeoff engines (engines of vertical thrust); 3—hemispherical discplane cabin; 4—longitudinal axis of the discplane; 5—the middle longitudinal axis of the engines of the vertical takeoff (of vertical thrust) of the discplane; 6—plane of the Earth's surface; 7—the angle between the axis of the engines of vertical takeoff (of vertical thrust) and the longitudinal axis of the discplane; 8—the angle between the longitudinal axis of the discplane and the plane of the Earth's surface; 9—telescopic supports (chassis) of the discplane.

In Fig. 1b schematically shows the proposed device using eight-bladed propellers (top view), where: 1—parabolic (dome-shaped) circular wing of the discplane in the form of a biconvex lens; 2—engines of vertical takeoff (of vertical thrust); 3—hemispherical discplane cabin.

To increase the functionality, maneuverability, fault tolerance, simplicity and smoothness of control and flight safety in the aircraft used eight-sector layout of

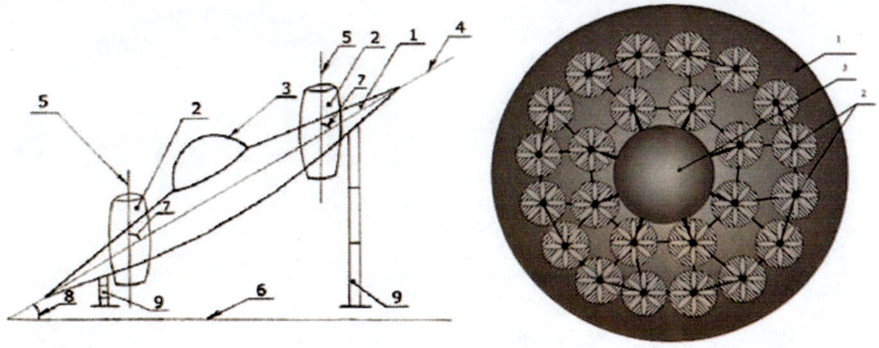

a) b)

Fig. 1 The discplane: **a** schematically shows the discplane at the time of its parking and during the 1st and 4th phases of flight (side view); **b** the discplane (top view)

vertical traction engines and decentralized system of separate parallel control of engines (Fig. 2a), the essence of which is as follows—vertical traction engines are located in 8 functional sectors of the wing 3 identical motors in each functional sector), each functional sector is an independent lever arm for tilting the disc in a certain direction (forward, backward, left, right, left-forward, left-back, right-forward and right-back), 3 autopilots, each of which has 8 independent Pulse Width Modulation outputs (PWM outputs), regulate the engine speed in such a way that each of these autopilots affects the operation of only one third of all aircraft engines, which does not result in the failure of one autopilot. to the loss of control over all aircraft engines, but only over one third of them part (only on engines that control this incorrectly working autopilot), additional duplicating synchronously working autopilots are also used, in case of incorrect operation of a part of the main (three) autopilots, the inputs for the Pulse Width Modulation signals (PWM inputs) on the stroke controllers are disconnected from the PWM outputs of the incorrectly operating autopilot and switched by transistor switches to the PWM outputs of the spare pulse-width modulation signals from a properly working autopilot to the PWM inputs of the regulators of the electric motors, hus, 3 main autopilots are on board and operate synchronously, each of which controls its one third of the engines, and another 6 (2 additional spare autopilots for each main, in case of failure), which provides the aircraft with a higher level survivability and fault tolerance.

In Fig. 2a shows the discplane (top view), an eight-sector layout of the engines of vertical traction and a diagram of a decentralized system of separate parallel control of its engines, where: A—autopilot, E—engine.

In Fig. 2b in comparison shows the dicplane (top view), an eight-sector layout of vertical traction engines and a diagram of a centralized control system for its engines by a single autopilot, where: A—autopilot, E—engine.

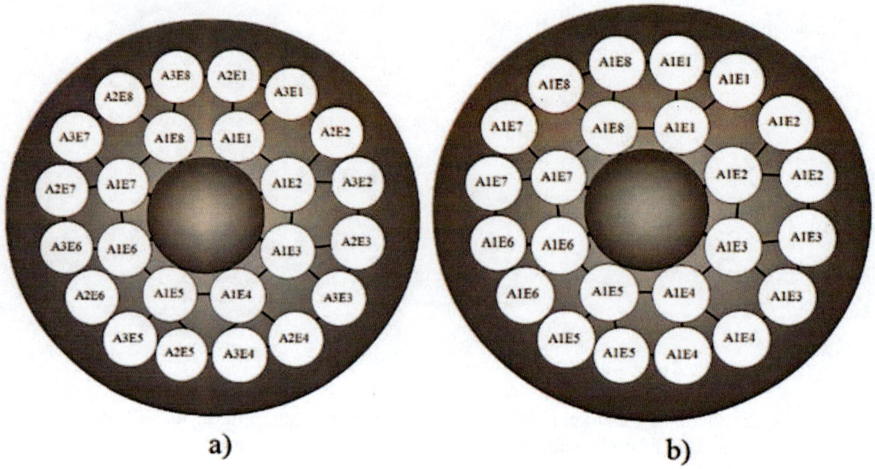

Fig. 2 The discplane (top view): **a** eight-sector layout of vertical traction engines and scheme of decentralized system of separate parallel control of its engines; **b** eight-sector layout of vertical traction engines and the scheme of the centralized control system of its engines by one autopilot

The discplane functions as follows. At the time of parking the discplane is in such a position that its front part (nose) is above its rear part (tail) and the engines of the vertical take-off in this position are directed upwards at an angle of 90 degrees to the Earth plane, the vertical take-off (1 phase of flight) in this position, after which the chassis is cleaned. Then to start moving forward (phase 2 of flight—see Figs. 3, 4b)—the speed of the rear engines increases and the discoplane leans forward and becomes in such a position that its longitudinal axis (4) becomes parallel to the Earth (6), and the axis of the engines of the vertical takeoff becomes at an angle 60° to the plane of the Earth (Figs. 3, 4b) (in other embodiments of the layout and purpose of the discplane, the axis of the engines of the vertical takeoff (5) may be located at an angle of 80°–10° to the plane of the Earth).

In Fig. 3 schematically shows the proposed device during the execution of the 2nd phase of the flight (side view), where: 1—parabolic (dome-shaped) wing of the discplane in the form of a biconvex lens; 2—engines of vertical takeoff (of vertical thrust); 3—hemispherical discplane cabin; 4—longitudinal axis of the discplane; 5—the middle longitudinal axis of the engines of the vertical takeoff (vertical thrust); 6—plane of the Earth's surface; 7—angle between the axis of the engines of the vertical take-off (of vertical thrust) and the longitudinal axis of the discplane.

After gaining a certain horizontal speed there is an additional lifting force of the discobolic parabolic circular wing and energy consumption is reduced. Next, the aircraft performs a horisontal movement forward from point A to point B, on approach to the point B equalize the speed of the rear and front engines and make a smooth vertical landing directly to the destination. After reaching the destination the joystick control is moved to the neutral position (in manual mode), or the autopilot stops moving forward on the pitch in automatic control mode (3 phase of

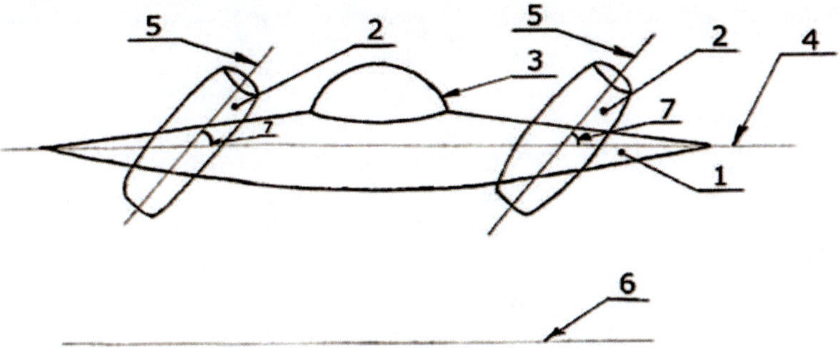

Fig. 3 Schematically shows the discplane during the 2st phase of flight (side view)

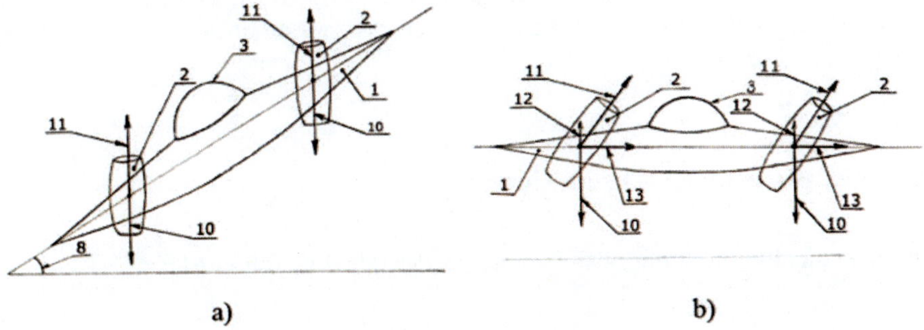

Fig. 4 The discplane: **a** schematically shows the discplane during the 3th phase of flight (side view); **b** schematically shows the discplane during the 2th phases of flight (side view)

the flight—see Fig. 4a), which leads to a decrease in the speed of rotation of the rear engines of vertical traction, tilting the discplane backward (at an angle of 30° to the Earth's plane in this discplane, or at an angle of 10°–80° in discplanes of another layout) and to a complete stop of the discplane in the horizontal direction and its hovering in the air (Fig. 4a), then there is a gradual decrease in speed of all engines of the vertical takeoff (4 phase of flight—see Fig. 1a) and the release of the chassis, which leads to a smooth vertical landing of the discplane on the desired site (destination).

In Fig. 4a schematically shows the proposed device during the execution of the 3rd phase of the flight, and during the vertical takeoff and landing with the chassis removed (side view), where: 1—parabolic (dome-shaped) wing of the discplane in the form of a biconvex lens; 2—engines of vertical takeoff (of vertical thrust); 3—hemispherical discplane cabin; 10—gravity vector; 11—vector of the thrust of engines of vertical takeoff (of vertical thrust).

In Fig. 4b schematically shows the proposed device during the execution of the 2nd phase of the flight (side view), where: 1—parabolic wing of the discplane in the form of a biconvex lens; 2—engines of vertical takeoff (of vertical thrust);

3—hemispherical discplane cabin; 10—gravity vector; 11—vector of thrust of the engines of vertical takeoff (of vertical thrust); 12—lifting vector of the parabolic wing of the discplane; 13—the resulting force vector, pushing the discplane in the horizontal direction.

If you need to fly to the left—increase the speed of the right engines and the aircraft tilts to the left and immediately flies to the left. If necessary, fly to the right—increase the speed of the left engines and the aircraft leans to the right and immediately flies to the right. If you need to fly back—increase the speed of the front engines and the aircraft leans back and immediately flies back. If it is necessary to turn the aircraft clockwise—increase the speed of the engines rotating clockwise. If it is necessary to turn the aircraft counterclockwise—increase the speed of the engines rotating counterclockwise. If necessary, fly up (if necessary, increase the height)—smoothly and evenly increase the speed of all engines of vertical takeoff (of vertical thrust). If necessary, fly down (if necessary, reduce the height)—smoothly and evenly reduce the speed of all engines of vertical takeoff (of vertical thrust).

5 Results and Its Discussion

Mathematically, the flight of the vehicle is described as follows [16]:

$$\begin{cases} x = Ax + B_u u + B_f f \\ y = C_y x + D_{yu} u + D_{yf} f \\ z = C_z x + D_{zu} u + D_{zw} f, \end{cases} \quad (1)$$

where x ϵ R^{11x1}—the state vector; u ϵ R^{4x1}—the control vector; y ϵ R^{11x1}—the output vector, to generate feedback; f ϵ R^{3x1}—vector of atmospheric excitations that act on an aircraft in horizontal and vertical planes (three-axis); z ϵ R^{3xl}—vector of the output variables to assess the quality of the system; A ϵ R^{11x11}, B_u ϵ R^{11x4}, B_f ϵ R^{11x3}, C_y ϵ R^{11x11}, D_{yu} ϵ R^{11x4}, D_{yw} ϵ R^{11x3}, C_z ϵ R^{3x11}, D_{ZU} ϵ R^{3x4}, D_{zf} ϵ R^{3x3}—matrix, which describe the model of the aircraft in the space of states. State vector of polikopter comprises the following components [16]:

$$x = [V_x, V_y, \omega_x, \omega_z, \gamma, \delta, V_z, \omega_y, r_{3,3}]^T \quad (2)$$

where V_x, V_y, V_z—longitudinal, lateral and vertical linear velocities; δ, γ—angles of pitch and roll; ω_x, ω_z, ω_y—angular velocity of the pitch, roll and yaw; $r_{3,3}$—the change of state feedback gyroscope.

The main driving force of the propeller—is a mechanical momentum, i.e. kinetic energy (Ek = mv2/2), which the propeller receives in contact with air molecules during its rotation, i.e. the nature of the driving (lifting) force of propeller—mechanical (gravitational).

The flight of such an aircraft (multirotor discplane) can be described by the following mathematical formulas: $G \cdot M \cdot m/R2 = mI \cdot v2/2$—for flights over a universal astronomical body with mass M of an aircraft of mass m, and $m \cdot g = mI \cdot v2/2$—for flights over the planet Earth of an aircraft of mass m (where m_I—is the mass of all air molecules (or other working body) with which the rotors (propellers, propellers) of the aircraft collide, G—is the gravitational constant). The mass of all air molecules (or other working body) with which the rotors (propellers, propellers) of the aircraft collide (mI) is in direct proportion to the surface of the rotor, to the density of the medium (air), and inversely proportional to the height above sea level.

$$m_I \approx S_r \cdot \rho_m / h, \tag{3}$$

where m_I—is the mass of all air molecules (or other working body) with which the rotors (propellers, propellers) of the aircraft collide, S_r—surface of the rotor, ρ_m—the density of the medium (air), h—height above sea level.

The speed of collision of the rotor with the molecules of the medium (air) (v) is in direct proportion to the diameter (radius) (D(r)) of the rotor and to the angular speed of rotation of the rotor (RPM)

$$v \approx S_r \cdot RPM, \tag{4}$$

where S_r—is a surface of the rotor, RPM—is a angular speed of rotation of the rotor.

In Fig. 5 schematically shows the mathematical relationship of the thrust of the engines of vertical takeoff (of vertical thrust) on the surface of the rotors.

Point of thrust reduction due to supersonic rotor thrust losses.

In Fig. 6 schematically shows the mathematical relationship of the thrust of the engines of vertical takeoff (of vertical thrust) on the diameter of the rotors.

In Fig. 7 schematically shows the mathematical relationship of the thrust of the engines of vertical takeoff (of vertical thrust) on the angular speed of rotation of the rotors (RPM).

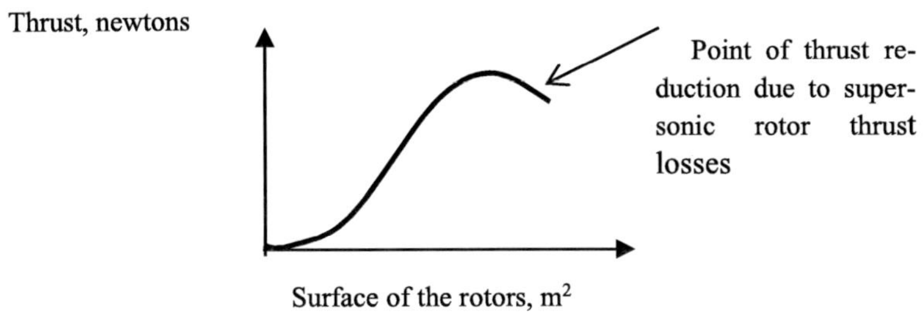

Fig. 5 Mathematical relationship of the thrust of the engines of vertical takeoff (of vertical thrust) on the surface of the rotors

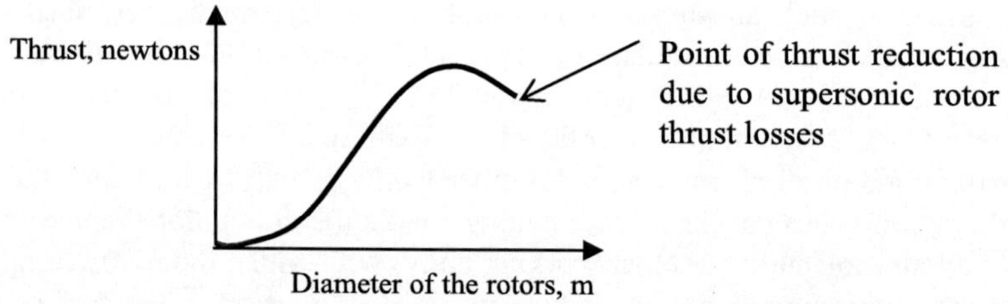

Fig. 6 Mathematical relationship of the thrust of the engines of vertical takeoff (of vertical thrust) on the diameter of the rotors

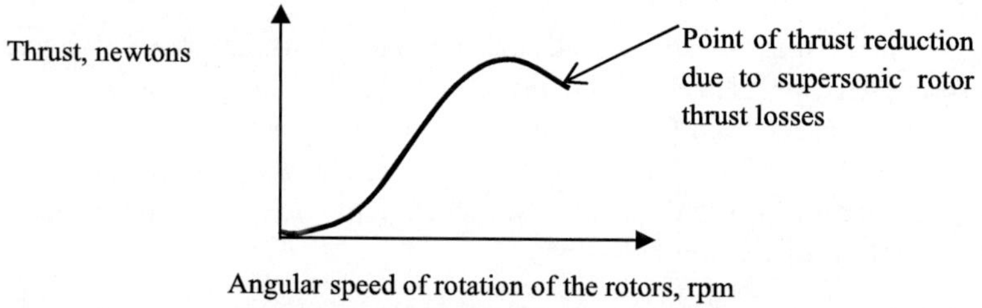

Fig. 7 Mathematical relationship of the thrust of the engines of vertical takeoff (of vertical thrust) on the angular speed of rotation of the rotors (RPM)

In Fig. 8 schematically shows the mathematical relationship of the thrust of the engines of vertical takeoff (of vertical thrust) on the number of rotor blades.

In Fig. 9 schematically shows the mathematical relationship of the thrust of the engines of vertical takeoff (of vertical thrust) on the height above sea level.

In Fig. 10 schematically shows the mathematical relationship of the total thrust of the all engines of vertical takeoff (total carrying capacity of the discplane) on the number of rotors.

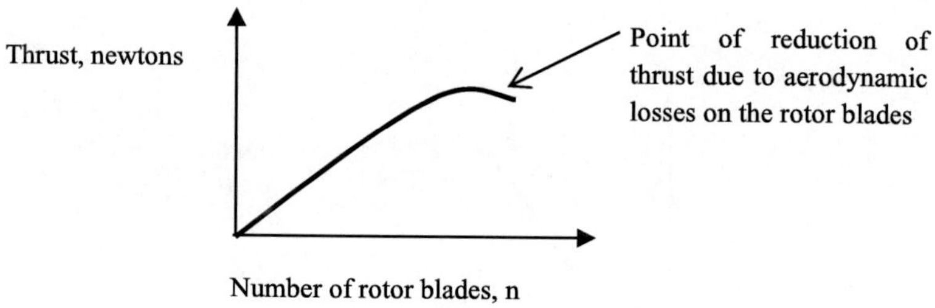

Fig. 8 Mathematical relationship of the thrust of the engines of vertical takeoff (of vertical thrust) on the number of rotor blades

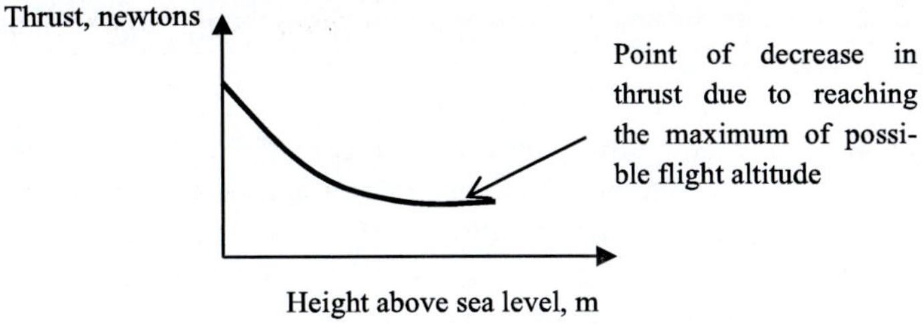

Fig. 9 Mathematical relationship of the thrust of the engines of vertical takeoff (of vertical thrust) on the height above sea level

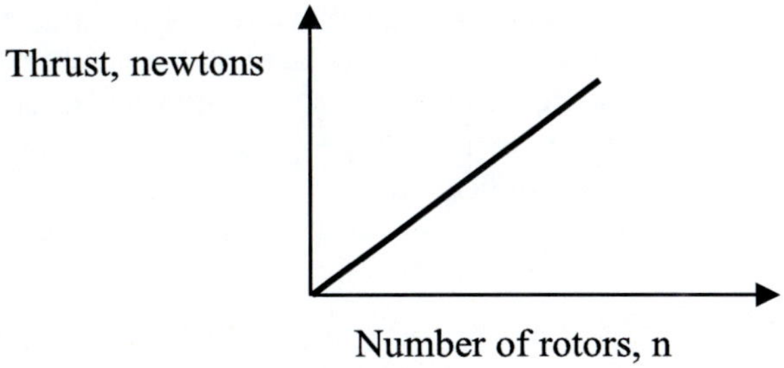

Fig. 10 Mathematical relationship of the total thrust of the all engines of vertical takeoff (total carrying capacity of the discplane) on the number of rotors

6 Conclusions

1. The technology of completely closed and hidden under the case and protective lattices of air screws allows to use this discplane safely in the conditions of densely populated city quarters.
2. The disc-shaped form of the aircraft allows you to fly at higher speeds in the dense layers of the atmosphere (due to the thinning of air above (in front of) the aircraft) and reduces energy consumption during flight by using the Coanda-effect—creating increased pressure (airbag) behind and under the discplane, and a simultaneous reduction in air pressure and aerodynamic drag at the front and top of the discplane.
3. The use of a decentralized system of separate parallel engine control and the use of backup autopilots and the system of switching PWM inputs to the PWM outputs of correctly operating autopilots along with the main autopilots allows to increase the failure resistance of the aircraft in the most dangerous emergencies and autopilot failure.

4. The use of a decentralized system of separate parallel engine control and the use of backup autopilots along with the main autopilots and use of the system of switching of PWM-inputs of the stroke regulators (ESC) to the PWM-outputs of correctly operating autopilots allows to increase the failure resistance of the aircraft in the most dangerous emergencies and autopilot failures.
5. The flight of such an aircraft (multirotor discplane) can be described by the following mathematical formulas: $G \cdot M \cdot m/R^2 = m_I \cdot v^2/2$—for flights over a universal astronomical body with mass M of an aircraft of mass m, and $m \cdot g = m_I \cdot v^2/2$—for flights over the planet Earth of an aircraft of mass m (where m_I—is the mass of all air molecules (or other working body) with which the rotors (propellers, propellers) of the aircraft collide, G—is the gravitational constant).
6. Directly proportional relationship of the total thrust of the all engines of vertical takeoff (total carrying capacity of the discplane) on the number of rotors allows to almost unlimited increase the carrying capacity,dimensions and weight of the discplane by increasing the number of its engines (rotors) of vertical takeoff (of vertical thrust) without a significant loss of the efficiency of the aircraft.

References

1. Eckhardt J, Nykänen L, Aapaoja A, Niemi P (2018) MaaS in rural areas—case Finland. Res Transp Bus Manag 27:75–83. https://doi.org/10.1016/j.rtbm.2018.09.005
2. ERTICO (2018) Mobility as a service—from modes to mobility. ERTICO, Brussels. Retrieved from https://ertico.com/wp-content/uploads/2018/09/Ertico_MaaS-vision-paper-2018_www.pdf
3. Gogola M, Sitányiová D (2020). Innovative mobility solutions in rural areas with case studies. Implications of mobility as a service (MaaS) in urban and rural environments: Emerg Res Opportunities 156–209
4. Stamos I (2018) What do data tell us? The story of the European logistics and road freight transportation sector. In: Proceedings of 7th transport research arena TRA 2018, 16–19 Apr 2018, Vienna
5. Trafa (2017) Shipping goods 2016. Trafik analysis. Statistik 2017:19, Stockholm
6. Tsiulin S, Reinau KH, Hilmola O-P, Goryaev N (2020) Blockchain in maritime port management: defining key conceptual framework. Rev Int Bus Strategy. 30(2):201–224
7. Blok A, Sharpanskykh A, Vert M (2018) Formal and computational modeling of anticipation mechanisms of resilience in the complex sociotechnical air transport system. Complex Adapt Syst Model 6:7. https://doi.org/10.1186/s40294-018-0058-2
8. Sato K, Chen Y (2018) Analysis of high-speed rail and airline transport cooperation in presence of non-purchase option. J Mod Transport 26:231–254. https://doi.org/10.1007/s40534-018-0172-z
9. Baranwal VC, Brönner M, Rønning J et al (2020) 3D interpretation of helicopter-borne frequency-domain electromagnetic (HEM) data from Ramså Basin and adjacent areas at Andøya, Norway. Earth Planets Space 72:52. https://doi.org/10.1186/s40623-020-01177-3
10. Christensen BR (2015) Use of UAV or remotely piloted aircraft and forward-looking infrared in forest, rural and wildland fire management: evaluation using simple economic analysis. N Z J For Sci 45:16. https://doi.org/10.1186/s40490-015-0044-9

11. Bagheinia R, Ghassemi A (2018) Simulated and experimental investigation of the airfoil contour forming of 301 austenitic stainless steel considering the springback. Int J Mech Mater Eng 13:1. https://doi.org/10.1186/s40712-017-0087-1
12. Ullah H, Abu-Tair M, McClean S et al (2020) Connecting disjoint nodes through a UAV-based wireless network for bridging communication using IEEE 802.11 protocols. J Wireless Com Network 142. https://doi.org/10.1186/s13638-020-01727-z
13. Raut S, Shimasaki K, Singh S et al (2019) Real-time high-resolution video stabilization using high-frame-rate jitter sensing. Robomech J 6:16. https://doi.org/10.1186/s40648-019-0144-z
14. Wu Y, Wang Y, Liu P et al (2019) Infrared LSS-target detection via adaptive TCAIE-LGM smoothing and pixel-based background subtraction. Photonic Sens 9:179–188. https://doi.org/10.1007/s13320-018-0523-8
15. Grigorev A, Tian Z, Rho S et al (2019) Deep person re-identification in UAV images. EURASIP J Adv Signal Process 2019:54. https://doi.org/10.1186/s13634-019-0647-z
16. Kharchenko VP, Chepizhenko VI, Tunick AA, Pavlova SV (2012) Avionika bezpilotnih lital'nih aparativ [Avionics of drones], 464

Obtaining Body Parts with Variable Wall Thickness Along the Perimeter

Yuliya Bessmertnaya, Alexander Malyshev, Vladimir Vikhorev, and Pavel Romanov

Abstract The article considers the production of housing parts with variable wall thickness along the perimeter. The use of this type of products is possible for various areas of industrial development, as the most important branch of the national economy. Examples of the use of cylindrical body parts of variable thickness in the electrical industry are given. The technological process of manufacturing body parts with variable wall thickness along the perimeter of sheet blanks consists of the operation of drawing without thinning and the operation of drawing with thinning and end support, as well as direct extrusion. The simulation of the described operations is carried out, the process diagrams and the semi-finished product obtained as a result of forming a round sheet billet are shown. The power modes of the first and second operations of the technological process of manufacturing body parts with variable wall thickness along the perimeter from sheet blanks are studied. Recommendations are given for the manufacture of body parts with and without a flange. The technological process of manufacturing body parts with variable wall thickness along the perimeter from sheet blanks presented in the article ensures high quality of products while saving material and energy resources, and also leads to a reduction in labor costs.

Keywords Wall thickness · Industrial development · Semi-finished product

Y. Bessmertnaya (✉) · P. Romanov
Tula State University, 92 Lenin Avenue, Tula 300012, Russia

A. Malyshev
Kaluga Branch of Bauman Moscow State Technical University (National Research University), 2 Bazhenova St., Kaluga 248000, Russia

V. Vikhorev
JSC "AvtoKom", 18 Azarovskaya St., Kaluga 248012, Russia

A. Bieliatynskyi and V. Breskich (eds.), *Safety in Aviation and Space Technologies*, Lecture Notes in Mechanical Engineering,
https://doi.org/10.1007/978-3-030-85057-9_39

1 Introduction

In various branches of mechanical engineering, body parts with variable wall thickness, which have high performance characteristics, are widely used. Shells with variable wall thickness along the perimeter are usually obtained either by hot or by cold die forging methods, for example, by forward or backward extrusion [1–5]. The disadvantage of the known methods is the significant energy consumption for the forming of the workpiece and the limited technological capabilities of the forming process. It is important to use a highly efficient resource and material saving technology for the manufacture of such products, which will allow obtaining high strength characteristics of finished products due to the strain hardening of the processed material.

One of the major consumers of various cylindrical body parts is the electrical industry. As part of the research work [6, 7] a series of experiments on body parts extrusion with variable wall thickness along the perimeter of the sheet rolled metal for the production of an experimental low-power electric motor with permanent magnets for the motor gearbox 19.3780.000.Э of the cable window regulator for cars VAZ-2172, VAZ-2123, VAZ-2192 (Fig. 1) was carried out with a satisfactory result on the production site of JSC "Autocom" in Kaluga JSC «KZAE» «Kaluga Plant of Automotive Electric Equipment».

Fig. 1 Electric motor body

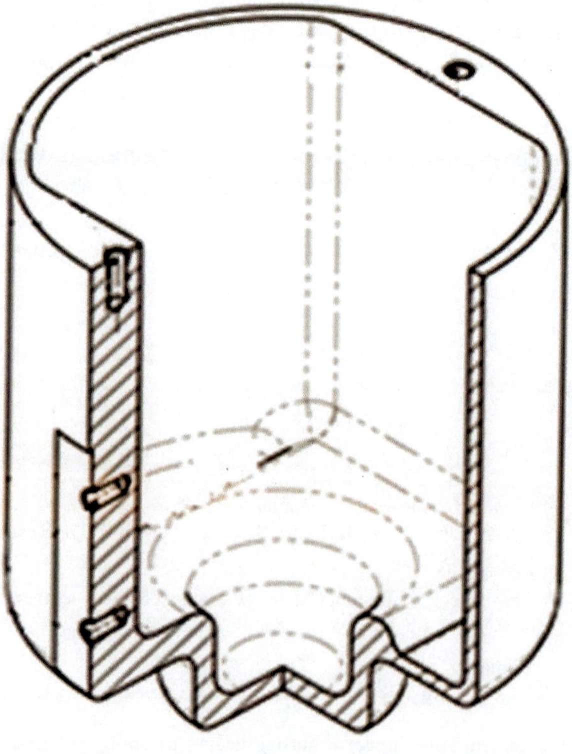

2 Materials and Methods

To reduce energy consumption when manufacturing a body part with variable wall thickness along the perimeter, extrusion operations with end support were studied. The research included the studies of the manufacture of semi-finished body parts with variable wall thickness along the perimeter of the blanks using extrusion without thinning and extrusion with thinning and the end support, as well as direct extrusion, that helped to get a body part with variable wall thickness along the perimeter. Bottom section forming has not been investigated.

Theoretical studies were carried out using the QForm2D/3D software package [8–10].

The first step involved the extrusion without thinning of the sheet round blank having a thickness equal to the maximum thickness of the finished workpiece (Fig. 2). Steel 1008 was used as the workpiece material. A hydraulic press with a nominal force of 50 MN was chosen as the equipment.

As a result, a semi-finished part was produced with outer diameter equal to the finished workpiece (Fig. 2b). The extrusion ratio in this case was 0.58.

The resulting semi-finished product was formed by extrusion with thinning and end support (Fig. 3). The use of a punch, the shape of which corresponds to the demanded size of the cavity of the finished part, creates an uneven gap at the die perimeter. Therefore, when forming in some areas, extrusion is carried out without the thinning of the wall, and in others—the extrusion involved the thinning of the wall. The extrusion ratio was 0.8.

During the extrusion, the height of the semi-finished part is changing unevenly. The height of the semi-finished part increases most intensively in the wall thinning areas. As a result, the end of the semi-finished part comes into contact with the circular protrusion of the punch. Therefore the end is supported, the force of which is in the opposite direction to the extrusion force. During the forming, the extrusion with a support is transformed into a direct extrusion (Fig. 4).

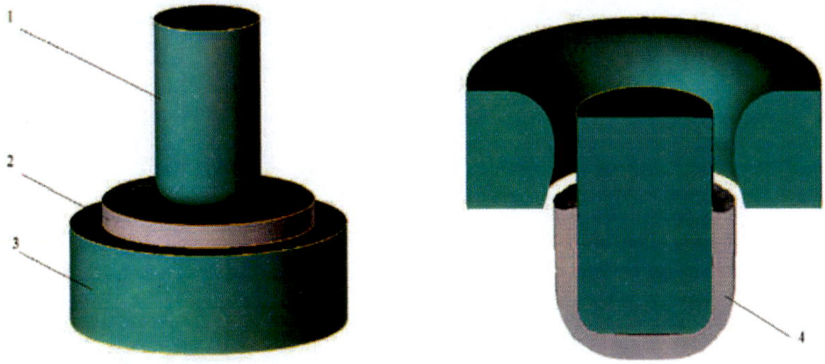

1 – punch, 2 – blank, 3 – die, 4 – semi-finished part

a b

Fig. 2 Diagram of the first operation: **a** the beginning of the process, **b** the end of the process

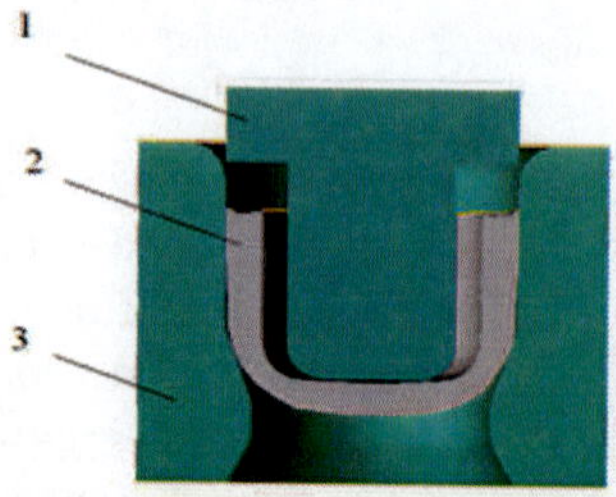

1 – punch, 2 – semi-finished part, 3 – die

a

b

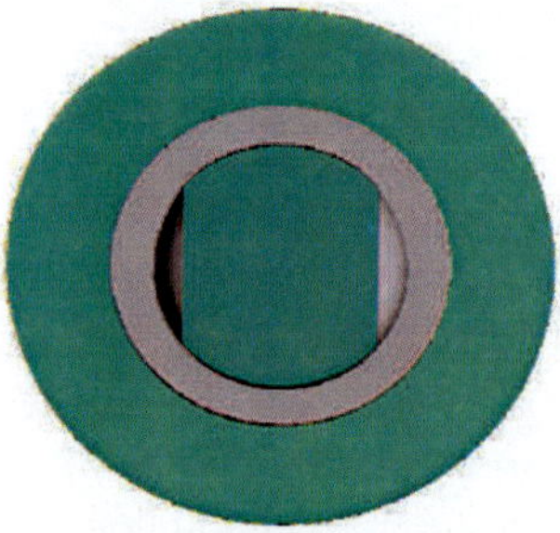

c

Fig. 3 Diagram of the second operation: **a** section 1, **b** section 2, **c** section of the working part

Fig. 4 Diagram of the completion of the second operation

3 Results

Figure 5 shows the obtained semi-finished products. It is clear that it is possible to obtain a semi-finished product with variable wall thickness along the perimeter from a round sheet blank in two operations.

The power modes of the first and second operations of the technological process of manufacture of body parts with variable wall thickness along the perimeter from sheet blanks have been studied.

Figures 6 and 7 present a graph of change of force according to the displacement during the first and second operations.

According to the analysis of the graphs in Fig. 6 it can be concluded that the force of the first operation increases proportionally relative to the displacement of the working tool and reaches the maximum value equal to 0.49 MN. After that, the force decreases.

Fig. 5 Semi-finished parts

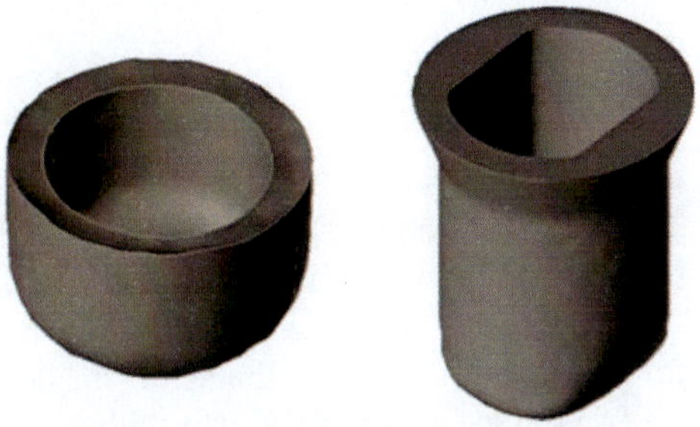

Fig. 6 Force plotted versus the displacement during the first operation

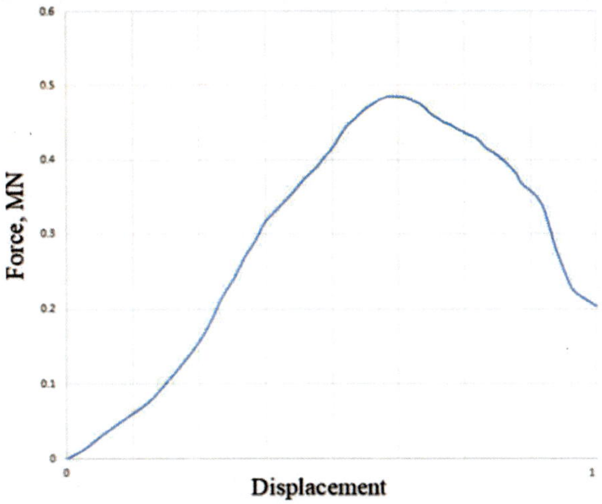

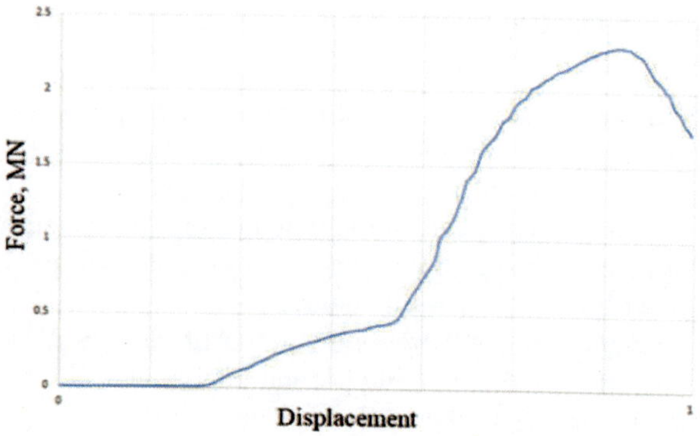

Fig. 7 Force plotted versus the displacement during the second operation

Figure 7 shows that the force at the first stage of forming increases insignificantly. Further, during the extrusion with thinning and direct extrusion, a sharp increase in the investigated parameter is observed. The maximum force of the process is reached practically at the end of the operation and it equals 2.3 MN.

4 Discussion

The semi-finished part produced as a result of the presented theoretical studies can be used for the manufacture of body parts both with and without a flange.

To produce parts with a flange it is necessary to form a flange at the funnel-shaped section of the semi-finished product.

To produce a cylindrical part without a flange (Fig. 8), it is necessary to perform the third extrusion operation with thinning of the funnel-shaped section and then trim the end of the semi-finished product.

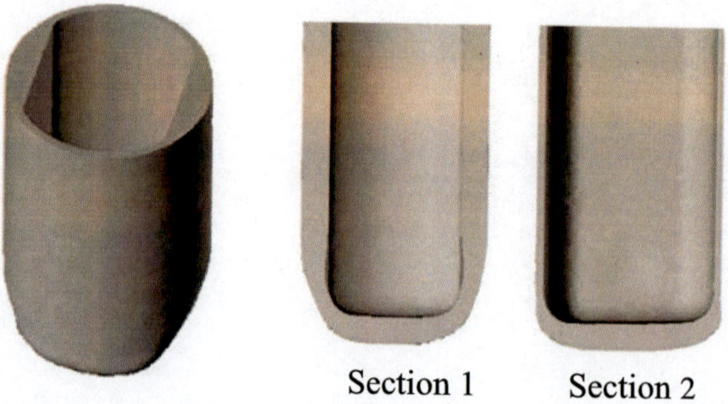

Fig. 8 Body part without a flange with variable wall thickness along the perimeter

5 Conclusion

The presented theoretical studies made it possible to solve the technological problems, which consist in designing a technological process with a minimum number of transitions, a high coefficient of material utilization, and precise geometric parameters. When comparing existing technologies for the production of body parts by means of backward extrusion with the proposed technology, it was found that energy consumption is reduced by almost 3 times. Moreover, the resulting product complies with the given performance parameters.

Acknowledgements This work was carried out within the framework of the grant NSh-2601.2020.8.

References

1. Semonov E (2010) Forging and stamping: handbook in 4 t. t. 3. Cold bulk stamping. Metal powder stamping. Mashinostroenie, Moscow
2. Zagirov N (2018) Theory of metal processing by pressure: a textbook. Siberian Federal University, Krasnoyarsk
3. Larin S, Platonov V, Korotkov V (2018) The design of a matrix for drawing of materials possessing planar anisotropy of mechanical properties. Tsvetnye Metally 7:83–87
4. Yaghoubi S, Fereshteh-Saniee F (2021) Optimization of the geometrical parameters for elevated temperature hydro-mechanical deep drawing process of 2024 aluminum alloy. J Process Mech Eng 2:151–216
5. Evin E, Daneshjo N, Mareš A, Tomáš M, Petrovčiková K (2021) Experimental assessment of friction coefficient in deep drawing and its verification by numerical simulation. Appl Sci (Switzerland) 6:2756
6. Malyshev A (2020) Application of variable thickness blanks in modern sheet-stamping production. Prospects for the development of processing technologies and equipment in mechanical engineering. In: Collection of scientific articles of the 5th All-Russian scientific and technical conference with international participation. Yugo-Zapad State University, Kursk, Russia, pp 161–163
7. Malyshev A, Bysov S, Orlik A, Bessmertnaya Yu (2020) Experimental study of the error in the height of the contour of box-shaped parts, obtained by drawing from welded blanks of different thickness. Chernye Metally 12:49–54
8. Vlasov A (2019) Finite element modeling of technological processes of forging and volumetric stamping. Bauman Moscow State Technical University Publishing House, Moscow
9. Pasynkov A, Larin S, Platonov V (2021) Analysis of the influence of the geometry of the input part of the matrix on the formation of the edge part of the cylindrical bill . In: 14th International scientific and technical conference on applied mechanics and systems dynamics, AMSD 2020, vol 1791(1). Omsk, Russian Federation
10. Schrek A, Svac P, Brusilova A, Gabrisova Z (2018) Simulation process deep drawing of tailor welded blanks DP600 and BH220 materials in tool with elastic blankholder. J Mech Eng 1:95–102

A Model of Electronic Document Management of Technical Documentation for Railway Automation

Dilshod Baratov, Nazirjon Aripov, Obidjon Muhiddinov, and Xasan Jumanov

Abstract The article discusses the features of electronic document management of technical documentation of railway automation and telemechanics. The article deals with the problems of synthesizing the mathematical description of the electronic document management of technical documentation for railway automation and telemechanics. For this, an overview of the actual processes of making, checking and applying technical documentation was carried out on the example of maintaining custom specifications for automation and telemechanics systems, which made it possible to identify document flow scenarios, as well as protocols of technical documentation properties. A method for creating a model of electronic document management of technical documentation based on the apparatus of graph theory has been advanced. Methods for determining sets for the proposed model are described, an algebra of workflow of TD using graphs is proposed. To set the matrix type of document flow representation, it is proposed to use a set of flat rectangular matrices, each of which represents the state of the system in some discrete unit of time. The resulting matrices of incidents determine the graph model of the document flow of the process under consideration, and the set of these matrices sets all possible scenarios for the movement of documents in the process, describes all possible states of documents and determines possible participants. The use of the proposed method for constructing a mathematical model based on graph theory makes it possible to develop reliable applied software for solving problems of operational document flow of technical documentation.

Keywords Electronic document management of technical documentation · Graph theory · Model of electronic document flow of technical documentation · Formal presentation of technical documentation · Technological process of maintaining custom specifications of railway automation and telemechanics · Matrix of incidents of electronic document flow of technical documentation

---

D. Baratov (✉) · N. Aripov · O. Muhiddinov · X. Jumanov
Tashkent State Transport University, Odilkhojayev 1, Tashkent, Uzbekistan100067

A. Bieliatynskyi and V. Breskich (eds.), *Safety in Aviation and Space Technologies*, Lecture Notes in Mechanical Engineering,
https://doi.org/10.1007/978-3-030-85057-9_40

1 Introduction

Currently, the most complex tasks are being solved in railway transport using modern information technologies. Microprocessor systems are used to manage and centralize stations and stages. Intensive work is continuing on the transfer to electronic document management in parallel with the traditional paper document management. In Uzbekistan, there is currently a definition of this term at the legislative level [1].

The issues of formalization of electronic document flow of technical documentation (TD) of railway automation are considered in works [2–12], however, the task of clearly formalizing the concept of "electronic document flow of technical documentation" and creating a model of technological processes of railway automation and telemechanics, as well as using existing information technology achievements to solve management problems remains urgent.

To create a model of electronic document flow of technical documentation (EDTD) for signaling, centralization and blocking devices, a survey of the actual processes of creating, checking and using technical documentation in automation and telemechanics systems was performed. This allowed us to identify document flow scenarios and property protocols for technical documents.

This article discusses a new approach to creating models of technical documentation workflow based on the graph theory apparatus. Methods for defining sets for the developed model are described, and a document flow algebra using graphs is proposed.

2 General Provisions

To build a graph model, need to defined a certain set of data that will be accepted by the basis. For a graph model, the data must be discrete and assume varying degrees of connectivity. The notation introduced in the paper is used as the basis for the graph model under consideration [13].

Formally, the EDTD process is represented as three finite sets and links of elements of these sets to each other. The mathematical notation of this process is represented as a triple

$$DT = \{Y, P, F\} \quad (1)$$

where DT—a formal model of electronic document flow of technical documentation; Y—a set of participants; P—a set of processes; F—a set of TD States with acceptable areas of values.

The set Y is defined as a finite set of actual document flow participants, and P is defined as a finite set of processes that are performed within the document flow system under consideration by participants from the set Y. F is a finite set of States that can accept TD after executing processes from the set P by participants from the set Y.

When building a graph model of the TD workflow, it is proposed to use the following method for displaying the TD workflow with graphs [14]. To set the vertex set of a graph use the number of possible States of Edges of F. The edges of the graph will be set by using a variety of processes P and installed this according so as to fulfill the following rules:

- one node of the graph corresponds to one and only one element of the set F;
- one edge of the graph corresponds to one and only one element of the set P;
- one element of the set F corresponds to one and only one vertex of the graph;
- one element of the set P corresponds to one and only one edge of the graph.

Such an identical mapping of the sets of States F to the set of vertices v and the sets of processes P to the set of edges e can be mathematically defined as follows: for any i, the statement is true.

$$v(i) \leftrightarrow F(i) \quad \text{and} \quad e(i) \leftrightarrow P(i), \tag{2}$$

where $i \in I,\ I = 1, 2, 3, \ldots, n$.

That is, two paired grammars are defined—the first grammar for establishing the translation of F to v, and the second grammar for establishing the translation of P to e.

Thus, the relationships between vertexes correspond identically to the relationships of States of the modeled workflow. After implementing the sets of processes P, the States F of the technical documentation change. In a workflow graph, the vertices of a graph connect edges if and only if the corresponding state vertices are connected by an action corresponding to the edge, i.e.

$$e = \begin{cases} e, & \text{if the edge exists;} \\ 0, & \text{if the edge is missing.} \end{cases} \tag{3}$$

The direction of the edges is set in such a way as to display the logic of the sequence of changing document flow States. A vertex i is an incoming vertex for a vertex j through an edge k if and only if the state i changes to the state j after the action k is performed. Thus, the States y1, y2,..., yn are mapped to the vertices of the graph v1, v2,..., vn and each pair of vertices vi and vj is connected by an edge eij going from vi to vj if and only if state vi is the input state for vj.

To get a clear description of various structural properties of document flow, it is useful to introduce a number of concepts defined and widely used in graph theory into the graph model [15–18]. The collection of a non-empty set V, an isolated set E (possibly empty), and a map F of a set $E(V^2)$ is a graph. The elements of the set V are called vertices, the elements of set E—graph edges, and F the mapping of the incidence graph.

3 Definitions of a Workflow Model Based on a Graph

To represent the document flow graph of technical documentation the following format is accepted

$$G = (V, E, R), \tag{4}$$

where V—set of graph vertices, E—the set of edges of a graph, R—the set of relations of incidence.

Thus, a graph G consists of a non-empty set of elements called vertices; a set of connected pairs from a set of vertices called edges; and a set of edge direction features [19].

A set consisting of vertices of a graph G is called the set of vertices of a graph and is denoted by V(G). Similarly, a set consisting of edges is called the set of edges and is denoted by E(G). If v_i and v_j are vertices of the graph G, then the edge of $v_i v_j$ is called the connection that connects v_i and v_j.

Two vertices v_i and v_j are boundary vertices of an edge *e* if v_i is the beginning of the edge and v_j is the end of the edge. Two vertices v_i and v_j are adjacent if they are distinct and exist, and there is an edge running from one of them to the other. It is assumed that an edge *e* comes from vertex v_i if v_i is the beginning but not the end of *e*, and that an edge goes to v_i if v_i is the end but not the beginning of *e*. In both cases, the edge *e* is called the incident vertex v_i, and the vertex v_i is called the incident edge *e*. The total number of edges incident to vertex v_i is a power of vertex v_i and is denoted $b(v_i)$.

4 The Matrix Form of Presentation of the Document Management of Technical Documentation

To represent the workflow of technical documentation, it is advisable to use the matrix form [20–24]. The matrix form uses three sets from the previously introduced triple {Y, P, F}.

It is believed that at the time of submission occurred updating the sets, i.e. all the States are represented by many forms, all actions leading to state changes lots of action, and the manufacturing steps in the form of roles in many of the participants. A matrix model is a model of the current state of the technical documentation workflow that operates with a finite number of technical documents.

To solve the above problem, it is proposed to use a set of flat rectangular document flow matrices, each of which represents the state of the system in a discrete unit of time. Columns in the workflow matrix are set to match the document States that are possible within the workflow lifecycle.

That is, the first column corresponds to the first element of set F, the second column the second element and so on, until the last element of the set F. The matrix

Rows of the document are associated with the process, the product of which leads to a change of state of at least one document. The first row corresponds to the first element of the set P, the second row to the second, and so on, for the entire set P.

Thus, it is got a rectangular matrix with columns, the number of which is equal to the dimension of the set F and rows in the dimension of the matrix P. this matrix is Filled with elements of the set of role participants in the modeled document flow Y. An element is filled into a matrix cell if and only if the corresponding participant performs an action corresponding to a row element, which results in a change in the state corresponding to the column element.

If the row action does not change the column state at this workflow step, the matrix element is filled with empty or null values. The criterion for the success of creating a matrix is its non-degeneracy in columns and rows. In other words, there is at least one column in the matrix that contains a non-empty element, and at least one row that contains a non-empty element. However, it is assumed that not all elements of the set of role participants in Y will be used for filling in.

Thus, a matrix of technical documentation workflow is obtained, the content of which uniquely corresponds to the state of document flow at the first step. After the first event occurs, namely, after the first action occurred that led to a change in at least one state, will be updated the workflow matrix. Namely, the matrix content will be brought in such a way that the matrix elements correspond to the current state—the second step of document flow. Thus, at the second step, get a matrix filled in by the participants of this step, who are at the intersection of the actions they perform and the States that these actions change.

Since the number of workflow steps may be large, but it is still finite, the workflow itself can be represented as a finite set of matrices described above. Each matrix represents the overall state of the entire composite document management system at a time when there is no change in document States.

5 Graph-Based Workflow Model

To illustrate the meaningful meaning of the concepts used, will be considered a document flow model based on the proposed graph model.

As a basis, will be considered the General technological process of maintaining custom specifications (MCS) for railway automation and telemechanics (Fig. 1), presented in the article [15].

The process consists of the following steps:

1. The Project organization makes custom specifications (CS) based on the completed project for the objects under construction or reconstruction.
2. The compiled CS is approved by the automation and telemechanics service.
3. After that, the CS is sent to the capital construction Department (CCD). This data is used to determine which equipment manufacturers need to order. Since there are several hundred items of equipment and devices in the CS, the

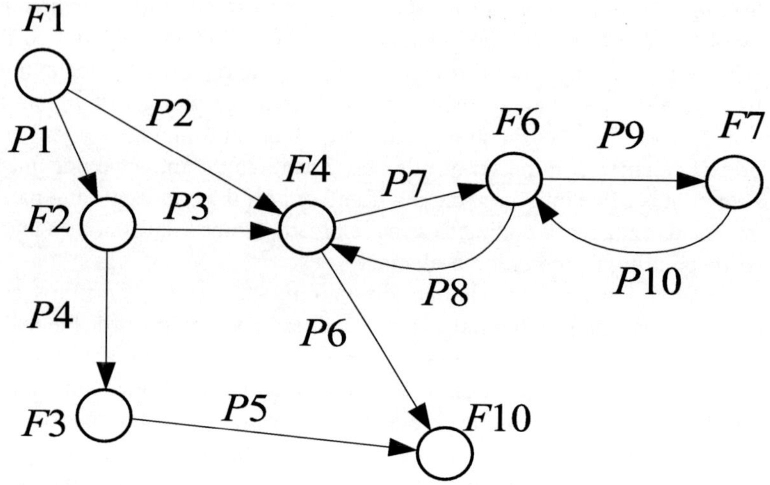

Fig. 1 Graph of the technological process for maintaining specifications of railway automation and telemechanics

probability that only one factory produces them is very small. CCD divides the equipment specified in the CS between suppliers.
4. Suppliers, in turn, give an order to the factory to complete the order.
5. At the end of the work, the plant sends the equipment to the warehouse signaling distance.

Will be considered the General technological process of technical characteristics CS of railway automation and telemechanics (Fig. 1), given in [19].

The following notation is used to set model parameters:

- documents are indicated by a set of forms used in the modeled process. Note these forms F1…F10;
- the process performed by documents that change States is defined by a set of processes P1…P10;
- performing processes P1…P10, are denoted by the set Y1…Y10.

Possible scenarios for the process of maintaining custom specifications that can be implemented by a given model will be considered. For this a table in the form of a matrix (Table 1) and possible transitions of the VSS process will be made (Table 2).

From table defining the set Y:

Y1 = P2F4∨P1F2.
Y2 = P1F2∨P3F4∨P4F3.
Y3 = P4F3∨P5F10.
Y4 = P2F4∨P6F6∨P9F4∨P3F4∨P4F10.
Y6 = P6F6∨P9F4∨P7F7∨P8F6.

Table 1 Matrix table {Y, P, F} for the MCS process

	F1	F2	F3	F4	F5	F6	F7	F8	F9	F10
P1	Y1	Y1 Y2								
P2				Y1 Y4						
P3				Y2 Y4						
P4			Y2 Y3							
P5										Y3 Y10
P6						Y4 Y6				
P7							Y6 Y7			
P8						Y7 Y6				
P9				Y6 Y4						
P10										Y4 Y10

Table 2 Table of possible transitions of the MCS process

#	P	F	Y1	Y2	Y3	Y4	Y5	Y6	Y7	Y8	Y9	Y10
1	P2	F4	1			1						
2	F6	F6				1		1				
3	P9	F4				1		1				
4	P7	F7						1	1			
5	P8	F6						1	1			
6	P1	F2	1	1								
7	P3	F4		1		1						
8	P4	F3		1	1							
9	P5	F10			1							1
10	P4	F10				1						1

$Y7 = P7F7 \vee P8F6.$
$Y10 = P5F10 \vee P4F10.$

From the set Y, those that have the same elements are combined $Y6 = P6F6 \vee P9F4 \vee Y7$.

Will be applied graph theory terminology to the workflow model of technical documentation. Then the possible workflow scenarios correspond to the graphical paths (Fig. 2).

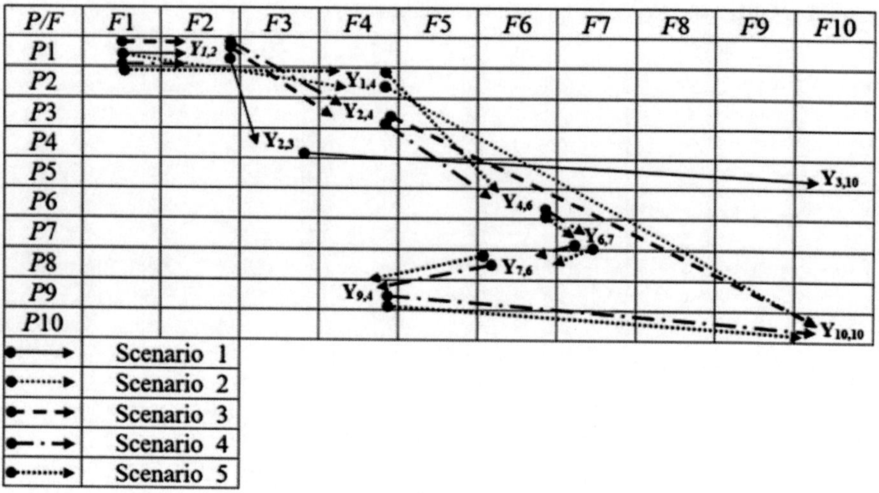

Fig. 2 Graph transitions of the technological process of maintaining specifications of railway automation and telemechanics

In the presented graph, there are five possible paths, which are denoted by edges. These paths correspond to the scenarios MCS.

Will be imagined the EDTD matrices corresponding to the considered scenarios. At each stage of the script, a workflow step is performed that corresponds to the work being done on the documents.

Scenario 1. At stage 1, the elements of the matrix P1, F1 and P1, F2 get the values Y1 and Y2 (Fig. 2). At stage 2, the matrix elements P1, F2 and P4, F3 get the values Y2 and Y3, and at stage 3, the elements P4, F3 and P5, F10 respectively get the values Y3 and Y10.

Scenario 2. At stage 1, the matrix elements P1, F1 and P2, F4 get the values Y1 and Y4, respectively. At stage 2, the matrix elements P2, F4 and P10, F10, respectively, get the values Y4 and Y10.

Scenario 3. At stage 1, the matrix elements P1, F1 and P1, F2 get the values Y1 and Y2. At stage 2, the matrix elements P1, F2 and P3, F4 get the values Y2 and Y4, and at stage 3, the elements P3, F4 and P10, F10 respectively get the values Y4 and Y10.

Scenario 4. At stage 1, the matrix elements P1, F1 and P1, F2 get the values Y1 and Y2. In step 2, the matrix elements P1, F2 and P3, F4 to get the values Y2 and Y4, for phase 3 elements P3, F4 and P6, F6 respectively receive the values of Y4 and Y6, in step 4, the elements P6, F6 and P7, F7 respectively receive the values of Y6 and Y7, in step 5 the elements P7, F6 and P8, F6 respectively receive the values of Y7 and Y6, and in step 6 the elements of P8, F6 and P9, F4 respectively receive the values of Y6 and Y4, in step 7, the elements P9,F4 and P10,F10 respectively receive the values of Y4 and Y10.

Scenario 5. At stage 1, the matrix elements P1, F1 and P2, F4 get the values Y1 and Y4, respectively. In step 2, the elements of the matrix P2, F4 and P6, F6

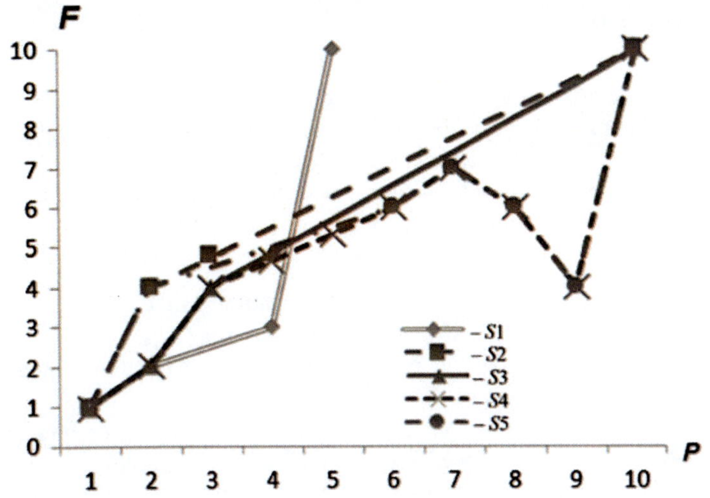

Fig. 3 Graph of changes in the state of the MCS process

respectively receive the values of Y4 and Y6, stage 3 elements Y6, Y7 and P6, F6 respectively receive the values of Y6 and Y7, the stage 4 elements P7, F7 and P8, F6 respectively receive the values of Y7 and Y6, in step 5 the elements of P8, F6 and P9, F4 respectively receive the values of Y6 and Y4, in step 6 the elements of P9, F4 and P10, F10 respectively receive the values of Y4 and Y10.

The resulting event matrices determine the model of the EDTD graph of the presented process.

The set of matrices specified above sets all possible scenarios of document flow in the process, describes all possible cases of document flow (Fig. 3) and defines possible participants.

In addition to the mapping matrix, it is convenient to display the graph by the contiguity matrix. Just as the matching matrix reflects the relationship between the ends and edges, so the ends of the contiguity matrix reflect the relationship between each other. In the proposed model, the contiguity matrix reflects the relationship of States (Fig. 4). the elements include actions that lead to changes in States.

Fig. 4 Graph of the ratio of states of the MCS process

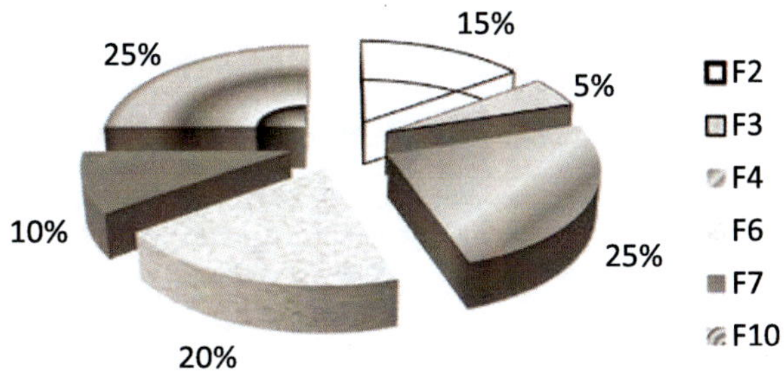

6 Conclusions

Based on the methodology for constructing a conceptual model of document flow systems for technical documentation [1] and a composite document flow model [10], this article presents a graph model of electronic document flow for technical documentation, which takes into account the decomposition of document flows into a set of process participants, a set of States, and a set of actions.

Obtained result:

- survey of real processes of creation, verification and use of technical documentation of automation and telemechanics systems;
- document flow scenarios and protocols of technical document properties are identified;
- developed a method for creating a model of electronic document flow of technical documentation based on the graph theory apparatus;
- it is proposed to use a set of flat straight-angle matrices to set the matrix form for the representation of technical documentation workflow;
- methods for defining sets for the proposed model are described, and an algebra of technical documentation workflow with the use of graphs is proposed.

Prospects:

- further improvement of the theoretical base of technical documentation workflow
- the ability to effectively identify all possible scenarios of document movement in the process, describe all possible document States, and identify possible participants
- development of computer application software based on the graph theory apparatus for solving problems of document management of technical documentation of railway automation and telemechanics systems.

References

1. Aripov N, Aliyev R, Baratov D, Ametova E (2016) Features of construction of systems of railway automatics and telemechanics at the organization of high-speed traffic in the Republic of Uzbekistan. Proc Eng 134:175–180. https://doi.org/10.1016/j.proeng.2016.01.057
2. Bentil W, Li LC, Chawner B (2021) An examination of electronic resource management in academic libraries in Ghana through the techniques of electronic resource management (TERMS) framework. J Acad Librarianship 47(1):102265. https://doi.org/10.1016/j.acalib.2020.102265
3. Nakayama M, Hustad E, Sutcliffe N (2021) Agility and system documentation in large-scale enterprise system projects: a knowledge management perspective. Proc Comput Sci 181:386–393. https://doi.org/10.1016/j.procs.2021.01.181

4. Baratov D, Astanaliev E (2021) Minimization of the automatic machine structure process of accounting and control of railway automation and telemechanics devices. In: E3S web of conferences, vol 244, p 08024. https://doi.org/10.1051/e3sconf/202124408024.
5. Pushpa AM (2020) Automatic source code documentation using code summarization technique of NLP. Proc Comput Sci 171:2522–2531. https://doi.org/10.1016/j.procs.2020.04.273
6. Oevermann J (2018) Semantic PDF segmentation for legacy documents in technical documentation. Proc Comput Sci 137:55–65. https://doi.org/10.1016/j.procs.2018.09.006
7. Sangeetha S, Hema P, Selvarani N, Geetha P, Karthikeyan P, Gowdham Kumar S (2021) Fuzzy graph theory in coloring graph. Mater Today: Proc. https://doi.org/10.1016/j.matpr.2021.01.797
8. Yavari R, Williams R, Riensche A, Hooper PA, Cole KD, Jacquemetton L, Halliday HS, Prahalada Krishna Rao (2021) Thermal modeling in metal additive manufacturing using graph theory—application to laser powder bed fusion of a large volume impeller. Additive Manuf 41:101956. https://doi.org/10.1016/j.addma.2021.101956
9. Davis Ian, McKay Andrew, Bennett Gareth J. (2021) A graph-theory approach to optimisation of an acoustic absorber targetinguns a specific noise spectrum that approaches the causal optimum minimum depth. J Sound Vibr 505:116135. https://doi.org/10.1016/j.jsv.2021.116135
10. Aripov N, Muxiddinov O, Baratov D, Jumanov X (2020) Block diagram of the automatic model of technical documentation. J Phys: Conf Ser 1614(1):012032. https://doi.org/10.1088/1742-6596/1614/1/012032
11. Ayaz A, Yanartaş M (2020) An analysis on the unified theory of acceptance and use of technology theory (UTAUT): acceptance of electronic document management system (EDMS). Comput Hum Behav Rep 2:100032. https://doi.org/10.1016/j.chbr.2020.100032
12. Alexey A, Kristina I, Evgeny T, Alexey T (2018) Electronic document processing operating map development for the implementation of the data management system in a scientific organization. Proc Comput Sci 145:248–253. https://doi.org/10.1016/j.procs.2018.11.053
13. Lee JY, Park H-A, Chung E (2018) Use of electronic critical care flow sheet data to predict unplanned extubation in ICUs. Int J Med Inform 117:6–12. https://doi.org/10.1016/j.ijmedinf.2018.05.011
14. Brandenburger N, Naumann A (2019) On track: a series of research about the effects of increasing railway automation on the train driver. IFAC-PapersOnLine 52(19):288–293. https://doi.org/10.1016/j.ifacol.2019.12.115
15. Muxiddinov O, Baratov D, Aripov N, Jumanov X (2020) Structurial schem diagram of the automatic model of technical documentation. J Phys: Conf Ser 1614(1):012032 https://doi.org/10.1088/1757-899X/918/1/012084
16. Ruziev D, Aripov N, Baratov D (2019) Formalized methods of analysis and synthesis of electronic document management of technical documentation. In: 2019 IEEE east-west design and test symposium, p 8884415. https://doi.org/10.1109/EWDTS.2019.8884415
17. Corman F, Trivella A, Keyvan-Ekbatani M (2021)Stochastic process in railway traffic flow: models, methods and implications. Transp Res Part C: Emerg Technol 128:103167. https://doi.org/10.1016/j.trc.2021.103167
18. Söylemez MT, Durmuş MS, Yıldırım U, Serhat T, Arcan S (2011) The application of automation theory to railway signalization systems: the case of turkish national railway signalization project. IFAC Proc Vol 44(1):10752–10757. https://doi.org/10.3182/20110828-6-IT-1002.03755
19. Baratov D, Aripov N (2019) Formalized scheme of technical documentation based on the accounting process and control of automatic and telemechanics devices. Int J Eng Adv Technol 8(3):479–484
20. David McGlade, Sandra S-H (2019) ML-based cyber incident detection for electronic medical record (EMR) systems. Smart Health 12:3–23. https://doi.org/10.1016/j.smhl.2018.05.001

21. Patrick R, Manfred G, Detlef G, Martin H, Stefan D (2020) The CASAD matrix method: introduction of a technique for the documentation, analysis, and optimization of context-aware systems. Proc CIRP 91:427–432. https://doi.org/10.1016/j.procir.2020.03.106
22. Khan Basheer A, Saha Arun K (2021)Direct numerical simulation of turbulent flow and heat transfer over a heated cube placed in a matrix of unheated cubes. Int J Heat Mass Transfer 171:121052. https://doi.org/10.1016/j.ijheatmasstransfer.2021.121052
23. Changjun L, Franz KD, Daeho L (2019) Capturing information on technology convergence, international collaboration, and knowledge flow from patent documents: a case of information and communication technology. Inf Process Manage 56(4):1576–1591. https://doi.org/10.1016/j.ipm.2018.09.007
24. Aoun J, Quaglietta E, Goverde RMP, Scheidt M, Blumenfeld M, Jack A, Redfern B (2021) A hybrid delphi-AHP multi-criteria analysis of moving block and virtual coupling railway signaling. Transp Res Part C: Emerg Technol 129:103250. https://doi.org/10.1016/j.trc.2021.103250

Parametric Analysis of the Boundary Layer of the Air Flow of the Ventilation Duct of the Brake Disc

Artem Litvinov, Ivan Yaitskov, Pavel Polyakov, Alexey Golikov, Evgeny Fedotov, and Nina Zadayanchuk

Abstract The air flow is used to cool the brake discs. Manufacturers of friction units equip a metal friction element of the unit with a ventilation device, assuming that it will intensively remove heat from heated surfaces. Macroparameters characterizing the efficiency of the brake disc ventilation apparatus are widely given in the studies of scientists from different countries. The speed and volume of air masses directly affect the amount of heat removed from heated surfaces. Surface parameters also affect the heat transfer process. Unfortunately, there are so few works devoted to the processes in the boundary layer of the air flow that directly affect the heat transfer from the surfaces of the ventilation apparatus. In a wide range of literary sources, the boundary layer has a heat-insulating effect, with an indistinctly expressed condition of movement and energy transfer.

Keywords Brake disc · Ventilation apparatus · Energy balance · Boundary layer · Heat and mass transfer · Loss of momentum · Displacement by air flow

1 Introduction

The research carried out provides an equation for the energy balance of the air flow in the boundary layer, taking into account all types of flow energy, from potential to heat. Analytical dependences of various parameters of the thickness of the boundary layer of the air flow are proposed. Based on the theory of air flow continuity, the thickness of the momentum loss in the boundary layer inside the ventilation

A. Litvinov (✉) · P. Polyakov · A. Golikov · E. Fedotov
Kuban State Technological University, Moskovskaya st., 2a, Krasnodar 350020, Russia

I. Yaitskov
Rostov State Transport University, Rostov Rifle Regiment of the People's Militiapi, 2, Rostov-on-Don 344038, Russia

N. Zadayanchuk
Kuban State University, Stavropolskaya, st. 149, Krasnodar 350040, Russia

A. Bieliatynskyi and V. Breskich (eds.), *Safety in Aviation and Space Technologies*, Lecture Notes in Mechanical Engineering,
https://doi.org/10.1007/978-3-030-85057-9_41

apparatus and the thickness of the displacement by the air flow from the boundary layer are determined.

According to theoretical studies of the energy balance in the boundary layer of the air flow in the ventilation apparatus, the parameter of the thickness of the heat energy loss in the boundary layer is given. In connection with the heating of air masses in the boundary layer, there is a change in the internal energy of the air, and as a result of its total enthalpy. Based on the same energy balance equation, the thickness of the enthalpy loss in the boundary layer of the air flow was determined.

As a result of analytical studies, a parametric analysis of the influence on the dynamic, thermal and energy components of the resistance to heat and mass transfer in the boundary layer was carried out. This allows in the future to take into account the parameters of influence on heat and mass transfer as a whole in the entire ventilation apparatus of the brake disc, and to improve the methodology for the development of thermal models of ventilated brake discs.

Literary search for sources devoted to cooling ventilated brake discs is reduced to the division of work into two directions. The first direction is the development of a thermal model of the brake mechanism. The second direction is to intensify the cooling of friction units.

The developed thermal models of friction units [1–3] are based on changing the parameters of the microgeometry of the surfaces of friction elements and the use of new materials in the manufacture of brake discs.

In the works of the second direction, various criteria are used to assess the intensity of cooling. Work [4] is devoted to the study of a heat pipe built into a ventilation device.

In studies [1, 5] changes in the geometry of the ribs of the ventilation apparatus of the brake disc are given. In [6], the angle of attack of forced air flow into ventilation ducts was proposed as a criterion. Work [7] is devoted to the volume of cooling air passing through a ventilation apparatus as a cooling criterion.

Modern studies of increasing the intensity of cooling of ventilated brake discs consider macroparameters without considering the parameters of the boundary layer of the air flow [8]. One of the criteria for the intensity of cooling is the thickness of the boundary layer of the air flow (Fig. 1). In modern studies, there are no criteria characterizing the size of the boundary layer and the processes occurring within it.

2 Materials and Methods

In studies [9], a system of differential equations for the motion of air masses in the boundary layer and a differential equation for the continuity of the flow in the boundary layer of the air flow in the ventilation duct of the brake disc (1) were given.

In studies [9], a system of differential equations for the motion of air masses in the boundary layer and a differential equation for the continuity of the flow in the

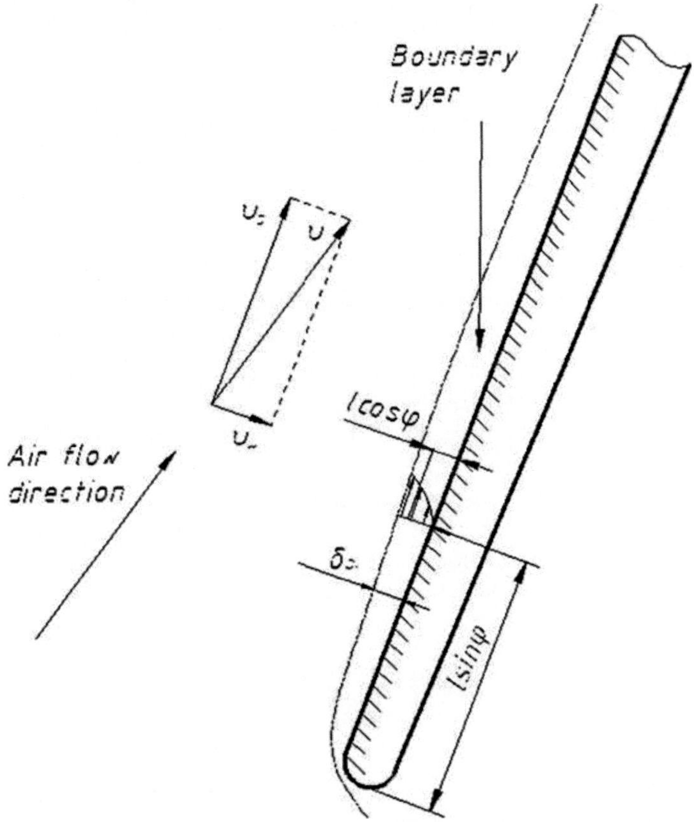

Fig. 1 Design diagram of the movement of the air flow washing the ribs of the brake disc ventilation apparatus

boundary layer of the air flow in the ventilation duct of the brake disc (1) were given.

$$\begin{cases} \rho_a\left(\upsilon_d \dfrac{\partial \upsilon_d}{\partial(l\sin\varphi)} + \upsilon_\omega \dfrac{\partial \upsilon_d}{\partial(l\cos\varphi)}\right) = -\dfrac{\partial p}{\partial(l\sin\varphi)} + \dfrac{\partial\left[\mu\left(\frac{\partial \upsilon_\omega}{\partial(l\sin\varphi)}\right)\right]}{\partial(l\cos\varphi)}, \\ \dfrac{\partial p}{\partial(l\cos\varphi)} = 0, \\ \dfrac{\partial(\rho_a \upsilon_\omega)}{\partial(l\cos\varphi)} + \dfrac{\partial(\rho_a \upsilon_d)}{\partial(l\sin\varphi)} = 0, \end{cases} \tag{1}$$

where

ρ_a is the air density in the ventilation duct of the brake disc, kg/m^3;
υ_d is the air flow velocity in the radial direction of the ventilation duct, m/s;

υ_ω air flow velocity in the transverse direction of the ventilation duct, m/s;
l is a vector characterizing the movement of the air flow in the ventilation duct, m;
φ is the angle between the vector of movement of the air flow in the ventilation duct and the horizontal, deg.
μ dynamic viscosity of air, Pa s;
p is the air flow pressure inside the ventilation duct, Pa.

Analyzing the equation of continuity of the flow, we can conclude that the term of the equation $\frac{\partial(\rho_a \upsilon_\omega)}{\partial(l \cos \varphi)}$ is responsible for the propagation of the velocity of air masses over the thickness of the boundary layer δ_{bl}. Due to the fact that the speed of the air flow across the ventilation duct will be relatively low and tend to zero.

To integrate the partial derivative, it is necessary to bring it to the usual differential equation. To take this assumption into account, we introduce the coefficient ψ into the equation [10–12].

Let's consider each term of the equation separately. If the sum is zero, then this may mean that each of the terms of the equation is infinitely small and can tend to zero. Let us assume that the left term of the equation is equal to zero, then equating $l \cos \varphi$ to the thickness of the boundary layer δ_{bl} we get:

$$\frac{\partial(\rho_a \upsilon_\omega)}{\partial(l \cos \varphi)} = \frac{\psi \rho_a d(\upsilon_\omega)}{d(\delta_{bl})} = 0, \tag{2}$$

By integrating expression (2) with respect to the value of the air flow velocity across the direction of movement of the air masses, we obtain:

$$\delta_{bl} = \int_0^\infty \psi \rho_a d\upsilon_\omega = \psi \rho_a (\upsilon_d - \upsilon_\omega) + C, \tag{3}$$

If we set the initial conditions: $\delta_{bl} = 0$, $\upsilon_\omega \to 0$, $\upsilon_d \to \upsilon_\infty$, then after transformation (3) we get:

$$\delta_{bl} = \int_0^\infty \psi \rho_a d\upsilon_\omega = \psi \rho_a \upsilon_d, \tag{4}$$

If we equate the right term of the equation to zero, we get:

$$\frac{\partial(\rho_a \upsilon_d)}{\partial(l \sin \varphi)} = \frac{d(\rho_a \upsilon_d)}{d(l \sin \varphi)} = \frac{d(\delta_{bl})}{\psi d(l \sin \varphi)} = 0. \tag{5}$$

When analyzing this equation of flow continuity, the air is calculated as homogeneous and its density ρ_a is assumed to be the same at all points of the ventilation duct. Further, when considering the boundary layer of the ventilation channel, the air density at the edge of the ventilation device ρ_{vd} and the air density at the boundary of the boundary layer and the free flow ρ_d will be different.

Provided that the movement of the air flow through the cross section of the boundary layer is preserved, we write the equations in the following form:

$$\frac{d(\delta_{bl})}{\psi d(l\sin\varphi)}+f(2+H)=\frac{\rho_{vd}\upsilon_{\omega}}{\rho_d\upsilon_d},\tag{6}$$

where f is a parameter characterizing the aerodynamic curvature of the rib of the ventilation apparatus by the air flow; H—parameter characterizing the boundary layer of the ventilation apparatus; ρ_{vd}—air density in the boundary layer at the edge of the ventilation duct, kg/m3; ρ_d is the air density in the boundary layer at the border with the free flow of the ventilation duct, kg/m^3.

The boundary layer thickness is an uncertain value and depends on a large number of unrelated parameters. In contrast to this value, there are parameters that can clearly characterize the boundary layer, such as the amount of displacement by the air flow in the boundary layer and the thickness of the momentum loss by the air flow in the boundary layer [13].

The parameter of the aerodynamic curvature of the rib of the ventilation apparatus by the air flow is:

$$f=\frac{\delta_l}{\upsilon_d}\frac{d\upsilon_d}{d(l\sin\varphi)},\tag{7}$$

where δ_l is the thickness of the momentum loss by the air flow in the boundary layer, m.

Parameter H is determined by the formula:

$$H=\frac{\delta_l}{\delta_d},\tag{8}$$

where δ_d is the thickness of displacement by the air flow in the boundary layer, m.

These two parameters are responsible for the change of air masses in the boundary layer in the ventilation duct of the brake disc.

Let us analyze each of the parameters of the boundary layer thickness:

$$\rho_d\upsilon_d\delta_d=\int\limits_0^{\infty}(\rho_d\upsilon_d-\rho_{vd}\upsilon_{\omega})d(l\cos\varphi),\tag{9}$$

The thickness of displacement by air flow in the boundary layer δ_d is determined by the segment through which the mass flow rate of the air flow will be equal to the loss of flow rate in the cross section of the boundary layer due to the braking effect of friction forces against the ribs of the ventilation apparatus during the flow of air flow in the ventilation duct.

The thickness of the displacement by the air flow in the boundary layer δ_d is determined by the formula:

$$\delta_d = \int_0^\infty \left(1 - \frac{\rho_{vd}\upsilon_\omega}{\rho_d\upsilon_d}\right) d(l\cos\varphi), \tag{10}$$

Thickness of momentum loss by air flow in the boundary layer:

$$\rho_d\upsilon_d^2\delta_l = \int_0^\infty \rho_{vd}\upsilon_\omega(\upsilon_d - \upsilon_\omega)d(l\cos\varphi), \tag{11}$$

This parameter characterizes the segment through which, during the air flow, the momentum will pass, equal to the loss of movement in the section of the boundary layer of the ventilation duct due to the braking effect of friction forces against the fins of the duct of the ventilation apparatus.

The thickness of the momentum loss by the air flow in the boundary layer δl is determined by the formula:

$$\delta_l = \int_0^\infty \frac{\rho_{vd}\upsilon_\omega}{\rho_d\upsilon_d}\left(1 - \frac{\upsilon_\omega}{\upsilon_d}\right) d(l\cos\varphi), \tag{12}$$

When the air flow is flushed with the rib of the ventilation duct, the differential equation of the continuity of the flow will have the following form:

$$\frac{\partial(\rho_a\upsilon_\omega R)}{\partial(l\cos\varphi)} + \frac{\partial(\rho_a\upsilon_d R)}{\partial(l\sin\varphi)} = 0, \tag{13}$$

where R is the radius of curvature of the curved rib of the ventilation duct, m.

The radius of curvature is much greater than the thickness of the boundary layer ($R \gg \delta_{bl}$), which is a condition for the application of this equation.

Taking into account expressions (7, 10), we change the Eq. (6), the form of which will have the form:

$$\frac{d(\delta_{bl})}{\psi d(l\sin\varphi)} + \frac{\delta_l}{\upsilon_d}\frac{d\upsilon_d}{d(l\cos\varphi)}(2+H) + \left(\frac{1}{\rho_d}\frac{d\rho_d}{d(l\sin\varphi)} + \frac{1}{R}\frac{dR}{d(l\sin\varphi)}\right)\delta_l = \frac{\rho_{vd}\upsilon_{vd}}{\rho_d\upsilon_d}, \tag{14}$$

For the curved section of the rib of the ventilation apparatus, the thickness of the displacement by the air flow in the boundary layer will have the following form:

$$\delta_d = \int_0^\infty \left(1 - \frac{\rho_{vd}\upsilon_\omega}{\rho_d\upsilon_d}\right)\left(1 \pm \frac{d(l\cos\varphi)}{R}\cos\beta\right)d(l\cos\varphi), \tag{15}$$

where β is the angle between the tangent to the meridian and the axis of the curved rib of the ventilation duct, deg.

For the curved section of the rib of the ventilation apparatus, the thickness of the impulse loss by the air flow in the boundary layer will have the following form:

$$\delta_l = \int_0^\infty \left(\frac{\rho_{vd}\upsilon_\omega}{\rho_d\upsilon_d}\right)\left(1 - \frac{\upsilon_\omega}{\upsilon_d}\right)\left(1 \pm \frac{d(l\cos\varphi)}{R}\cos\beta\right)d(l\cos\varphi), \tag{16}$$

Let us write the energy balance equation for the boundary layer of the air flow in the channels of the ventilation apparatus:

$$\rho_a\upsilon_d\frac{\partial(\upsilon_d^2 + c_pT)}{\partial(l\sin\varphi)} + \rho_a\upsilon_\omega\frac{\partial(\upsilon_d^2 + c_pT)}{\partial(l\cos\varphi)} = \frac{\partial\left[\lambda\frac{\partial\left(T\cdot\left[1 - \frac{2H}{3\mathrm{Pr}}\upsilon_\rho\right]\right)}{\partial(l\sin\varphi)}\right]}{\partial(l\sin\varphi)} - \frac{2}{3}\mu\left[\frac{\partial\upsilon_d}{\partial(l\sin\varphi)}\right] + 2\mu\left[\frac{\partial\upsilon_d}{\partial(l\sin\varphi)}\right]^2 + \upsilon_d\frac{\partial\left[\mu\left(\frac{\partial\upsilon_\omega}{\partial(l\sin\varphi)}\right)\right]}{\partial(l\cos\varphi)}. \tag{17}$$

where

c_p is the specific isobaric heat capacity of the boundary layer of the air flow.
T is the average temperature of the boundary layer of the air flow in the ventilation apparatus of the brake disc, K.
H is the total enthalpy of the boundary layer;
P_r is the Prandl number.

On the left side of the energy balance equation, the numerator denotes the enthalpy of deceleration of the air flow at the junction of the boundary layer and free flow:

$$H_d - \upsilon_d^2 + c_pT = c_pT_d. \tag{18}$$

T_d is the average temperature at the boundary between the boundary layer and the consolidated flow in the brake disc ventilation apparatus, K.

The enthalpy value at the boundary of the boundary layer is considered constant.

$$\rho_a\upsilon_d\frac{\partial(\upsilon_d^2 + c_pT_d)}{\partial(l\sin\varphi)} + \rho_a\upsilon_\omega\frac{\partial(\upsilon_d^2 + c_pT_d)}{\partial(l\cos\varphi)} = 0. \tag{19}$$

From the left-hand side of Eq. (17), we subtract term-by-term Eq. (19) and integrate over the thickness of the boundary layer:

$$c_p \int_0^{\delta_{bl}} \frac{\partial \rho_a \upsilon_d (T - T_d)}{\partial (l \sin\varphi)} d(l\cos\varphi) + c_p \int_0^{\delta_{bl}} \frac{\partial \rho_a \upsilon_\omega (T - T_d)}{\partial (l\cos\varphi)} d(l\cos\varphi)$$

$$= \int_0^{\delta_{bl}} \left(\frac{\partial \left[\lambda \frac{\partial T \cdot \left[1 - \frac{2H}{3\mathrm{Pr}} \upsilon_d\right]}{\partial (l\sin\varphi)} \right]}{\partial (l \sin\varphi)} \right) d(l\cos\varphi)$$

$$- \frac{2}{3}\mu \int_0^{\delta_{bl}} \left[\frac{\partial \upsilon_d}{\partial (l\sin\varphi)} \right] d(l\cos\varphi) \tag{20}$$

$$+ 2\mu \int_0^{\delta_{bl}} \left[\frac{\partial \upsilon_d}{\partial (l\sin\varphi)} \right]^2 d(l\cos\varphi)$$

$$+ \int_0^{\delta_{bl}} \left(\upsilon_d \frac{\partial \left[\mu \left(\frac{\partial \upsilon_\omega}{\partial (l\sin\varphi)} \right) \right]}{\partial (l\cos\varphi)} \right) d(l\cos\varphi).$$

Boundary conditions at $d\cos\varphi = 0$, $\upsilon_\omega = \upsilon_{vd}$, $\upsilon_d = 0$, $T = T_{vd}$, $\rho_a = \rho_{vd}$, at the value $d\cos\varphi = \delta_{bl}$, $\upsilon_\omega = 0$, $T = T_d$, $\rho_a = \rho_d$.

Due to the boundary conditions:

$$\int_0^{\delta_{bl}} \frac{\partial \rho_a \upsilon_d (T - T_d)}{\partial (l\sin\varphi)} d(l\cos\varphi) = \rho_a \upsilon_{vd} (T - T_d)\Big|_0^{\delta_{bl}} = -\rho_{vd} \upsilon_{vd} (T_{vd} - T_d);$$

$$\int_0^{\delta_{bl}} \left(\frac{\partial \left[\lambda \frac{\partial \left(T \cdot \left[1 - \frac{2H}{3\mathrm{Pr}} \upsilon_d\right] \right)}{\partial (l\sin\varphi)} \right]}{\partial (l\sin\varphi)} \right) d(l\cos\varphi) = \lambda \frac{\partial T_{vd}}{\partial (l\sin\varphi)}\bigg|_0^{\delta_{bl}} + \frac{2}{3}\lambda \frac{\partial T_{vd} \cdot H \cdot \upsilon_d}{\partial (l\sin\varphi)}\bigg|_0^{\delta_{bl}}. \tag{21}$$

where T_{vd} is the temperature of the rib of the brake disc ventilation apparatus, K.

Let us write the energy balance equation at the surface of the rib of the ventilation apparatus:

$$-c_p \rho_{vd} \upsilon_{vd} (T_{vd} - T_d) = \lambda \frac{\partial T_{vd}}{\partial (l\sin\varphi)}. \tag{22}$$

Taking into account the heat removal from the fins of the ventilation apparatus, the parameter characterizing the boundary layer is the thickness of the heat energy loss in the boundary layer:

$$\delta_{le} = \int_0^\infty \left(\frac{\rho_{vd}\upsilon_\omega(T - T_d)}{\rho_d\upsilon_d(T_{vd} - T_d)}\right) d(l\cos\varphi) = \int_0^\infty \frac{\rho_{vd}\upsilon_\omega}{\rho_d\upsilon_d}\left(1 - \frac{(T_{vd} - T)}{(T_{vd} - T_d)}\right) d(l\cos\varphi). \tag{23}$$

The heat flux from the rib of the ventilation duct is:

$$q_{vd} = \lambda \frac{\partial T_{vd}}{\partial(l\sin\varphi)}\bigg|_0^{\delta_{bl}} \tag{24}$$

The energy balance equation at the surface of the rib of the ventilation apparatus is written in the following form:

$$\frac{d(\delta_{le}\rho_d\upsilon_d\Delta T)}{d(l\sin\varphi)} = \frac{c_p\rho_{vd}\upsilon_{vd}\Delta T + q_{vd}}{c_p}, \tag{25}$$

where $\Delta T = T_{vd} - T_d$ is the temperature difference between the rib of the ventilation duct and the free air flow in the ventilation duct, °C.

When the air flow reacts with the ribs of the brake disc ventilation apparatus, the energy balance equation looks as follows:

$$\frac{d(\delta_H\rho_d\upsilon_d\Delta H)}{d(l\sin\varphi)} = \frac{c_p\rho_{vd}\upsilon_{vd}\Delta H + q_{vd}}{c_p}, \tag{26}$$

where $\Delta H = H_{vd} - H_d$ is the difference between the enthalpy between the edge of the ventilation duct and the enthalpy of deceleration of the free air flow at the boundary with the boundary layer in the ventilation duct; δ_H—thickness of enthalpy loss by air flow in the boundary layer, m [14].

The obtained parameters of the boundary layer, namely, δ_d, δ_l, δ_{le}, δ_H, characterize the boundary layer of the air flow in the brake disc ventilation apparatus (Fig. 2).

From the point of view of heat and mass transfer of air masses in the ventilation apparatus, the parameters δd, δl can determine the effect of the change of matter in the boundary layer of the ventilation ducts and characterize the amount of potential energy required to renew it in the boundary layer. The parameters δ_{le}, δ_H determine the heat losses in the boundary layer and the energy consumption for the diffusion of the air flow with the ribs of the ventilation duct. These parameters of the boundary layer thickness make it possible to formulate the dependences of the Re numbers for the boundary layer:

Reynolds number at loss of airflow momentum in the boundary layer:

$$\mathrm{Re}_d = \frac{\rho_{vd}\upsilon_{vd}\delta_d}{\mu_{vd}}. \tag{27}$$

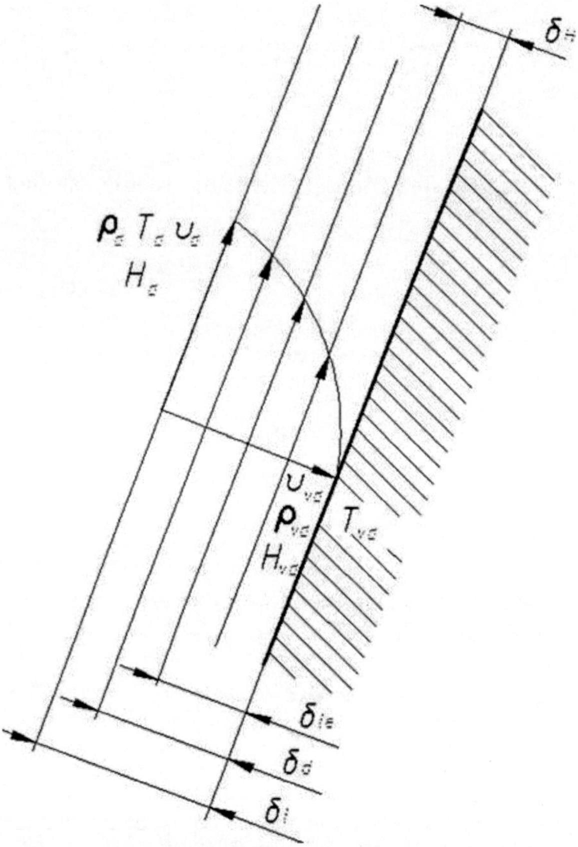

Fig. 2 Parameters of the boundary layer of the air layer in the ventilation apparatus of the brake disc

Reynolds number for air flow displacement in the boundary layer:

$$\mathrm{Re}_l = \frac{\rho_{vd} \upsilon_{vd} \delta_l}{\mu_{vd}}. \tag{28}$$

Reynolds number with loss of air flow energy in the boundary layer:

$$\mathrm{Re}_{le} = \frac{\rho_{vd} \upsilon_{vd} \delta_{le}}{\mu_{vd}}. \tag{29}$$

Reynolds number at loss of enthalpy of air flow in the boundary layer:

$$\mathrm{Re}_H = \frac{\rho_{vd} \upsilon_{vd} \delta_H}{\mu_{vd}}. \tag{30}$$

From the conducted theoretical research, it can be concluded that the resistance of heat and mass transfer processes in the boundary layer of the air flow in the ventilation apparatus of the brake disc is influenced by two groups of parameters. The first group is the design parameters (d, R, φ), the second group is more multiparameter and includes the following parameters (υ_d, υ_{vd}, ρ_d, ρ_{vd}, δ_d, δ_l, δ_{le}, δ_H, Re_d, Re_l, Re_{le}, Re_H, q_{vd}, T_{vd}, T_d, H_{vd}, H_d). The laws of heat and mass transfer resistance are characterized by dynamic losses due to surface friction of the air flow (C_{fd}, C_{fl}), thermal St_{le} and energy St_H enthalpy losses are as follows:

Dynamic impulse losses of the air flow in the boundary layer:

$$C_{fd} = f(d,\ R,\ \varphi, \upsilon_d,\ \upsilon_{vd},\ \rho_d,\ \rho_{vd},\ \delta_d,\ \mathrm{Re}_d). \tag{31}$$

Dynamic losses during air flow displacement in the boundary layer:

$$C_{fl} = f(d,\ R,\ \varphi, \upsilon_d,\ \upsilon_{vd},\ \rho_d,\ \rho_{vd},\ \delta_l,\ \mathrm{Re}_l). \tag{32}$$

Heat loss of air flow in the boundary layer:

$$St_{le} = f(d,\ R,\ \varphi, \upsilon_d,\ \upsilon_{vd},\ \rho_d,\ \rho_{vd},\ \delta_{le},\ q_{vd}, \mathrm{Re}_{le}, T_{vd},\ T_d). \tag{33}$$

Energy loss of enthalpy of air flow in the boundary layer:

$$St_H = f(d,\ R,\ \varphi, \upsilon_d,\ \upsilon_{vd},\ \rho_d,\ \rho_{vd},\ \delta_H,\ \mathrm{Re}_H, H_{vd},\ H_d). \tag{34}$$

Heat and energy losses of enthalpy are estimated by the Stanton number.

3 Conclusion

According to the laws of heat and mass transfer resistance, the processes occurring in the boundary layer and its parameters depend primarily on the air flow regime. As a result of the laminar movement of air masses, the parameters of the boundary layer will be lower than in the turbulent mode of the air flow. These dependences and laws of resistance to heat and mass transfer can explain the nature and mechanisms of heat transfer to the washing air flow from the ribs of the brake disc ventilation apparatus. In addition to the laws, these studies clarify the parameters of the boundary layer of the air flow in the ventilation ducts of the brake disc.

References

1. Indira R, Bharatish A (2020) Optimization of ventilated brake disc rotor geometry for enhanced structural characteristics. J Meas Eng 8:98–106. https://doi.org/10.21595/jme.2020.21399
2. Yanab H, Zhangbc Q, Lubc T (2016) Heat transfer enhancement by X-type lattice in ventilated brake disc. Int J Therm Sci 107:39–55. https://doi.org/10.1016/j.ijthermalsci.2016.03.026

3. Polyakov PA, Tagiev RS Fedotov ES, Polyakova EA, Moskalenko MB (2020) Development of a thermal model of the brake disc of the friction unit. Bull Irkutsk State Tech Univ 24(1 (150)): 64–76. https://doi.org/10.21285/1814-3520-2020-1-64-76
4. Polyakov PA, Litvinov AE, Polyakova EA, Fedotov ES, Tagiev RS (2020) Design of surface profile of pairs of friction unit. IOP Conf. Series: Materials Science and Engineering. 6. DOI: https://doi.org/10.1088/1757-899X/843/1/012001.
5. Qifei J, Wangb L, ShuiaY (2020) Thermal analysis of ventilated brake disc based on heat transfer enhancement of heat pipe. Int J Therm Sci 155. https://doi.org/10.1016/j.ijthermalsci.2020.106356
6. Nejat A, Aslani M, Mirzakhalili E, Najian Asl R (2011) Heat transfer enhancement in ventilated brake disk using double airfoil vanes. J Therm Sci Eng Appl 3:10. https://doi.org/10.1115/1.4004931
7. Litvinov AE, Polyakov PA, Polyakova EA, Tagiev RS, Fedotov ES, Golikov AA (2021) Development of a model of pressure distribution in the ventilation ducts of the brake disc during forced cooling. Bull ISTU named after M. T. Kalashnikov. 24(1):19–30. https://doi.org/10.22213/2413-1172-2021-1-19-30
8. Litvinov AE, Polyakov PA, Polyakova EA, Tagiev RS, Fedotov ES, Golikov AA (2020) Development of a methodology for assessing the cooling system of brake discs. M. T. Kalashnikov. 23(1):14–22. https://doi.org/10.22213/2413-1172-2020-1-14-22
9. Jiang F, Xu W, Deng J, Wei W (2020) Flow Structures of the air-water layer in the free surface region of high-speed open channel flows. Math Probl Eng. https://doi.org/10.1155/2020/5903763
10. Mamtaz F, Hossain A, Sharmin N (2018) Solution of boundary layer and thermal boundary layer equation. J Asian Re J Math 11(4):1–15. https://doi.org/10.9734/ARJOM/2018/45267
11. Kolesnikov VI, Chebakov MI, Danilchenko SA (2019) Analysis of the effect of the hardened surface layer properties of the tribocontact parts on their wear and stress-strain state. Adv Mate. Springer Proc Phys 224:637–642
12. Kolesnikov VI, Chebakov MI, Danilchenko SA (2019) The nonstationary contact problem of thermoelasticity for a ring in the presence of a reinforcing layer on the friction surface. J Friction Wear 4:447–455
13. Volchenko NA, Volchenko DA, Polyakov PA, Fedotov ES, Evchenko AS (2019) Pulse-contact frictional interaction of microprotrusions of friction pairs of brake devices. In: IOP conference series: materials science and engineering, vol 560(10)
14. Kolesnikov IV (2003) Thermophysical processes in metal-polymer tribosystems, 279

Vector Analysis of Strategic Sustainability Airport Operators and Airlines

Aigul Pankratova, Svetlana Borodulina, and Alexandr Gubenko

Abstract This article describes approaches to assessing strategic sustainability. The results of interaction between airport operators and airlines are important for understanding strategic sustainability of market players (operators) in the air transport sector. Strategic sustainability of airport operators and airlines is greatly influenced by the specific nature and focus of strategic interaction. Strategic sustainability can be presented as a vector of market players' development using a number of parameters, which reflect sustainability, and reliability of a long-term functioning of the mentioned market players in the particular sphere of their interaction. The article describes the spheres of strategic interaction, focus objects in the spheres chosen, system of indicators reflecting the results of strategic interaction between airport operators and airlines. The article offers to use a method, which makes it possible to create a vector description of strategic sustainability and interaction potential. Using a case study, the article describes the factors and the process of analyzing strategic sustainability of market players in the air transport sector. It also presents a vector description of strategic sustainability and the interaction potential be-tween an airport operator and an airline.

Keywords Strategic sustainability · Airport operators · Airlines · Potential · Strategic interaction

1 Introduction

The strategic development of modern airport operators and airlines has been taking place against the background of the mobile environment, which is determined by geopolitical, macroeconomic and social progress, as well as by changes in the industry itself. In the current conditions, corporate strategies will significantly define the company's sustainable functioning in the future. Due to the specific nature of the air transport sector there is a necessity of active interaction between

A. Pankratova (✉) · S. Borodulina · A. Gubenko
University of Civil Aviation, Pilotov str. 38, 196210 St. Petersburg, Russia

A. Bieliatynskyi and V. Breskich (eds.), *Safety in Aviation and Space Technologies*, Lecture Notes in Mechanical Engineering,
https://doi.org/10.1007/978-3-030-85057-9_42

two important players in the market, namely airport operators and airlines. This interaction be-tween them greatly influences the competitiveness of domestic carries and defines the amount and structure of passenger traffic and transport availability of Russian Federation regions.

Aviation companies should forecast the factors influencing the results of their interaction as well as find the most rational strategies and solutions taking into consideration the motives, strategic goals and interests of the related parties.

2 Literature Review

Modern approaches to understanding strategic sustainability, including process approach, profit margin approach, systems approach, and market approach, are presented in papers by Yashin, Grigorian and others [1–3]. The systems approach seems to be the most interesting. According to this approach, strategic sustainability is defined as a combination of controllable dynamic components which, taken together, ensure the object's stable performance and development. The authors [4–7] say that the lack of a common method of along with the excess of tools for assessing strategic sustainability are significant obstacles in the development of this kind of analysis. It is therefore important to mention the authors who have developed effective tools for the strategic management of corporate sustainability in different industries and sectors of the economy. Paper [8] contributes to the successful implementation of the sustainability corporate strategies, summarizes the theoretical background of the sustainability, determines the objectives, rules of the sustainability strategic alternatives and the critical factors of the strategic choices in a chemical industry, describes the framework of the conventional Balanced Scorecard (BSC) introduces the proposals for the elements of the BSC for the different sustainability strategies. We identified the branch focus is the authors' studies [2, 4, 5, 9, 10]. We agree that the branchers specifics and indicators should be taken into account in strategic development. However, we think that it can be difficult to develop a common method of assessing corporate sustainability in the context of strategic planning and management. The most important results of the study [11] revolve around the idea that the impact of applying a strategic vision is greatly related to the creation of a new, modern and sustainable management, capable of adapting to the necessities of the current economic and social environment.

The authors [12] link strategic sustainability and leadership to the emergence of senior positions in corporate management. The most important results of the study [13] revolve around the idea that the impact of applying a strategic vision is greatly related to the creation of a new, modern and sustainable management, capable of adapting to the necessities of the current economic and social environment.

3 Problem Statement

From our point of view, all systems within a company which participate in the shaping of its sustainability in the market should be described by a set of parameters representing the most important positions of the company. When assessing strategic sustainability, it is important to take into consideration parameters describing the company's target position it the market over the long term. Two kinds of companies—airport operators (AO) and airlines (A)—are responsible for producing end products in air transport. It is important to note that activities and operations which they perform can be inseparable from each other. In this context, in order to understand the nature of either current or strategic sustainability of market players in the air transport sector it is vital to assess the quality of corporate interaction. The variety of interaction activities between airport operators and airlines means that it is necessary to solve such tasks as security management, quality control, growth in passenger traffic and volume of transportation. The high level of their interaction will have a positive impact on current and strategic sustainability of both airport operators and airlines. Strategic sustainability in the air transport market are:

- the use of carrying capacity over the long term;
- compliance with market requirements regarding technologies and carrying capacity of transport infrastructure over the long term;
- the quality of customer service management.

Strategic sustainability of airport operators and airlines will be significantly influenced by the specific nature and focus of strategic interaction. Their optimization will further lead to a particular result, which will determine future positions of air transport companies interacting with each other. Strategic sustainability can be presented as a vector of market players' development using a number of parameters which reflect sustainability and reliability of a long-term functioning of the mentioned market players in the particular sphere of their interaction. In its turn, strategic interaction (SI) as a factor and cause of any level of strategic sustainability can be described using a set of parameters which will be different for different formats and spheres of a company's interaction in the air transportation market. It is therefore necessary to describe the set of parameters which will reflect the results of strategic interaction between airport operators and airlines in the chosen sphere of strategic interaction as well as to choose a method which will make it possible to create a vector description of strategic sustainability depending on the parameter chosen and the vector optimal for the chosen pair "airport operator—airline". The analyzing process in the aviation market can be presented as a diagram (Fig. 1).

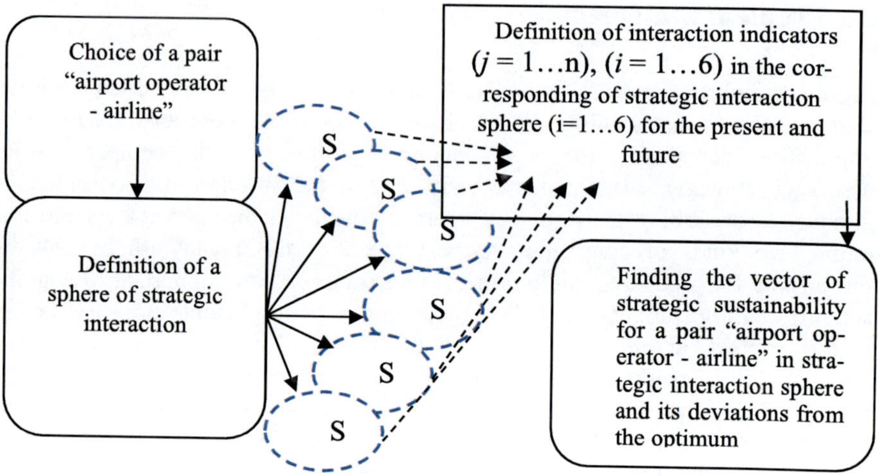

Fig. 1 Analysis of strategic sustainability of the subjects in the aviation market

4 Methodology

In determining the directions of interaction between airport operators and airlines, earlier studies [14] identified areas of strategic interaction which are described by the following conditions and formats (see Fig. 2).

The concept of the sphere of strategic interaction, which we research, includes the area of functioning of airport operators and airlines in which their strategic goals are achieved through the implementation of joint decisions, projects and actions taking into account the vector of strategic development. Indicators reflecting the results of strategic interaction in the spheres SI1–SI6 (Fig. 2) and describing their impact on strategic sustainability in Table 1. In order to identify indicators to describe the results of strategic interaction an expert assessment was performed in the study. Experts (middle and senior managers) of airport operators and airlines were involved in the expert assessment. The formalization of the target vector and, in general, the methodology for the systems analysis of corporate sustainability has been studied in detail by several authors Gorjunov E.V., Shalanov V.N. and others.

According to the method used in the works mentioned, the vector x1 is a target vector and a radius vector, the beginning of which is the zero values of the indicators, and the end is the target values of the indicators. Since the strategy, according to the authors Gorjunov E.V., is the direction of development of the object, and the direction can be formally given by a vector with coordinates, then the vector x0 will be the strategy currently being implemented, and the strategy for implementation in the future will be x1. The potential of the strategy being formed is expressed as an indicator of the relationship between the vectors.

In our study, we will use the vector model of relationship visualization to estimate the potential of strategic interaction for airport operators and airlines,

SI1	Technological focus: business processes, technological processes, operations, information technology, etc.	SI2	Marketing focus: ticketing system, non aviation revenues, advertising campaign, customer base
SI3	Organizational focus: parameters of routes, flights, geography of flights, aircraft, fares, benefits, discounts	SI4	Organizational and economic focus: personnel, level of service and quality of services, range of services
SI5	Technical focus: airfield, air terminal, business terminal, aircraft maintenance facilities, aircraft, special motor transport, equipment, communications	SI6	Resource focus: resources, equipment, personnel, suppliers, aircraft safety management system

Fig. 2 Characteristics of the areas of strategic interaction between airport operators and airlines

and then, after finding the cosine we will find the angle which describes the relationship between the vector of the target and the vector of the current position (strategy). The method of assessing the potential is based on the calculation of the weighted average of the growth potentials of significant indicators, so the potential of strategic interaction between airport operators and airlines can be described by the following equation:

$$\mathrm{ISI} = \sum_{v=1}^{ps} D_v \cdot x\,ISi_v \tag{1}$$

where ISI—an indicator of assessing the potential of strategic interaction between airport operators and airlines; Dv—an indicator of the significance of each indicator IISI within the SI sphere (v = 1…PSIs).

However, if the indicator tends to decrease with its improvement, the inverse value of the indicator is used in the calculations. The ISI indicator provides an approximate assessment of the extent to which the strategy being developed is better than the one implemented at the moment judging by the most important indicators of resource support and the possibilities of realizing the potential of interaction between airport operators and airlines chosen for assessment. The chart for estimating the potential for strategic interaction is presented in Fig. 3. The estimates obtained can be used to forecast the potential of strategic interaction.

Table 1 Indicators which describe strategic interaction between airport operators and airlines in the spheres mentioned and affect the strategic sustainability of companies

SIi	Indicators (I)
SI1	Wear rate of the terminal building; profit; passenger traffic; capacity of the airport; expenses for one passenger; rankings of social and economic development of the region where the airport is located; delay time; the level of flight safety, the number of runways; the number of employees; the share of transfer passengers; staff performance (revenue per employee)
SI2	Revenues from aviation services; takeoffs and landings; non-aviation revenue; passenger traffic; seat occupancy rate; passenger traffic; flight geography index; rankings of social and economic development of the region of airport location; revenue per passenger; passenger satisfaction; the destinations; revenue per employee; aviation mobility in the region
SI3	Revenues from aviation; transfer passengers; flight geography index (the ratio of passenger traffic volume to the number of flights per week to different airports); passenger traffic; rankings of social and economic development of the region; the destinations; takeoffs and landings
SI4	The non-aviation revenue; revenues from aviation services; flight geography index; wear rate of the terminal building; passenger traffic; the number of flights per week; rankings of social and economic development of the region; average annual revenue per passenger; connection time; total delay time; flight safety, the incidents at the airport per 1000 takeoffs and landings; the destinations; the share of transfer passengers; staff performance (revenue per employee)
SI5	The number of aircraft; the length of the runway; investments in upgrading airport infrastructure; wear rate of the terminal building; wear rate of the runway; capacity of the airport; passenger traffic; expenses per passenger; runways; flight safety level; the accidents at the airport for 1000 takeoffs and landings; transfer passengers
SI6	The number of takeoffs and landings per employee; seat occupancy rate; wear rate of the runway; wear rate of the airport building; passenger traffic volume; the number of passengers per employee; the annual average number of flights per week; the amount of expenses per passenger; profitability of services; average annual revenue per passenger

In order to provide a vector description of the strategic sustainability of the pair 'airport operator—airline', we selected a method of quantifying assessment of the parameters of the economic development vector which we adapted to long-term sustainability calculating calculations depending on the sphere of their interaction, the focus object, and indicators reflecting the sustainability of such interaction given in Table 1 and in Fig. 2.

As the main mathematical tools of research, we use the elements of the theory of vector algebra given in [15], in which vector description in multidimensional Euclidean space is made using the parameters of the length of the vector LV, as well as the angles of the directions between the vector and its projections on the axis.

According to this method, the length of the vector LV is determined by the equation:

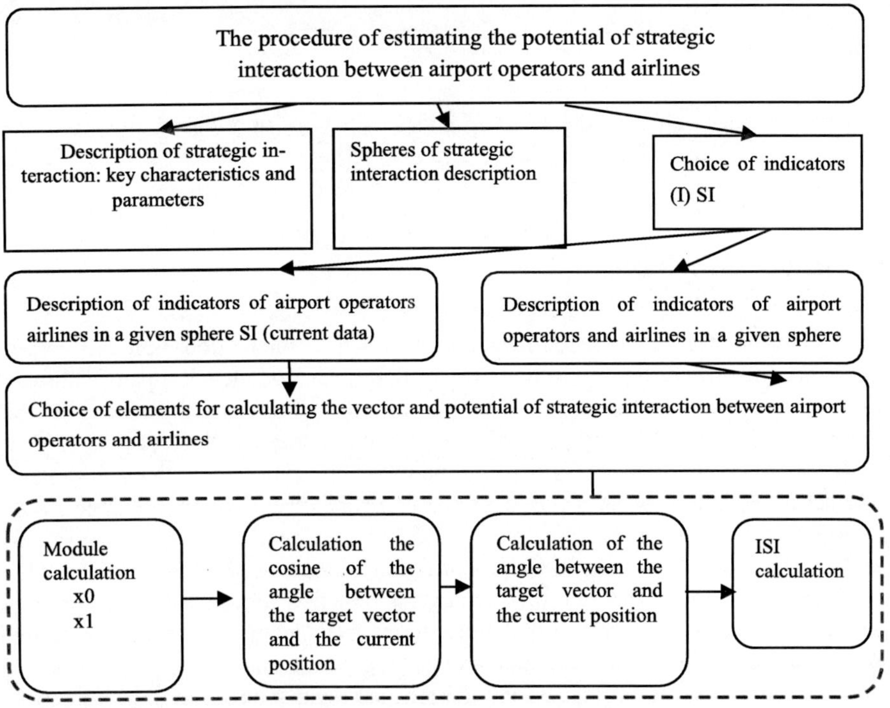

Fig. 3 The sequence of estimating the potential for strategic interaction

$$\mathrm{LV} = \sqrt{\sum (R_{Iij})^2} \tag{2}$$

where R_{Iij} the growth rate (either real or forecast) of the indicators I i j describing the result of the strategic interaction between airport operators and airlines leading to a certain level of strategic sustainability of a pair of subjects.

In order to find the angles $\alpha_{i\ j}$ between the projections of the R_{Iij} on the axis of the vector V and the resulting angle α_0 we will use the following models:

$$\alpha_{ij} = \frac{\arccos R_{Iij}}{LV} \tag{3}$$

$$\alpha_{ij} = \sum \alpha ij \tag{4}$$

Provided that the components of the V vectors are equally directed to the coordinate axes and that the R_{Iij} parameters are equal, the values of bisector angles are found as follows:

$$\alpha_{ij} = \frac{\arccos 1}{j} \tag{5}$$

The reference values of the bisector angles of the vector V based on the number of parameters j according to author Gorjunov [15].

5 Research Results

The analysis of formats recommended of strategic interaction between airport operators and airlines as well as that of current projects for such companies allowed us to assess the potential of strategic interaction and forecast the results of this interaction for the following pairs of airport operators and airlines (Table 2).

Using the methodology proposed in this paper, the following values of strategic interaction potential for airport operators and airlines were obtained (Table 3):

In order to increase the effectiveness of strategic interaction, airport operators and airlines need to choose development options based on the formation of interconnected strategies. The result of strategic interaction between airport operators and airlines largely depends on the ability to choose adequate solutions taking into consideration the current positions and strategic orientations. The potential of implementing different strategies in different strategic interaction spheres can be described for each company. For example, data for Khabarovsk Airport are shown in Fig. 4. Good visualization of the differences between the basic strategy and the strategy being developed generated is reflected by the angle between the target vector, which is a formalized expression of the strategy being developed, and the established vector, which is a formalized expression of the strategy implemented over the period research.

As the figures show, the strategy of interaction with Ural Airlines which is being developed has the highest potential for Khabarovsk Airport. It is 30.6% higher than the potential of the strategy implemented by this pair of companies in the current period, the angle between the target vector and the established vector of development of the current interaction being 2.8°, which indicates a slight difference in the format of the strategy being formed from the current one.

For Avrora Airlines, the indicator of interaction with Khabarovsk Airport in the sphere SI2 is 1.240, the angle of the vector being 1.9°. In order to assess the strategic sustainability of the pair "airport operator—airline" in the sphere of strategic interaction SI3, the data given in Table 4 were used. This table also shows the results of calculations for the models being described and interaction indicators at j = 7. The deviation from the standard in the table was calculated by comparing the reference angle (for i = 3 at j = 7 aref = 67.7 in Fig. 4) with the angles of the vectors for current and forecast. The results of calculation show that the range of growth rates (vector V projections) in the current period is 1.001–1.080 (8%). In the forecast period, if the airport operator and the airline interact in the organizational

Table 2 The description of strategic solutions for different spheres of strategic interaction

Pair of airport operator and airline	Spheres of strategic interaction	Projects	Focus of strategic interaction	Object of strategic interaction	Strategic goal
Krasnodar Airport—Rossiya Airlines	*SI*1 *SI*4	Creation of an aviation training center	Technological	Personnel management, IT	To improve the efficiency of business processes
			Organizational and economic	Personnel, revenue, service quality	To increase revenue from non-aviation services
Yakutsk Airport—Yakutia Airlines	*SI*2	Construction of a business terminal	Marketing	Business terminal, special offers for VIPpass	To increase revenue from non-aviation services
Kurumoch—Rossiya Airlines	*SI*4	Opening of a business Hall	Organizational and economic	VIP segment, quality	To increase focus on non-aviation services
Khabarovsk—Avrora Alns	*SI*6	Construction of an aircraft hangar	Resource	Investment	To minimize MRO costs, to ensure stable operation
Sochi—Rossiya Alns	*SI*5		Technical	Hangar	To modernize infrastructure
Khabarovsk Airport—Ural Airlines	*SI*2 *SI*3	Transit route Yekaterinburg—Khabarovsk—Harbin—Khabarovsk	Marketing Organizational	Service promotion Route	To increase Traffic, routes
Sochi Airport—Rossiya Airlines	*SI*4	Development of a route Sochi—Beijing—Sochi	Organizational and economic	Route, expansion of flight geography	To expand flight geography, to increase routes, to increase revenue
Khabarovsk—Avrora Airlines	*SI*2	Flexible tariff policy	Marketing	Service promotion	To increase the Flights and passenger traffic
Vladivostok Airport—Avrora Airlines	*SI*2	Increase flights on the route Vladivostok—Yangzi—Vladivostok	Marketing	Service promotion	To increase flights and passenger traffic

(continued)

Table 2 (continued)

Pair of airport operator and airline	Spheres of strategic interaction	Projects	Focus of strategic interaction	Object of strategic interaction	Strategic goal
Vladivostok Airport—Avrora Airlines	*SI2*	Increase flights on the route Vladivostok—Sapporo—Vladivostok	Marketing	Service promotion	To increase the number of flights and passenger traffic
Vladivostok—Avrora Airlines	*SI2*	Development of a route Vladivostok—Jiamusi—Vladivostok	Marketing	Service promotion	To increase flights and passenger traffic
Khabarovsk—Avrora Airlines	*SI3*	Development Khabarovsk—Jiamusi—Khabarovsk route	Organizational	Route	To increase routes and flights
Khabarovsk Airport—Avrora Airlines	*SI3*	Development of a route Khabarovsk—Sanya—Khabarovsk	Organizational	Route	To increase routes and flights

area SI3, where focus objects are parameters of routes, flights, geography of flights, aircraft, fares, and discounts, the range of values is 1.070–1.291 (21%). The corresponding angular parameters in degrees are 66.9–68.6 (2.5%) for the current period and 62.0–67.1 (8.2%) for the forecast period.

The biggest changes in the level of strategic sustainability in the forecast period are caused by such parameters as an increase in the number of takeoffs and landings as well as the number of routes, which lead to an increase in passenger traffic and incomes of the subjects of the air transport market. The accepted reference parameter—the bisector angle under the number of parameters $j = 7$ equal to 67.7 degrees (see Fig. 4)—has bigger deviations in the forecast period (0.9–9.2) from parameters SI3 than in the current period (−1.4 to 1.3), which indicates at the same time an improvement in the stability of the interaction between the airport operator and the airline in a positive direction of the vector of strategic sustainability and increased risk of failure to achieve a goal.

So total angle of parameters αj—degrees (450 in tabl.4) from the optimal total angle $\alpha 0 = 474$ (Fig. 4) in the current period is much less than in the forecast. The angular values of the growth rate of the input parameters reflect strategic sustainability. All of them have positive values, which means that they indicate the significance of the parameter in increasing the length of the V vector of strategic stability, i.e. in reinforcement the stability of the functioning of the pair "airport

Table 3 The values of strategic interaction potential for airport operators and airlines

Project number	Subjects of strategic interaction	Spheres of strategic interaction	ISI	ISI (%)
1	Krasnodar Airport—Rossiya Airlines	*SI*1	1.162	16.2
2		*SI*4	1.131	13.1
3	Yakutsk Airport—Yakutia Airlines	*SI*2	1.044	4.4
4	Kurumoch Airport—Rossiya Airlines	*SI*4	1.088	8.8
5	Khabarovsk Airport—Avrora Airlines	*SI*6	1.150	15.0
6	Sochi Airport—Rossiya Airlines	*SI*5	1.137	13.7
7	Khabarovsk Airport—Ural Airlines	*SI*2	1.243	24.3
8		*SI*3	1.306	30.6
9	Sochi Airport—Rossiya Airlines	*SI*4	1.145	14.5
10	Khabarovsk Airport—Avrora Airlines	*SI*2	1.240	24.0
11	Vladivostok Airport—Avrora Airlines	*SI*2	1.185	18.5
12	Vladivostok Airport—Avrora Airlines	*SI*2	1.129	12.9
13	Vladivostok Airport—Avrora Airlines	*SI*2	1.134	13.4
14	Khabarovsk Airport—Avrora Airlines	*SI*3	1.176	17.6
15	Khabarovsk Airport—Avrora Airlines	*SI*3	1.073	7.3

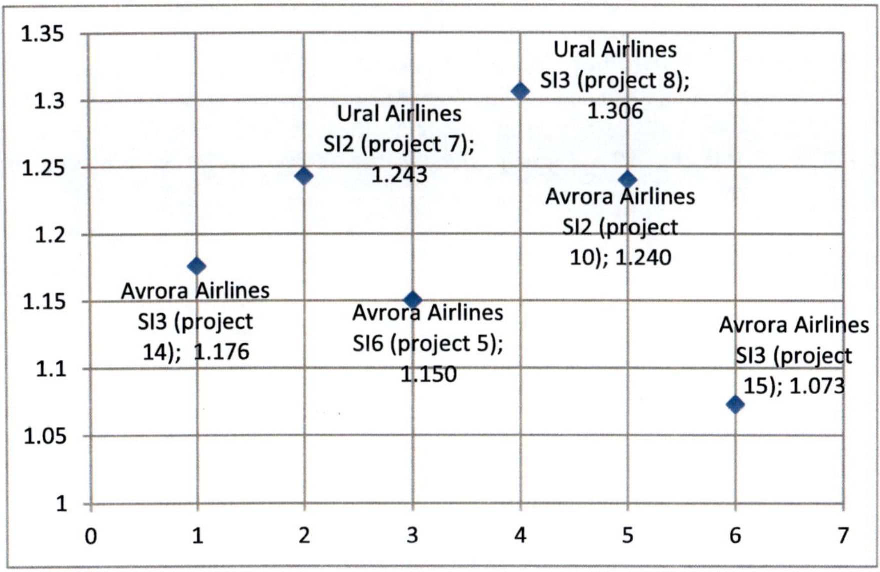

Fig. 4 The potential of Khabarovsk Airport for implementing strategies of interaction with different airlines in different strategic interaction spheres

Table 4 Data and results of calculations of the vector of strategic sustainability of the pair "airport operator—airline" in the sphere SI3 (i = 3)

I_i/estimated value	Current period			SI3 forecast			Change in direction (°)
	Growth rate (R_I)	α_j (°)	Deviation (%)	Growth rate (R_I)	α_j (°)	Deviation (%)	
1. Revenues from aviation services	1.071	67.1	1.0	1.230	63.4	6.8	−3.7
2. The take-offs and landings per year	1.033	67.9	−0.3	1.291	62.0	9.2	−5.9
3. Flight geography	1.001	68.6	−1.4	1.070	67.1	0.9	−1.6
4. The share of transfer passengers	1.003	68.6	−1.3	1.100	66.4	2.0	−2.2
5. Passenger traffic	1.068	67.1	0.9	1.290	62.0	9.2	−5.1
6. The number of routes (in average)	1.010	68.4	−1.1	1.210	63.9	6.0	−4.6
7. The number of destinations	1.080	66.9	1.3	1.150	65.3	3.7	−1.6
Vector length *LV*	2.748	–	–	3.160	–	–	–
The sum of angles α_o	–	474.6	–	–	450.0	–	−24.6

operator—airline" in a given sphere of strategic interaction. So they indicate that the goal can be reached more quickly. The strategic stability of the interaction between the airport operator and the airline in the SI3 area increased by 1.15 times (3.160/2.748). The deviation from the bisector angle will describe the possible unbalance of the system from its current position (Fig. 5).

As a result of strategic interaction between companies in the SI3 sphere, the length of the strategic sustainability vector in the forecast period increased from 2.748 to 3.160 (by 15%) largely due to the efforts made in terms of the

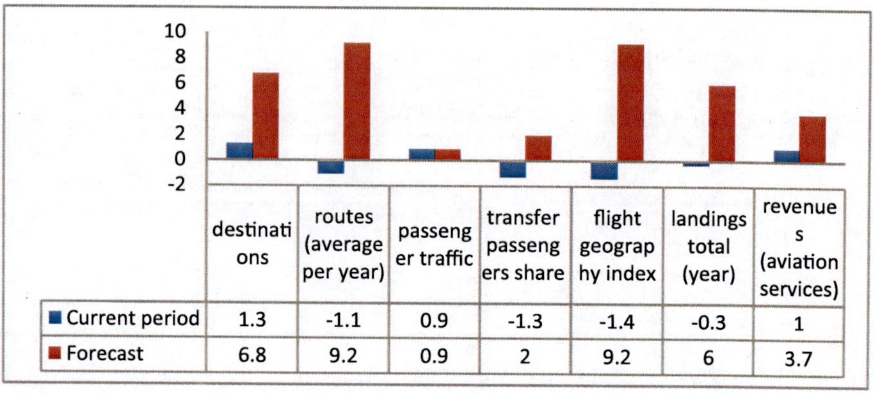

Fig. 5 Deviations of the angles of the vectors of input parameters in the composition of the strategic stability vector from the reference angle, %

organizational coordination of routes, flight geography, fares and shipping tariffs. In order to improve interaction indicators and achieve the required level of strategic sustainability, the formats and conditions for the strategic interaction between airport operators and airlines and the sphere by which they are determined must be adjusted to account for possible changes in the strategies of each participant.

6 Conclusion

The results of analyzing ISI show that their values will differ depending on the sphere and that they depend on the method and the possibility of realizing the potential in terms of the resources used and the capabilities of the interacting parties. Airport operators and airlines receive tools for modeling and forecasting results and selecting the optimal interaction parameters in different formats. In determining the goals and prospects of strategic interaction, airport operators and airlines analyze each other's strategies to assess the possibilities for building effective relationships in the future and balancing their individual development strategies in the chosen sphere. Thus, airport operators and airlines, analyzing their positions and having an idea of what future strategies can be formed to ensure strategic sustainability, can, using this approach, reasonably select the most appropriate solutions.

References

1. Yashin N, Grigoryan E (2014) Methodological aspects of enterprise sustainability. Bull Saratov State Socio-econ Univ 5(54):113–117. https://doi.org/10.0000/cyberleninka.ru/article/n/metodologicheskie-aspekty-obespecheniya-ustoychivosti-predpriyatiya
2. Bushuyev AN, Gugelev AV, Yashin NS (2014) Standard regulation of corporate management. Mediterr J Soc Sci 5(24):442–447. https://doi.org/10.5901/mjss.2014.v5n24p442
3. Rastova YI, Rastov MA (2018) Strategic management of a modern organization: concept synergy effect. Manag Sci 8(3):20–31. https://doi.org/10.26794/2404-022X-2018-8-3-20-31
4. Kuznetsov SV, Rastova YI, Sushcheva NV (2018) The role of public companies in creating a platform for economic growth in Saint Petersburg. Baltic Region 10(1):37–55. https://doi.org/10.5922/2079-8555-2018-1-3
5. Teece JD (2007) Explicating dynamic capabilities: the nature and microfoundations of (sustainable) enterprise performance. Strateg Manag J 28(13):1319–1350. https://doi.org/10.1002/smj.640
6. Goetz BE (2019) Avoiding managerial obsolescence. Calif Manag Rev 1965 7(3):91. https://doi.org/10.2307/41165637
7. Freeman E (2015) Strategic management: Cambridge. Cambridge University Press Publ. https://doi.org/10.1017/CBO9781139192675
8. Borodulina S, Pantina T (2021) Model of sustainable economic development in the context of inland waterway transport management. Adv Intell Syst Comput 1258. http://doi-org-443.webvpn.fjmu.edu.cn/https://doi.org/10.1007/978-3-030-57450-5_68

9. Tikhonov AA (2013) The choice of the model of strategic management in the development of agricultural companies. Trends Manag 3(3):82–90
10. Shrivastava P, Ivanaj S, Ivanaj V (2016) Strategic technological innovation for sustainable development. https://doi.org/10.1504/IJTM.2012.049104
11. Keleinikova SV, Samygin DY, Bulycheva YV (2017) Models of the organizational and economic development mechanism of the regional vegetables market. Finan Manag 2:11–23. https://doi.org/10.25136/2409-7802.2017.2.21175
12. Fülöp G, Hódi Hernádi B (2012) Corporate sustainability: strategic alternatives and methodology. Implementation. https://www.issbs.si/press/ISBN/978-961-6813-10-5/papers/ML12_041.pdf
13. Morarescu C (2017) Applying a strategic vision to create a sustainable management for the 21'st century http://conference.management.ase.ro/archives/2017/pdf/3_13.pdf .
14. Pankratova AR (2017) Strategic cooperation between airport enterprises and airline companies. Econ Manag 8:48–53
15. Gorjunov EV (2010) Vector method in the theory of economic analysis. Econ Anal Theory Pract 27(182):39–44

Fault Detection Technique for Electromechanical Actuator of the Aircraft Using Neural Networks

Georgy Veresnikov, Valentin Lebedev, Oleg Ogorodnikov, and Artem Golev

Abstract The paper concerns the problem of development of the technique for rapid fault detection for electromechanical actuator (EMA) of the aircraft related to dissipative torque change. Increased dissipative torque is a common cause of EMA failure. To address this problem, a data processing scheme has been suggested for EMA's operation using neural networks. The work describes the feature filtering methods used for reduction of neural network input vector dimension. Trained neural networks are designed to develop algorithms for determining the EMA's technical state. The algorithms are proposed to be used for flight safety in the health usage monitoring systems (HUMS) of the aircraft. The paper shows the computing experiment results based on the data obtained by using computational model of unmanned aircraft's EMA while processing the take-off cyclogram. The data includes test and training sets for neural network. The computing experiment results shows efficiency of fault detection related to dissipative torque change.

Keywords Neural network · Select feature · Wrapper methods · Filter methods · Fault detection · Electromechanical actuator · Aircraft

1 Introduction

With the advancements in electrical aircraft production technologies, the relevance of techniques development has been enhanced for detection of EMA's failures affecting substantially the flying safety [1]. Herewith, a promising exploration trend is the use of neural networks in searching for regularities that may be used successfully in early EMA diagnostics. The data for neural networks training may be obtained as a result of mathematic modelling and operation of the aircraft's EMA. It is often appropriate to use neural networks in alliance with methods that allow for reducing the feature space dimension in order to avoid the "curse of dimension-

G. Veresnikov (✉) · V. Lebedev · O. Ogorodnikov · A. Golev
Institute of Control Sciences of Russian Academy of Sciences, Profsoyuznaya street, Moscow 117997, Russia

A. Bieliatynskyi and V. Breskich (eds.), *Safety in Aviation and Space Technologies*, Lecture Notes in Mechanical Engineering,
https://doi.org/10.1007/978-3-030-85057-9_43

ality" [2, 3]. Usually in reducing of the feature space dimension, the initial variety of features undergoes feature selection procedures. Feature selection means formation of the relevant features for developing classification model/technique. Using the "wrapper methods" [4] is the most cross-cutting approach for feature selection. These methods optimize based on information obtained as a result of iterative accomplishment of comprehensive search and formalisation regularities using data mining techniques, in particular, neural networks. The key feature of the "wrapper methods" is that in each iteration of optimization, technique models/techniques are developed that can be further used for integrated application. For each subset of features generated by "wrapper methods, the neural network training is performed, decision tree and growing pyramidal network are built and etc. General optimization techniques used for solving of combinatorial problems of optimal subsets formation based on the given initial set, provide a basis for "wrapper methods". In developing of the "wrapper methods" the following optimization techniques are used: genetic algorithms [5], scatter search [6], particle swarm optimization [7], simulated annealing [8] and etc.

When there is a large number of initial features and the volume of data analysed, the use of the "wrapper methods" might prove unacceptable due to substantial computational costs for models/techniques development in each iteration of optimisation calculations. The approach to selection of features based on the "filter methods" application allows for computational costs reduction common to "wrapper methods". The key feature of "filter methods" is that the feature selection is performed via designing the simplified models. Such methods are based on correlation/regressive dependences, information gain (IG), statistical tests (t-test, chi-square-test) and so on [9–12]. The efficiency of "filter methods" application is limited inasmuch as there is no opportunity to describe complex regularities with simple models, however, these methods require less computational costs against the "wrapper methods" [13, 14].

To address the development of fault detection technique for aircraft's EMA related to the dissipative torque change, the article features:

- Data processing scheme on EMA operation using neural networks;
- Description of features filtration technique for increasing the efficiency of neural networks application in EMA fault detection task;
- The research findings concerning the purposed data processing scheme for the data obtained by the use of the computational model of unmanned aircraft's EMA.

2 Data Processing Scheme

We propose the data processing scheme on EMA operation (Fig. 1) that yields the technique for EMA fault detection.

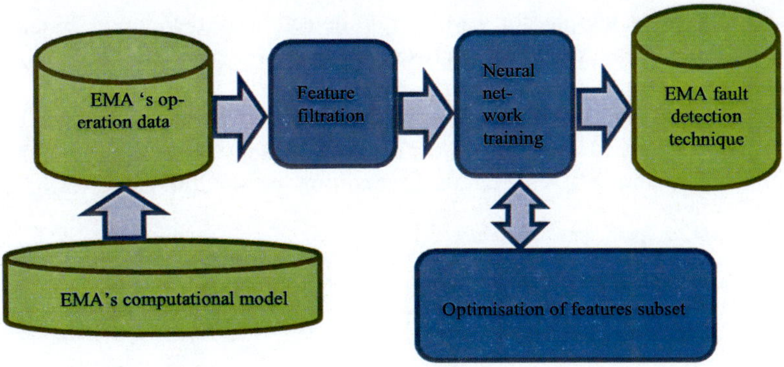

Fig. 1 Data processing scheme

The EMA fault detection technique development related to the dissipative torque change, as part of the proposed scheme, is formalised as the task for classification in two classes. Based on the EMA operation data $\{(\bar{x}_1, \bar{y}_1), \ldots, (\bar{x}_n, \bar{y}_n)\}$, $\bar{\boldsymbol{x}}_i \in$ X, $\bar{\boldsymbol{y}}_i \in$ Y, $i = 1,\ldots,n$, we need to build the EMA fault detection technique Φ:X → Y, where X—features values set—controlled and controlling EMA parameters, Y—a set composed of two class labels reflecting operative and pre-accident condition of EMA. Φ technique allows to determine the class label $\bar{\boldsymbol{y}} \in$ Y for the arbitrary vector $\bar{\boldsymbol{x}} \in$ X. In teaching the neural network using which the technique Φ is implemented, the marks of classes are used that are presented in Table 1.

When the trained neural network if used for EMA's technical condition determination, the maximal element of the vector $\bar{\boldsymbol{y}}$ shall be set to 1, the least element shall be set to 0 ("the winner takes all" principle).

The initial data on EMA operation are generated based on EMA's computational model, which takes into account the influence of the dissipative torque on the controlled EMA's parameters. Later on, the data obtained using the EMA's computational model are replenished by statistical data accumulated as a result of the aircraft's flights.

As part of the scheme in Fig. 1, the data on EMA operation are primarily processed using the "filter methods", which enables to reduce the dimension of neural network input vector (features space) and the quantity of EMA's controlled parameters respectively necessary for EMA fault detection without significant computational costs.

Further, the optimization of remained features subset is performed using the "wrapper methods". At each optimization step, the neural network is trained, a

Table 1 Presentation of class labels determining the EMA's technical condition

No	EMA's technical condition class	Class label $\bar{y}$
1	Normal EMA's technical condition	(0.1)
2	EMA's pre-accident technical condition	(1.0)

cross-cutting Φ technique for EMA fault detection is developed based on the generated features subset. Neural network precision based on the test set is used as the target function.

The resulting Φ technique is applied for determination of EMA's technical condition. For that end, the controlled parameter values vector is given to the neural network input, and EMA's technical condition estimation is performed at the output.

3 Features Filtration

To reduce the features space dimension—the controlled parameters vector $\bar{\boldsymbol{x}} = (x_1, \ldots, x_k)$ a three-step procedure is used.

Step 1. Building non-linear models of multiregression to determine the relationships between vector elements $\bar{\boldsymbol{x}}$:

$$x_i(\bar{x}_i') = \sum_{d=1}^{N} a_d B_d\left(\bar{x}_i'\right), \tag{1}$$

where x_i—feature dependent from the vector $\bar{x}$, a_d—estimated model coefficient, $B_d\left(\bar{\boldsymbol{x}}_i'\right)$—basis function, $\bar{\boldsymbol{x}}_i' = \{x_1, \ldots, x_k\} \backslash \{x_i\} = (x_1', \ldots, x_{k-1}')$, $i = 1,\ldots k$, N—number of basis functions.

If the number of controlled parameters is more than 1, the formation of basis functions may be difficult for data exploration. Hence, the approach for non-linear multiregression model is used that offers the implementation of technique 1.

Technique 1.

1. Formation of a set of elementary functions $b_j\ x'$, x'—spontaneous feature from vector $\bar{\boldsymbol{x}}_i'$, $j = 1\ldots z$, z—number of elementary functions (polynominals till the third degree are used as basis functions).
2. Selection of function integration method $b_j(x')$ for basis functions $B_d\left(\bar{\boldsymbol{x}}_i'\right)$ formation. For example, the set is formed based on elementary functions $\{b_1(x'), \ldots, b_z(x')\}$ chosen by the researcher: $G =$, where m—the number of G elements used for one basis function formation. Then the basis functions $B_d\left(\bar{\boldsymbol{x}}_i'\right)$, $d = 1, \ldots, N$, where $N = (k+m-1)!/(m!(k-1)!)$, are a multiplication of combination elements from the set G from m. An additional restriction may also be imposed on use of the elementary functions of common parameters in one basis function.
3. Composition and quality assessment for a non-linear multiregression model.

End of Technique 1.

To estimate the quality of the multiregression model obtained (estimation of features' correlation ratio), the standard determination coefficient R^2 that varies from 0 to 1. If R^2 is above the threshold value set by the data researcher (usually close to 1), the dependent parameter may be excluded from the features subset (vector elements $\bar{\boldsymbol{x}}$) used for EMA's technical condition estimation).

Step 2. Features sort based on statistical chi-square-test. This test helps to calculate the criterion χ^2 for each feature from EMA's controlled parameters vector $\bar{\boldsymbol{x}}$. The most informative for solving classification task is the feature for which χ^2 is maximum.

Step 3. If there are features with a high regression interdependence, the dependent features with the least significant χ^2 shall be excluded from vector $\bar{\boldsymbol{x}}$.

4 Experimental Findings

The scheme provided in Fig. 1 has been measured using EMA operation data obtained with EMA's computational model described in [15]. While modelling, the control input was used (Fig. 2) corresponding to the take-off cyclogram of the unmanned 400 kg aircraft.

To obtain the data on EMA's operation corresponding to operative and pre-accident EMA's condition while processing the take-off cyclogram, the value of Mcf dissipative torque varied. Mcf value for EMA's operative technical condition varied from 0.225 to 0.375, and the EMA's pre-accident technical condition varied from 0.875 to 1.025. The data obtained were split into test and training sets (Table 2).

The use of *Mcf* values provided in Table 2 resulted in generation of data on EMA operation while the take-off cyclogram processing that include the following EMA's parameters (features): x_1—input signal (grade), x_2—rotor's turn (grade), x_3—output signal (grade), x_4—supply voltage (V), x_5—supply current (A), x_6—supply power (Watt), x_7—shaft speed (rot./min), x_8—shaft torque (Nm), x_9—shaft power (Watt), x_{10}—output shaft speed (rot./min), x_{11}—output torque (Nm), x_{12}—power output (Watt).

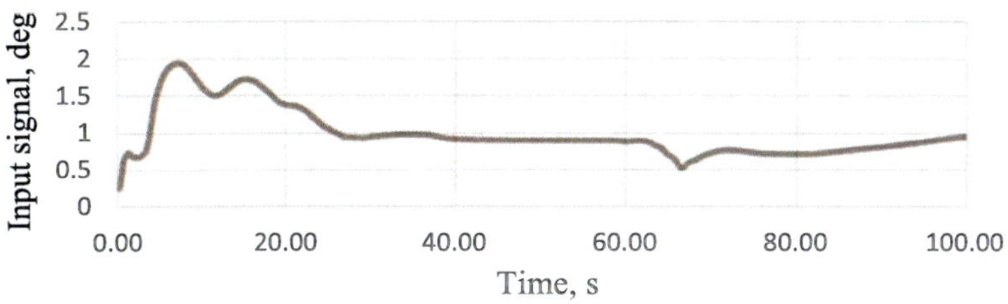

Fig. 2 Control input of EMA's computational model in the take-off cyclogram processing

Table 2 Splitting the data to training and test sets

Mcf dissipative torque	Set type	Class
0.225	Training	EMA's operative condition
0.275	Test	EMA's operative condition
0.325	Test	EMA's operative condition
0.375	Test	EMA's operative condition
0.875	Test	EMA's pre-accident technical condition
0.925	Test	EMA's pre-accident technical condition
0.975	Test	EMA's pre-accident technical condition
1.025	Training	EMA's pre-accident technical condition

Each cyclogram of unmanned aircraft's 100 s take-off in one *Mcf* value from Table 2 accounts for 10,000 parameter measurements x_1–x_{12}. Consequently, the training set contains 20,000 cases, the test set contains 60,000 cases.

Without data processing scheme provided in Fig. 1, the neural network precision with the maximum number of 16 neurons in hidden layer based on the test set is 96.7% at most. 50 neuron networks in total have been trained because neural network training is not a deterministic algorithm; 4–16 neurons are used in the hidden layer, activation functions—the hyperbolic tangent (tan h). Broyden-Fletcher-Goldfarb-Shanno (BFGS) method was used for neural work training. The parameters of the best 5 neural networks and the obtained preciseness based on test and training sets are provided in Table 3.

Precision means the percentage vectors classified correctly from training and test sets.

We will apply the data processing scheme provided in Fig. 1. Figure 3 demonstrates the results of use of the feature filtration techniques (see Sect. 2).

Figure 3 demonstrates that the parameters form 4 groups of connected components. According to the chi-square-test, in every selected group one parameter that is the most significant (informative) is selected as neural network inlet. Consequently, x_2, x_5, x_9, x_{11} parameters are considered as potential neural network inlets.

The results of the parameters subset optimization by direct enumeration method show that it is possible to increase the precision on test and training sets only by

Table 3 Neural networks with input parameters x_1–x_{12}

Number of neurons			Classification precision	
Input layer	Hidden layer	Output layer	Training set (%)	Test set (%)
12	11	2	98.7300	93.96167
12	16	2	99.3800	96.70167
12	12	2	95.0700	92.41167
12	11	2	98.3000	95.02167
12	10	2	99.9800	94.79000

Parameters related to regression dependence (step 1)

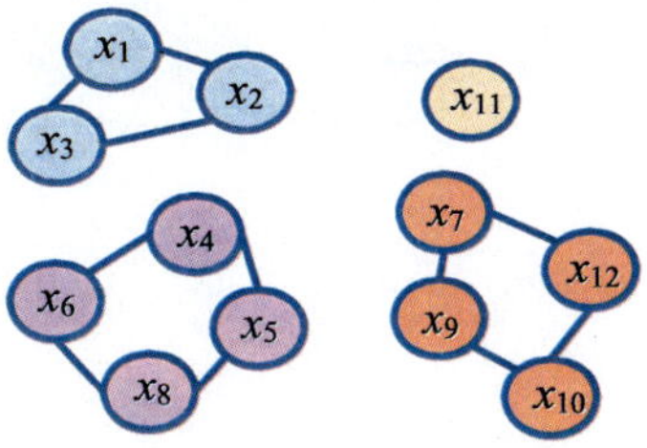

The result of using chi-square-test (step 2)

Variable name	Importance of the parameter (by chi-square-test)	p-level
x_2	41807.30	0.000000
x_3	41807.30	0.000000
x_1	40590.56	0.000000
x_5	19452.68	0.000000
x_8	19452.68	0.000000
x_4	19272.22	0.000000
x_6	14806.20	0.000000
x_9	1811.61	0.000000
x_7	115.80	0.000000
x_{10}	115.80	0.000000
x_{12}	26.56	0.005362
x_{11}	10.22	0.009165

Fig. 3 Feature filtration results for the neural network training

using all the four parameters. 50 neural networks in total have been trained (with the input parameters x_2, x_5, x_9, x_{11}), 4–16 neurons are used in the hidden layer, the hyperbolic tangent is used as the activation functions of hidden layer neurons. The parameters of the best 5 neural networks and the obtained preciseness based on test and training sets are provided in Table 4.

Thus, by reducing the number of neural network input parameters, we managed to somewhat increase the classification precision and decrease the number of neurons. It is also worth mentioning that the speed of the neural networks training has increased that is related to the decrease in the number of neural connections.

Figure 4 demonstrates the result of neural network that has shown 98.01% precision in test set while processing the take-off cyclogram for dissipative torque $Mcf = 0.225$ (training set, class 1—operative condition).

Figure 5 demonstrates the result of the same neural network use while processing the take-off cyclogram for dissipative torque Mcf = 0.375 (test set, class 1—operative condition). The worst result demonstrated is related to the landmark value use from Mcf variation interval.

Table 4 Neural networks with input parameters x_2, x_5, x_9, x_{11}

Neurons number			Classification precision	
Input layer	Hidden layer	Output layer	Training set (%)	Test set (%)
4	14	2	99.98500	98.01000
4	15	2	99.04500	95.14333
4	11	2	96.68500	95.40500
4	16	2	99.79000	96.70667
4	11	2	96.68000	94.00167

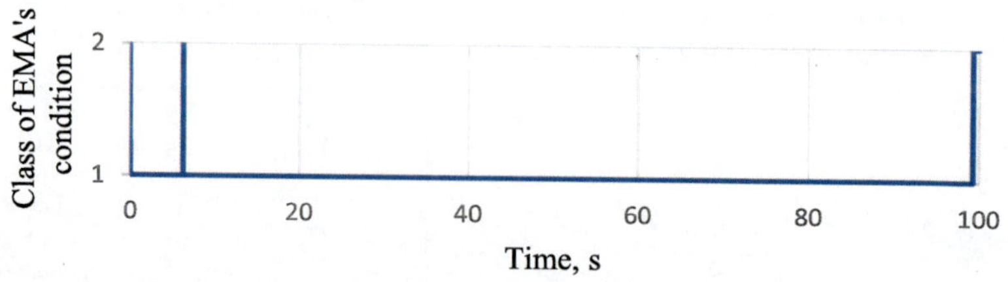

Fig. 4 The result of the neural network use while processing the take-off cyclogram (training set, class 1—EMA's operative condition)

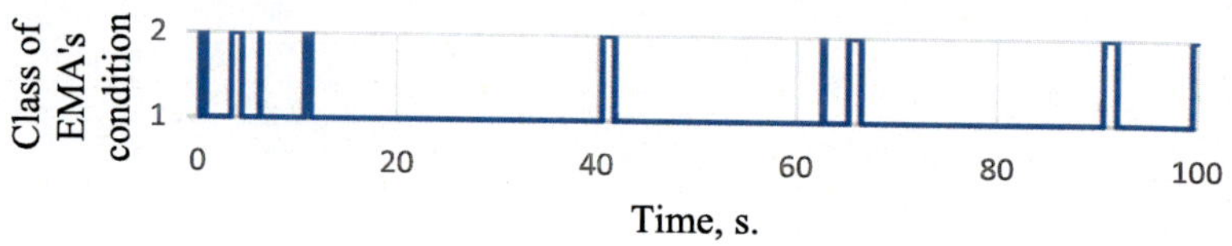

Fig. 5 The result of the neural network use while processing the take-off cyclogram (training set, class 1—operative condition)

Figures 4 and 5 demonstrate that in determining of EMA's operative condition in the test set against the training set there are more time spans, on which the neural network determines EMA's technical condition class incorrectly—it takes the operative condition for the pre-accident condition.

Figure 6 demonstrates the result of the neural network use is demonstrated while processing the take-off cyclogram for dissipative torque $Mcf = 1.025$ (training set, class 2—pre-accident condition).

Figure 7 demonstrates the result of the neural network use is demonstrated while processing the take-off cyclogram for dissipative torque $Mcf = 0.875$ (test set, class 2—pre-accident condition). The worst result demonstrated is also related to the landmark value use from *Mcf* variation interval.

Figures 6 and 7 demonstrate that in determining EMA's invalid condition in the test set against the training set there are more time spans, on which the neural network determines EMA's technical condition class incorrectly—it takes the pre-accident condition for the operative condition.

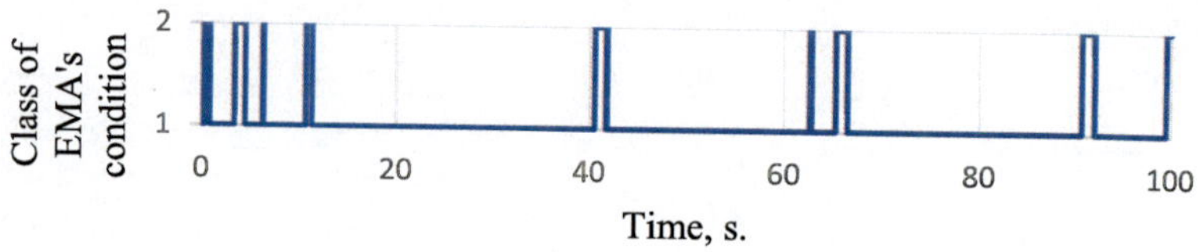

Fig. 6 The result of the neural network use while processing the take-off cyclogram (training set, class 2—pre-accident condition)

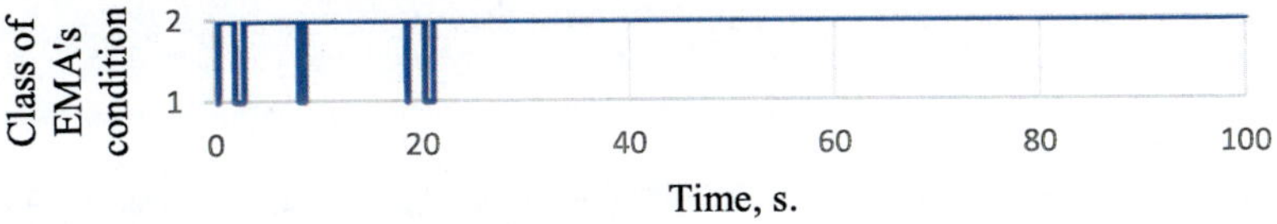

Fig. 7 The result of the neural network use while processing the take-off cyclogram (training set, class 2—pre-accident condition)

Thus, the following results of data analysis have been obtained while processing the take-off cyclogram using the EMA's computational model: the number of parameters has reduced from 13 to 4, the number of neurons in the hidden layer of the neural network has reduced, here with the precision of detecting EMA's technical condition in the test set has been extended.

5 Conclusion

The paper describes the task of the aircraft's EMA fault detection connected to the dissipative torque change. A solution to this problem has been offered using the data processing scheme, in which the methods of decreasing the feature space dimension and neural networks are integrated. The research has been performed concerning the purposed scheme on EMA's operation data obtained by using the computational model of unmanned aircraft's EMA. Consequently, we managed to reduce the number of inputs and to increase the neural network precision in training and test sets. Later on, it is planned to form the extended cases base including the EMA's cases to solve the classification tasks and forecast the development of various types of EMA's degradation based on specified EMA's computational model, resource bench tests and operation of control EMAs in various operating modes and external effects factors. In this case, it is recommended to consider the time in the input vector and apply convolution neural networks.

References

1. Kuvshinov V, Steblinkin A, Erofeev E, Skryabin A (2015) Safety, energy efficiency research on advanced more electrical flight control actuation systems for short/middle range passenger aircraft. More Electric Aircraft Proceedings. Toulouse, France
2. Alpaydin E (2014) Introduction to machine learning, 3rd edn. The MIT Press, Massachusetts, p 612
3. Verleysen M, François D (2005) The curse of dimensionality in data mining and time series prediction. In: The international work-conference on artificial neural networks, pp 758–770
4. Panthong R, Srivihok A (2015) Wrapper feature subset selection for dimension reduction based on ensemble learning algorithm. Procedia Comput Sci 72:162–169

5. Soufan O, Kleftogiannis D, Kalnis P, Bajic VB (2015) DWFS: a wrapper feature selection tool based on a parallel genetic algorithm. PLoS ONE 10(2):1–23
6. Gendreau M, Potvin J-Y (2019) Handbook of metaheuristics, 3rd edn. Springer International Publishing
7. Xue B, Zhang M, Browne WN (2013) Particle swarm optimization for feature selection in classification: a multi-objective approach. IEEE Trans Cybern 43:6
8. Jeong I-S, Hong-Ki Kim H-K, Kim T-H, Lee DH, Kim KJ, Kang S-H (2016) A feature selection approach based on simulated annealing for detecting various denial of service attacks. J Softw Netw 1:173–190
9. Yu L, Liu H (2003) Feature selection for high-dimensional data: a fast correlation-based filter solution. In: Proceedings of the 12'th international conference on machine learning (ICML-2003), pp 856–863
10. Greenwood PE, Nikulin MS (1996) A guide to chi-squared testing Wiley
11. Pereira RB, Plastino A, Zadrozny B (2015) Merschmann information gain feature selection for multi-label classification. J Inf Data Manag 6(1):48–58
12. Piramuthu S (2004) Evaluating feature selection methods for learning in data mining applications. Eur J Oper Res 156:483–494
13. Chuang L-Y, Ke C-H, Yang C-H (2008) A hybrid both filter and wrapper feature selection method for microarray classification. In: Proceedings of international multiconference of engineers and computer scientists, pp 146–150
14. Ke C-H, Yang C-H, Chuang L-Y, Yang C-H (2008) A hybrid filter/wrapper approach of feature selection for gene expression data. In: Systems, man and cybernetics, IEEE international conference, pp 2664–2670
15. Veresnikov GS, Skryabin AV (2020) The development of electromechanical actuator mathematical model for fault identification using data mining methods. In: Proceedings of the 13th international conference "management of large-scale system development" (MLSD)

Subminiature Eddy Current Transducer for Inspection of Welded Joints Obtained by Friction Stir Welding

Alexey Ishkov and Vladimir Malikov

Abstract This paper presents a measuring system focused on the study of welded joints obtained by friction stir welding (FSW). The system is based on an eddy current transducer. The study shows key technical information that describes the principles of operation of an eddy-current transducer, as well as a method for searching for violations of the internal structure of welded joints of aluminum plates. The proposed measuring system incorporates the functionality to automatically change the main operating frequency and cutoff frequency of the filtration system. Testing of the apparatus was carried out on aluminum-magnesium alloy plates assembled together by friction stir welding. The results obtained in the course of a series of experiments allow drawing conclusions regarding the qualitative characteristics of the obtained welded joints.

Keywords Friction stir welding · Eddy current transducer · Introduced voltage · Electrical conductivity · Flaw detection

1 Introduction

Structures based on aluminum alloys are now widely used due to the combination of consumer characteristics. The use of such alloys in mechanical engineering, the production of aircraft and space vehicles requires special attention to the strength and mechanical properties of parts and components. In this case, it is necessary to take into account several significant aspects, namely: one of the widespread methods of creating structures is welding; however, fusion welding has some limitations of applicability to aluminum parts. In this regard, it is necessary to use other technological methods of creating joints, among which friction stir welding (FSW) is recognized as the most effective [1]. At the same time, it should be borne

A. Ishkov (✉)
Altay State Agricultural University, Krasnoarmeysky, 73, Barnaul 656049, Russia

V. Malikov
Altay State University, Lenina, 61, Barnaul 656049, Russia

A. Bieliatynskyi and V. Breskich (eds.), *Safety in Aviation and Space Technologies*, Lecture Notes in Mechanical Engineering,
https://doi.org/10.1007/978-3-030-85057-9_44

in mind that due to significant temperature and deformation gradients, welded seams are susceptible to the occurrence of internal structure violations and heterogeneity. External mechanical action on such defects in welded joints can lead to early formation of cracks and premature failure. Considering that the geographic habitat and functioning of structures made by the FSW-method includes regions with extreme temperature regimes, it becomes necessary to carefully check the quality of the joints [2].

FSW-method opens up prospects for creating permanent joints of aluminum parts from such combinations of alloys, which under normal conditions cannot be subjected to electric arc welding or argon-arc welding. For example, in the manufacture of housing elements with increased strength, small dimensions and weight [3]. Speaking about the scale of the use of FSW, one can cite as examples large foreign aircraft concerns, such as Airbus and Boeing, as well as domestic manufacturers, such as CJSC Cheboksary Enterprise «Sespel», which create sealed containers for auto transportation, including dangerous goods.

It is important to understand that there is a specific list of possible violations of the internal structure of welded joints during FSW, which is different from defects that can occur during fusion welding. This circumstance became the reason for the development of an integrated approach to diagnostics of the quality of the obtained compounds [4].

There are methods for detecting violations of the internal structure in the volume of a welded joint using ultrasonic and X-ray equipment, one of them is presented in the following work [4]. It should be noted that the ultrasonic testing method, provided that a phased antenna array is used, allows a high degree of reliability of the results of scanning violations both on the surface and in the thickness of the welded joint. However, in addition to defects localized in the bulk of the joint, the strength characteristics of the obtained parts are significantly affected by surface disturbances in the structure of the substance, such as cracks on the surface, pores and lack of fusion of the joint root. With regard to these defects, the eddy-current method of non-destructive testing showed high efficiency and accuracy of their detection. Aspects of the application of this method and the limits of its applicability are considered in the works of a number of authors.

In particular, Mandache et al. [5] considered an ultrasound method for diagnostics within the framework of structures obtained by means of FSW, and also showed the effectiveness of the approach in relation to defects that are located in depth. At the same time, the lack of accuracy of this method in relation to surface defects is noted. A. Lamarre, together with co-authors [6], is considered a pioneer in the application of a matrix of several sensors for ultrasonic scanning of joints created by FSW. Johnston [7] carried out a versatile study of the limits of applicability of the ultrasonic method to search for internal violations associated with oxidation at the boundaries of a welded joint.

The eddy current method, due to its effectiveness in relation to aluminum alloys, is also in the range of interests of the scientific community [8]. So N. Goldfine in co-authorship [9] has developed a sensor that uses the principles of eddy currents and is aimed at detecting violations of welded joints of aluminum parts by friction stir welding, with a size of more than 150 μm. A. Lamarre, noted earlier, also studied the possibility of combining ultrasonic and eddy current sensors [6], which, according to calculations, could increase the efficiency of scanning welded joints within the framework of the FSW-method. With all this, the key aspect of the application of the eddy current approach to nondestructive testing of welded joints is the change in the electrical conductivity of the substance that occurs during the direct FSW [10].

A number of authors have suggested ways to get around this circumstance. So Rosado [11] considered the possibility of subtracting indicators from two ECT and computer analysis of the signal received from the measuring circuit to improve the accuracy of the results. The method developed by them made it possible to detect structural violations with a dimension of 60 μm. Li et al. [12] presented a combination of different methods. The authors were able to significantly increase the rate of measurement of defects on the welded joint of the 2219-T6 aluminum alloy with the addition of copper, silicon, manganese, iron, etc.

Hu [13] analyzed the application of magnetic field sensors when scanning joints created by friction stir welding. In the course of the study, the fact of a change in the magnetic permeability of such welds was established, which in turn opened up prospects for the application of methods for searching for typical defects of the order of 13–130 mm, based on eddy-current effects.

A different vision of the measuring system using eddy current principles is described in [14]. The proposed system included an ECT sensor, a component generating the required signal, a subsystem for processing the output signal, and an automated subsystem for moving the sensor. This scanning complex was focused on the search for microscopic violations of the internal structure of the substance connected by the FSW-method. Testing was carried out on parts made of AA2024 alloy and there was the applicability of the system for detecting defects with dimensions of 50 μm or more.

In paper [15] we discussed the application of the eddy current testing (ECT), a rather new technique of motion induced eddy current testing (MIECT), to the non-destructive testing of friction stir welding. The use of static magnetic fields enables a much higher field penetration into the object under test. If the magnetic flux density and the electrical conductivity of the specimen are high enough, even deep internal defects can be detected. It has been shown that MIECT is a promising technique to evaluate the weld zones of friction stir welding.

Thus, we understand the relevance of the topic of developing an eddy current transducer aimed at scanning welded joints created by the FSW-method and the corresponding measuring system. This work is aimed at revealing violations of the internal structure of welded joints of aluminum alloys obtained by the FSW-method and developing an integrated approach for eddy current testing of such materials.

2 Experimental Technique

The ultra-compact eddy-current transducer was created for the purpose of local non-destructive diagnostics of the physical parameters of parts made of aluminum alloys [16, 17]. The electrical conductivity of the substance and its distribution both over the surface and in the thickness of the scanned material were chosen as a key parameter. The chosen method is based on the correlation between the intensity and distribution of eddy currents in the scanned sample and the electromagnetic (and related) parameters of the sample, as well as the relative position of the scanning sensor and the controlled object.

Within the framework of the software and hardware-measuring complex, the eddy current transducer is connected to the audio interface of a personal computer running Windows OS, on which specialized software is installed. The software is based on the C++ language and is aimed at controlling the voltage parameters on the excitation circuit of the eddy-current transducer, as well as registering (in conventional units) and processing the output signals from the measuring circuit, as a result of which the obtained data are converted into values of electrical conductivity.

The hardware and software complex includes pyramidal cores made of 2000 HM3 ferrite with a magnetic permeability of 500, on the basis of which the exciting, measuring and compensation circuits of the system were assembled from copper wire 5 μm thick. The latter is connected to the measurement winding and serves the purpose of eliminating the influence of the excitation circuit on the resulting indicators of the output signal. The excitation and measurement circuits have diameters of 0.3 and 0.05 mm, respectively, and contain 200 turns, while the compensation circuit has from 160 to 180 turns.

The eddy current transducer is implemented in the form of a transformer that includes three circuits (windings of excitation, measurement and compensation) and a magnetic circuit installed inside the cylindrical base. The platform has special slots for winding tracks, which are impregnated with a compound at a temperature of 200 °C. This procedure protects the windings from destruction when a ferrite shield is subsequently applied to them, which is necessary to improve the accuracy of localizing the electromagnetic field on the scanned sample. To prevent the contact of the transducer with the object under study, the ECT is placed in a corundum washer. The presented transducer is designed for effective concentration of the magnetic field within the boundaries of a zone with an area of 2500 μm^2 and providing a significant penetration depth into the test object at sufficiently low operating frequencies.

The hardware and software complex operates according to the following conceptual scheme. The software installed on the PC controls the source of rectangular voltage pulses with a repetition rate f1 and an amplitude U. Depending on the set research task, the repetition rate and amplitude of the pulses are subject to the necessary correction. The complex includes two integrators connected in series, which convert the signal form from rectangular to sinusoidal. Next, the signal is amplified in a special unit (the gain is 20 ± 2), after which it is directed to the excitation circuit of the converter. Passing through the windings of the transducer, the sinusoidal current causes an electromagnetic field, which in turn induces eddy currents in the electrically conducting material. The existing inhomogeneities of the internal structure of the substance in the area of the transducer affect the parameters of the eddy current field, which induces an electromotive force (EMF) in the measuring winding. The resulting EMF is sent for amplification to a special selective amplifier, the action of which is based on a two-stage modified Delyann filter with multi-loop feedback. A second selective amplification stage was added to the standard filter model to increase the amplitude of the signal at the desired frequency. Modification of the filter made it possible to increase the stability of the system and reduce the sensitivity to a small spread in the parameters of circuit elements. The operating frequency of the filter is 300–700 Hz.

At the next stage, the signal passes through an amplitude detector and is sent to an analog-to-digital converter of PC. In view of the fact that the frequency control of the generated signal on the excitation circuit occurs simultaneously with the control of the cutoff frequency of the filtering system, the scanning complex extracts the signal carrying useful information. In particular, the nature of the distribution of electrical conductivity inside the substance indicates the presence of internal structural disturbances.

The control of this scanning unit is carried out in an automated mode and is implemented at the digital level. The functionality of the VDSS-8 software and hardware complex includes both direct management of the scanning process and subsequent analysis of the collected information. Visualization of the recorded data is preceded by a three-dimensional stage of signal preprocessing using the electronic unit of the meter-transformer, which interacts with the spectrum analyzer of the read signal. Such a scheme of the installation device makes it possible to display data in graphical form, which in turn increases the convenience of information perception since such a visual representation often has a positive effect on the perception of complex phenomena and processes. The software and hardware complex was designed to provide the operator with a wide range of options for adjusting parameters and fine-tuning individual characteristics of the installation to improve the accuracy of measurements in each specific case. Adjustment is understood as entering the values of the amplitude and phase into the computing block, which is necessary to determine the coefficient of the inverse transformation function. It is also necessary to take into account such essential parameters as the thickness of the scanned object, as well as the size of the technological gap between the measuring sensor and the surface of the control object. Next, the system registers the signal indicators in two significant states, namely, in the absence of the scanned object in the sensor range, as well as in the absence of a zero technological gap between it and the ECT. These data are used as reference points insofar as their boundaries will contain the true values obtained during the measurement of the control object. After completing all the debugging procedures, the reading sensor is positioned above the object in such a way as to allow the converter to move freely on the surface and at a constant speed. The latter condition is of critical importance in the context of transmitting indicators in the graphical form concerning the points of the welded joint. The location of the transducer was discrete with a step of 0.5 mm and was considered as the distance from the edge of the scanned sample to the current location of the transducer.

Within the framework of this study, a series of practical measurements was made. However, the system has previously passed the stage of sensor calibration at a frequency of 100–1000 Hz. The stage consisted in registering the voltage introduced from the area where there were no defects. Then the procedure was repeated on the area of the material, which could have violations of the internal structure. In this case, the operating frequency varied in the range of 100–1000 Hz with a step of 100 Hz. The purpose of this step was to determine the frequency at which the applied voltage deviated the most in comparison with the area without defects. The voltage that corresponded to this frequency was taken as a key parameter that indicated the presence of a defect.

Then, using the ECT, direct scanning of aluminum alloys was carried out, the connection of which was obtained by the FSW-method at a field strength of 800 A/m. The scanning area on the surface of the object was 0.1 mm, the duration of one measurement at one frequency was 0.1 s, when scanning all frequencies—1 s, the calibration time was 0.5 s. The introduced voltage, created by the field of eddy currents in the scanned object, was chosen as a variable characteristic. The measurement was carried out while moving the sensor across the welded joint or across the area of defects.

3 Research Results and Their Discussion

SAMPLE No. 1.

Sample No. 1 is two plates (plate 1 and plate 2), each 8.3 mm thick, made of an alloy of copper and aluminum, assembled by the FSW-method with a welded joint width of 1.5 cm. During scanning, the sensor moved perpendicular to the weld with distance measurement from the calibration area to the measurement area. In the course of measurements, differences in the values of the inserted voltage were revealed, which indicates the difference in materials. In Fig. 1 data on the results of scanning are presented: when the sensor approaches the boundaries of the welded joint, areas with a significant decrease in the insertion voltage are observed. Thus, the correspondence between the stress drop and the boundaries of the welded joint is established.

SAMPLE No. 2

Sample No. 2 is two plates each 5 mm thick with technological inhomogeneities at the welded joint (the study areas are marked with numbers 1–8). The movement of the sensor during scanning was perpendicular to the axis of the welded joint over the area containing defects, with the measurement of the distance from the calibration area to the study area.

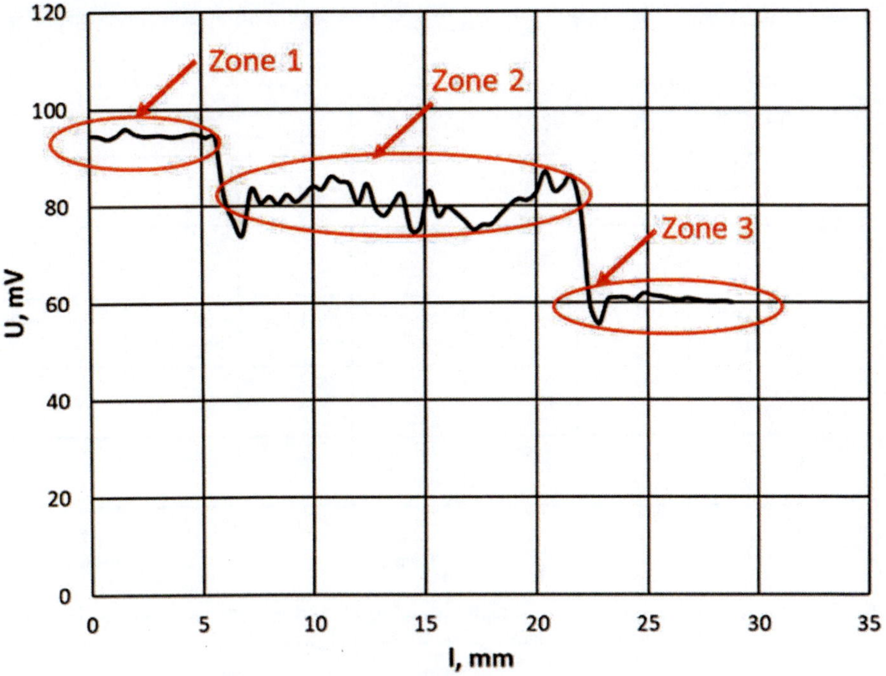

Fig. 1 The results of measurements of a sample No. 1: U—introduced voltage, l—sensor position, zone 1—plate 1, zone 3—plate 2, zone 2 - welded joint

Region 1 is a cavity having a diameter of 2 mm and located inside the scanned object at a depth of 3 mm from the measurement surface. The defect was detected by the scanning system at an operating frequency of 700 Hz (Fig. 2).

The defect area is marked with a dotted line. The decrease in the voltage on the measuring circuit within the boundaries of the cavity was 25 mV compared to the calibration section and 7 mV compared to the zone of the welded joint, the boundaries of which are also marked by a drop in the insertion voltage.

Marks 2 and 3 indicate the zones of the surface crack in the welded joint, 0.2 mm wide. This defect was registered by a scanning system at almost any frequency. Differences in the insertion voltage were about 35 mV compared to the calibration section and about 15 mV compared to the welded joint zone (Fig. 3a, b).

On area 4, a cylindrical cavity with a diameter created by milling the plate from the inside from the opposite side is marked. The cavity has the following parameters: depth—4 mm, diameter—1.5 mm, distance to the scanning surface—1 mm (Fig. 4).

The transmitter has accurately identified the presence of a defect. The change in the insertion voltage was 30 mV compared to the calibration section and 10 mV compared to the welded joint.

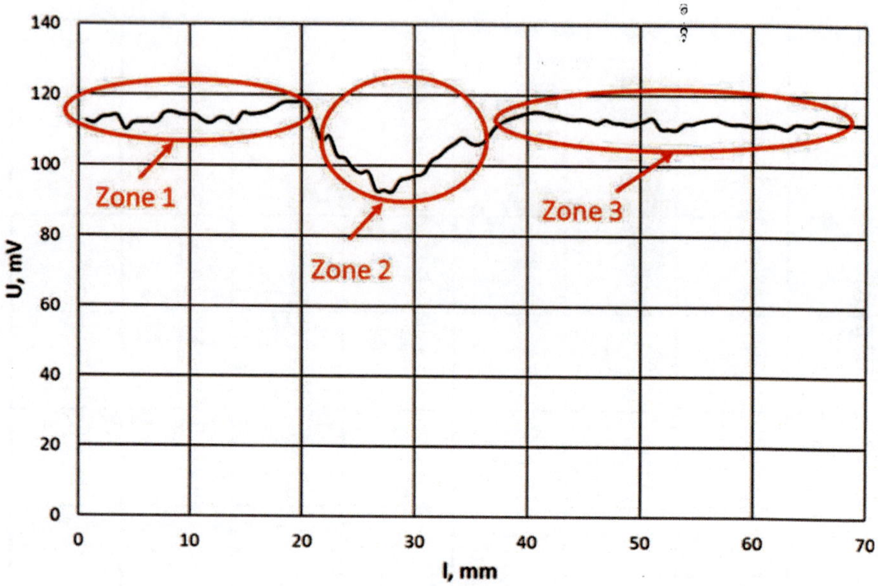

Fig. 2 The results of measurements of sample No. 2 (region 1): U—introduced voltage, l—sensor position, zone 1—plate 1, zone 3—plate 2, zone 2—welded joint

On area 4, a cylindrical cavity with a diameter created by milling the plate from the inside from the opposite side is marked. The cavity has the following parameters: depth—4 mm, diameter—1.5 mm, distance to the scanning surface—1 mm (Fig. 4).

The transmitter has accurately identified the presence of a defect. The change in the insertion voltage was 30 mV compared to the calibration section and 10 mV compared to the welded joint.

Areas 5–8 mark the areas of the welded joint with internal cavities (holes). They are marked by characteristic dips at the boundaries of the connection (Figs. 5, 6, 7 and 8).

The change in the insertion voltage was 5–10 mV relative to the region without defects. In view of the fact that similar data were obtained at facility No. 1 (Fig. 1), there are reasons to speak of this as a feature of this type of welded joint.

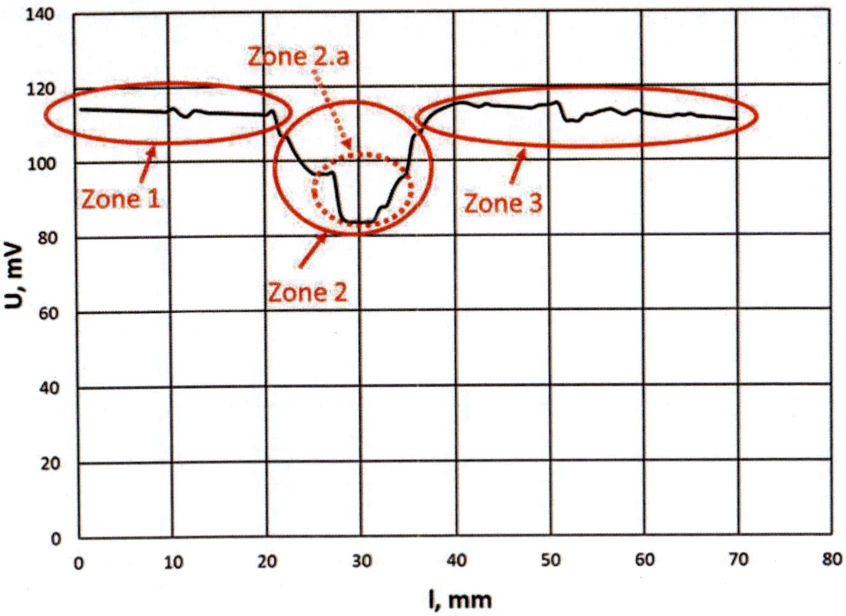

a)

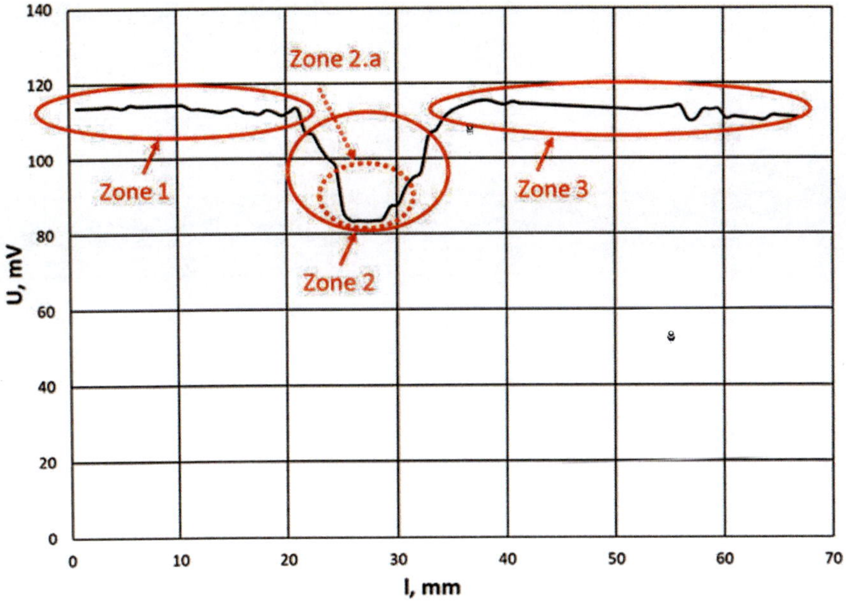

b)

Fig. 3 The results of measurements of a sample No. 2, region 2 (**a**), region 3 (**b**): U—introduced voltage, l—sensor position, zone 1—plate 1, zone 3—plate 2, zone 2—welded joint, zone 2a—surface crack

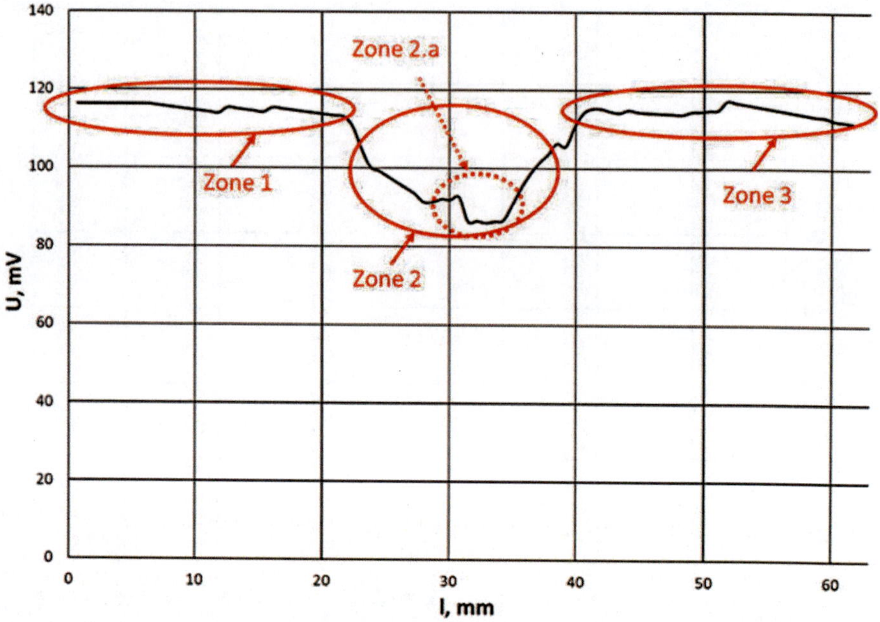

Fig. 4 The results of measurements of sample No. 2, region 4: U—introduced voltage, l—sensor position, zone 1—plate 1, zone 3—plate 2, zone 2—welded joint, zone 2a—cylindrical cavity from the inside

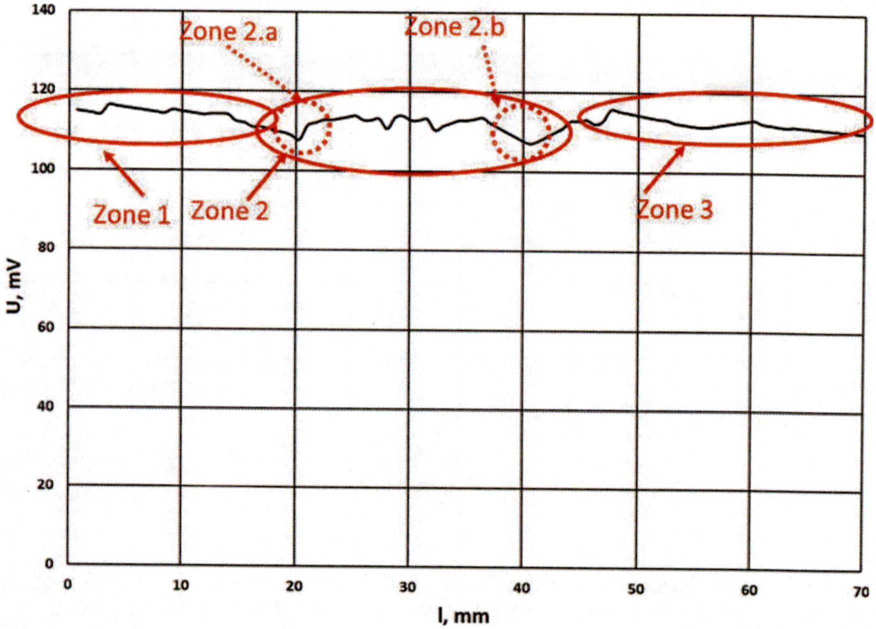

Fig. 5 The results of measurements of sample No. 2, region 5: U—introduced voltage, l—sensor position, zone 1—plate 1, zone 3—plate 2, zone 2—welded joint, zone 2a, zone 2b—internal cavities (holes)

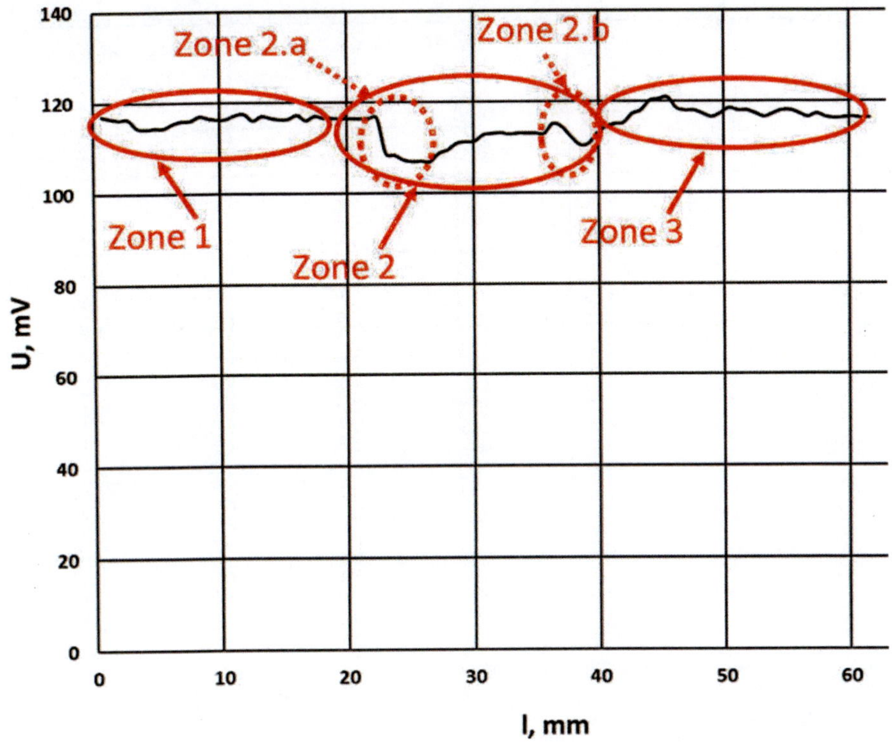

Fig. 6 The results of measurements of sample No. 2, region 6: U—introduced voltage, l—sensor position, zone 1—plate 1, zone 3—plate 2, zone 2—welded joint, zone 2a, zone 2b—internal cavities (holes)

4 Conclusion

The presented flaw detection hardware and software complex is designed for scanning plates made of copper and aluminum alloys, fastened by friction stir welding. The paper shows the results of a series of experiments demonstrating the possibility of detecting ultra-small defects in welded joints by the eddy current method of non-destructive testing. Ultra-compact transducers assembled on the basis of pyramidal cores make it possible to detect, with a high degree of accuracy and efficiency, cracks and cavities with a diameter of 0.2–2 mm and located at a depth of 3 mm. In addition, the paper shows the applicability of the scanning complex for monitoring the region of the boundaries of the welded joint.

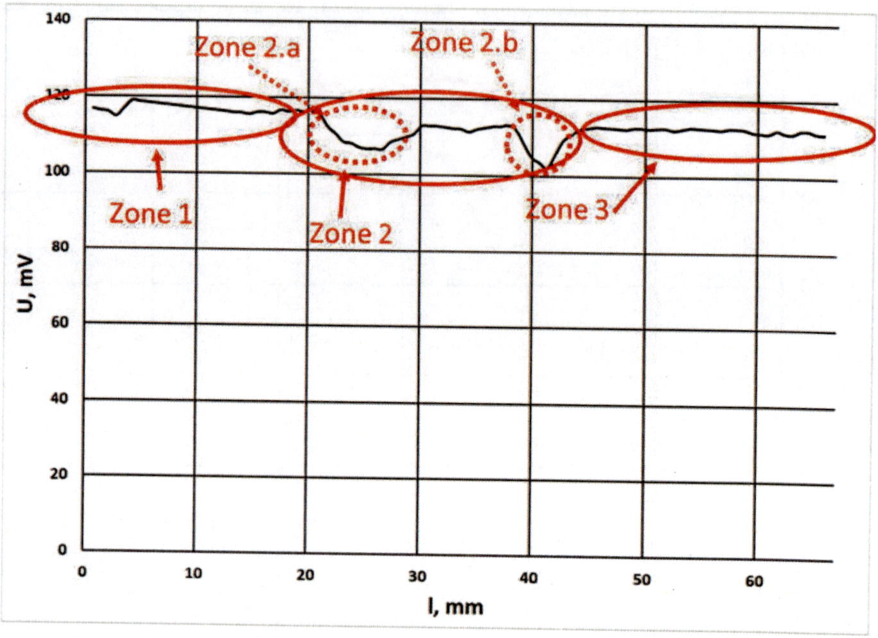

Fig. 7 The results of measurements of sample No. 2, region 7: U—introduced voltage, l—sensor position, zone 1—plate 1, zone 3—plate 2, zone 2—welded joint, zone 2a, zone 2b—internal cavities (holes)

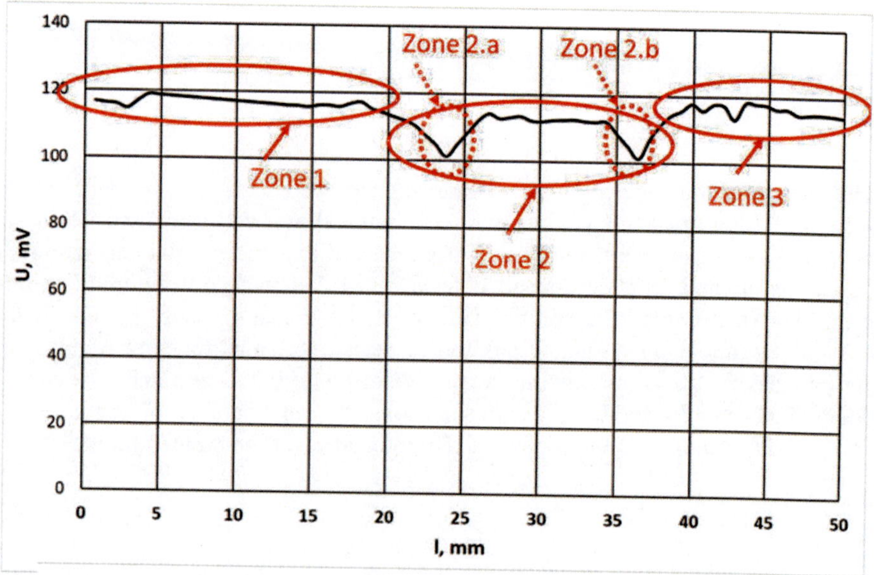

Fig. 8 The results of measurements of sample No. 2, region 8: U—introduced voltage, l—sensor position, zone 1—plate 1, zone 3—plate 2, zone 2—welded joint, zone 2a, zone 2b—internal cavities (holes)

References

1. Mishra RS, Ma ZY (2005) Friction stir welding and processing. Mater Sci Eng R Rep 50: 1–78
2. Egorov AV, Polyakov VV, Salita DS, Kolubaev EA (2015) Inspection of aluminum alloys by a multi-frequency eddy current method. Defence Technol 11(2):99–103
3. Bakshaev VA, Vasilev PA (2014) Friction stir welding in manufacturing of large-sized aluminum alloy products. Non-ferrous Metals 1:14–18
4. Rubtsov VE, Kolubaev EA, Tarasov YuS (2014) Thermography inspection of friction stir welding. AIP Conf Proc 1623:535–538
5. Mandache C, Dubourg L, Merati A, Jahazi M (2008) Pulsed eddy current testing of friction stir welds. Mater Eval 66(4):382–386
6. Moles M, Lamarre A, Dupuis O (2004) Complete inspection of friction stir welds in Aluminum using ultrasonic and Eddy currents arrays. In: Proceedings 16th WCNDT 2004. World conference on NDT, pp 87–92
7. Johnston PH (2008) Addressing the limit of detectability of residual oxide discontinuities in friction stir butt welds of aluminum using phased array ultrasound. In: Proceedings—35th annual review of progress in quantitative nondestructive evaluation, pp 1–26
8. Klimova TA, Il'inkova TA, Shigapov AI (2015) Non-destructive control of the weakening of profiles made of V95OChT2 aluminum alloy. Metal Sci Heat Treat Metals 5:37–41
9. Grundy D, Zilberstein V, Goldfine N (2005) MWM®-array inspection for quality control of friction stir welded extrusions. In: Proceedings 7th international conference on trends in welding research, pp 227–232
10. Santos T, Ramos PM, Vilaça P (2008) Nondestructive testing of friction stir welding: comparison of planar eddy current probes. In: Proceedings, IMEKO TC4 symposium, pp 507–512
11. Rosado LS, Santos TG, Piedade M, Ramos PM, Vilaca P (2010) Advanced technique for non-destructive testing of friction stir welding of metals. J Measur 43:1021–1030
12. Li B, Shen Y, Hu W (2011) The study of defects in aluminium 2219–T6 Thick butt friction stir welds with the application of multiple nondestructive testing methods. Mater Des 32:2073–2084
13. Hu B, Yu R, Zou H (2012) Magnetic non-destructive testing methods for thin plate aluminum alloys. NDT&E Int 47:66–69
14. Vilaça P, Santos TG, Rosado L, Miranda RM (2013) Innovative concept and application of ec probe for inspection of friction stir welds. In: Proceedings. The 12th international conference of the Slovenian society for non-destructive testing application of contemporary non-destructive testing in engineering, pp 329–338
15. Brauer H, Otterbach JM (2019) Friction stir weld inspection using the motion induced eddy current testing technique. In: AIP conference proceedings, 080004
16. Dmitriev SF, Ishkov AV, Katasonov AO (2017) Measurement system for studying flaws in alloy slabs by means of subminiature eddy-current transducers. Meas Tech 60(4):372–375
17. Dmitriev SF, Ishkov AV, Malikov VN, Sagalakov AM (2017) Scanning the welded seams of titanium alloys by using subminiature eddy current transducers. Mater Sci Forum 906: 147–152

Application of Machine Learning Algorithms in Ensuring Information Security of the Internet of Things in Water Transport

Izolda Li, Artem Arbuzov, Vladimir Shilkin, Irina Shilkina, and Nadezhda Baryshnikova

Abstract IoT technologies are gaining increasing influence in all areas of human activity from wearable devices to industrial applications. With the rapid development of water transport infrastructure, as well as with the growth of commissioned port and berthing mechanisms, serious prerequisites have arisen for the use of control and monitoring tools. IoT systems allow you to flexibly and efficiently solve such problems. The IoT architecture includes various levels of communication and information transfer. There are many protocols and applications running at each layer, which can be vulnerable to attacks of intruders. Nowadays, research related to solving IoT security issues using various technologies and concepts, such as blockchain, statistical analysis, machine learning, etc., is actively underway. This paper provides an overview of machine learning methods used at various layers of the Internet of Things architecture. In addition to the layers of the IoT architecture, it also discusses the machine learning methods used to counter the increasingly widespread attacks of social engineering that lead to the loss of integrity, confidentiality and availability of information. At the end of the paper, the machine learning methods that best match the architectural and hardware capabilities of the IoT were selected.

Keywords IoT · Information security · Internet of ships · Smart shipping · Machine learning (ML) · Social Engineering · Accidents · Risk

1 Introduction

Nowadays, the Internet of Things (IoT) is one of the promising technologies that is being actively implemented in many areas, from healthcare to agriculture. The integration of any new technology opens up new vectors for attacks against the

I. Li (✉) · A. Arbuzov · V. Shilkin · I. Shilkina · N. Baryshnikova
Admiral Makarov State University of Maritime and Inland Shipping,
Dvinskaya str 5/7, Saint-Petersburg 198035, Russia
e-mail: liiv@gumrf.ru

A. Bieliatynskyi and V. Breskich (eds.), *Safety in Aviation and Space Technologies*,
Lecture Notes in Mechanical Engineering,
https://doi.org/10.1007/978-3-030-85057-9_45

system. Such incidents can have serious negative consequences, ranging from access to personal data of users and ending with threats to human life [1].

With the integration of the IoT into local and global information systems integrated into various branches of human activity, a huge amount of vital information has actually come under threat. This problem has also affected the transport system as a whole, and water transport is not an exception. The industry is conducting research on the integration of IoT technology.

In the field of water transport, the use of IoT systems is possible:

- in the infrastructure of ports, including the management of cranes, containers and other mechanisms;
- in ship management and monitoring;
- in the process of loading and unloading containers;
- in the management of buoys in the aquatic environment;
- in the monitoring and control of ship machinery;
- when monitoring the mental and physical condition of the crew members during the performance of their duties;
- when monitoring the condition of cargo, etc.

The implementation of such an IoT system will lead to the need to develop applications that process large amounts of information collected by IoT devices. These applications should solve such problems as:

- planning and optimization of shipping routes;
- intelligent decision making on physical and information security issues, including monitoring of incidents and failures;
- tracking the condition and movement of goods;
- monitoring of the environment and ecological state;
- providing energy efficient and energy saving solutions for the operation of equipment and ships;
- automatic docking.

The work [2] introduces the definition of the Internet of Ships (IoS): "The Internet of Ships is the interconnection of sensory objects such as ships, crews, cargo, on-board equipment, aquatic environment, water structures, coastal structures and other navigation elements that are embedded in various sensory and heterogeneous networking technologies that enable these entities to collect and exchange data".

The relevance of the integration of IoT systems in the field of water transport is no less relevant than for the transport industry as a whole. Historically, one of the most demanded modes of transport in the field of cargo transportation is water transport. In 2019 alone, seaports handled about 4 million tons of cargo in Europe inland waterways—about 522 million tons of cargo.

These indicators will grow steadily due to the development of the global maritime economy. This will lead to an even greater increase in ships, which means to an increase in congestion of various waterways, which will negatively affect both economic and environmental indicators. In addition, the maneuverability of ships in traffic jams will be limited, which will create additional risks of accidents and other incidents. To solve these problems on the basis of the Internet of Things (IoT), the concept of the Internet of Ships (IoS) was proposed, adapted to the specifics of maritime transport, and various problematic aspects of using the IoT system on "smart ships" were considered [3].

The concept of the maritime Internet of Things was originally proposed by the United Nations International Maritime Organization under the name "Electronic Navigation" to harmonize and digitize maritime navigation information and operations, and to support a wide range of maritime services. The paper [4] discusses the practical problems of the implementation of the maritime Internet of Things related to the marine environment, service requirements and radio frequency spectrum. Researchers propose solutions for the architecture of the maritime Internet of Things networks, the radio interface and the radio frequency spectrum of this specific communication system.

In the field of ecology, work is also underway to prevent damage to the marine ecosystem caused by leaks of hydrocarbon-type oil. To recover oil from the surface of the water, a system that can provide a high recovery rate in a moving environment is required. In paper [5], an intelligent oil skimmer with a mechanism of two round discs and the IoT (Internet of Things) monitoring system based on a smartphone controlled by a user is proposed. The oil gathering system is controlled taking into account weather conditions (temperature and humidity) from any geographic location through a server. Experiments have shown an overall average efficiency of about 89.87%.

The work [6] proposes the development and implementation of an Internet of Maritime Things (IoMT) platform that supports long-distance and high-speed communications for remote and online monitoring of sea water quality. Initially, the IoMT system and gateway that interact with software platforms of the international IoT standard for universal use in the field of marine monitoring is being developed.

As an example of research in the field of automated detection of marine objects, we can cite work [7] using Internet of Things (IoT) surveillance systems. Traditional methods based on machine learning have had some success in specific object detection problems. However, ships are considered to be relatively small objects, so they are difficult to detect with high accuracy. An algorithm for the detection of marine objects based on electroencephalography (EEG) is proposed for intelligent surveillance systems using IoT technology.

Research to develop special protocols designed to improve the security of data transmission in IoT systems is also underway. The work [8] examines the issues of ensuring the security of networks of the Internet of Maritime Things (IoMaT). A secure mobile Ad-hoc/Delay Tolerant (S-MADNET) routing protocol is proposed for the marine environment using VHF equipment available on most ships. The proposed secure network is designed to utilize the existing Automatic Identification System (AIS), which ships use for positioning and navigation assistance. The IoMaT routing module is also presented, which sends data from marine sensors using the proposed secure protocol, taking into account information from the AIS system level. In addition, a new IoMaT-enabled AIS message format is proposed that meets the requirements of the secure routing protocol.

The use of IoT in water transport is not just a theoretical possibility, it is already being translated into practical application. According to the results of the Inmarsat report [9], a fairly large number of shipping companies are open to implement solutions based on IoT technologies, since IoT offers tremendous opportunities in the maritime industry. As can be seen from the above works, the introduction of IoT technologies is becoming more widespread. However, at the same time, as in all industries related to information technology, aspects of information security are often ignored until significant incidents occur that lead to economic or environmental consequences and also pose a threat to human life.

With such a high level of interest in the use of this technology, it is necessary to solve the question of how to protect IoT systems from the actions of intruders at all levels of IoT interaction. At the moment, a large number of concepts have been proposed aimed at increasing the security of the Internet of Things. The most interesting of these are blockchain [10], statistical analysis of vulnerabilities [11], a list of major OWASP vulnerabilities [12], and machine learning (ML) algorithms [13], the latter will be discussed in this paper.

To date, there are already several review papers on ML-based IoT security. This paper proposes its own approach to the classification of existing approaches and algorithms. The level of social engineering, as one of the most vulnerable levels today, is also added.

The structure of this paper is as follows. The main levels of the architecture of the Internet of Things are considered. Vulnerabilities and attacks that are possible at each level are identified. Machine learning solutions that can counter each type of attack are provided. Machine learning solutions capable of countering social engineering attacks are considered.

In conclusion, the paper summarizes and selects the most promising machine learning methods. The selection criterion is universality at different levels of architecture, as well as the frequency of use in previous studies.

2 IoT Architecture

The IoT architecture can be described by a gateway that provides communication between hardware applications and physical IoT devices using various communication protocols, as shown in Fig. 1. Each layer employs a wide range of different connection protocols for receiving and transmitting information/data [14]. The specificity of building the IoT architecture lies in the fact that the range of services provided to the user expands at each level.

The standard architecture of the Internet of Things is subdivided into three layers: the application layer, the network layer, and the physical (perception) layer [15].

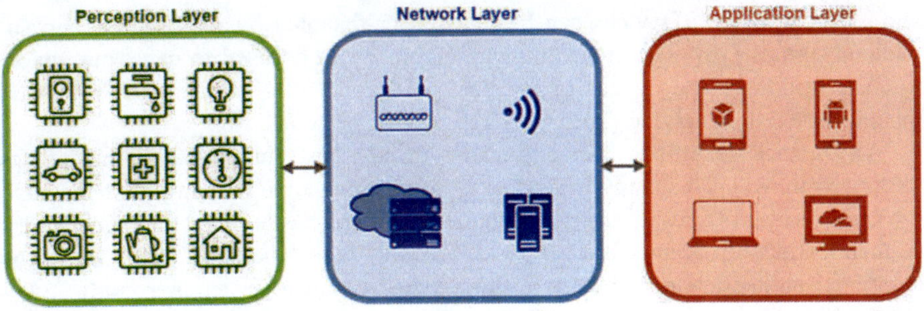

Fig. 1 IoT standard architecture

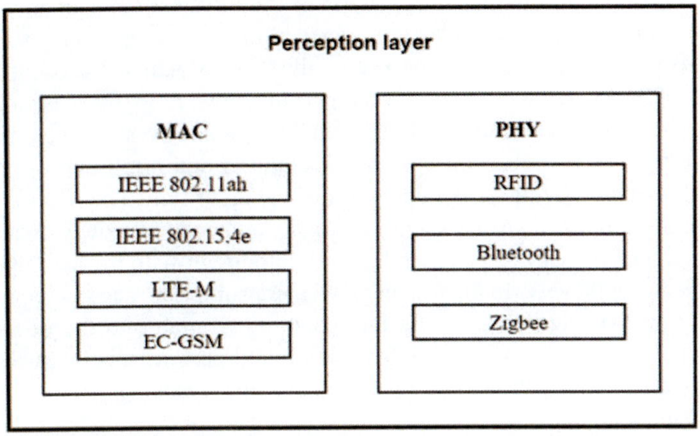

Fig. 2 Basic perception layer transfer protocols

2.1 Perception Layer

The perception (physical) layer is the first in the IoT architecture. This layer includes physical devices of the Internet of Things. This layer is divided into two sublayers: medium access control (MAC) and physical (PHY). The MAC sublayer is responsible for establishing communication between physical devices, Fig. 2 To establish a connection, the MAC uses the IEEE 802.11ah and IEEE 802.15.4e protocols, as well as the LTE and GSM cellular standards [15]. The PHY is responsible for devices aimed at receiving information from the physical medium and transmitting it, i.e. sensors. This sublayer uses such protocols as Bluetooth, RFID, etc. [16].

Physical Layer Vulnerabilities and Attacks That Use Them. The main vulnerability of the physical layer manifests itself at the stage of device authentication. IoT devices at the physical layer exchange information using the MAC sublayer and have limited computing resources [17], so traditional authentication methods are not applicable here. Devices can be blocked or disabled by interference and suspicious signals [18]. Therefore, this vulnerability can be used to carry out attacks, such as spoofing and attacks using electromagnetic interference [19], which can result in denial of service [20].

Anti-attack Solutions. An alternative to standard authentication means was proposed in [21]. This study looked at supervised machine learning methods based on Q-learning. As a result, the use of these algorithms has reduced the percentage of authentication errors by about 64%. Moreover, these methods differ from conventional authentication methods in higher performance. In another study [22], a method based on combining reinforcement learning (RL) and deep convolutional neural networks—Convolutional Neural Network (CNN) was proposed, which successfully counteracted interference.

When choosing machine learning tools at the physical level of the IoT architecture, the first thing to consider is the capabilities of computing resources. The solution to this problem can be the use of the Tiny Machine Learning (TinyML) algorithm [23] for devices with low computing power. In the case of the IoT, examples of such devices are motion and/or thermometry sensors, smart kettles, stereos, refrigerators, and so on. The ML method is optimized due to the detailed tuning of the microcircuit for the operation of the algorithm. The ML method itself is maximally simplified to work with a small number of narrowly focused tasks. With the help of this approach (flashing the microcircuit with the ML algorithm), it is possible to achieve a reduction in power consumption and the possibility of implementing ML algorithms using the resources available to the device. In our opinion, this technology solves the issue of limited performance of such primitive devices as sensors, which are the basis of many IoT systems.

2.2 Network Layer

The network layer in IoT systems is the link between devices and applications. Figure 3 shows some of the possible network layer protocols. At this layer, a huge number of communication protocols are used, such as 3–5G, WiFi, IPv6, etc. [14]. This layer includes a local cloud and servers that store and process intermediate data. The main task of this level is to work with big data [24]. The fact is that every second, physical IoT devices generate huge amounts of data that are stored and processed in the storages of IoT systems. Nowadays, systems based on deep machine learning algorithms—Deep Learning (DL) [25] are used to process big data. The computing power of these algorithms allows processing information in up-to-date time, which makes it possible to provide users with services in real time.

Network Layer Vulnerabilities and Attacks. The surface of the network layer can be conditionally divided into the surface of network transportation and cloud systems. The IoT network uses a huge number of devices using both wired and wireless connections. Due to the abundance of protocols, information is transmitted openly, which is especially clearly visible when transmitting information over a

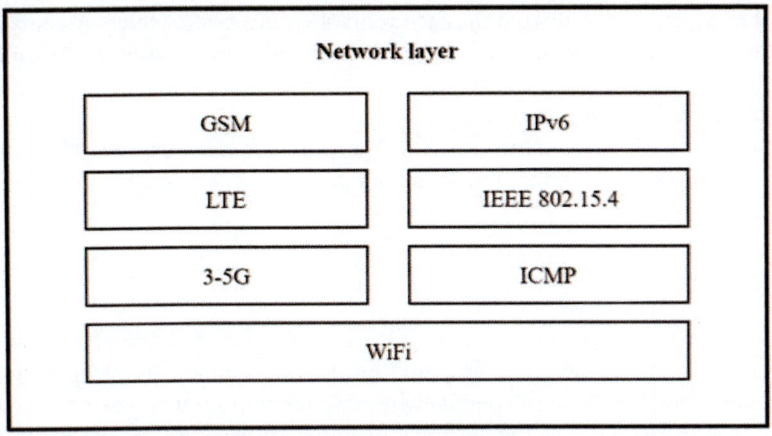

Fig. 3 Basic protocols and standards used at the network layer

sensor network (SN). The sensor network connects sensors in IoT systems that exchange information and transmit it to servers and the cloud. With this openness, the sensor network needs a reliable security protocol [9].

The network layer is susceptible to such attacks as man-in-the-middle, DoS attacks, traffic analysis, black hole attacks (dropping incoming/outgoing packets, which can lead to denial of service). In the case when information is exchanged between sensor devices, the system is vulnerable to the Sybil attack (a type of attack in a peer-to-peer network, as a result of which the victim connects only to nodes controlled by the attacker) [26].

The IoT system operates with a colossal amount of data. According to forecasts, by 2025, 175 zettabytes of information will be annually processed by devices [27]. With such an increase in the amount of data, there may be a shortage of standard information stores. The solution to this problem was cloud storage systems, the potential of which is theoretically limitless. In addition to storing big data, cloud technologies have provided the ability to perform computing directly in the cloud. With this technology, IoT systems can be updated in real time. There is a risk of unauthorized access to cloud users' data through DoS attacks [28].

ML Based Solutions to Counter Network Attacks. In study [29], work for comparing existing machine learning algorithms aimed at detecting DoS attacks in the Internet of Things was carried out. Such algorithms as Neural Network (NN), Random Forest (RF) and Support Vector Machine (SVM) were considered. Another work [30] proposed a solution for detecting DoS attacks in a network based on the SVM algorithm.

In addition to looking for solutions aimed at countering DoS attacks, it is worth considering methods aimed at detecting anomalies/intrusions. One of the most commonly used intrusion detection solutions is traffic analysis. To solve this problem, ML algorithms are actively used [31], the essence of which is as follows: there are rules on the basis of which packets that do not correspond to these rules are selected and blocked from the incoming/outgoing traffic. Such systems do not work perfectly, false alarms often occur, which calls into question the reliability of this method. Therefore, in addition to analyzing traffic, a behavior-based model is used to detect intrusions.

In [32], a trust-based approach aimed at intrusion detection was investigated. Here, when analyzing traffic, the following parameters are considered:

- Level of trust in the device;
- Traffic type;
- Traffic class.

Intrusions that exploit the 0-day vulnerability are not detected by traditional intrusion prevention methods. The solution to this problem is machine learning algorithms, which are currently actively used in modern intrusion detection systems (IDS). In order to increase the accuracy of intrusion detection, a method based on artificial intelligence in a network controlled by the IoT was proposed [33]. The essence of this method is that two ML algorithms are used to capture traffic:

1. Bat algorithm based on the Particle Swarm Optimization method,
2. Differential Mutation algorithm.

The flow is then classified using a random forest (RF) algorithm, and potentially dangerous anomalies (intrusions) are identified.

There is also a development [34] aimed at intrusion detection based on the ML classification algorithm—Optimum-Path Forest (OPF).

The work [35], during which a method for detecting DDoS attacks on local IoT devices was developed, should be noted. It uses such machine learning methods as SVM with a linear kernel–Lagrangian Support Vector Machine (LSVM), NN, RF with an estimation of Gini impurity scores, the k-Nearest Neighbors algorithm (KNN). These algorithms perfectly solve the problem of classification when detecting attacks. The accuracy of detecting attacks based on a test suite for 5 algorithms was over 99%.

2.3 Application Layer

The application layer is the highest layer in the IoT architecture. This level contains custom applications that allow controlling the IoT system. A good example of human interaction with the IoT system is Apple's Smart Home. All new models of IPhone, MacBook, iPad come with a built-in IoT control application called "Home". This application allows you to control HomeKit accessories from any device from Apple [36].

At this level, two types of protocols are mainly used (Fig. 4). Some are focused on messaging, such as MQTT, DDS, CoAP, and so on. Others focus on service searching, examples of which are SSDP and mDNS.

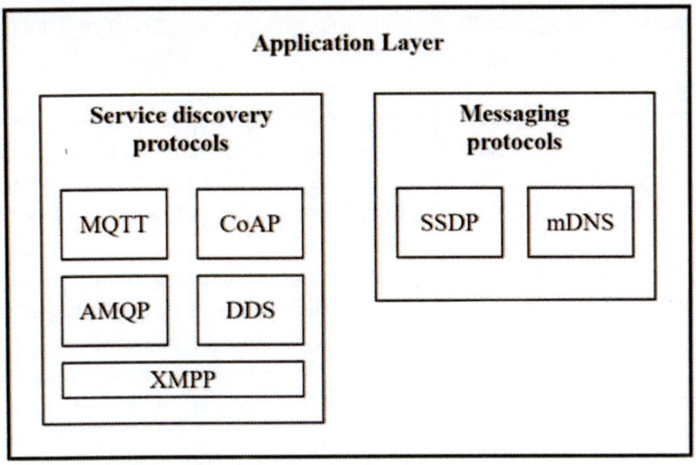

Fig. 4 Application layer protocols

The application layer is no less important than the perception and network layer, since it largely depends on it whether the consumer will use the IoT system. Convenience is paramount here, so user interfaces are becoming more and more simplified over time and more often than usual become the target of cyberattacks.

Application Layer Vulnerabilities and Attacks. At the application level of IoT systems, the most common are 7 protocols: MQTT, CoAP, AMQP, DDS, XMPP, mDNS, SSDP. These protocols are used for the purpose of transmitting information, as well as for searching for services. The choice of protocols depends on what functions the IoT system should perform. For example, CoAP and MQTT are great for collecting data, while mDNS and SSDP are preferred in IoT systems for service searching.

These protocols are an excellent solution for transporting data, but in terms of security, they are not so perfect. Some protocols have weak protection mechanisms, while other protocols lack such mechanisms at all [37]. Let's consider the security of each protocol separately.

The MQTT protocol is a lightweight transmission protocol that makes it relevant to the IoT. In terms of security, this protocol supports TLS-based encryption [38] and also various authentication mechanisms. However, risks for the MQTT protocol often arise from misconfiguration of TSL and the use of weak ciphers [38, 39]. These risks make the MQTT protocol susceptible to denial of service attacks. An example of such an attack is a stack overflow, which can be achieved by sending a single message containing more than 65,400 "/" characters [40].

The DDS protocol is used to exchange data, and it also uses a wide variety of security mechanisms. Also, a special security specification has been developed for this protocol [41], thanks to which an architecture is built based on special plug-ins. However, this specification uses the handshake protocol to confirm the permission of plugins, which transmits information about the capabilities of the participants in the open form. This vulnerability allows an attacker to gain access to confidential information [42]. Also, individual plugins can carry vulnerabilities, which in turn can be used to unauthorized or unintentionally connect to another participant.

AMQP is an open protocol that performs messaging functions. Nowadays, this protocol is popular in networks where there is a need for asynchronous communication. Vulnerabilities in this protocol are primarily related to misconfiguration, as well as user interfaces. Open commands to disable this protocol can be used for denial of service purposes.

In the CoAP protocol, it is necessary to find a balance between the performance of the protocol and its security. The higher the security level, the greater the load on computing resources. Therefore, neglecting some protection services for the sake of performance can make this protocol vulnerable to attacks. The protocol uses DTLS, an implementation of TLS based on the UDP protocol, which opens up the possibility of a denial of service attack.

XMPP supports a wide variety of security mechanisms. This protocol has problems with authentication, which gives rise to vulnerabilities like CVE-2019-1845 [43]. The essence of this vulnerability is that an attacker can remotely cause a service to crash without being authenticated, ultimately leading to a Denial of Service (DoS).

SSDP and mDNS are geared towards finding IoT services. SSDP aims to find services and advertise them on private networks. The mDNS protocol performs a similar service lookup function and also deals with the resolution of host names on local links, i.e. the protocol allows the device to select a name in the local zone (pseudo top-level domain), which is used to identify hosts. These protocols do not have any built-in defense mechanisms [44], which makes them completely unarmed in front of an attacker. The main part of attacks on these protocols, as well as on all other protocols, is denial of service (DoS).

It is worth considering that applications that are installed to interact with the IoT system can also be used as a springboard for deploying attacks on the IoT application itself. At the moment, most of the market share is made up of devices based on the Android OS, which, in turn, have an open architecture and are equipped with an application programming interface. This openness enables attackers to create applications with undeclared capabilities and display such software on various resources. As a result, such a modified system can get on the devices of the user of the IoT system, thus authorizing an attack on the IoT network.

ML Based Solutions to Counter Network Attacks. In our opinion, application layer attacks are heterogeneous enough to select the best ML solution for each type of attack. Therefore, it would be a good option to consider algorithms aimed at detecting various types of attacks.

The study [45] compared several machine learning algorithms—classifiers that can be used to detect attacks. This study was primarily conducted in order to identify the most appropriate machine learning algorithms for detecting Botnet attacks. Namely, the Naive Bayes classifier (NB), the J48 algorithm and Artificial Neural Networks (ANN) algorithms. The results showed that J48 and ANN algorithms outperform NB in terms of speed and lightness. Also, in another study [46], it was concluded that Q-Learning algorithms show excellent results in terms of the accuracy of attack detection and the delay obtained in the detection of attacks.

It should also be taken into account that attackers have the ability to create malware and deliver it to the user's device. Therefore, it is necessary to consider ML-based solutions for detecting malware on mobile devices. In [47], several types of machine learning were considered, namely NB, supervised learning (SL), learning using a Decision Tree (DT), and others. As a result, a parallel classification model was proposed, in which these algorithms were used together. The accuracy of the proposed model is about 97.5%.

2.4 Ml Solutions Countering Social Engineering Threats in IoT

Social engineering (SE) is an impact using manipulation methods aimed at both one person and a group of people, the purpose of which is to gain access to confidential information or to control the behavior of the target [48]. Examples of SEs are phishing emails, fake web pages, fake calls, etc. Note that the indicator of protection against attacks using social engineering methods mainly depends on the person himself. In [49], the author conducts a survey, as a result of which it turns out that even with the most advanced means of protection, a person still remains vulnerable to attacks. Here, the degree of awareness of the person himself about the methods of social engineering matters.

Existing ML mechanisms are capable of helping users to protect themselves from such manipulation methods as phishing. The study [50] developed a mechanism for identifying phishing sites using classifier algorithms: Naive Bayesian classifier (NB), Support Vector Machine (SVM), Random Forest (RF), and Neural Network (NN). As a result, the authors came to the conclusion that SVM has the best performance in terms of selecting and working with heterogeneous datasets.

If we talk about phishing emails, then the same solution algorithms as in the study [50], with the exception of NB, were used by the authors in another study [51]. In addition to the methods listed above, we used Logistic regression, which, as a result, showed the highest accuracy in predicting phishing emails. Also, the authors of this study showed the possibility of improving the logical regression method by using feature selection methods.

3 Results

IoT systems are improving every year, including from the point of view of information security, but today this technology has a fairly large number of vulnerabilities, due to which systems are susceptible to a wide variety of attacks. As a result, there are serious consequences, ranging from the compromise of any confidential information and ending with the disabling of the system itself.

Machine learning offers a myriad of solutions to the security problems of IoT systems, but not all of them can be put into practice. The most versatile were the neural networks (NN). So, at the perception layer, a solution was proposed using convolutional neural networks (CNN).

Research has also been carried out at the application and network layers, and solutions have been proposed using neural networks (NN). But neural networks are not able to cope with the entire spectrum of attacks aimed at IoT, these algorithms show their effectiveness in a bid with other ML methods. So, for example, at the network layer, NNs are used in conjunction with SVM and RF. At the application layer, in addition to NN, botnet attack detection solutions can also use the J48 algorithm. At the perception layer, in addition to CNN, RL methods are often used in the solution.

In the field of countering social engineering, namely phishing, one of the studies also used NN, although this method is not the best option. The best results in this study were shown by SVM, which again indicates that there is no one ML reference method for solving all the security problems of IoT systems.

4 Conclusions

As a result of this work, the following conclusions can be drawn. In the field of water transport, IoT technologies are increasingly used. There is a growing interest of the scientific and industrial community in the implementation of IoT systems. However, the issues of information security during data transfer between IoT devices and control terminals, issues of safe storage and access to information on water transport are insufficiently studied. This paper discusses one of the most effective classes of methods for preventing attacks and methods for detecting vulnerabilities at various levels of the IoT system operation—machine learning methods.

Among the methods of machine learning, the most common and effective methods at various levels of the IoT security system are methods using neural networks. When choosing machine learning methods, it is necessary to take into account the limited computational capabilities of the devices used in the system. Incidents of recent years have clearly shown the need to attract methods aimed at influencing not only the system, but also its direct consumers using social engineering. Therefore, along with methods of countering attacks on the system, machine learning methods were considered that identify phishing emails.

References

1. Chacko A, Hayajneh T (2018) Security and privacy issues with IoT in healthcare. EAI Endorsed Trans Pervasive Health Technol 4(14):155079. https://doi.org/10.4108/eai.13-7-2018.155079

2. Liu G, Perez R, Munoz JA, Regueira F (2016) Internet of ships: the future ahead. World J Eng Technol 4(03):220
3. The internet of ships: a new design for smart ships. https://www.rina.org.uk/TheInternetofShipsanewdesignforSmartShips.html. Accessed February 2020
4. Xia T, Wang MM, Zhang J, Wang L (2020) Maritime Internet of things: challenges and solutions. IEEE Wirel Commun 27(2):188–196. https://doi.org/10.1109/MWC.001.1900322
5. Ansar Khan Md, Golam Mostafa SK, Md RA, Hoque Md, Mizanul, (2021) Implementing and improving the IoT based weather monitoring and controlling double discs type oil skimmer. Electr Signal Process Tech. https://doi.org/10.1109/ICREST51555.2021.9331192
6. Song Y, Shin H, Koo S, Baek S, Seo J (2021) Internet of maritime things platform for remote marine water quality monitoring. IEEE Internet Things J 1–113. https://doi.org/10.1109/JIOT.2021.3079931
7. Duan Y, Li Z, Tao X, Li Q, Shuzhan Hu, Jianhua Lu (2020) EEG-based maritime object detection for IoT-driven surveillance systems in smart ocean. IEEE Internet Things J 7 (10):9678–9687. https://doi.org/10.1109/JIOT.2020.2991025
8. Al-Khalidi M, Al-Zaidi R, Woods J, Reed M, Pereira E (2020) Securing marine data networks in an IoT environment. In: 7th international conference on future internet of things and cloud (FiCloud). https://doi.org/10.1109/FiCloud.2019.00025
9. Inmarsat Report "Inmarsat Research Programme offers unique maritime IoT insights". https://www.inmarsat.com/en/news/latest-news/maritime/2018/inmarsat-research-programme-offers-unique-maritime-iot-insights.html. Accessed 06 June 2018
10. Fremantle P, Aziz B, Kirkham T (2017) Enhancing IoT security and privacy with distributed ledgers—a position paper. https://doi.org/10.5220/0006353903440349
11. Ferrara P, Mandal AK, Cortesi A, Spoto F (2020) Static analysis for discovering IoT vulnerabilities. Int J Softw Tools Technol Transfer 23(2):1–18. https://doi.org/10.1007/s10009-020-00592-x
12. OWASP top internet of things. https://owasp.org/www-pdf-archive/OWASP-IoT-Top-10-2018-final.pdf (Available online)
13. Androcec D, Vrcek N (2018) Machine learning for the internet of things security: a systematic review. In: Conference: 13th international conference on software technologies. https://doi.org/10.5220/0006841205630570
14. Bandyopadhyay S, Sengupta M, Maiti S, Dutta S (2011) Role of middleware for internet of things: a study. Int J Comput Sci Eng Surv 2(3). https://doi.org/10.5121/ijcses.2011.2307
15. Saadeh M, Sleit A, Sabri KE, Almobaideen W (2018) Hierarchical architecture and protocol for mobile object authentication in the context of IoT smart cities. J Netw Comput Appl 121:1–19
16. Razzaque MA, Milojevic-Jevric M, Palade A, Clarke S (2016) Middleware for internet of things: a survey. IEEE Internet Things J 3(1):70–95. https://doi.org/10.1109/JIOT.2015.2498900
17. Anand P, Singh Y, Selwal A, Alazab M, Tanwar S, Kumar N (2020) IoT vulnerability assessment for sustainable computing. Threats Curr Solutions Open Challenges 8:168825–168853. https://doi.org/10.1109/ACCESS.2020.3022842
18. Yao X, Farha F, Li R, Psychoula I, Chen L, Ning H (2020) Security and privacy issues of physical objects in the IoT: challenges and opportunities. Digital Commun Netw 2–3. https://doi.org/10.1016/j.dcan.2020.09.001
19. European Telecommunications Standards Institute. Machine-To-Machine Communications (M2M) (2019) Smart Metering Use Cases. European Telecommunications Standards Institute; Valbonne, France. ETSI TR 102 691 V1.1.1
20. Staniec K, Kowal M (2020) On vulnerability of selected IoT systems to radio jamming—a proposal of deployment practices. 20(21):6152. https://doi.org/10.3390/s20216152
21. Xiao L, Li Y, Han G, Liu G, Zhuang W (2016) PHY-layer spoofing detection with reinforcement learning in wireless networks. IEEE Trans Veh Technol 65(12):10037–10047. https://doi.org/10.1109/TVT.2016.2524258

22. Han G, Xiao L, Poor HV (2017) Two-dimensional anti-jamming communication based on deep reinforcement learning. In: IEEE int. conf. acoustics speech and signal processing, pp 1–5. https://doi.org/10.1109/ICASSP.2017.7952524
23. Doyu H, Morabito R, Brachmann M (2021) A TinyMLaaS ecosystem for machine learning in IoT: overview and research challenges. In: 2021 international symposium on VLSI design, automation and test (VLSI-DAT). https://doi.org/10.1109/VLSI-DAT52063.2021.9427352
24. Venkatraman S, Overmars A, Fahd K, Parvin S, Kaspi S (2020) Security challenges for big data and IoT, In: BDET 2020: proceedings of the 2020 2nd international conference on big data engineering and technology, pp 1–6. https://doi.org/10.1145/3378904.3378907
25. Savic M, Lukic M, Danilovic D, Bodroski Z, Bajovic D, Mezei I, Vukobratovic D, Skrbic S, Jakovetic D (2021) Deep learning anomaly detection for cellular IoT with applications in smart logistics, pp 59406–59419. https://doi.org/10.1109/ACCESS.2021.3072916
26. Rajan A, Jithish J, Sankaran S (2017) Sybil attack in IOT: Modelling and defenses. In: Conference: 2017 international conference on advances in computing, communications and informatics (ICACCI). https://doi.org/10.1109/ICACCI.2017.8126193
27. Reinsel D, Gantz J, Rydning J (2018) The digitization of the world from edge to core. Available online: https://www.seagate.com/ru/ru/our-story/data-age-2025/
28. Masdari M, Jalali M (2016) A survey and taxonomy of DoS attacks in cloud computing: DoS attacks in cloud computing. Secur Commun Netw 9(16). https://doi.org/10.1002/sec.1539
29. Doshi R, Apthorpe N, Feamster N (2018) Machine learning ddos detection for consumer internet of things devices. IEEE Security and Privacy Workshops (SPW), pp 29–35. https://doi.org/10.1109/SPW.2018.00013
30. Ye J, Cheng X, Zhu J, Feng L, Song L (2018) A ddos attack detection method based on svm in software defined network. Secur Commun Netw 4:8. https://doi.org/10.1155/2018/9804061
31. Mehta A, Kaur Sandhu J, Sapra L (2020) Machine learning in wireless sensor networks: a retrospective. In: 2020 sixth international conference on parallel, distributed and grid computing (PDGC). https://doi.org/10.1109/PDGC50313.2020.9315767
32. Meng W (2018) Intrusion detection in the era of iot: Building trust via traffic filtering and sampling. Computer 51:36–43. https://doi.org/10.1109/MC.2018.3011034
33. Li J, Zhao Z, Li R, Zhang H, Zhang T (2019) Ai-based two-stage intrusion detection for software defined iot networks. IEEE Internet Things J. https://doi.org/10.1109/JIOT.2018.2883344
34. Samie F, Bauer L, Henkel J, Chen Z, Ma N, Liu B (2019) From cloud down to things: an overview of machine learning in internet of things. IEEE Internet Things J (IoT-J):4921–4934. https://doi.org/10.1109/JIOT.2019.2893866
35. Mohammadi S, Desai V, Karimipour H (2018) Multivariate mutual information-based feature selection for cyber intrusion detection. In: IEEE electrical power and energy conference (EPEC), pp 1–6. https://doi.org/10.1109/EPEC.2018.8598326
36. Apple inc., HomeKit. (2021). https://developer.apple.com/homekit/
37. Nebbione G, Carla Calzarossa M (2020) Security of IoT application layer protocols: challenges and findings. Future Internet 12(3):55, 20. https://doi.org/10.3390/fi12030055
38. LiJinshu P, Wang S (2019) iTLS/iDTLS: lightweight end-to-end security protocol for IoT through minimal latency. In: The ACM SIGCOMM 2019 conference posters and demos. https://doi.org/10.1145/3342280.3342347
39. Ivanov O, Ruzhentsev V, Oliynykov R (2018) Comparison of modern network attacks on TLS protocol. In: Proceedings of the 2018 international scientific-practical conference problems of infocommunications. Science and Technology (PIC S&T), pp 565–570. https://doi.org/10.1109/INFOCOMMST.2018.8632026
40. NATIONAL VULNERABILITY DATABASE. CVE-2019-11779 Detail, 19 Sept. 2019. Available online: https://nvd.nist.gov/vuln/detail/CVE-2019-11779
41. DDS SECURITY SPECIFICATION VERSION 1.1, Jul. 2018. Available online: https://www.omg.org/spec/DDS-SECURITY/

42. White R, Caiazza G, Jiang C, Ou X, Yang Z, Cortesi A, Christensen H (2019) Network reconnaissance and vulnerability excavation of secure DDS systems. In: Proceedings of the 2019 IEEE European symposium on security and privacy workshops (EuroS&PW), pp 57–66. https://doi.org/10.1109/EuroSPW.2019.00013
43. CVE-2019-1845 (2019). Available online: https://cve.mitre.org/cgi-bin/cvename.cgi?name=CVE-2019-1845
44. Kim TH, Douglas Reeves A survey of domain name system vulnerabilities and attacks. J Surveill Secur Safety. Available online: https://jsssjournal.com/article/view/3660
45. Soe YN, Feng Y, Santosa PI, Hartanto R, Sakurai K (2020) Machine learning-based IoT-Botnet attack detection with sequential architecture. https://doi.org/10.3390/s20164372
46. Xiao L, Li Y, Huang X, Du XJ (2017) Cloud-based malware detection game for mobile devices with offloading. IEEE Trans Mobile Comput 16(10):2742–2750. https://doi.org/10.1109/TMC.2017.2687918
47. Muthuswamy S (2019) Malware detection through machine learning techniques. Int J Adv Trends Comput Sci Eng 8(5):2408–2413 (October 2019). https://doi.org/10.30534/ijatcse/2019/82852019
48. Irani D (2011) Reverse social engineering attacks in OnLine social networks. Detection of intrusions and malware, and vulnerability. In: Assessment 8th international conference DIMVA. https://doi.org/10.1007/978-3-642-22424-9_4
49. Kumar A, Chaudhary M, Kumar N (2015) Social engineering threats and awareness: a survey. Eur J Adv Eng Technol 15–19
50. Kiruthiga R, Akila D (2019) Phishing websites detection using machine learning. Int J Adv Comput Sci Appl 10:7. https://doi.org/10.35940/ijrte.B1018.0982S1119
51. Meenu SG (2019) Phishing detection using machine learning techniques. Int J Eng Adv Technol 9:2

Enroute Alternate Planning Minima Determination Method Based on METAR Database Analysis

Konstantin Kuts and Gennadii Kovalenko

Abstract The random processes analysis of meteorological conditions changes is carried out: the ceiling and visibility range at the enroute alternates Nizhnevartovsk and Bratsk, selected when planning Russian airlines Far Eastern flights. The random processes analysis is based on the meteorological aerodrome half-hour observations database in 2017 and 2018 for the direct and return flights periods to the Russian Far East. Based on the analysis, the ceiling and visibility regression forecast on direct and return flights, necessary to determine the enroute alternate planning minima, is made. The existing methods shortcomings for the enroute alternate planning minima determining used in the Russian Federation, the United States, the European Union and other countries are indicated. A new enroute alternate planning minima determination method based on a meteorological condition changes regression forecast is proposed. The new method also can be used for North Atlantic, Pacific and other long-range routes and helps the airlines increase their flights regularity by increasing the amount of enroute alternate available days in year. The ceiling and visibility increments calculated using the new method must be used regardless of the enroute alternate area of operation radius.

Keywords Planning minima · Ceiling · Visibility · Enroute alternate · Regression forecast · EDTO

1 Introduction

During flight planning, the flight crew and the flight dispatcher select enroute alternates to ensure flight safety. The enroute alternate is determined based on the predicted meteorological conditions correspondence at these alternates to the

K. Kuts (✉) · G. Kovalenko
Saint Petersburg State University of Civil Aviation, Pilotov str., 38, Saint Petersburg 196210, Russian Federation
e-mail: pilotkuts@ya.ru

A. Bieliatynskyi and V. Breskich (eds.), *Safety in Aviation and Space Technologies*, Lecture Notes in Mechanical Engineering,
https://doi.org/10.1007/978-3-030-85057-9_46

planning minima during the period of its possible use in flight. Planning minima—the minimum ceiling and visibility values that the aerodrome must meet when choosing it as an alternate.

According to the International Civil Aviation Organization (ICAO) requirements, the enroute alternate planning minima are defined as the sum of the operating minima and the ceiling and visibility increments. According to the ICAO standard, the increments are set by the state of the operator [1–3].

2 Existing Methods for Planning Minima Determination

An important condition for ensuring flight safety is high-quality pre-flight training, one of the main elements is the weather conditions assessment along the flight route, in particular on the enroute alternates. The enroute alternate planning minima determination for airlines in the Russian Federation is carried out in accordance with the Federal Aviation Regulations (FAR-128) as shown in Table 1. This enroute alternate planning minima determination method corresponds to FAA

Table 1 FAA AC120-42 and FAR-128 enroute alternate planning minima

Approach facility configuration	Alternate airport IFR weather minimum ceiling	Alternate airport IFR weather minimum visibility
For airports with at least one operational navigational facility providing a straight in non-precision approach procedure, or Category I precision approach, or, when applicable, a circling maneuver from an instrument approach procedure	Add 400 ft to the minimum descent altitude (height) MDA (H) or decision altitude (height) DA(H), as applicable	Add 1600 m to the landing minimum
For airports with at least two operational navigational facilities, each providing a straight-in approach procedure to different suitable runways	Add 200 ft to the higher DA (H) or MDA(H) of the two approaches used	Add 800 m to the higher authorized landing minimum of the two approaches used
One useable authorized Category II ILS IAP	300 ft	1200 m
One useable authorized category III ILS Instrument Approach Procedure (IAP)	200 ft	800 m or RVR 550 m

Table 2 EASA AMC20-6 enroute alternate planning minima

Approach facility	Ceiling	Visibility
Precision approach	Authorised DA(H) plus an increment of 200 ft	Authorised visibility plus an increment of 800 m
Non-precision approach or circling approach	Authorised MDA(H) plus an increment of 400 ft	Authorised visibility plus an increment of 1500 m

Advisory Circular AC120-42B [4]. In the European Union (European Aviation Safety Agency—EASA), a simpler method of planning minima determining is currently used (Table 2) [5]. Australian civil aviation authorities set own minima determination method, which differs from FAA's and EASA's [6].

The existing methods for the planning minima determining have two drawbacks:

1. The planning minima as the sum of ceiling and visibility increments and operating minima are determined only for an extended diversion time operation rule (EDTO). This means that the methods specified in Tables 1 and 2 are used only for the enroute alternate area of operation radius more than 60 min, and for the 60 min radius, increments do not need to be used [1–6].
2. Studies [7–9] show that the forecast validity depends on the time interval size between the forecast preparation and the forecast using time—the actual enroute alternate passion time. Therefore, the planning minima determining method should take this factor into account in one way or another, but none of the existing methods takes into account the forecast justification degree.

Studies [7–12] on the problems of the ceiling and visibility forecast do not address the enroute alternate planning minima determining issues, which would guarantee with a certain probability that during the flight the enroute alternate will be available to the crew according to weather conditions. Most of the studies [10–12] are devoted to the aerodrome climatic characteristics, that is, the meteorological parameters characteristics changes over a long-term period without relation to the aerodrome operating minima are determined. Statistical data show the average seasonal percentages of the selected enroute alternates non-compliance with the already set planning minima by extended diversion time operation rules (EDTO) [13]. It should be noted that the meteorological parameters variability is already taken into account in the weather forecasts used by the flight crew and the flight dispatcher when choosing the enroute alternate, but their justification, according to regulatory documents, does not exceed 80%, therefore, the need for increments does not disappear [14–18].

3 The Enroute Alternate Planning Minima Determining Method Based on the Ceiling and Visibility Increments Regression Forecast

The choice of weather-appropriate enroute alternates is most relevant for airlines operating long-range flights in regions where the enroute alternate choice is limited. For example, the airlines of the Russian Federation operating flights to the Far East often choose the same enroute alternates (Fig. 1). Similarly, the airlines operating flights in other regions, for example, in the North Atlantic, do the same.

Another feature of long-range flights is the fixed departure and arrival time of unidirectional long-range aircraft (LRA) flights. The time may have a slight variation of ± 1 h, which is primarily due to economic reasons—the daily release of day and night tracks in the North Atlantic.

Consider the airlines of the Russian Federation Far Eastern routes are operated on the LRA Boeing 777-300ER (category D, area of operation radius 180 min) from Sheremetyevo (ICAO-code: UUEE) to Vladivostok (UHWW), Khabarovsk (UHHH), Yuzhno-Sakhalinsk (UHSS), Petropavlovsk-Kamchatsky (UHPP) and back. The average direct flights departure time from Sheremetyevo is 15.00 UTC. The average return flights departure time from Far Eastern cities is 09.00 UTC. The passion time for Nizhnevartovsk USNN and Bratsk UIBB on direct flights is 18.00 UTC and 21.00 UTC, respectively. The average total flight time is 9 h, so the time period of the flight is divided into 3 equal intervals.

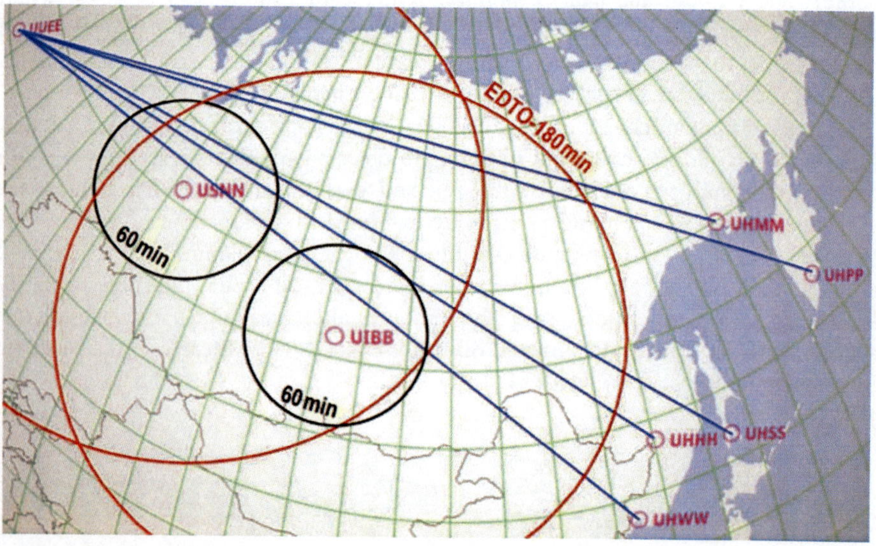

Fig. 1 Enroute alternates Nizhnevartovsk USNN and Bratsk UIBB location, selected when planning Far Eastern flights on the Boeing 777

Consider the increments changing processes in $\Delta h = \Delta h(t)$ and visibility $\Delta l = \Delta l(t)$, required to determine the planning minima during the LRA flights period. To determine the increments, we use the worst enroute alternate operating minima (*hэ*, *lэ*)—the circling minima [19, 20]. In practice, the flight crew will use the approach with a lower minimum, so using the Circle-to-Land minima in the calculations will put an additional margin in the increments. Circling minima for category D LRA *hop* = 700 ft for USNN, *hop* = 1100 ft for UIBB, *lop* = 3600 м for USNN и UIBB. For the purposes of developing the method, we will consider the weather conditions for years 2017 and 2018. The ceiling and visibility increments will be determined by the following formulas:

$$\begin{aligned} \Delta h(t_i) &= \begin{cases} 0, & if h(t_i) \geq h_{op} \\ h_{op} - h(t_i), & if h(t_i) < h_{op} \end{cases}, \\ \Delta l(t_i) &= \begin{cases} 0, & if\ l(t_i) \geq l_{op} \\ l_{op} - l(t_i), & if\ l(t_i) < l_{op} \end{cases}, \end{aligned} \tag{1}$$

where h_{op}—operating minima MDA(H), l_{op}—operating minima visibility, *h*(*ti*)—ceiling in a half-hour METAR, *l*(*ti*)—visibility in a half-hour METAR.

Using Microsoft Excel 2019 and the meteorological aerodrome report (METAR) archive at the USNN and UIBB aerodromes for 2017 and 2018 [21], we will build tables for analyzing a random processes ensemble of increment changes at each enroute alternate separately for 2017 and 2018 (Table 3). The interval for registering the process value is 0.5 h, similar to METAR. For each ensemble section, we determine the mean-mathematical expectation $M[\Delta h(ti)]$, $M[\Delta l(ti)]$ and the standard deviation (SD) $\sigma[\Delta h(ti)]$, $\sigma[\Delta l(ti)]$. The process review interval corresponds to the time interval from departure to the corresponding enroute alternate passion.

Consider a sample random process ensemble cross-section of changing the ceiling and visibility increments. The increments distribution increments in the cross section are performed by Excel using the Scott formula, which takes into account the sample SD ratio. The ensemble cross-section typical histogram, shown in Fig. 2 has the highest frequency in the interval where the average values taken as

Table 3 The Excel 2019 table fragment for analyzing the ceiling increments changing processes in feet for 2018 at the Nizhnevartovsk aerodrome USNN

Date	Direct flight (Tdirect + 3 h)			Date	Return flight (Treturn + 6 h)		
	15:00	18:00	9:00		9:30	14:30	15:00
19.10.18	200	100	19.10.18	0	200	100	200
20.10.18	0	0	19.10.18	0	0	200	0
Average $M[\Delta h(ti)]$	11.2	15.9		5.5	7.4	12.1	11.2
SD $\sigma[\Delta h(ti)]$	60.3	80.3		46.6	51.1	64.8	60.3

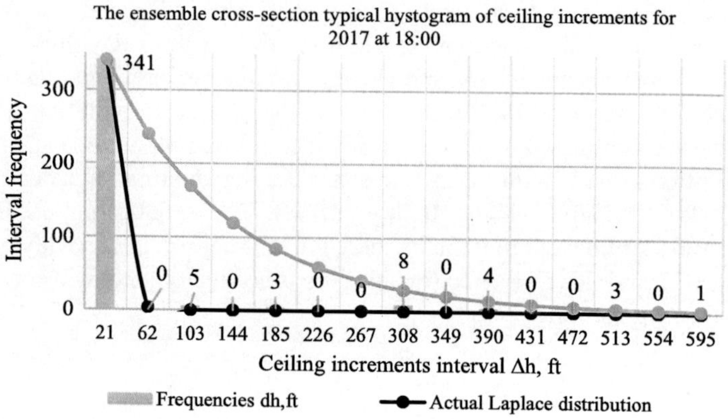

Fig. 2 The random process ensemble cross section typical histogram of ceiling and visibility increments changing

mathematical expectations fall. The distribution density in the cross section can be defined as normal with an axis on the sample average value. According to the Kolmogorov and Pearson agreement criterions, it can be represented as Exponential Laplace law at the significance level of 10^{-5} [22].

Exponential Laplace distribution for ceiling increment is represented by formulas:

$$f(\Delta h) = \frac{\lambda}{2} e^{-\lambda|\Delta h - m_{\Delta h}|} \tag{2}$$

$$\lambda = \frac{\sqrt{2}}{\sigma_{\Delta h}} \tag{3}$$

Similar formulas can be written to visibility increment. As shown in the Fig. 2, the estimated SD ($\lambda = 0.0085$) will be at least 10 times greater than the actual SD ($\lambda = 0.11$) which makes the forecast more reliable. Therefore, to build a linear forecast, we take 2SD for Laplace law, which will give 0.963 probability for a deviation in single direction. Thus, the ceiling and visibility increments conditional mathematical expectation for building linear forecast can be calculated using the following formulas:

$$\Delta hcond(ti) = M[\Delta h(ti)] + 2\sigma[\Delta h(ti)] \tag{4}$$

$$\Delta lcond(ti) = M[\Delta l(ti)] + 2\sigma[\Delta l(ti)] \tag{5}$$

A 0.963 probability will be sufficient, since from the point of view of the determining increments safety, deviations are estimated only for the worse. In addition, most of the values in the ensemble are 0, so if we consider the real increments negative values, the density axis will go far into the negative values area. Zero increment values provide even more protection when determining their values.

In order to predict a random process, it is necessary to prove its stationarity in the narrow sense—the presence of identical distribution densities in the ensemble cross section. We perform the Fisher and Student tests to check the changing increments process stationarity in the broad and narrow senses. The tests were performed for the time series ensemble cross-section average values at a significance level of 0.01 (Table 4).

The results of the Fisher and Student tests for ceiling and visibility increments changing process can be represented as stationary only in a narrow sense, therefore, there is a trend in the processes and it is possible to perform the process linear forecast according to a conditional mathematical expectation, calculated for a probability of 0.963 [23, 24]. Using Excel 2019, we will perform regression forecasts of the ceiling and visibility increments changes processes. The obtained increment values for direct and return flights will be reflected in the forecast graphs (Figs. 3, 4, 5, 6, 7, 8, 9 and 10).

The increments should be rounded up to 100 ft and 100 m, respectively. It is advisable to use the obtained increment values to determine the planning minima for the expected landing system at the aerodrome. This is because often the most accurate landing system may not work or the wind direction does not allow the most accurate landing system to be used. It is also advisable to set the increments for the next year based on the current year data and adjust the increments for the next year based on the current year results (Table 5).

Let us evaluate the effectiveness of the obtained increment values for determining the planning minima by comparing the number of days when the aerodrome is unavailable according to the planning minima calculated according to the current method given in Table 1 and the minima calculated according to the new method in total for ceiling and visibility for 2017 and 2018, noting that the effectiveness was evaluated based on the current values, and the minima are set for the "next" year (Table 6). The planning minima calculation will be made for the most accurate instrument landing system (ILS).

4 Conclusion

A new enroute alternate planning minima determination method based on the meteorological aerodrome reports database analysis will allow us to establish reasonable increments values, thereby increasing the enroute alternate choice reliability, which will help prevent aviation accidents associated with the closure of the selected enroute alternate [25]. According to the current planning minima determination method (FAR-128, FAA AC120-42), the average number of days per year when the enroute alternate is unavailable can reach 30% in the North Atlantic and even 70% in the North Pacific region [13]. The new increments determination method will increase the efficiency of using the enroute alternate due to the fact that

Table 4 Fischer's F-test and Student's t-test with the same variances results in Excel 2019 for ceiling increments time series

Parameter	Nizhnevartovsk USNN				Bratsk UIBB			
	2017		2018		2017		2018	
	Direct	Return	Direct	Return	Direct	Return	Direct	Return
F-value	0.0104	0.0096	0.0027	0.0004	0.0008	0.0005	0.0004	0.0017
F-critical area	±0.1181	±0.2407	±0.1181	±0.2407	±0.2407	±0.1181	±0.2407	±0.1181
t-value	−41.33	−22.2	−17.52	−6.37	−18.23	−14.38	−18.23	−23.713
t-critical area	±3.0545	±2.7969	±3.0545	±2.7969	±2.7969	±3.0545	±2.7969	±3.0545

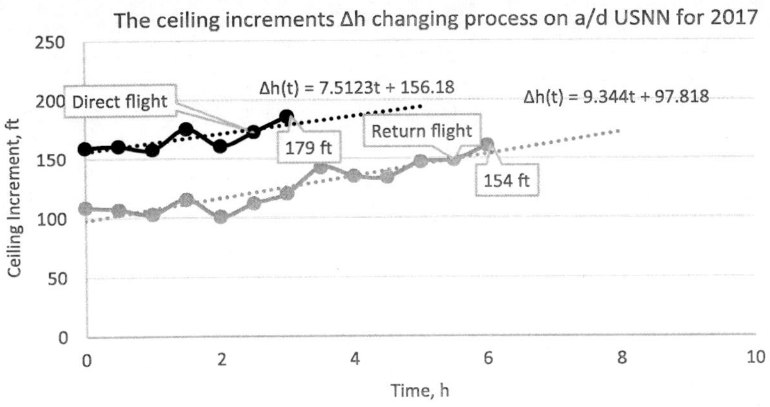

Fig. 3 The ceiling increments Δh changing process on a/d USNN for 2017

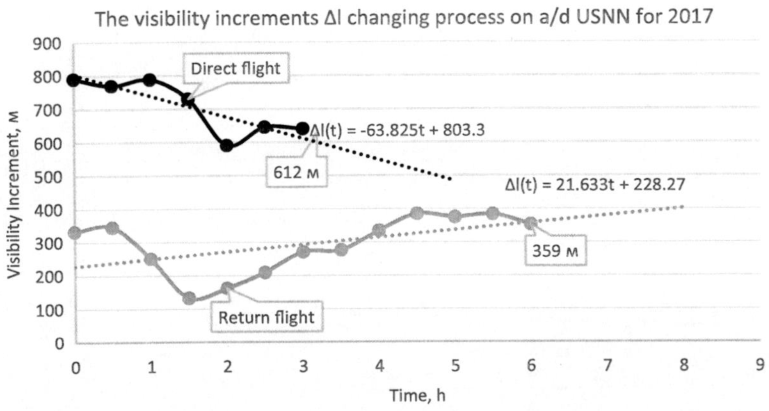

Fig. 4 The visibility increments Δl changing process on a/d USNN for 2017

the number of days in a year when the enroute alternate is not available will be reduced by at least 1.5 times (see Table 6), so it is advisable to use it not only on the Far Eastern routes of Russian airlines, but also on the North Atlantic, Pacific and other long-range routes. It is important to note that the increments calculated in the article must be used when planning LRA flights, regardless of the enroute alternate area of operation radius.

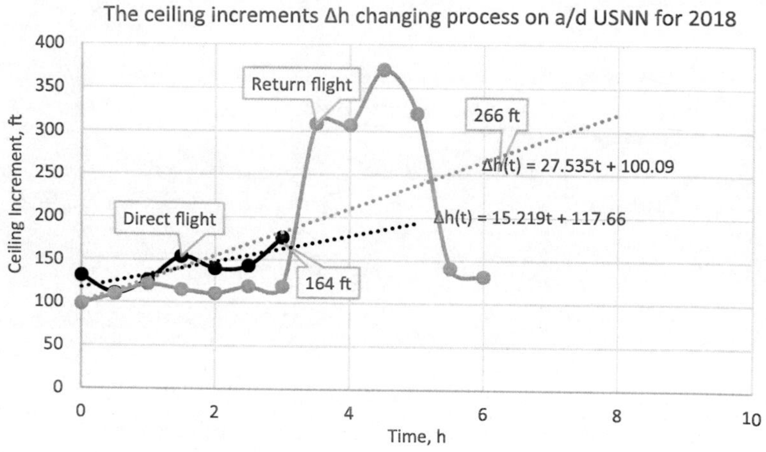

Fig. 5 The ceiling increments Δh changing process on a/d USNN for 2018

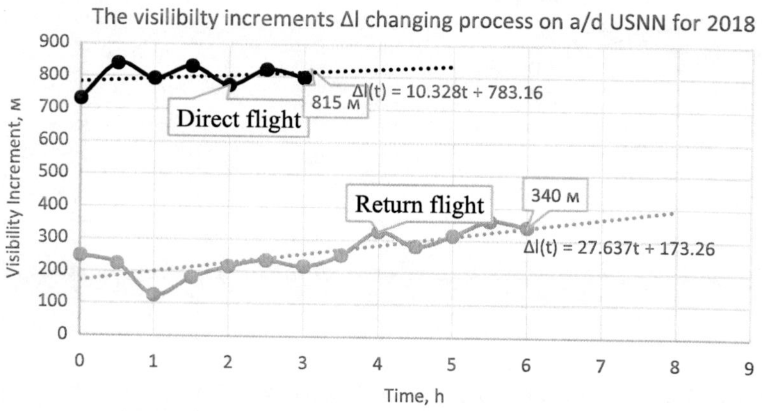

Fig. 6 The visibility increments Δl changing process on a/d USNN for 2018

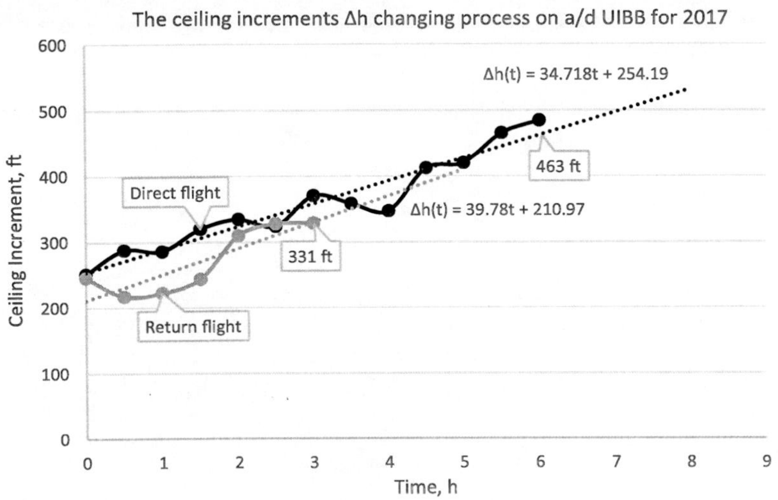

Fig. 7 The ceiling increments Δh changing process on a/d UIBB for 2017

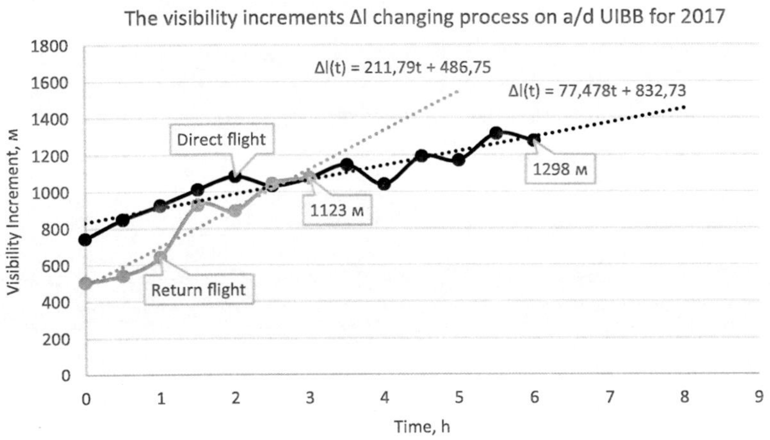

Fig. 8 The visibility increments Δl changing process on a/d UIBB for 2017

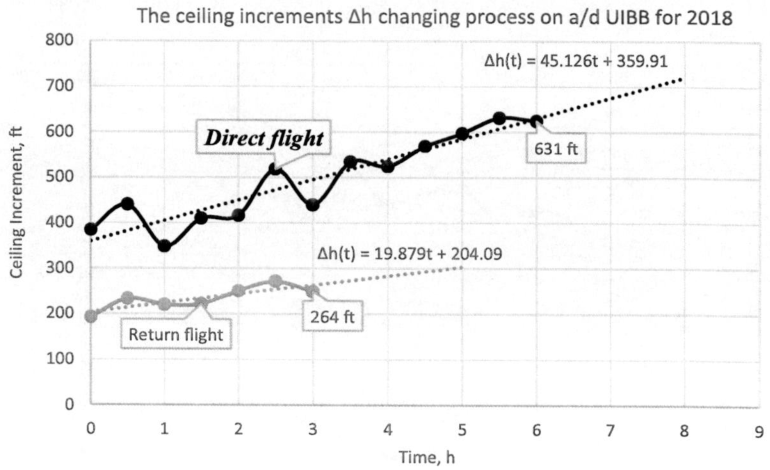

Fig. 9 The ceiling increments Δh changing process on a/d UIBB for 2018

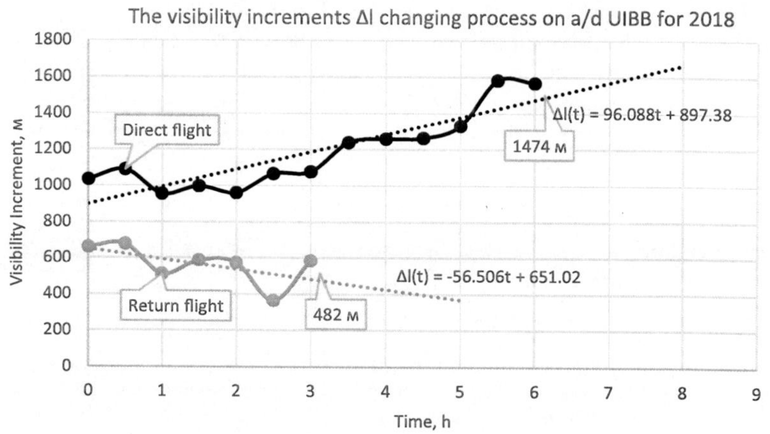

Fig. 10 The visibility increments Δl changing process on a/d UIBB for 2018

Table 5 Enroute alternate USNN and UIBB planning minima

Approach facility	Nizhnevartovsk USNN				Bratsk UIBB			
	Direct flight		Return flight		Direct flight		Return flight	
	Ceiling	Visibility	Ceiling	Visibility	Ceiling	Visibility	Ceiling	Visibility
2018 (based on 2017 data)	+200 ft to DA (H) or MDA (H)	+700 m to visibility minima	+200 ft to DA (H) or MDA (H)	+400 m to visibility minima	+500 ft to DA (H) or MDA (H)	+1300 m to visibility minima	+400 ft к DA (H) or MDA (H)	+1200 m to visibility minima
2019 (based on 2018 data)	+200 ft to DA (H) or MDA (H)	+900 m to visibility minima	+300 ft to DA (H) or MDA (H)	+400 m to visibility minima	+700 ft to DA (H) or MDA (H)	+1500 m to visibility minima	+300 ft к DA (H) or MDA (H)	+500 m to visibility minima

Table 6 The effectiveness evaluation of the new enroute alternate planning minima determination method for 2017 and 2018 on direct and return flights

2017–2018	USNN 200 ft × 550 m + new increments	USNN 400 ft × 2150 m FAR-128	UIBB 200 ft × 720 m + new increments	UIBB 400 ft × 2320 m FAR-128
Average amount of days in a year when enroute alternate is unavailable	9	24	27	40

References

1. Annex 6 to Convention on international civil aviation organization (2016) Operation of aircraft. Part I. International commercial air transport. Aeroplanes. Tenth edition. Montreal: ICAO. ISBN 978-92-9249-971-6. p 64. p 4.3.5. Meteorological conditions
2. Doc 9976 (2015) Flight planning and performance manual (FPFM). Montreal: ICAO, First edition. ISBN 978-92-9249-838-2. p 63. p 4.14. Alternate aerodrome planning minima—establishing incremental values for ceiling and visibility. https://unitingaviation.com/livecycle/Documents/ICAO_Doc_9976-1_EN.pdf
3. Doc 10085 (2017) Extended Diversion Time Operations (EDTO) manual. Montreal: ICAO, first edition. ISBN 978-92-9258-327-9. p 74. 3.5.2.4 EDTO alternate aerodrome weather minima. https://www.icao.int/MID/Documents/2020/EDTO%20Workshop/10085_cons_en.pdf
4. FAA Advisory Circular AC120-42B (2008) «Extended Operations (ETOPS and Polar Operations)». U. S. Department of Transportation, Federal Aviation Administration. https://www.faa.gov/regulations_policies/advisory_circulars/index.cfm/go/document.information/documentID/73587

5. EASA Acceptable Means of Compliance AMC20-6 (2010)«Extended range operation with two-engine aeroplanes ETOPS certification and operation». European Aviation Safety Agency. https://www.easa.europa.eu/sites/default/files/dfu/Annex%20II%20-%20AMC%2020-6.pdf
6. Civil Aviation Order 82.0 (2014) Air Operators' Certificates—applications for certificates and general requirements Instrument 2014—Amendment Order. Australian Government, Civil Aviation Safety Authority, Federal Register of Legislative Instruments F2015C00204. p 7. Standard EDTO alternate aerodrome planning minima. https://www.legislation.gov.au/Details/F2015C00204/7154b981-4ee4-40a1-8bbc-a705cd57b096
7. Chaabani H, Werghi N, Kamoun F, Taha B, Outay F, Yasar A-U-H (2018) Estimating meteorological visibility range under foggy weather conditions: a deep learning approach. Procedia Comput Sci 141:478–483. https://doi.org/10.1016/j.procs.2018.10.139
8. Rossa A, Liechti K, Zappa M, Bruen M, Germann U, Haase G, Keil C, Krahe P (2011) The COST 731 action: a review on uncertainty propagation in advanced hydro-meteorological forecast systems. Atmos Res 100(2–3):150–167. https://doi.org/10.1016/j.atmosres.2010.11.016
9. Hogan RJ, O'Connor EJ, Illingworth AJ (2011) Verification of cloud-fraction forecasts. Q J R Meteorol Soc 135(643):1494–1511. Part B. https://doi.org/10.1002/qj.481
10. Vetrova EI, Skriptunova EN, Shakina NP (2013) Low clouds and their forecast at the airports of the European part of the former USSR. Russ Meteorol Hydrol 1(38):6–19. https://doi.org/10.3103/S1068373913010020
11. Ivanova AR, Skriptunova EN (2018) Variations in several climatological characteristics at aerodromes in the Russian Federation in 2001–2015. Russ Meteorol Hydrol 5(43):302–312. https://doi.org/10.3103/S1068373918050047
12. de Oliveira MW, Eufrásio ABR, Guterres MX, Murça MCR, de Arantes Gomes R (2021) Analysis of airport weather impact on on-time performance of arrival flights for the Brazilian domestic air transportation system. J Air Transp Manag 91:101974. https://doi.org/10.1016/j.jairtraman.2020.101974
13. ETOPS Guide Volume III «Operational Guidelines and Methods» (2009) Flight Operations Engineering Boeing Commercial Airplane Group Seattle, Washington. USA, rev. 2, 4.33
14. Annex 3 to Convention on international civil aviation organization (2016) Meteorological service for international air navigation, 19th edn. Montreal. ICAO. 64. 55. p. 6.2. Aerodrome forecasts
15. Doc 8896 (2011) Manual of aeronautical meteorological practice. Montreal: ICAO, ninth edition. 43. p. 3.2. Accuracy of aeronautical meteorological forecast. https://www.skybrary.aero/bookshelf/books/2506.pdf
16. Günter Mahringer (2012) The role of meteorological forecast verification in aviation. https://www.skybrary.aero/bookshelf/books/2331.pdf
17. Hostache R, Matgen P, Montanari A, Montanari M, Hoffmann L, Pfister L (2011) Propagation of uncertainties in coupled hydro-meteorological forecasting systems: A stochastic approach for the assessment of the total predictive uncertainty. Atmos Res 100(2–3):263–274. https://doi.org/10.1016/j.atmosres.2010.09.014
18. Buxi G, Hansen M (2013) Generating day-of-operation probabilistic capacity scenarios from weather forecasts. Transp Res Part C: Emerg Technol 33:153–166. https://doi.org/10.1016/j.trc.2012.12.006
19. Doc 9365 (2017) Manual of all-weather operations. Montreal. ICAO, 4th edn. 86. p 6.4.7. Circling approach minima
20. David Owens (2011) The circling approach. Safety first. https://safetyfirst.airbus.com/app/themes/mh_newsdesk/documents/archives/the-circling-approach.pdf
21. Metar/Speci/Taf reports selection query. Professional information about meteorological conditions in the world. https://ogimet.com/metars.phtml.en
22. Hussain SA, Ahmad I, Saghir A, Aslam M, Almanjahie IM (2021) Mean ranked acceptance sampling plan under exponential distribution. Ain Shams Eng J. https://doi.org/10.1016/j.asej.2021.03.008

23. Qiqing Yu (2020) The maximum likelihood estimator under the exponential distribution with right-censored linear regression data. Statist Probab Lett 165:108846. https://doi.org/10.1016/j.spl.2020.108846
24. Shalabh (2013) A revisit to efficient forecasting in linear regression models. J Multivar Anal 114:161–170. https://doi.org/10.1016/j.jmva.2012.07.017
25. INCIDENT Air China #CA983 made an emergency landing Siberia due to a fire alarm, all passengers evacuated via emergency slides. https://www.airlive.net/incident-air-china-ca983-made-an-emergency-landing-siberia-due-to-a-fire-alarm-all-passengers-evacuated-via-emergency-slides/

Development of a Car Door Hinge with a Temperature-Sensitive Element

Maria Mironova, Ivan Senin, Etibar Balaev, Tatiana Konovalova, and Artem Litvinov

Abstract The article shows the study of the influence of the anthropometric characteristics of drivers and passengers on the time of evacuation from the passenger compartment of the vehicle, through the window opening of the vehicle door in an accident with fire in the event of a jammed door lock. It was determined that a large category of drivers and passengers could not be evacuated through the vehicle window opening and was determined as the most optimal way of evacuation directly from the doorway. Based on the data obtained, a new design of the car door hinge was proposed using materials with shape memory effect, which makes it possible to break the connection between the door and the body. The principles of operation of the proposed design of car door hinges are shown.

Keywords Accident · Door lock · Materials with shape memory · Car door hinges · Transport infrastructure

1 Introduction

The constantly growing number of vehicles, as well as the uneven loading of the road network, contribute to the increase in the number of accidents. The analysis of statistics in the Russian Federation showed that about a thousand people die in road accidents every year due to technical malfunctions of vehicles. A large number of health consequences of road traffic accidents and the number of deaths as a result of road accidents depends on the level of active safety of vehicles, but recently the number of victims has been growing due to insufficient provision of post-accident safety of vehicles. It should be noted that the design of the vehicle, in contrast to other components of the VADS complex, is undergoing continuous modernization, therefore, practical opportunities appear for a fairly prompt implementation of measures to improve safety.

M. Mironova (✉) · I. Senin · E. Balaev · T. Konovalova · A. Litvinov
Kuban State Technological University, Moskovskaya st., 2a, Krasnodar 350020, Russia

A. Bieliatynskyi and V. Breskich (eds.), *Safety in Aviation and Space Technologies*, Lecture Notes in Mechanical Engineering,
https://doi.org/10.1007/978-3-030-85057-9_47

The design of modern cars today is being developed taking into account the needs of users, many aspects are taken into account, including ergonomics, but still one of the main tasks of evacuating a person from the passenger compartment of a vehicle without the use of specialized devices remains unresolved (in special cases in case of an accident, door locking, vehicle fire, etc.). The collection and analysis of anthropometric characteristics for various groups of drivers and passengers is the most important task of the design process of the structure, both of the vehicle itself and of its individual elements, which explains the geometric dimensions, shapes, and materials from which certain structural elements or units of the vehicle are made.

The design of the car also provides for the evacuation of people from the vehicle interior in the event of an accident with fire when doorways are jammed, through a window opening, the shape and dimensions of which allow for evacuation. However, the analysis showed that for a large category of people in view of their age, anthropometric physical and other features, evacuation through the window opening is significantly difficult, and sometimes not possible. Moreover, such evacuation from the vehicle in the event of a fire is much more effective through the doorway [1–5].

Having studied the design features of assemblies and assembly units that ensure the connection between the door and the body, it can be concluded that such a connection is provided by means of door hinges and a door lock, which are fixation points. The existing structures do not allow to destroy this connection without human intervention using special equipment. However, in the event of an accident, when a vehicle fires, when the door lock is blocked, the existing time for evacuating people from the vehicle is limited, and evacuation through glass openings is not always possible due to the anthropometric and physical characteristics of a person, thus the doorway is the most desirable area. for emergency evacuation of people from the vehicle interior. To provide such a possibility, it is necessary to break the connection between the door and the vehicle body in the places of their fixation. So one of the solution methods is the development of a new design of automobile hinges, allowing to destroy this connection, which is due to the hinge axis connecting the movable and fixed link of the hinge, by means of the interface surface of the cylindrical surface of the axis with the cylindrical holes of the movable and fixed elements, providing one degree of freedom of rotation around the axis of the hinge, which allows you to open and close the vehicle door. Thus, solving the problem of destroying the links between the moving link and the fixed link of the automobile loop, a more efficient possibility of evacuating people from the vehicle interior, through the doorway, is provided, which will increase the likelihood of saving people and reduce the degree of injury.

It should be noted that in the Russian Federation, as well as throughout the world as a whole, the national priority is the life and health of people. Therefore, the focus of this project is very relevant, since the socio-economic damage from road accidents has a negative impact not only on the demographic indicators of Russia, but also on indicators in the production sector. All categories of the population die and get injured in road accidents, which entails a decrease in the population of the country and the level of health, life expectancy, length of the working period, distortion of the age pyramid. The safety of vehicles affects road safety in general, and the continuously modernizing design of the vehicle allows you to quickly implement measures to improve road safety [4–8].

2 Materials and Equipment

To determine the most optimal evacuation routes from the vehicle interior in the event of an emergency (in the event of a vehicle fire), anthropometric data were collected. The collection of anthropometric data was carried out for car drivers in Russia.

The study was carried out taking into account the requirements of GOST R ISO 15535-2012, GOST R ISO 7250-1-2013, GOST R 56620.2-2015/ISO/TR 7250-2: 2010, GOST R ISO 7250-3-2019.

This study involved 153 people aged 16–65 from Russia. Data was collected using a set of structured questionnaires that were completed by study participants and a control group.

The measurements were carried out on three cars of three brands, namely Volkswagen, Kia and Lada with different parameters of doors and window openings of vehicles. The data were analyzed using a two-sample t-test to determine the relationship between the time it takes to evacuate the vehicle and human anthropometry. It should be noted that the studies were carried out under normal environmental conditions.

3 Analysis of Simulation Results and Experimental Data

According to the data obtained on the basis of information from the emergency response services, the time scale of the vehicle burning, presented in Fig. 1, was established and the effect of temperature exposure on the driver and passengers is shown in Table 1.

So, according to the data in Table 1 and Fig. 1, the optimal time and temperature for evacuating people from the vehicle is the time at which the temperatures should not exceed 60–70 °C, at these temperatures the harm to health will be minimal, and therefore the desired time for evacuation is 40–50 s. At the same time, according to studies of the influence of anthropometric, age and physical characteristics of people on the speed of self-evacuation through the window opening of the vehicle presented in Table 2, there are categories of people who do not fit into this period of time. It is also worth noting that the sample being formed is preliminary and does not take into account all categories of people in terms of anthropometric and physical characteristics of people. Plus, the time spent for evacuation does not take into account possible injuries resulting from an accident, as well as the state of a

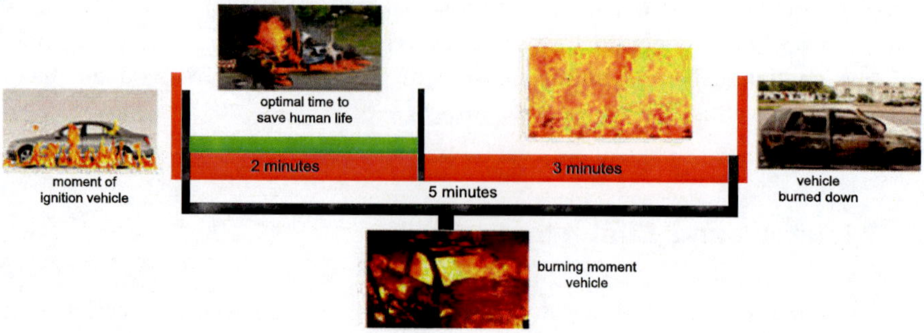

Fig. 1 Timeline of vehicle burning

Table 1 Influence of temperature exposure on the driver and passengers in the vehicle

Heating temperature of body elements (°C)	Human condition
Up to 60	Erythema (redness of the skin)
Up to 70	Vesiculation (blistering)
Up to 100	Destruction of the skin with partial preservation of capillaries
100 and more	Muscle burn. Loss of consciousness and death of a person

person, including mental, which affects the time that will be spent for taking measures, significantly increasing it.

Figure 2 shows a histogram of the time dependence of evacuation from the vehicle interior, on age through the window opening of the vehicle door and through the body door opening. According to which it can be concluded that the most effective evacuation is possible through the vehicle doorway.

At the same time, based on the analysis, it can be seen that the total time for the evacuation of people from the passenger compartment, before receiving damage to the body incompatible with life, is 100–120 s, in which the injuries and burns received are of moderate severity. At the same time, evacuation through the window opening of a trained person with a height of 183 cm, weighing 78 kg, aged 35 years of athletic build takes 35–50 s from the passenger seat and from 50 to 80 s from the driver's seat is 15–30 s, while through the window opening is impossible [8–10].

After analyzing the data obtained, it can be concluded that the evacuation of people is possible through the window opening, however, it is very difficult if we take into account the speed of flame propagation during a fire in an emergency, since the time spent on evacuation from the vehicle exceeds the time at which loss of consciousness occurs from high temperature. The vehicle doorway is the most optimal and desirable area for evacuation in an emergency. To provide such an

Table 2 Influence of anthropometric data of the driver and passengers on potential escape routes from the passenger compartment of a motor vehicle

	Gender	Body weight (kg)	Height (person's height, m)	Chest width (cm)	Hips width in sitting position (cm)	Shoulder joints width (shoulder width, cm)	Waist width (cm)	Time of evacuation from the vehicle through the window opening (s)	Time of evacuation from the vehicle interior through the doorway (s)	Time of evacuation from the vehicle interior through the sunroof (s)
16–25	Male	60–71	165–170	76.5–108	85–95.9	48.5–54.5	69.8–84	35.3	3.25	Not investigated
16–25	Female	49–65	158–169	86.3–94.6	92,8–102.3	44.8–46.3	63.6–78.9	35.1	3.17	Not investigated
25–35	Male	68–76	173–179	96–113	95,9–108.9	49–52.5	77.6–92	38.6	3.51	Not investigated
25–35	Female	52–65	161–173	91,7–100.2	98.9–105.6	45.9–51.2	69.8–84	38.3	3.56	Not investigated
35–45	Male	68–81	168–175	108–118	103.6–121.9	54.8–56	83.7–96	38.9	4.09	Not investigated
35–45	Female	52–68	161–169	98.3–113	103.6–108.9	46.7–51.2	77.8–87	40.02	4.00	Not investigated
45–55	Male	70–85	168–170	112–124	118.1–124.4	55.3–58	89.6–96	43.02	5.83	Not investigated
45–55	Female	52–68	161–169	100.3–115.6	118.6–125.7	48.9–55,2	81,8–89.9	43.11	5.76	Not investigated
55–65	Male	70–85	165–169	96–113.6	124.5–128.4	55.9–58	91.3–101.7	Difficult or not possible	8.14	Not investigated
55–65	Female	67–83	161–165	115.3–120	125.3–130	48.9–55.2	87.6–90.9	Difficult or not possible	8.06	Not investigated

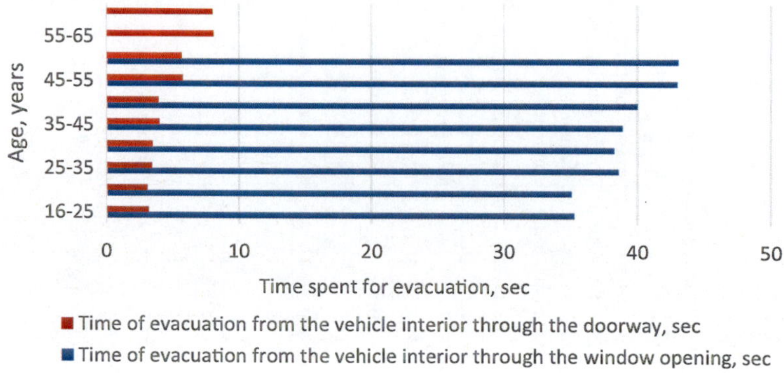

Fig. 2 Time dependence of evacuation from the vehicle interior, on age

opportunity, it is necessary to destroy the bonds between the door and the vehicle body in the places of their fixation. So, one of the solutions is the development of a new design of automobile hinges, where, in order to ensure the loss of communication between the movable and fixed link and, as a consequence, the vehicle body and door, it was decided to remove the axle from the coupling with one of the hinge links. For this, it was decided to use materials with thermoelastic phase transformations, which are also called shape memory materials. This is due to the fact that this class of materials during phase transformations retains short-range order and the proximity of atoms, which, in turn, allows restoring the shape of an article characteristic of a given phase state. Thus, this material can be both a sensor that records the temperature change inside the car, and the actuator (drive) itself, which will deform and provide an exit from the hinge axis mating from the opening of the movable link, after which the door can be knocked out from the side of the hinges. Combining these registration and execution functions improves the reliability of the system and greatly reduces the likelihood of error. To ensure the exit from the mating of the axis and the hole of the fixed link, it is necessary to make the axis composite, ensuring the possibility of retraction of the axle shafts towards the moving link, thereby breaking the connection between the fixed and the moving link by means of the axis. However, in this case, it is necessary to provide a structure that guarantees constant mating of the axle shafts with the holes of the fixed link in normal operation. The solution to this design problem is shown in Figs. 3 and 4.

So Fig. 3 shows one of the variants of a two-link door hinge of a car with a temperature-sensitive element made in the form of a spring, which, in the event of a fire, will deform, compressing and tightening the two axle shafts, taking them out of conjugation with a fixed link, thus, the connection between with a fixed and movable link, which will allow with the application of little effort to knock out the car door, thereby freeing the most effective escape route.

In this case, in normal operation, the axle shafts have a position that ensures constant coupling of the axle shafts and the holes of the fixed link. Thus, the

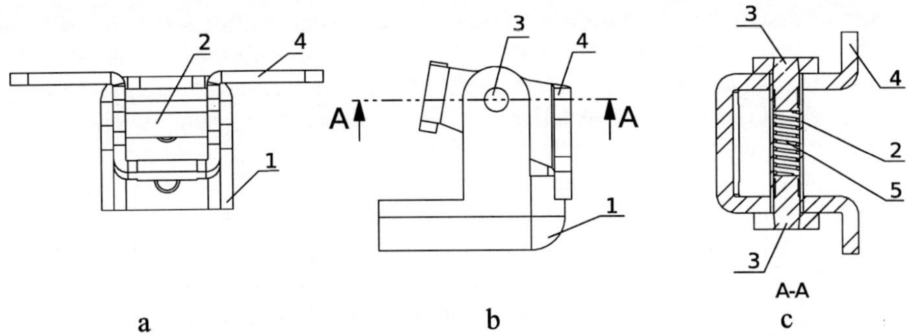

Fig. 3 Two-link car door hinge with a temperature-sensitive element made in the form of a spring: **a** top view, **b** side view, section A-A. 1—fixed link; 2—tube; 3—axle shafts; 4—movable link; 5—spring

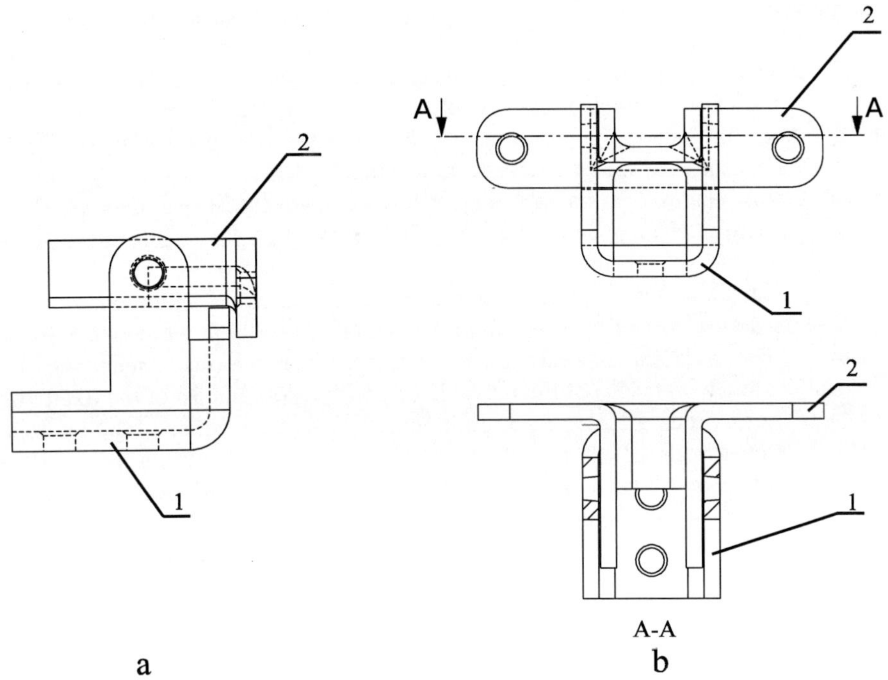

Fig. 4 Two-link car door hinge with a temperature-sensitive element made in the form of a spring: **a** side view, **b** front view and its section A-A. 1—fixed link; 2—movable link

semiaxes have two extreme positions in the tube—initial and working. In the initial position, the tips of the semiaxes are mated along the entire length of the tapered holes of the fixed link, while the fixed and movable links of the two-link car door hinge are connected. In the working position, the ends of the semiaxes do not protrude above the end planes of the tube and the outer end planes of the movable link, becoming flush with them, while the fixed and movable link of the two-link car door hinge are not connected. Both positions of the axle shafts are provided and fixed by a spring made of material with a high-temperature shape memory effect, the ends of which are inserted into the holes of the axle shafts head. Possible deformation of the spring during heating as a result of martensitic-austenitic transformation exceeds the total length of the tapered parts of the tips of the semiaxes, which ensures their guaranteed removal from the mating with the moving link. The conical shape of the tips of the pins along the length of the mating with the hole of the fixed link and the holes of the fixed link ensures their quick removal from the mating, since there will be no sliding friction forces [8, 9].

So Fig. 4 shows the second version of a two-link door hinge of a car with a thermosensitive element, which is a movable link, but a combination of manufacturing both a fixed link and both links is also possible, where the axle shafts are made on the surface of the movable link mating with the fixed link. Manufacturing the movable link of a two-link car door hinge or the movable and fixed links of a two-link car door hinge from a material with a high-temperature shape memory effect allows it to be instantly disconnected in the event of a fire in the car, regardless of its position in an accident. A link made of materials with a high-temperature shape memory effect at ordinary temperature has a martensitic state, when heated, a martensitic-austenitic transformation occurs, as a result of which a link made of a material with a high-temperature shape memory effect begins to deform, taking on a predetermined shape corresponding to the austeite state of the material. In the event of a fire in the car, a moving link made of a material with a high-temperature shape memory effect or a fixed and moving link of a two-link door hinge of a car made of a material with a high-temperature shape memory effect as a result of an increase in temperature deforms, disengaging the conical bosses (half-shafts) of the movable link with tapered holes of the fixed link of the two-link car door hinge. As a result, regardless of the position of the car body (it will be on wheels or overturned on its side or roof), the movable link and the fixed link of the two-link door hinge of the car cease to be connected, and the car door, as a result, ceases to be connected to the car body and, by applying minimal effort, it will be fashionable to knock out the door, after which the timely evacuation of passengers and the driver of this vehicle is possible [10–12].

4 Conclusion

According to ongoing research, evacuation through a doorway in the event of a car fire is most effective with other escape routes. However, in the event of a jammed door lock, the preferred escape location is not available. At the same time, the sooner people are evacuated from the vehicle, the less damage caused to their health. To ensure such a possibility of evacuation in the event of jamming of the vehicle door lock, hinges have been developed containing an element in the structure of a material with a shape memory effect, which makes it possible to provide instant disassembly of the car door hinge, by removing the elements performing the function of an axis (semi-axis) from the hole of the fixed link from the interface, when temperature exposure. Thus, a violation of the connection between the door and the vehicle body from the hinge side is ensured and when a slight force is applied, the door is knocked out, since the door lock design does not work well for bending and tensile loads in the transverse direction. All this allows you to quickly clear the doorway for the timely quick evacuation of people from the vehicle [13–15].

References

1. Balaev EY, Konovalova TV, Litvinov AE, Nadiryan SL, Senin IS (2020) Optimization of freight traffic in cities using the forecast transport model of freight traffic. In: IOP conference series: materials science and engineering, 012045
2. Konovalova TV, Nadiryan SL, Litvinov AE (2018) Transport systems of cities in the processes of new industrialization. Adv Soc Sci Edu Hum Res 590–593
3. Konovalova TV, Kotenkova IN (2013) Transport infrastructure. Kuban State Technological University, Krasnodar
4. Sukharev OS (2015) Industrial policy and the development of transport infrastructure in Russia. National interests: priorities and security. L, 11, 1, 286, 2–20
5. Kravchenko P, Oleshchenko E (2016) Mechanisms of functional properties formation of traffic safety systems. In: 12th international conference organization and traffic safety management in large cities, pp 367–372
6. Konovalova T, Zarovnaya L (2016) The assessment model for economic efficiency of traffic safety improvements. In: 12th international conference organization and traffic safety management in large cities, pp 311–315
7. Kurakina E, Evtyukov S (2018) Results of studying road construction parameters condition. Archit Eng 3(1):29–37
8. George R, Jana I, Joseph K (2013) International student's traffic engineering project seminar meps. World Transp Technol Mach 4(43):106–111
9. Balaev EY (2018) Patent 184114, IPC E05D 7/00 (2006.01), SPK E05D 7/00 (2018.08)
10. Balaev EY (2018) Patent 184565, IPC E05B 7/00 (2006.01), SPK E05D 7/00 (2018.08)
11. TR CU 018/2011. Technical Regulations of the Customs Union. On the safety of wheeled vehicles Uniform provisions concerning the approval of vehicles with regard to door latches and door retention components, Addendum 10: Regulation, 11, Revision 2
12. Uniform provisions concerning the approval of vehicles with regard to door latches and door retention components, Addendum10: Regulation, 11, Revision 3
13. Pugachev I, Kulikov Y, Markelov G, Sheshera N (2016) Factor analysis of traffic organization and safety systems. In: 12th international conference "Organization and traffic safety management in large cities", SPbOTSIC-2016, pp 529–535
14. Antonyan ON, Kostryukova NV (2011) Motivation as a factor in increasing labor productivity in a crisis. In: Materials of the XXIII intra-university scientific and practical conference. Ministry of Education and Science of the Russian Federation; Volgograd State University of Architecture and Civil Engineering, pp 469–472
15. Notchenko VV, Lineva EL (2015) Assessment of the transport infrastructure of the region as a necessary condition for entrepreneurship development. St. Petersburg State Polytechnic university. J Econ 211:172–181. https://doi.org/10.5862/JE.211.1

Mathematical Model of Multi-item Spare Parts Reservation

Veniamin Shilovsky, Igor Skobtsov, and Evgeniy Pitukhin

Abstract A mathematical model for determining the optimal number of reserved spare parts in the warehouse of an enterprise operating transport and technological machines is proved and tested. The solution of this problem is important in improving the efficiency of technical operation of machines and equipment used in any enterprise. The market conditions of management require a higher level of spare parts demand planning, a new level of spare parts supply system management. The purpose and tasks of economically justified spare parts reserves are to maintain the rolling stock in working condition in the absence or minimal downtime of machines waiting for repair due to the lack of spare parts. To do this, it is necessary to have the optimal number of required spare parts in the warehouse of the company that operates or repairs the machines. The paper is devoted to the rationale for making optimal economically sound management decisions to provide spare parts for any equipment by creating and implementing the procedure and methodology for reserving spare parts in the form of a mathematical model that takes into account a sufficient number of factors and relationships between controlled and uncontrolled variables. An example of calculating the optimal number of spare parts to reserve, de-pending on the distance to the source of spare parts, is given. The results of the calculations confirm the adequacy of the proposed mathematical model to the real production conditions.

Keywords Spare Parts · Reserve · Optimization target function

1 Introduction

One of the reasons for the increase in the cost of technical operation of machines is the imperfection of the spare parts supply system [1].

V. Shilovsky · I. Skobtsov · E. Pitukhin (✉)
Petrozavodsk State University, Lenin Ave. 33, Petrozavodsk 185910, Russia
e-mail: eugene@petrsu.ru

A. Bieliatynskyi and V. Breskich (eds.), *Safety in Aviation and Space Technologies*, Lecture Notes in Mechanical Engineering,
https://doi.org/10.1007/978-3-030-85057-9_48

The role of the supply and creation of a reserve fund of spare parts (SP) in the organization of technical service and reduction of downtime due to the lack of SP along with the work [1] are considered in [2–5]. It is proposed to determine the economically justified parameters of the multi-item reserve of the SP by minimizing the costs of reserving and using spare parts and developing an analytical expression of the target function and an algorithm for its implementation in relation to any equipment [5–8]. The issues of creating and operating a reserve fund of spare parts, taking into account logistics and their life cycle are actively being developed in the scientific and technical communities [7–9]. The organization of a rational supply of spare parts in order to reduce downtime and material losses during the technical operation of equipment is discussed in sufficient detail in [10–14]. Despite many works of domestic and foreign authors [1–17] concerning economically justified supplies and stocks of SP, insufficient attention is paid to automating the monitoring of this process and making appropriate management decisions through the creation of a set of mathematical models describing the course of real events and algorithms for their implementation. The solution of these issues is especially important for foreign expensive equipment and the corresponding cost of their SP [5, 12].

The purpose of this work is to develop and test a mathematical model for determining the parameters of optimal multi-item reservation of spare parts for deterministic and fully satisfied stationary demand.

2 Materials and Methods

The material for the study is scientific works in the field of technical operation, marketing and management of technical service and supply of spare parts for transport and technological machines, as well as the authors' own works.

2.1 *Task Statement*

There are two approaches [1, 2, 12] for determining the optimal amount of spare parts reservation (SP).

The first approach is to determine the scope of the *i*th name of the SP, at which the condition is fulfilled:

$$C(X_i) \leq C_o, \tag{1}$$

where $C(X_i)$—cost of a set of reserve elements; C_o—limit on the value of the reserve; $i = 1, 2, \ldots, n$—the number of typo nominal of SP.

The second approach is to determine the volume of the $\{X_i\}$ SP *i*th name at which the condition is met:

$$R(X_i) \geq R_o, \tag{2}$$

where $R(X_i)$—system reliability indicator; R_o—limit on the minimum acceptable level of system reliability.

The efficiency of reserving SP in the warehouse of the enterprise can be estimated by the following cost function:

$$H(X_i^*) = \sum_{i=1}^{p} \min\{C_{Xi}\}, \tag{3}$$

where X_i^*—the optimal number of SP ith names corresponding to the minimum total cost for eliminating machine failures, i.e. losses from machine downtime and the creation of a reserve of SP, pcs.; C_{Xi}—the total expected costs for eliminating failures of the fleet of machines, creating a reserve X_i of SP ith name and losses from downtime of equipment due to the lack of SP, rub.

To create a mathematical model for determining the optimal number of reserved SP, we form all the elements of economic costs into groups:

- the cost for eliminating machine failures in the presence of the SP ith name in the warehouse of the enterprise, R_{1i};
- the cost for eliminating machine failures in the absence of SP ith name in the warehouse of the enterprise, R_{2i};
- economic losses of the enterprise from machines downtime, D_i;
- the cost for reserving the X_i of the SP ith name, S_i.

As a result, the total expected costs are represented by the following target function

$$C_{Xi} = R_{1i} + R_{2i} + D_i + S_i \rightarrow \min. \tag{4}$$

The time intervals between the occurrence of demand (operating time of parts up to resource failure) are independent and random variables, which, according to the work [16], have one of the distributions: normal, lognormal, Weibull or exponential, the most common in the theory of modeling indicators of the durability of parts [12, 15].

The simulation of the operating time of the part up to resource failure in the case of the normal distribution law of a random variable can be carried out using the expression [2, 5]:

$$t = M + \sigma \cdot X, \tag{5}$$

where t—part resource value, engine hours; M—mathematical expectation of the part resource, engine hours; σ—standard deviation of the part resource, engine hour; X—random variable of the standard normal distribution law $N(0, 1)$ with parameters $M = 0$ and $\sigma = 1$.

3 Results

3.1 Model Components and Their Interaction

At the beginning of the modeling process, the model sets the initial period of the stock of the RFP for each of their names and the estimated time period T, within which the simulation will be carried out.

The estimated simulation time period T is assigned in calendar units (days), and the time intervals between part failures are generated in engine hours. To combine the named time streams, the conversion factor K_c is used, which is equal to:

$$K_c = \frac{24}{T_c}, \tag{6}$$

where T_c—average daily operating time of the machine, engine hours/day.

The simulated operating time of the parts up to replacement (resource failure), translated into calendar time, is equal to:

$$t_{rf,i,j,l} = K_c \cdot t'_{rf}, \tag{7}$$

where i—name of SP $(i = \overline{1,P})$; P—number of SP under consideration, pcs; j—machine number $(j = \overline{1,N})$; N—number of machines, provided with SP, pcs; l—part number (SP) $(l = \overline{1,L})$; L—number of parts, installed on the one machine, pcs; t'_{rf}—simulated operating time of parts up to failure, engine hours.

The operating time values can be set in accordance with the actual operating time for each machine (part), if there is information from the beginning of the operation, or modeled using the formula:

$$t_{i,j,l} = M\left(-\frac{U_i}{2}\right), \tag{8}$$

where M—mathematical expectation of the part resource, engine hours; U_i—independent and uniformly distributed random numbers on the interval [0…1].

If there is a stock of the SP ith name $X_i \geq 1$ in the warehouse, the resulting failure is eliminated at the expense of this reserve, as a result of which the reserve is reduced by one and takes the value $X_i - 1$. In this case, the cost for eliminating the failure is equal:

$$R_{1i} = C_{oi} + \Delta C_i + R_{dlv,j}, \tag{9}$$

where C_{oi}—the cost of the SP ith name, rub; ΔC_i—the added value of the SP ith name, rub; $R_{dlv,j}$—the cost of delivery of the SP ith name from the warehouse to the jth machine, rub.

The added value (warehouse rent, insurance, working capital freeze, maintenance and security costs) is equal to:

$$\Delta C_i = c \cdot C_{oi}\left(t_{rf,i,j,l} - t_{i,j,l}\right), \tag{10}$$

where c—the coefficient of the specific increase in the cost of one SP located in the warehouse of the enterprise, h^{-1}.

The cost of delivery of the SP from the warehouse of the enterprise to the jth machine is equal to:

$$R_{dlv,j} = Q \cdot L_j, \tag{11}$$

where Q—the unit cost of delivery of one SP ith name from the warehouse to jth machine, rub/km; L_j—the distance from the warehouse of the enterprise to the jth machine that requires a SP, km.

The time spent on the delivery of the SP from the warehouse of the enterprise is equal to:

$$t_{dlv,j} = \frac{L_j}{V_{av}}, \tag{12}$$

where V_{av}—average vehicle speed, km/h.

The costs for eliminating the failure in the absence of a SP in the warehouse of the enterprise and delivering it from the warehouse of the dealer (external source) will be:

$$R_{2i} = C_{odi} + R_{dlv,i} + R_{dlv,j}, \tag{13}$$

where C_{odi}—the cost of the SP ith name in the dealer's warehouse, rub; $R_{dlv,i}$—the cost of delivery of the SP from the dealer's warehouse to the enterprise's warehouse, rub.

The cost $R_{dlv,i}$ is determined by the formula:

$$R_{dlv,i} = Q_i \cdot L_i, \tag{14}$$

where Q_i—unit cost of delivery of the SP from the dealer's warehouse, rub/km; L_i—distance from the dealer's warehouse to the enterprise's warehouse, km.

The time spent on delivery of SP from the dealer's warehouse is equal to:

$$t_{dlv,i} = \frac{2L_i}{V_{avi}}, \tag{15}$$

where V_{avi}—average vehicle speed, km/h.

The economic losses of the enterprise from the downtime of the jth machine in the event of the failure of the part of the ith name are equal to:

$$D_i = (Z - Z_n) \cdot t_{mdt,j}, \tag{16}$$

where Z—economic losses of the enterprise from machine downtime, rub/h; Z_n—variable costs associated with a reduction in the consumption of material resources (fuel-containing masses, salary) during the machine downtime, rub/h; $t_{mdt,j}$—downtime period, within which the failure is eliminated, h.

$$t_{mdt,j} = t_{dlv,i} + t_{dlv,j}. \tag{17}$$

The costs of the enterprise for reserving the SP ith name, which were not used during the calculation period, are equal to:

$$S_i = C_{oi} \cdot x_i \cdot T \cdot c, \tag{18}$$

where x_i—the number of unused SP the ith name that are in the warehouse of the enterprise at the end of the calculation period T.

Finally, the target function that determines the efficiency of the process of reserving the SP ith name for the fleet of machines takes the following form:

$$\{C_{Xi}\} = \sum_{i=1}^{p}\sum_{k=1}^{K}(R_{1i} + R_{2i} + D_i + S_i)_k \rightarrow \min, \tag{19}$$

where k—the number of terms of the target function (cycles) during a single run of the model $(k = \overline{1,K})$.

3.2 Estimation of the Required Number of Model Runs

To achieve the required results in simulation modeling, it is necessary to use statistical methods for analyzing the simulation model runs [5, 10, 16].

To obtain an estimate of C_{Xi} with a given error, it is necessary to determine the number of model runs using the following algorithm [5]:

1. We perform independent repeated runs of the model of the multi-item reserve of the spare parts, as a result of which we get ten values of independent estimates

$$(C_{Xi})_f, f = \left(\overline{1,F}\right). \tag{20}$$

2. Determine the sample mean of the estimate

$$\overline{C_{Xi}}(F) = \frac{\sum_{f=1}^{F}(C_{Xi})_f}{F}. \tag{21}$$

3. Determine the sample variance of the estimate $S^2(F)$:

$$S^2(F) = \frac{\sum_{f=1}^{F}\left[(C_{Xi})_f - \overline{C_{Xi}}(F)\right]^2}{F-1}. \tag{22}$$

4. Set the relative error γ of the mean estimate C_{Xi} and the confidence level α.
5. Determine the adjusted relative error γ' required to obtain the specified relative error γ:

$$\gamma' = \frac{\gamma}{1+\gamma}. \tag{23}$$

6. Determine the required number of model runs $M(\gamma)$ for the average estimate C_{Xi} with a given relative error γ and confidence level α by the following expression:

$$F(\gamma) = \min\left\{ i \geq \left[10 - \left[\frac{t_{i-1,1-\alpha/2} - \sqrt{\frac{S^2(F)}{i}}}{\left|\overline{C_{Xi}}(F)\right|}\right]^{-1}\right] \leq \gamma' \right\}. \tag{24}$$

where $t_{i-1,1-\alpha/2}$ is the quantile of the Student function.

The number of repeated model runs $F(\gamma)$ is defined as the smallest integer i at which the above condition is satisfied.

A general list of controllable and uncontrollable variables of the target function is given in the list below:

- Controllable variables

  X_i—the number of SP ith names;

- Uncontrollable variables

  P—number of SP under consideration;
  T—estimated time period;
  N—number of machines, provided with SP;
  K—the number of parts installed on the machine;
  T_c—average daily operating time of the machine;
  C_{oi}—the cost of the SP ith name in the warehouse of the enterprise;
  C_{odi}—the cost of the SP ith name in the dealer's warehouse;
  c—the coefficient of the specific increase in the cost of one SP;
  Q—the unit cost of delivery of one SP ith name from the warehouse of the enterprise;
  Q_i—unit cost of delivery of the SP from the dealer's warehouse;
  L_j—the distance from the warehouse of the enterprise to the jth machine;
  L_i—distance from the dealer's warehouse to the enterprise's warehouse;
  V_{av}—average speed of the vehicle for the delivery of SP within the enterprise;

V_{avi}—average speed of a vehicle for delivery of spare parts from the dealer's warehouse;
Z—economic losses of the enterprise from machine downtime;
Z_n—variable costs associated with a reduction in the consumption of material resources during the machine downtime;
F—number of terms of the target function in one model run.

The features of the mathematical model are as follows:

- the multi-item of the optimized reserve of the SP;
- the full set of controllable and uncontrollable variables in the model;
- structure of the model, built on the basis of discrete-event simulation modeling. The functioning of the model is based on random events—the operating time of the part up to the first failure.

4 Discussion

4.1 Checking the Operability and Adequacy of the Model

The task of checking the operability and adequacy of the mathematical model will be carried out when creating extreme situations in the initial data, namely, at different distances of delivery of spare parts.

Let's consider different situations for an industrial complex that has fourteen transport and technological machines of the same brand, which are distributed across four enterprises or workshops as follows:

- 1st enterprise—five cars;
- 2nd enterprise—three cars;
- 3rd enterprise—two cars;
- 4th enterprise—four cars.

The average daily operating time of machines $T_c = 12$ engine hours/day. The distance between the enterprises (consumers) and the central warehouse is given in Table 1.

The results of the calculations of the optimal number of reserved SP depending on the distance to the source of the SP are given in Table 2.

The given example of specific calculations fully confirms the logical conclusions that follow from purely qualitative considerations and till the calculations are carried out.

Table 1 The value of the distances between the consumers of spare parts

No	Variant of distance	Value of distance, km
1	Central warehouse (store)—consumer No 1 (C1)	70
2	Central warehouse (CW)—consumer No 2 (C2)	60
3	CW—consumer No 3 (C3)	30
4	CW—consumer No 4 (C4)	50
5	C1—C2	130
6	C1—C3	100
7	C1—C4	120
8	C2—C3	90
9	C2—C4	110
10	C3—C4	80

Table 2 Results of calculations of the optimal number of reserved SP depending on the distance to the external source of SP

Distance to the external source of the SP, km	The number of reserved SP at the enterprise, pcs	Total costs of the enterprise, rub
The cost of the part (SP)—C_{oi} = 1100 rubles and C_{odi} = 1000 rubles; the type of the law of distribution of the resource of the part—normal (L_{av} = 1500 engine hours; σ_L = 750 engine hours); the number of parts on the machine (applicability)—K = 2; the calculation period T = 100 days		
5	5	17,148.91
500	14	19,608.88
3000	18	37,561.64
Spare part: C_{oi} = 5100 rubles and C_{odi} = 5000 rubles; the law of resource distribution—Weibull (L_{av} = 1000 engine hours, σ_L = 300 engine hours, K = 1, T = 100 days)		
5	3	64,278.47
500	11	74,493.21
3000	13	86,572.86

5 Conclusions

The mathematical model of multi-item spare parts reservation in relation to transport and technological machines is proved.

An algorithm for estimation of the required number of model runs is proposed.

The given examples of specific calculations (Table 2) fully confirm the logical conclusions that follow from purely qualitative considerations and before the calculations.

The results of testing the model showed its efficiency and adequacy to real production conditions.

References

1. Veritelnik EA (2019) Technical and economic indicators of the work of the warehouse of spare parts as an element of a motor transport enterprise. World Transp Technol Mach 2 (65):115–123
2. Pitukhin AV (2018) Methodology for organizing the supply of a multi-item reserve of spare parts. Bull Moscow State Univ For 2(22):61–67
3. Veritelnik EA (2016) Optimization of the warehouse of spare parts of the motor transport enterprise. Motor Transp Enterp 8:51–54
4. Zaprudnov VI (2017) The need for a park of forest machines in technical maintenance. For Bull 21(2):76–79
5. Salivonik AV (2014) Improving the efficiency of logging machines by optimal reservation of spare parts and materials, 80
6. Aldoshin NV (2015) The choice of strategies for creating a secondary fund of spare parts. Bull Goryachkin VP Moscow State Agroengineering Univ 1(65):7–11
7. Veretelnik EA (2018) Improving the efficiency of motor transport enterprises by improving the efficiency of the warehouse of spare parts. World Transp Technol Mach 3(62):137–145
8. Baranov N (2003) Risk and optimum in the formation of reserves. Resour Inf Supply Competition 3:35–40
9. Arifullin IV (2016) Complex assessment of the quality of delivery of spare parts for the maintenance of the motor vehicle fleet. Bull Moscow Automobile Road State Tech Univ 3 (56):218–220
10. Hansmann F (1966) Application of mathematical methods in the management of production and sales, 184
11. Anily S (1991) Capacitated two-stage multi-item production model with joint setup const. Oper Res 3:244–254
12. Veretelnik EA (2016) Study of disruption cars-trucks Mercedes-Benz Actros 1844 LS to optimize the fund of spare parts auto-transport enterprise. Bull Transp Inf 8(254):30–34
13. Arifullin IV (2014) System of the lifecycle of spare parts of rolling stock. Transp Sci Technol Manage 1:56–58
14. Ryzhikov YI (2001) Inventory management, 384
15. Shilovskiy VN (2015) Marketing and management of technical service of machines and equipment, 272
16. Shilovskiy VN (2001) Theoretical bases and strategies of the organization of marketing and management of technical service of geographically distributed machines and equipment, 324
17. Shilovskiy VN (2012) Development of marketing and management of technical service of logging machines, 268

Analysis of Methods and Design Elements of Docking and Un-docking in Orbit

Daria Malyh, Ruslan Peshkov, Sergey Vaulin, Victor Fedorov, and Mikhail Shalashov

Abstract At the current level of spacecraft development, orbital docking is increasingly used. The requirements for docking devices, as well as their complexity, are increasing. The purpose of the article is to analyze the methods and design elements of docking and undocking in orbit. Based on the results of the review, a conclusion was made about the structural elements necessary for the docking and undocking of the demonstrator modules of the universal space platform.

Keywords Docking · Undocking · Docking device · Spacecraft · Service in space

1 Introduction

The docking system is designed to perform three main tasks: docking, forming a sealed joint with sufficient load-bearing capacity, and undocking [1]. On the spacecraft to be docked, docking units are installed, which provide the entire volume of docking and undocking operations. The combination of such devices and mechanisms that provide physical docking and undocking is called a docking device.

An active docking vehicle is a spacecraft that has a device on board that performs the basic operations of docking and undocking in orbit. A passive docking unit is a device that provides interaction with an active docking unit.

The docking unit elements, whose task is to form a sealed joint, consist of a body, a docking mechanism, a joint sealing mechanism, a detaching mechanism, and additional components and elements.

D. Malyh (✉) · R. Peshkov · S. Vaulin · V. Fedorov · M. Shalashov
South Ural State University, Lenin Avenue, 76, Chelyabinsk 454000, Russian Federation
e-mail: malyhda@susu.ru

S. Vaulin · V. Fedorov
Autonomous Nonprofit Organization «Industrial Innovation Development Center» (ANO «Industrial Innovation Development Center»), Chelyabinsk 454000, Russian Federation

A. Bieliatynskyi and V. Breskich (eds.), *Safety in Aviation and Space Technologies*, Lecture Notes in Mechanical Engineering,
https://doi.org/10.1007/978-3-030-85057-9_49

In this paper, consider possible methods of docking and undocking during missions in orbit, which can be used in a reusable small upper stage as part of a universal space platform (USP).

2 General Information and Approaches

The USP task is the delivery and creation of satellites for various purposes to the LEO (Low Earth Orbit) and GEO (Geosynchronous Equatorial Orbit) of the Earth, the creation of spacecraft. The USP can include three modules (depending on the purpose): the service systems module, the payload module and the reusable upper stage module (if necessary, the transfer of the satellite from the LEO to the GSO, the delivery of the spacecraft to a given point in outer space). Another task of the USP may be to land on a body with a small gravitational field. To approach the asteroid, a reusable upper stage module is used, from which the service systems module and the payload module are then separated and the landing is carried out.

To solve the tasks assigned to the USP, in particular the reusable upper stage, the most important element is the system of docking and undocking in orbit.

There are two classes of docking devices: docking on the principle of combining cones ("pin-cone") and docking with a peripheral arrangement of docking devices [2].

A distinctive feature of the APAS (androgynous peripheral docking assembly) type peripheral device (androgynous-peripheral docking unit) is that two spacecraft equipped with such devices can easily dock with each other (APAS can perform both active and passive roles).

Another method of docking is the use of a single docking mechanism (Common Berthing Mechanism) [3]. Its distinctive feature is that after connecting the two sides of the mechanism, the pressure seals the joint to tightness.

In 2010, the International Docking System Standard was developed to help create compatible docking mechanisms [4].

The docking methods described above are mostly used on the International Space Station, but the docking and undocking of spacecraft or satellites has a number of features.

The development of such mechanisms for docking and undocking is closely related to the solution of issues of orbital maintenance of satellites, or the fight against space debris, one of the main sources of which are non-functioning spacecraft and their debris. At the same time, if the spacecraft to which the docking takes place is working during the orbital maintenance, then the objects of space debris are non-cooperative targets, that is, they may not have docking or fixing nodes. Therefore, it is possible to use contactless docking, that is, using a network, self-opening membranes, or bags [5].

Docking devices with the help of robotic manipulators are being actively studied. The use of such mechanisms allows to simultaneously solve the tasks of docking and fixing, while several robotic manipulators can be used (Fig. 1).

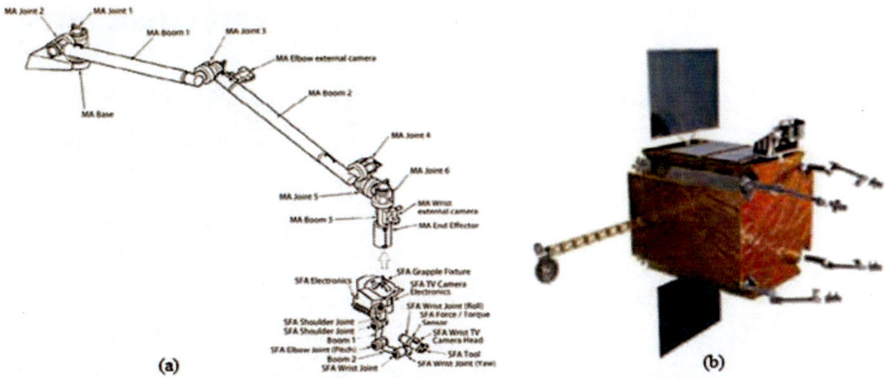

Fig. 1 Diagram of a robotic device **a** JAXA's grip on the engine nozzle **b** capture by multiple manipulators (NM1 spacecraft) [3]

When docking spacecraft, the already described method according to the "pin-cone" scheme is also promising. For example, when approaching the docking object, a pin is extended from the reusable upper stage and inserted into the nozzle part of the rocket engine device, followed by fixation (Fig. 2).

In addition, the developments of Astrium, the largest manufacturer of spacecraft, are interesting. In particular, the possibility of using a sliding ring element, which is gradually tightened until the object of space debris is captured, is being studied (Fig. 3).

There are several projects for docking and undocking in orbit of varying degrees of elaboration, both for commercial companies and space agencies [6].

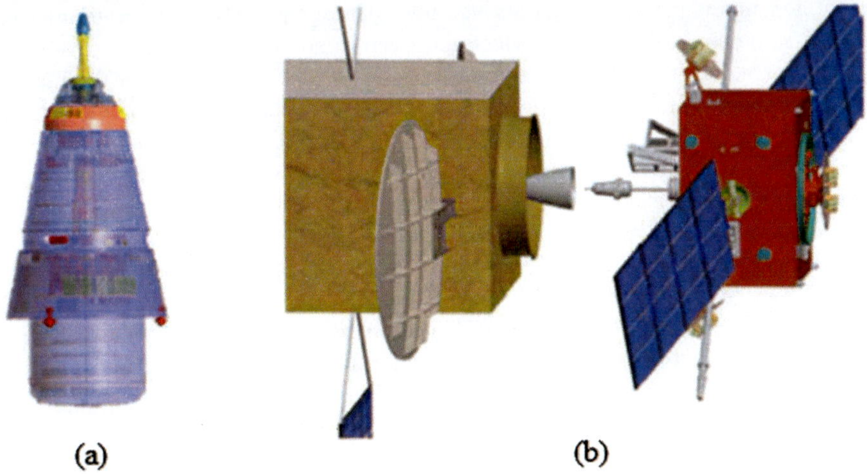

Fig. 2 «Pin–cone» docking system **a** docking device **b** start of the docking stage [2]

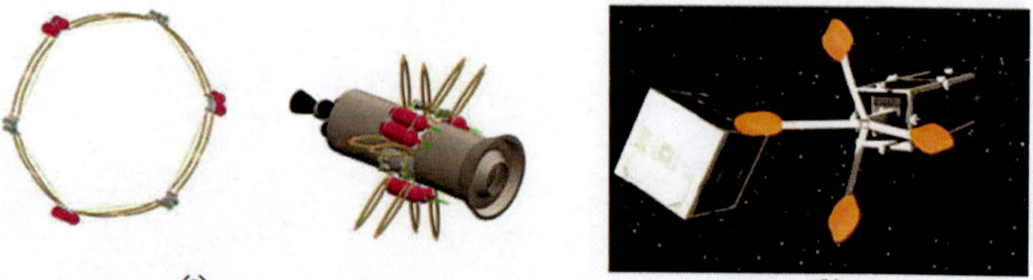

(a) (b)

Fig. 3 Astrium developments **a** sliding ring element **b** using «tentacles» [5]

3 Results of Review Main Projects for Docking and Undocking in Orbit

Airbus is developing a project to extend the service life of satellites in low-Earth and geostationary orbits, including the removal of debris (using nets, harpoons and sails). Spacecraft equipped with such systems will belong to the new SpaceTug class. The practical implementation of the project is planned for 2023.

It is expected that this will be possible with the help of the ASSIST system. The central mechanical part will first perform a soft docking, and then a mechanical retraction, which is completed during the hard docking phase using the centering pins. The docking is fully controlled by a robotic arm with a working body (end-effector), which is equipped with a light source to provide a clear indication of the markers placed on the mechanisms of the docking devices. On the working body there are hydraulic and electrical connections and a gripping mechanism that connects to the mooring device (Fig. 4). It also includes three hydraulic couplings (fuel, oxidizer, and xenon) for connection to the elastomeric O-rings. One of the main elements is the tip—probe, driven by a pantograph. The pantograph mechanism uses a central drive shaft, which is driven by a stepper motor.

The main stages of docking are shown in Fig. 5. With the help of the manipulator and the cameras, the alignment is carried out so that the center of the probe is located within the receiving cone, and the mounting pins are combined with the guides that have a conical shape to eliminate minor distortions. During the docking process, the working body enters the cavity of the step shaft through the "neck". Information about the docking process is monitored using force sensors. As soon as the probe is expanded, the clamping sleeve moves along the cylindrical part of the probe in the direction of the hub, thereby coming into contact with the "neck". Thus, a rigid docking will be achieved. To compensate for the displacement relative to the main axis, the liquid plane can rotate relative to the central cylinder, the pantograph and the bushing have a degree of freedom of rotation relative to the main shaft. The refueling surface begins to move in the direction of the mated satellite until it stops with the installation of pins in the holes and the automatic connection of three hydraulic couplings.

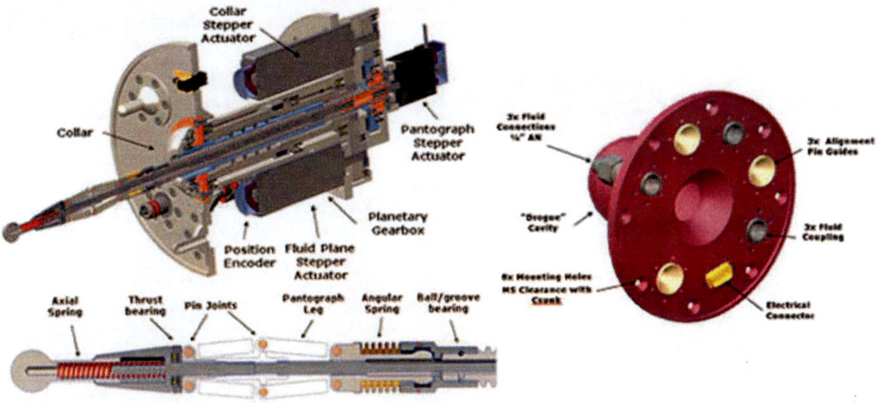

Fig. 4 Elements of the docking system [7]

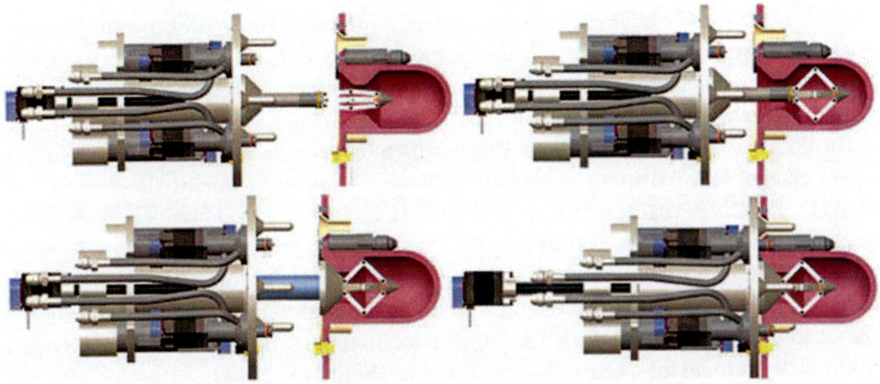

Fig. 5 The main stages of docking [7]

OneWeb has now begun building its global Internet system, which is based on a grouping of nearly 600 satellites in low-Earth orbit. One of the issues is to prevent the creation of space debris in the event of a satellite failure. For this purpose, for several years, together with the company Altius Space Machines, a universal docking mechanism DogTag was developed—an inexpensive, lightweight and versatile interface for small satellites [8]. Figure 6 shows the DogTag components. Dimensions of the DogTag mechanism 150 mm × 150 mm × 65 mm, weight 250 g.

In Japan, Astroscale, under the ELSA-d program, will launch a mission on March 22, 2021, demonstrating the basic technologies necessary for the docking and undocking of two satellites (with masses of 175 (dimensions 660 × 640 1100 mm) and 17 kg (dimensions 480 × 500 × 225 mm)) as part of the use of the received solutions for space cleaning [9, 10].

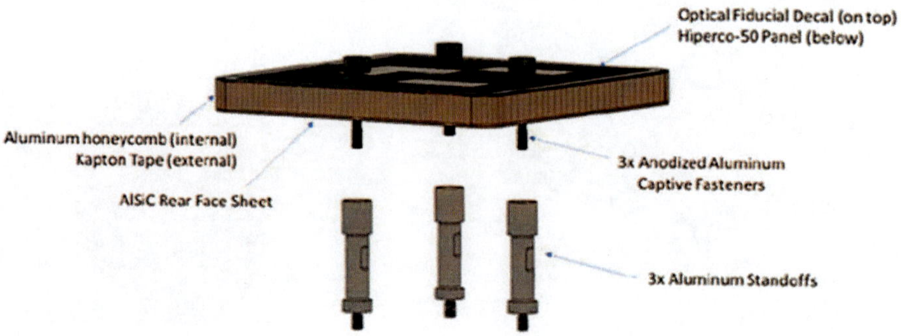

Fig. 6 DogTag components [8]

For docking and undocking vehicles, it is also possible to use a different approach, for example, using an autonomous space robot-a spacecraft. A similar device is being developed in the company Busek and is called SOUL (Satellite On Umbilical Line) [11]. This is a small (weighing about 10 kg) robot with the necessary equipment, depending on the purpose of its use, which connects satellites through a hose (a "robotic arm" with infinite degrees of freedom, which allows access to places inaccessible to ordinary robotic arms).

In 2020, with the help of the Russian Proton launch vehicle with the Briz-M upper stage, the British company Effective Space planned to launch two SPACE DRONE spacecraft into orbit. SPACE DRONE is equipped with a universal system consisting of four robotic manipulators, which allows to dock with a satellite (weighing from 1500 to 4000 kg.) in geostationary orbit and using its own electric power plant (four ion engines on the manipulators) to change the location of the "captured" satellite, adjust its orbit or inclination. The weight of the device is 400 kg. Its dimensions are $1 \times 1 \times 1.25$ m (Fig. 7) [12–14].

One of the modern approaches to creating new next-generation spacecraft is the use of standard cubic modules, especially when assembled in orbit (the iBoss—Intelligent Building Blocks for On-Orbit Satellite Servicing program). The base element of the module had a cubic shape with a side length of 40 cm, which met the installation requirements of most types of instruments for spacecraft. In addition, the size of the module can be increased from the main element to $40 \times 40 \times 80$ cm or $80 \times 80 \times 80$ cm. An example of the use of iBoss is the modular design of the ADM-AEOLUS satellite, which was integrated with many cubic iBoss elements, except for the payload chamber, solar panels and the central tube (Fig. 8) [15–17].

In 2007, two satellites-ASTRO (Autonomous Space Transfer and Robotic Orbitek) and NEXTSat (Next generation of Satellite) were launched under the Orbital Express program to test the technology of autonomous rendezvous, docking, refueling (with hydrazine), replacing individual elements (in particular, a special ORU (Orbital Replacement Unit) module, which symbolized the satellite's batteries) and undocking (Fig. 9) [18, 19].

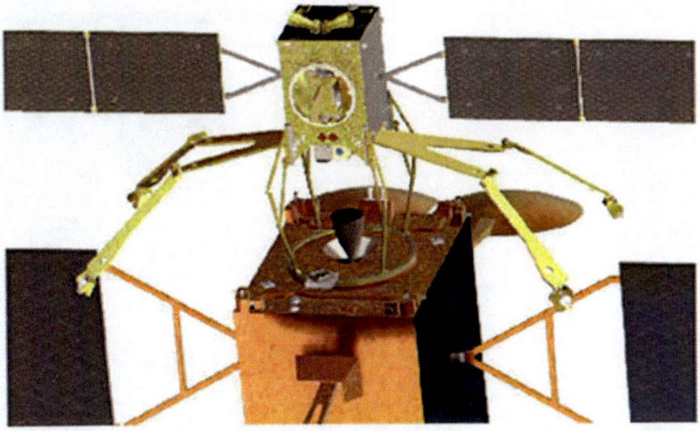

Fig. 7 The process of docking a SPACE DRONE with a satellite [12]

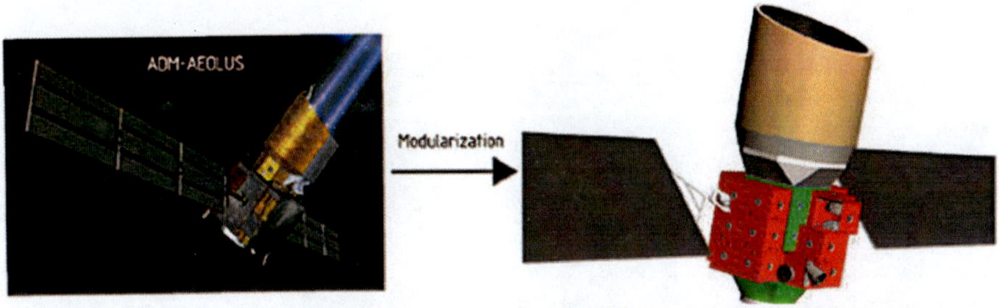

Fig. 8 Modification of the ADM-AEOLUS satellite using iBoss [15–17]

The docking and undocking system consists of the active and passive sides. On the active side there are grippers and the drive system, on the passive side there are gripper locks and a sensor indicating the correct activation of the gripping levers [20].

The docking process consists of the following basic operations (Fig. 10):

1. The docking of the active mechanism begins in the deployed (open) state. The passive structure is held on the station within the capture capabilities of the active mechanism.
2. After receiving the command, the engine starts to drive the ball-screw transmission, which moves the inner piston along the length of the body of the active mechanism. The piston moves the three separate four-link levers that make up the coupling devices. The levers are then moved down the roller element. The cam action of the roller causes the levers to contract, entering the passive structure. The wedge-shaped design of the passive structure guides the ends of the levers into the central grooves. Capture is achieved because the passive structure is limited by the framework of connections.

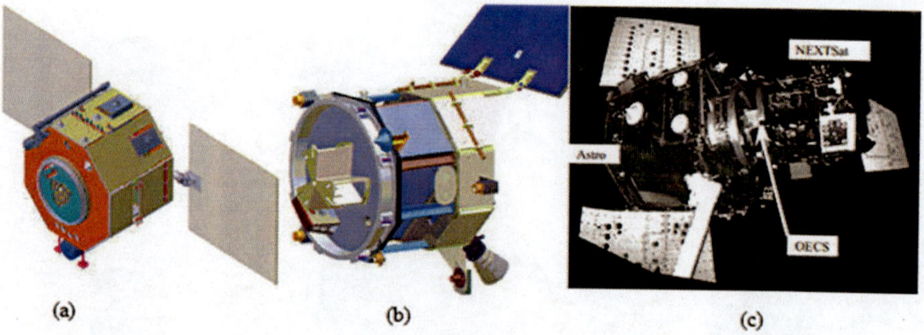

Fig. 9 Satellites of the orbital express program **a** ASTRO **b** NEXTSat **c** photo in orbit [19]

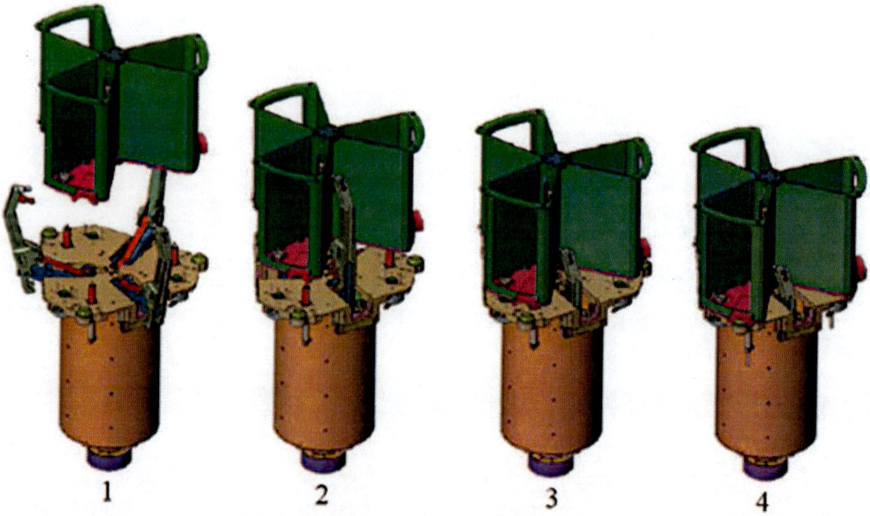

Fig. 10 Orbital express capture system [20]

3. As the connecting tips move down the grooves, they come into contact with the shelf, allowing the active and passive structures to be combined. As you draw in, the alignment occurs in stages. The continued movement of the lever mechanisms leads to the fact that the interface plate of the passive structure contacts the posts of the ejector rods, which act as a support for alignment at three points on the active mechanism. Each alignment element consists of a spring-loaded pin with tapered ends, which is inserted into the tapered cups on the passive structure. These features provide a rough alignment. With further tightening, the additional elements align the electrical connectors within the permissible deviation.

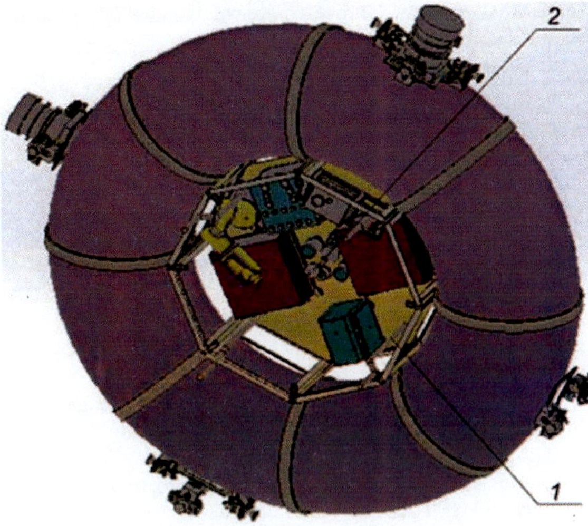

Fig. 11 Docking elements of the upper stage module that is part of the universal space platform demonstrator (1) capture manipulator (2) pin

4. In the final stages of tightening, the engine creates the necessary preload to ensure the required rigidity of the interface.

In addition to the listed projects for docking and undocking, there are many others, in particular, the project of the company Made in space-Archinaut [21], the CIRAS project of Northrop Grumman [22], the projects of Robotic Servicing of Geosynchronous Satellites (RSGS) [23], Phoenix [24], etc., but many of them are either already closed or are in the initial stages of development and most of the technical solutions that were planned or are planned to be used in already implemented projects, which were described above.

4 Conclusion

Based on the results of the analysis of the methods and structural elements of docking and undocking in orbit, it can be concluded that the most promising element for docking and undocking is the use of the classical "pin-cone" scheme with a retractable element (Fig. 11), while the engine nozzle acts as a cone, and manipulators to create a tight engagement with the ring of the spacecraft launch adapter, which is typical for most geostationary satellites.

To ensure versatility, it is necessary to place additional elements: lidar, color, infrared cameras for recognizing the docking location, a force and torque sensor in the manipulator.

Acknowledgements The research was funded by Ministry of Education and Science of the Chelyabinsk Region.

References

1. Abramov IP, Aldashkin EV, Alekseev IV (2016) Rocket and space technology. 548
2. Syromyatnikov VS (1984) Docking devices of spacecraft. 216
3. McLaughlin RJ, Warr WH (2001) The Common Berthing Mechanism (CBM) for international space station. Soc Autom Eng 34(4):8
4. International Docking System Standard (2021) Interface definition document. https://www.internationaldockingstandard.com/
5. Trushlyakov VI, Utkin EA (2013) Review of means of docking and capturing objects of large-sized space debris. Omsk Sci Bull 2:56–61
6. Davis JP, Mayberry JP, Penn JP (2019) On-orbit servicing: inspection, repair, refuel, upgrade, and assembly of satellites in space. Aerospace 14
7. Medina A, Tomassini A, Suatoni M, Avilés M, Solway N, Coxhill I (2017) Towards a standardized grasping and refuelling on-orbit servicing for geo space-craft. Acta Astronaut 134:10. https://doi.org/10.1016/j.actaastro.2017.01.022
8. Maclay T, Goff J, Sheehan JP, Han E (2020) The development of commercially viable ADR services: introduction of a small-satellite grappling interface. J Space Saf Eng 7(3):364–368. https://doi.org/10.1016/j.jsse.2020.08.002
9. Blackerby C, Okamoto A, Iizuka S, Kobayashi Y, Fujimoto K, Seto Y (2019) The ELSA-d end-of-life debris removal mission: preparing for launch. Paper presented at the Proceedings of the International Astronautical Congress, IAC, 8
10. ELSA-d Launch Press Kit—Astroscale. ELSA-d-Launch-Press-Kit-2021-ENG.pdf. Last accessed 29 March 2021
11. Hruby V, DeLuccia C, Williams D (2018) Tethered robot for spacecraft self inspection and servicing. Paper presented at the AIAA SPACE and astronautics forum and exposition, 9. https://doi.org/10.2514/6.2018-5342
12. Barnett DL, Campbell D, Azriel O, Halsband A (2019) Space drone adaptable servicing spacecraft. Paper presented at the Proceedings of the International Astronautical Congress, IAC. https://indico.esa.int/event/234/contributions/3749/attachments/3072/3776/2018_CSID_LinnBarnett_Spacedrone.pdf. Last accessed 29 March 2021
13. Halsband A, Taaseh N, Pariente M, Michael R (2016) Docking system and method for satellites. https://patentimages.storage.googleapis.com/59/ca/3e/2048bb8b784d11/WO2016030890A1.pdf. Last accessed 29 March 2021
14. Reitman M, Spitzer A, Halsband A, Azriel O (2021) Service satellite for providing in-orbit services using variable thruster control. https://patentimages.storage.googleapis.com/47/02/67/1365c736b567a4/US20180251240A1.pdf. Last accessed 29 March 2021
15. Jansen F, Brandt T, Dafnis A, Detsis E, Ferraris S, Findlay JAP (2019) In PPS flagship with IBOSS building blocks. Paper presented at the Proceedings of the International Astronautical Congress, IAC, 7
16. Kortmann M, Rühl S, Weise J, Kreisel J, Schervan TA, Schmidt H, Dafnis A (2015) Building block-based "iBOSS" approach: fully modular systems with standard interface to enhance future satellites. Paper presented at the Proceedings of the International Astronautical Congress, IAC, 11
17. Kreisel J, Schervan TA, Schroeder KA (2019) Game-changing space system inter-face enabling multiple modular and building block-based architectures for orbital and exploration missions. Paper presented at the Proceedings of the International Astronautical Congress, IAC, 7
18. Li WJ, Cheng DY, Liu XG (2019) On-orbit service (OOS) of space-craft: a review of engineering developments. Prog Aerosp Sci 108:32–120

19. Shoemaker J, Wright M (2004) Orbital express space operations architecture program. Proc SPIE—Int Soc Opt Eng 5419:57–65. https://doi.org/10.1117/12.544067
20. Chrisstiansen S, Nilson T (2008) Docking system mechanism utilized on orbital express program. In: Proceedings of the 39th Aerospace Mechanisms Symposium, pp 207–220
21. Archinaut. http://madeinspace.us/archinaut/. Last accessed 29 March 2021
22. Orbital ATK to develop "Tipping Point" Orbital Assembly Technologies. http://www.spaceflightinsider.com/organizations/orbital-sciences-corp/orbital-develop-tipping-point-orbital-assembly-technologies/. Last accessed 29 March 2021
23. Robotic servicing of geosynchronous satellites. https://www.darpa.mil/program/robotic-servicing-of-geosynchronous-satellites. Last accessed 29 March 2021
24. The DARPA phoenix spacecraft servicing program: overview and plans for risk reduction

On the Issue of Estimating Errors in the Operation of the Interactive System for Improving Road Safety Outside Settlements

Vladimir Zeer, Evgeniy Grazhdantsev, Ruslan Glukhikh, Pavel Rastovcev, and Alexey Elagin

Abstract The article presents the result of the analysis of statistical data of accidents in the Krasnoyarsk Territory on sections of roads outside settlements with their greatest concentration. The main types of vehicle collisions and their accompanying conditions were identified. A mathematical apparatus for determining the minimum safe distance between vehicles located closely in an emergency zone on a section of a road outside settlements is proposed, in which it is necessary to notify the drivers of these vehicles about the danger in conditions of insufficient and/or limited visibility in various road situations, taking into account errors in positioning and determining the speed of the vehicle, the response time of the system, the effective coefficient of adhesion of the tires to the road surface. The results of experimental studies of vehicle positioning and speed errors determined by the driver's early warning system about hazards on the road and their impact on the operability of the designed system are presented.

Keywords Road safety · Interactive driver warning system · Positioning error · Safe distance · Rest of track · Braking distance

1 Introduction

Among the main mortality factors in the Russian Federation, road accidents account for 1% of the total. At the same time, the predominant proportion of dead people is in the age group from 30 to 40 years, that is, the able-bodied population.

In the Krasnoyarsk region, as in the whole Russian Federation, over the past few years there has been a downward trend in the number of accidents in road transport, including the number of dead and wounded. However, the whole country, the Krasnoyarsk Territory is among the regions with a high share of accidents outside

V. Zeer · E. Grazhdantsev (✉) · R. Glukhikh · P. Rastovcev · A. Elagin
Sibir Federal University, Polytechnic Institute, pr. Svobodnyy, 79,
Krasnoyarsk 660041, Russia
e-mail: egrazhdantsev@sfu-kras.ru

A. Bieliatynskyi and V. Breskich (eds.), *Safety in Aviation and Space Technologies*,
Lecture Notes in Mechanical Engineering,
https://doi.org/10.1007/978-3-030-85057-9_50

settlements [1, 2], where the speed of vehicles is high and as a result the severity of the consequences increases.

An analysis of the number of accidents injured and killed in them per 1000 km of roads in the Krasnoyarsk region (Fig. 1) shows a significant risk of getting into accidents, which is 110 accidents for every 1000 km of roads. At the same time, the number of victims and deaths for every 1000 km is 132 and 12 people, respectively, which is also a significant negative indicator. At the same time, the value of the number of accidents and deaths in them per 1000 km in the Krasnoyarsk region corresponds to the average value of these indicators in the country.

100% victims of road accidents annually account for 8–9% dead, which puts the Krasnoyarsk Territory in tenth place in terms of accidents among 85% constituent entities of the Russian Federation and in eleventh percent-in terms of the number of deaths in accidents.

It should be noted that nine out of ten accidents occur due to violations of traffic rules (violation of the speed limit [3–5], violation of the priority of intersections, etc.), and the most frequent type of accidents outside settlements are collisions. Also important factors affecting the occurrence of accidents outside settlements are weather conditions (snow, rain, fog), time of day, complex terrain of the road, which determine insufficient and limited visibility [6, 7], and road conditions (ice, snow roll, wet road surface), which determine the braking properties of cars, their handling and stability [according to Russian State Standard ODM 218.6.015–2015].

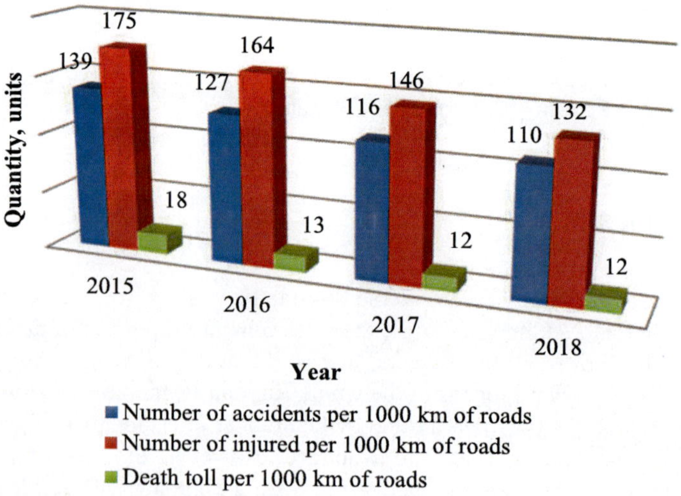

Fig. 1 Dynamics of the number of accidents suffered and killed in them per 1000 km of roads in the Krasnoyarsk region from 2015 to 2018

2 Data Definition Methods

Based on the analysis of statistical data on accidents that occurred on regional and federal roads outside the settlements of the Krasnoyarsk Territory for 2016–2019, several sections with the largest concentration of accidents, the number of wounded and killed in them were identified. To determine the causes of accidents, the following were used: road section plans in order to determine areas with limited visibility; accident cards for determining the coordinates, type of accident, weather and road conditions, time of day at the time of the accident, in order to determine the occurrence of insufficient visibility, the state of the road surface. The results of theoretical studies showed that more than 94% of accidents are due to unfavorable weather conditions, time of day and terrain of the road (62% of accidents were accompanied by limited and (or) insufficient visibility, 34% of accidents were accompanied by unsatisfactory pavement), while all the listed factors together have 20% of accidents (Fig. 2).

Equipment used. In order to improve road safety outside settlements, an interactive early warning system for vehicle drivers on the occurrence of an emergency situation in conditions of limited and insufficient visibility is proposed, using cellular communications, satellite navigation systems, the Internet, a remote server and additional information sources [8].

We will consider dangerous situations those situations that, provided that no response measures are taken, will lead to a traffic accident.

In fact, response measures aimed at preventing accidents are reduced either to a decrease in speed, or to a change in the direction of movement of vehicles, or to a combination of these actions.

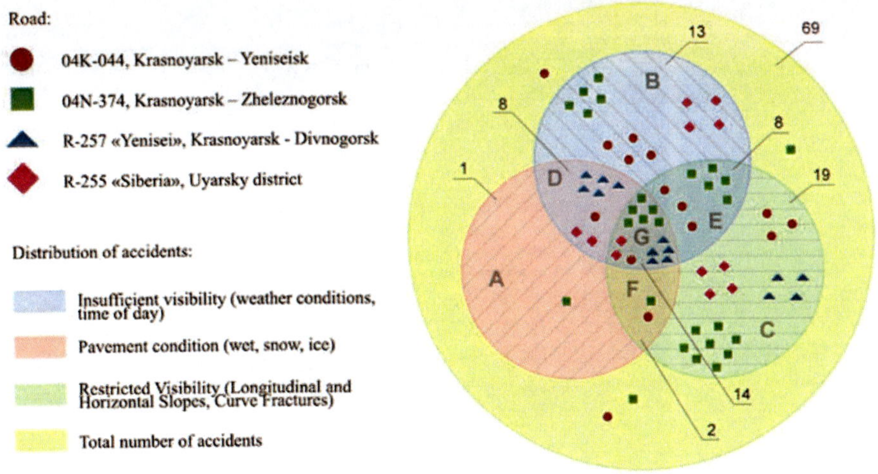

Fig. 2 Diagram of accident distribution according to the conditions accompanying their occurrence

As a rule, speed reduction is carried out by braking and occurs when the driver estimates the distance to the nearest object on the way and determines the safe distance to it. At the same time, the safe distance is determined by the driver subjectively on the basis of incoming information, available knowledge and acquired experience.

At the same time, roads and their elements are designed in such a way as to ensure the timely receipt of information on changes in road parameters and other road users, mainly by providing visibility parameters in which the driver can take the necessary actions to prevent an accident when an obstacle occurs on the way.

The system makes it possible to warn about the possible occurrence of vehicle collisions by determining critical values of their traffic parameters in real time on an interactive map of a section of the road outside the settlement, taking into account weather conditions, the type of road surface [9], and, if necessary, timely informing specific road users by means of sound and visual signals corresponding to the type of emergency. This system involves working with the 3 most common types of collisions: collision with an oncoming vehicle; collision with the vehicle in the same direction; vehicle collision at intersections [10–12] (Fig. 3).

The above-mentioned regulations adopt the following concepts of visibility:

- Visibility during overtaking is the minimum visibility distance to an oncoming car moving at a design speed, necessary for a safe overtaking maneuver.

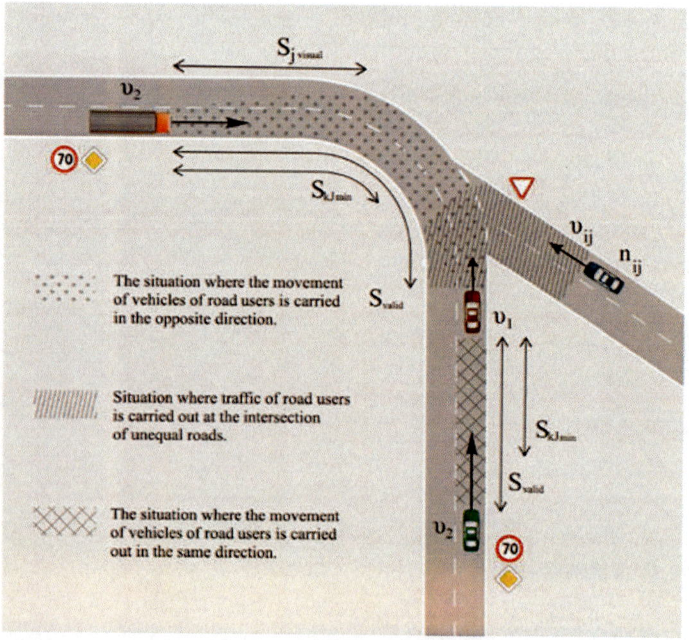

Fig. 3 Situations of movement of a pair of vehicles of road users

- Visibility of oncoming car is the minimum visibility distance to the oncoming car moving at a calculated speed, providing a safe interruption of overtaking from the oncoming lane.
- Insufficient visibility—the visibility of the road is less than 300 m in conditions of fog, rain, snowfall and the like, as well as at dusk.
- Limited visibility—the driver's visibility of the road in the direction of movement, limited by the terrain, geometric parameters of the road, vegetation, buildings, structures or other objects, including vehicles.
- Visibility distance is the distance from the front of the passenger vehicle at which the road design elements and other objects in the direction of movement are distinguished from the driver's seat, the orientation of which allows the vehicle to lead along the corresponding lane.

Based on the above definitions of visibility, it can be concluded that visibility is the distance from a vehicle to an object in the path of movement of that vehicle, allowing for a safe maneuver.

The conditions of warning signals on possible occurrence of an emergency situation for the *k*th pair of vehicles located in the *J*th emergency hazardous area on the *j*th section of the road outside settlements are determined as follows:

$$S_{kJvalid} \leq S_{kJmin} \text{ or } S_{jvisual} \leq S_{kJmin},$$

where

$S_{kJvalid}$ is the actual distance between the *k*th pair of vehicles, which according to the interactive map of the *j*th section of the road outside settlements are close to each other in the *J*th emergency hazardous area, m, determined taking into account the terrain;

S_{kJmin} is the minimum safe distance between the *k*th pair of vehicles, which according to the interactive map of the *j*th section of the road outside settlements are close to each other in the *J*th emergency danger zone, meters;

$S_{jvisual}$ distance of visibility on the *j*th section of the road outside settlements, meters, determined according to weather conditions.

For situations when the movement of the *k*th pair of road users' vehicles is carried out in the opposite, opposite direction or at the intersection of unequal roads, SkJmin is determined in the form:—differences of tracks travelled by the *k*th pair of HW during the time required to completely stop the catching-up HW taking into account the remaining path between them (1);—the sum of tracks (2);—the path travelled by the approaching HW to the intersection of unequal roads along the main road (3), respectively, meters:

$$S_{kJ\,\min} = (\upsilon_{2J} - \upsilon_{1J}) \cdot (t_{reac} + t_{ac.dr.} + t_{jresp}) + \frac{(\upsilon_{2J} - \upsilon_{1J})^2}{2\delta_{\max}} + S_{ost}, \qquad (1)$$

$$S_{kJ\,\min} = (\upsilon_{1J} + \upsilon_{2J}) \cdot (t_{reac} + t_{ac.dr.} + t_{jresp}) + \frac{(\upsilon_{1J} + \upsilon_{2J})^2}{2\delta_{\max}} + S_{ost}; \quad (2)$$

$$S_{kJ\,\min} = \upsilon_{1J} \cdot (t_{reac} + t_{ac.dr.} + t_{jresp}) + \frac{\upsilon_{1J}^2}{2\delta_{\max}} + S_{ost}, \quad (3)$$

where

$\upsilon_{1J}, \upsilon_{2J}$ is the speed of the *k*th pair of the vehicle in the *J*th emergency hazardous area at the time of determining the emergency hazardous situation, m/sec;

t_{reac} average reaction time of the *i*th road user who is the vehicle driver (varies from 1 to 1.5 s [according to Russian State Standard ODM 218.4.005-2010]), sec;

$t_{ac.dr.}$ average time of actuation of the brake drive of the vehicle of the *i*th road user (taken from 0.6 to 0.8 s [according to Russian State Standard GOST 33997-2016]), sec;

t_{jresp} response time of the vehicle collision prevention system on the *j*th section of the road outside settlements, sec;

$\delta_{\max}$ maximum possible deceleration of the road user's vehicle taking into account the coefficient of adhesion of tyres to the road surface of the *j*th section of the road outside settlements, determined by the formula, m/sec^2;

S_{ost}—the rest of the path between the *k*th pair of road users vehicles, which, according to the formed interactive map of the *j*th section of the road outside settlements, are close to each other in the *J*th emergency danger zone, m, determined for situations of the associated (4), oncoming direction (5) and at the intersection of unequal roads (6), respectively meters:

$$S_{ost} = (\upsilon_{2J} - \upsilon_{1J}) \cdot t_{jresp}; \quad (4)$$

$$S_{ost} = (\upsilon_{1J} + \upsilon_{2J}) \cdot t_{jresp}; \quad (5)$$

$$S_{ost} = \upsilon_{1J} \cdot t_{jresp} \quad (6)$$

A mobile phone application has been developed to test the technical capabilities of the system, which provides transmission of the coordinates of the vehicle (s) location and determination of their speeds and directions of movement on the remote server, as well as reception of service data in the form of periodic messages to the phones indicating the distances between the vehicle and the vehicle speed determined by the system, when limiting parameters occur with the possibility of variation of these parameters: maximum vehicle speed; minimum specified distance between vehicles; data polling rate.

3 Research Results

The results of experimental studies showed:—the response time of the system, taking into account various factors, such as communication quality, in places with different landscapes, roadside infrastructure and distance from Krasnoyarsk, is within 1 s, which satisfies the correct operation of the system [13];—the difference between the measured vehicle speed using the VIZIR-2 M speed meter and the determined system was from 1 to 2 km/h with a speed variation from 20 to 90 km/h, the relative error of the vehicle speed does not exceed 3%, which also satisfies the correct operation of the system. The low speed error is explained by the practical invariability of the positioning error at the nearest design points (determining the coordinates of the vehicle location), due to the absence of sharply changing factors of the external impact of weather conditions, changes in landscape and infrastructure impacts.

The results of experiments to determine the positioning error of the vehicle are presented in Table 1. The actual distance between the participants of experiment Sd, equipped with telephones with the developed application, was measured by the mechanical road wheel WheelPro CONDTROL between the marks left on the road by the participants at the time of delivery of the first message to the phone (the restriction conditions set by the system are reached), taking into account the errors of label drop depending on the speed of the vehicle.

Based on the obtained experimental data, the relationships of relative positioning errors relative to the given limit distance Sd/Sz and the distance between vehicles determined by the Ss/Sd system, depending on the speed of the vehicle, are constructed (Fig. 4). Approximating the obtained dependencies (Fig. 4), it can be concluded that the relative positioning error with increasing vehicle speed decreases, which is relevant for the purposes of the developed system.

In order to verify the adequacy of the proposed version of calculating the remainder of the path between the *k*th pair of vehicles of road users located close to each other in the *J*th Sost emergency danger zone, dependencies were built: an experimental-averaged positioning error in the form of a distance difference determined by the system and the actual ΔSs-d; Sost path residue (according to (6)); minimum safe distance between *k*th pair of SkJmin vehicle (refer to (3) excluding the rest of track Sost) at system response times equal to 1 s, driver response equal to 1 s, actuation of vehicle brake drive equal to 0.6 s and various vehicle decelerations (maximum possible deceleration taking into account the coefficient of adhesion of tyres with the road surface $\varphi = 0.8 = 7.84$ m/s^2 and working deceleration = 2.6 m/s^2) from the speed of the vehicle (Fig. 5).

Analyzing the obtained dependencies, it can be concluded that the value of the positioning error is practically commensurate with the calculated remainder of the path between the vehicle, while at high vehicle speeds the calculated remainder of the path will be more than the value of the positioning error. This provides at high speeds a reliable safe interval between vehicles, the drivers of which apply emergency braking corresponding to the current coefficient of adhesion of the tires to the

Table 1 Results of experiments on determination of vehicle positioning error

No and conditions of the experiment	Distance between participants on the phone (defined by the system), meters		Actual distance between participants (labels) (measured based on mark bounce), meters	Difference between distances defined by the system and valid $\Delta Ss\text{-}d$, meters	Average distance between participants per experiment on phone *Ss*, meters	Average actual distance between participants *Sd*, meters
	1 participant	2 participant				
1. Towards both participants, set distance between participants 50 m, set speed 3 km/h	47.67	47.67	42.6	5.07	46.8	36.3
	49.48	45.92	49.5	0.02		
	47.86	47.86	31.2	16.66		
	42.22	38.29	21.8	20.42		
	46,23	46.44	46	0.21		
2. Towards both participants, set distance between participants 100 m, set speed 30 km/h	83.43	83.42	56.3	27.12	83.4	55.4
	63.94	63.94	27.1	36.84		
	85.44	85.44	54.9	30.54		
	90.19	90.19	31.6	58.59		
	94.22	79.54	56.7	22.84		
3. Towards both participants, set distance between participants 200 m, set speed 50 km/h	185.33	201.06	104.8	96.26	185.5	119.5
	182.58	182.58	84.3	98.28		
	188.41	168.73	113.4	55.33		
	191.8	163.5	119.3	44.2		
	179.37	179.37	79.3	100.07		
4. One participant standing, set distance between participants 200 m, set speed 90 km/h	189.63	–	111.42	78.21	185.7	157.6
	187.09	–	136.3	50.79		
	180.34	–	128.9	51.44		

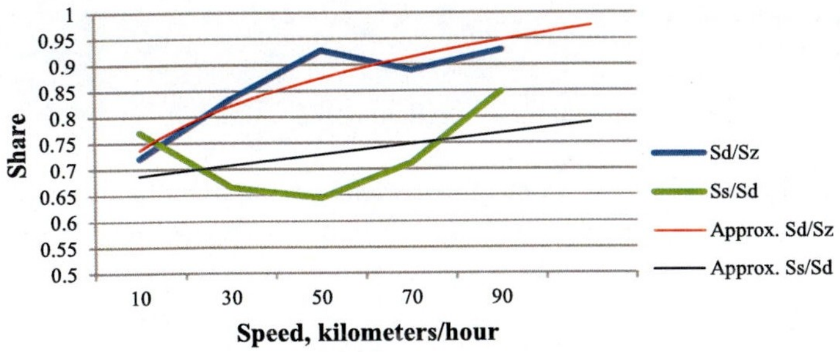

Fig. 4 Dependencies of relative positioning errors relative to the specified limit distance Sd/Sz and the distance between vehicles determined by the system Ss/Sd, depending on the vehicle speed

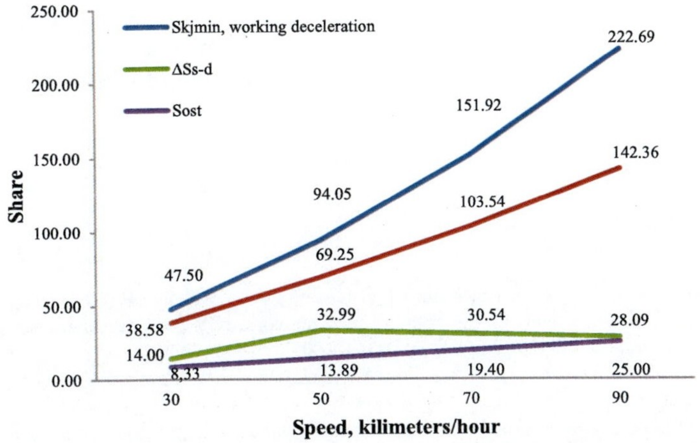

Fig. 5 Dependencies of the position error averaged per experiment in the form of the distance difference determined by the system and the actual $\Delta Ss\text{-}d$, the remainder of the *Sost* path, the minimum safe distance between the *k*th pair of the *SkJmin* vehicle (according to (3) without taking into account the rest of the path *Sost*) at system response times equal to 1 s, driver reaction equal to 1 s, actuation of the vehicle brake drive equal to 0.6 s and various vehicle decelerations (maximum possible deceleration taking into account the coefficient of adhesion of tyres with the road surface equal to 0.8 = 7.84 m/s^2 and working deceleration = 2.6 m/s^2) from the vehicle speed

road surface. At low vehicle speeds, drivers already make visual contact with emerging "obstacles" and accordingly have time to act according to the current road situation, so the action of the developed system is not advisable in this case. To ensure road safety, as well as the reliability of the system, a two-step hazard warning should be applied to drivers, the first stage corresponds to the minimum safe distance between the vehicle calculated during the working deceleration, the second—at the maximum possible deceleration, taking into account the existing coefficient of adhesion of tyres to the road surface on emergency dangerous sections of roads outside settlements.

4 Conclusions

The analysis of statistical data on accidents that occurred on roads outside the settlements of the Krasnoyarsk Territory [14] made it possible to identify the main types of vehicle collisions for which a mathematical apparatus for determining the minimum safe distance between vehicles close to the emergency zone is proposed, in which it is necessary to notify the drivers of these vehicles about the danger in conditions of insufficient and/or limited visibility. According to the obtained experimental data on positioning errors, vehicle speed determined by the driver early warning system about hazards on the road, and their analysis, it can be concluded that the designed system provides reliable operation at high vehicle speeds [15].

Thus, timely notification of a possible emergency in conditions of insufficient and limited visibility, implemented through the designed system, will increase road safety, especially on emergency-dangerous sections of roads outside settlements, where vehicle speeds are usually high and, accordingly, the severity of the consequences of an accident is high.

References

1. Baklanova KV, Voevodin ES, Fomin EV, Kashura AS, Cheban EP (2021) Identification of factors affecting accidents on the intercity road network. IOP Conf Ser Mater Sci Eng 12005. https://doi.org/10.1088/1757-899X/1061/1/012005
2. Voevodin ES, Pulyanova KV, Kashura AS, Askhabov AM, Fomin EV, Golub NV (2018) Reduction of accidents on public roads by increasing informativeness of drivers. In: Proceedings of the international conference: aviamechanical engineering and transport, pp 443–447. https://doi.org/10.2991/avent-18.2018.85
3. Khakzar M, Bond A, Rakotonirainy A, Oviedo O, Trespalacios S, Dehkordi G (2021) Driver influence on vehicle trajectory prediction. Accid Anal Prev 157:106165
4. Duan K, Yana X, Li X, Hang J, Yang J (2020) Analysis of driver's decision distance and merging distance in work zone area based on parametric survival models: with the aid of a driving simulator experiment. Transp Res Part F: Traffic Psychol Behav 71:31–48
5. Yuab Bo, Chena Y, Baob S (2019) Quantifying visual road environment to establish a speeding prediction model: an examination using naturalistic driving data. Accid Anal Prev 129:289–298
6. Yoneda K, Suganuma N, Yanase R, Aldibaja M (2019) Automated driving recognition technologies for adverse weather conditions. IATSS Res 43(4):253–262
7. Yasanthi RGN, Babak M (2020) Modeling free-flow speed variations under adverse road-weather conditions: case of cold region highways. Case Stud Transp Policy 8(1):22–30
8. Glukhikh RS, Grazhdantsev EV, Zeer VA, Rastovtsev PA, Elagin AI (2020) Drivers alert system in the conditions of information shortage. J Phys Conf Ser 1515:052009
9. Reinmuellera K, Steinhauserb M (2019) Adaptive forward collision warnings: the impact of imperfect technology on behavioral adaptation, warning effectiveness and acceptance. Accid Anal Prev 128:217–229
10. Augustsson N, Wilhelmsson M (2010) Verification of collision avoidance functionality. A mileage multiplier approach to verify future collision avoidance systems. Master of Science Thesis in the Master Degree Programme Systems, Control and Mechatronics (Department of

Signals and Systems. Division of Automatic Control, Automation and Mechatronics Chalmers University of Technology, Goteborg, Sweden) EX034/2010
11. Abboud K, Zhuang W (2009) Modeling and analysis for emergency messaging delay in vehicular ad hoc networks. In: Proceedings of the IEEE GLOBECOM
12. Cheng H-T, Shan H, Zhuang W (2011) Infotainment and road safety service support in vehicular networking: from a communication perspective. Mech Syst Signal Process 25:2020–2038
13. Zeer VA, Grazhdantsev EV, Glukhikh RS, Avdeev RM, Elagin AI (2020) Justification of the system parameters for alerting drivers about the emergency situation occurrence. IOP Conf Ser Mater Sci Eng 1061, 012043
14. Pulyanova KV, Voevodin ES, Fadeev AI, Fomin EV, Askhabov AM, Kashura AS (2019) Analysis of road safety assessment methods. Int Conf Innovations Automot Aerosp Eng 12015. https://doi.org/10.1088/1757-899X/632/1/012015
15. Winkler S, Kazazi Ju, Vollrath M (2018) How to warn drivers in various safety-critical situations—different strategies, different reactions. Accid Anal Prev 117:410–426